寰宇文献 Universal Library | SINOLOGY 系列

SELECTED WORKS OF BERTHOLD LAUFER

劳费尔著作集

第七卷

[美] 劳费尔 著

黄曙辉 编

中西书局

ZHONGXI BOOK COMPANY

图书在版编目(CIP)数据

劳费尔著作集 / (美) 劳费尔著；黄曙辉编. —上
海：中西书局，2022
 (寰宇文献)
 ISBN 978-7-5475-2015-4

 Ⅰ. ①劳… Ⅱ. ①劳… ②黄… Ⅲ. ①劳费尔 – 人类
学 – 文集 Ⅳ. ①Q98-53

中国版本图书馆CIP数据核字（2022）第207067号

第 7 卷

1

098

西藏六十干支的运用

T'OUNG PAO

通報

ou

ARCHIVES

CONCERNANT L'HISTOIRE, LES LANGUES,
LA GÉOGRAPHIE ET L'ETHNOGRAPHIE
DE
L'ASIE ORIENTALE

Revue dirigée par

Henri CORDIER
Membre de l'Institut
Professeur à l'Ecole spéciale des Langues orientales vivantes
ET
Edouard CHAVANNES
Membre de l'Institut, Professeur au Collège de France.

———

VOL. XIV.

———

LIBRAIRIE ET IMPRIMERIE
CI-DEVANT
E. J. BRILL
LEIDE — 1913.

THE APPLICATION OF THE TIBETAN
SEXAGENARY CYCLE

WITH REFERENCE TO

P. PELLIOT, *Le cycle sexagénaire dans la chronologie tibétaine*
(*Journal asiatique*, Mai-Juin, 1913, pp. 633—667)

BY

BERTHOLD LAUFER.

Ch'ang-an cannot have seen any brighter days than Paris when
M. Pelliot, a second Hüan Tsang, with his treasures of ancient
books, manuscripts, scrolls and statues, returned from his journey in
Central Asia which will ever be memorable in the annals of scientific exploration. His archæological material bearing on the languages,
literature and history of almost all nations of Central Asia has
naturally led him to transgress the boundary stones which were
set up by the commonly accepted Monroe doctrine of sinology, and
to take deep plunges into Turkish, Mongol, Tungusian, Tibetan,
and kindred subjects. In studying the work of previous scholars in
these fields, M. Pelliot encountered a great deal that could not
pass muster before his scrutinizing eagle eye, and that he was able
to enlighten considerably with the solid fund of his superior Chinese
and historical knowledge. In the present investigation he turns his
searchlight on the prevailing methods of computing the Tibetan
years of the sexagenary cycle into our system of time-reckoning;
he x-rays the father of this system, ALEXANDER CSOMA, who, in
his famous Tibetan Grammar (Calcutta, 1834), expounded a calculation of Tibetan years which ever since has been a sanctified dogma
of Tibetan philology (with two exceptions which escaped the atten-

39

tion of M. Pelliot), and discovers in it two fundamental errors of calculation which gave rise to all subsequent misunderstandings. After careful examination of M. Pelliot's deductions and conclusions, and after testing them also from Tibetan works of chronology and numerous examples of dates furnished by Tibetan books, it is the foremost and pleasant duty of the reviewer to acknowledge without restraint that the results obtained by M. Pelliot are perfectly correct, and that the rectifications proposed and conveniently summed up by him on p. 663 must be generally and immediately adopted.

The nerve of the whole matter is the date of the first year of the first Tibetan cycle. Csoma had calculated it at the year 1026, and M. Pelliot justly reveals the fact that he committed an error of calculation, and that this date must be fixed at 1027.[1]) This year as the starting-point of the Tibetan reckoning after cycles is moreover confirmed by the working of the system. It should be pointed out that this discovery of M. Pelliot is not entirely original. It was Father A. DESGODINS of the *Missions Étrangères* who as far back as 1899 proposed to fix the beginning of the first year of the Tibetan cycle at the year 1027. In his "Essai de grammaire thibétaine pour le langage parlé," p. 87 (Hongkong, Imprimerie de Nazareth, 1899) Father DESGODINS says literally: "Nous avons fait le tableau complet des cycles de 60 ans, en partant de l'année, telle qu'on la compte au Thibet: et nous avons trouvé que la première année du premier cycle thibétain était l'an 1027 de l'ère chrétienne, et non l'an 1026, comme disent Csoma et Mr. Foucaux.

---

1) Despite his wrong calculation, CSOMA has converted correctly at least one date. In his translation of a Tibetan passport which was published in Hyde's *Historia Religionis Veterum Persarum* (*J. A. S. B.*, Vol. II, 1833, p. 202, or *J A. S. B*, N. S., Vol. VII, N° 4, 1911 [containing a reprint of Csoma's papers], p. 26) the date *earth-dragon* (*sa abrug*) is justly reduced to 1688, also the Chinese cyclical signs *Vú Dhín* (*wu ch'en*) being correctly added in a footnote; but then immediately follows the sentence: "The Tibetan reckoning commences from February, 1026".

Quoi qu'il en soit, lorsqu'on est en pays thibétain, rien de plus facile que de savoir l'année que les Thibétains comptent actuellement; et, partant de là, on se fait un petit tableau pour les années suivantes. Cela suffit pour l'usage ordinaire." In the "Dictionnaire thibétain-latin-français par les Missionnaires Catholiques du Thibet" (Hongkong, 1899) edited by Father DESGODINS, to whom is due also a large share in the collection of the material, particularly from the native dictionaries, the same statement is repeated twice, —first on p. 932 under the word *rab abyun* "Cyclus 60 annorum (1[us] annus 1[i] cycli incoepit 1027 post X[um])," secondly on p. 976 where it is said: "La 1re année du 1er cycle de 60 ans *me mo yos* correspond à l'an 1027 de l'ère chrétienne." The cyclical determination indicated by the Tibetan words is *fire-hare*, and this is identical with the one revealed by M. PELLIOT (p. 651) from the *Reu mig*. The writer can himself vouchsafe the correctness of the fact that the first year of the first cycle is designated *fire-hare*, as he found this indication in Tibetan works on chronology. It is thus obvious, that Father Desgodins, toward the end of the last century, through a process of calculation similar to that of M. Pelliot and through an actual knowledge of the Tibetan chronological system, had arrived at the same result. The merit of M. Pelliot is certainly not lessened by the fact of priority which his countryman may justly claim, for the rectification of the humble missionary, couched in such a modest form, passed unnoticed and did not stir up those concerned in the case. There is not any doubt either that M. Pelliot, independent of his predecessor, has been led to his result by sheer commonsense and the exertion of his own brainpower. As the facts are, Desgodins and Pelliot are the only ones to be incarnations of Mañjuçrī, while all the others, the present writer among them, have been deluded by a temptation of Māra.

While Father Desgodins, as far as I know, never gave in his writings any practical examples of Tibetan dates, there is another scholar who, though he has never stated his opinion on the Tibetan cycle and its application, proves by his method of conversion that he understood it well, -- and this is V. VASILYEV. M. Pelliot would have himself traced this fact easily, had he consulted Tāranātha together with the translation of Schiefner in that of Vasilyev (and it is always safe to consult the two), or VASILYEV's "Vorrede zu seiner russischen Übersetzung von Tāranātha" (translated by SCHIEFNER and published as a separate pamphlet, St. Petersburg, 1869). M. PELLIOT (p. 648 note) attributes the correct calculation of the year 1608 as the date of the composition of Tāranātha to SCHIEFNER, but this feat is plainly to be credited to VASILYEV (p. XVIII). It turns out that VASILYEV was acquainted with the *Reu mig* of our friend Chandra Das, styled by him (Vasilyev) the Chronological Tables of Sumba Chutuktu (= Sum-pa mk'an-po).[1]). It is Vasilyev who correctly identifies the *earth-monkey* year with our year 1608 as the date of Tāranātha's work, and (this is the salient point) the *wood-pig* year with our year 1575 as the date of Tāranātha's birth, -- both data being taken from the *Reu mig* where in fact they are so given. Schiefner copied from Vasilyev the date 1608, but changed the other date into 1573. M. Pelliot, who without the knowledge of Vasilyev's indication correctly arrived at the date 1575 by utilizing the statement of the colophon that Tāranātha wrote his work in his thirty-fourth year, very generously excuses Schiefner on the ground that 1573 is a simple misprint; I could wish to share this point of view, but to my regret I can not.

---

1) The work Kalpasuvriksha referred to by SCHIEFNER, in which the same dates are said to be contained as those pointed out by VASILYEV, is nothing but the Sanskritized title of the *dPag bsam ljon bzan* of Sum-pa mk'an-po; and as the chronological table *Reu mig* forms a portion of the latter work, so also Schiefner indeed speaks of the *Reu mig*. This seems to have escaped M. Pelliot.

Before me is a copy of the Tibetan text of *Reu mig* written by Schiefner's unmistakable hand (already referred to by G. Huth, *Z. D. M. G.*, 1895, p. 280); in this copy, Schiefner has entered after the cyclical determinations the dates of Csoma in each case of a coincidence of events. Of course, this does not prove that Schiefner absolutely believed in the dates established by Csoma; but if we notice that he marked the datum of the journey of bSod-nams rgya-mts'o into Mongolia (*fire-ox*) = Csoma 1575, he is liable to the suspicion that he found the date for *wood-pig* two lines above by deducting 2 from 1575, and thus arrived at his date 1573. Taking further into account that Schiefner, as already shown by Pelliot, fell a victim to Schlagintweit, there is good reason to believe that prior to this time he was victimized by Csoma; the one almost necessarily implies the other. It is there-fore impossible to assume that the correct calculation 1608 is due to Schiefner whom M. Pelliot will have to put down on his black list.[1]) Vasilyev, who had made his Tibetan studies among the Lamas during a ten years' seclusion at Peking, had the advantage of being removed from the European contagion which had spread from India. There is no means of ascertaining what opinion was upheld by Vasilyev in regard to Tibetan chronology, and for lack of evidence I should hesitate to confer upon him any posthumous title. The two examples mentioned are the only ones traceable in his works and clearly stand out as exceptions in the history of

---

1) Schiefner has seldom had the opportunity of dealing with dates, and as far as pos-sible kept aloof from the translation of colophons. But to his honor it should not be pas-sed over in silence that in his *Eine tibetische Lebensbeschreibung Çâkjamuni's* (St. Peters-burg, 1849, p. 1) he has correctly reduced the date of the authorship of the work, *wood-tiger*, to 1734 (while the date of the print, 40th year of K'ien-lung is not, as stated, 1776 but 1775). The days and months given in both dates are carefully avoided, and the colophon is untranslated. The fact is overlooked that the year of the Jovian cycle *dmar ser* (Skr. *piṅgala*) given in correspondence with *wood-tiger* does not correspond to it but to *fire-serpent*, accordingly to 1737; one of the two dates must be wrong.

Russian scholarship. The repetition of Csoma's errors on the pages of our Russian colleagues goes to prove that Vasilyev did not bequeath to them any substantial lesson bearing on this question. O. Kovalevski (Монгольская Хрестоматія, Vol. II, p. 271, Kasan, 1837), without quoting Csoma, implicitly shows that he believed in his chronology by lining up three dates for the lifetime of bTson-k'a-pa, first the fanciful statement of Georgi 1232—1312, secondly the correct date of Klaproth 1357—1419, and thirdly the date 1355—1417 after *Vaiḍūrya dkar-po*, as given by Csoma in 1834; his very manner of expressing himself on this occasion bears out his endorsement of Csoma's dates.[1])

Prof. N. Küner at the Oriental Institute of Vladivostok, in his thorough and conscientious work "Description of Tibet"

---

1) It should not be forgotten that Mongol philology was developed in Europe on lines entirely different from Tibetan philology. Russia counted Mongols among her subjects, and Russian Mongolists always plodded along under the auspices of Mongol assistants. If Schmidt and Klaproth were correct in their conversion of Mongol cyclical dates into occidental years, this was by no means an heroic deed but simply due to information received from their Mongol interpreters. Tibet was always secluded and far removed from us, our workers had to push their own plough, and had to forego the privilege of consulting natives of the country. The opportunity and temptation of forming wrong conclusions were thus far greater. It is necessary to insist upon this point of view, in order to observe a correct perspective of judgment. Also the subjects treated on either side were different. In the Mongol branch of research, history was uppermost in the minds of scholars; in Tibetan it was the language, the problems of Sanskrit literature, and the religious side of Lamaism by which students were chiefly attracted, while history was much neglected. Certainly, students of Tibetan *did* always notice the divergence of their calculations from those of Schmidt and Klaproth (also, as will be shown below, Dr. Huth, contrary to the opinion of M. Pelliot), but what did Schmidt and Klaproth know about Tibetan chronology? They never stated that Tibetan and Mongol year-reckoning agreed with each other, nor that *their* system of computation should hold good also for the Tibetan cycle. Nor is there reason to wonder that Lama Tsybikov converted correctly the cyclical dates given in the Tibetan text of *Hor c'os byuṅ* edited by Huth; as a Mongol, he simply adopted the Russian mode in vogue of recalculating Mongol cyclical dates into the years of our era, but there is no visible proof forthcoming that he proceeded on the basis of an intelligent insight into the workings of Tibetan chronology, or on an understanding of the mutual relations of the two cycles. The result of a mathematical problem may often be guessed, or found by means of intuition or imagination; it is the demonstration on which everything depends.

(in Russian, Vol. II, 1, p. 107, Vladivostok, 1908), devotes a brief chapter to time-reckoning, and (invoking Rockhill) states as a fact (p. 108) that the first year of the first Tibetan cycle of sixty years appears in Tibet as late as the year 1026, so that the year 1908 appears as 43rd year of the 15th cycle.[1]) It is certainly easy to talk about Bu-ston, and to refer the reader to his "Histoire de la religion f. 23 et suiv. de l'édition xylographe tibétaine (Lhasa)" without giving any chronological and bibliographical references, as M. Th. de Stcherbatskoi does.[2]) The "wrong" dates which are made after a "system" are then still better than such a blank.

M. Pelliot passes on from Csoma to Huc, Koeppen and Schlagintweit. At this point M. Pelliot does not seem to me to do full justice to the facts in his attempt to trace the history of the case. If the history of this error must be written (and the history of an error is also a contribution to truth and one capable of preventing similar errors in the future), equal justice should be dealt out to all with equal measure. The propagator of Csoma's ideas in Europe was M. Ph. Éd. Foucaux (1811—1894) who published his "Grammaire de la langue tibétaine" in 1858 at a time when he was "professeur de langue tibétaine à l'École Impériale des Langues Orientales". Foucaux was decidedly a superior man, of keen intelligence, of bright and fertile ideas,[3]) commanding a full mastery of Sanskrit and Tibetan; and whoever has worked through his edition and translation of the Tibetan version of the Lalitavistara, will cherish the memory of this hard and patient worker with a

---

1) Despite this wrong statement, 1906 is correctly identified with *fire-horse*, and 1907 with *fire-sheep*.

2) *La littérature Yogācāra d'après Bouston* (Extrait du *Muséon*, Louvain, 1905).

3) One should peruse, for instance, his Discours prononcé à l'ouverture du cours de langue et de littérature tibétaine près à la Bibliothèque Royale, dated at the end Paris, 31 janvier 1842, the preface to his *Spécimen du Gya-tcher-rol-pa* (Paris, 1841), and the introduction to the translation of Lalitavistara (Paris, 1848). The present writer is proud of owning a copy of the latter work dedicated by Foucaux with his own hand to Jäschke.

profound feeling of reverence and admiration. His grammar, though based on the researches of Csoma, is an original work revealing the independent thinker on almost every page and, up to the present time, is the most useful book for the study of the Tibetan literary language.[1]) In fact, every student of Tibetan has made his juvenile start from this book which always enjoyed the highest authority in our academic instruction. Now while Foucaux in this work has carefully considered and sifted all statements and opinions of Csoma, he has embodied in it, without a word of criticism or any re-examination, Csoma's "manière de compter le temps" (p. 146) in its whole range; in particular, he has authorized and sanctioned "le commencement du premier cycle à partir de l'an 1026 de l'ère chrétienne" (p. 148). This step was decisive for the further development of this matter in European science; M. Foucaux had impressed on it the seal of his high academic authority, and since this legalization, the error has been raised into the rank of a dogma and believed to be a fact.

The correctness of this point of view of the matter is corroborated by two facts, — first by a long successive line of illustrious scholars in France following in the trail of Foucaux and all unreservedly accepting his teaching in matters of Tibetan chronology up to recent times (even after the rectification of Father Desgodins), and second by the fact that it was from France that the germ of the error was carried to America. For our great authority on subjects Tibetan, Mr. W. W. ROCKHILL, was a student of Tibetan under M. Foucaux, and in his fundamental work "Notes on the Ethnology of Tibet" (*Report U. S. Nat. Mus. for 1893*, p. 721, Washington, 1895) stated: "The first year of the first cycle of sixty years is

---

1) The same judgment was pronounced by the writer in 1900 (*W. Z K. M*, Vol. XII, p. 297). — *The Manuel de tibétain classique* of Dr. P. Cordier announced for some time is expected with great interest.

A.D. 1026, consequently 1894 is the twenty-ninth year of the fifteenth cycle, or the 'Wood Horse' (*shing ta*) year of the fifteenth cycle". Here, the year 1026 is plainly laid down as a fact [1]). The further remark of Mr. Rockhill shows where the root of the evil really lay, for his indication of the year 1894 as being a *wood-horse* year is perfectly correct and in harmony with the table drawn up by M. Pelliot [2]). If Mr. Rockhill had had M. Pelliot's table at his disposal at the time when he wrote that paragraph, he would have doubtless noticed that, if the year 1894 was a *wood-horse* year, the

---

1) Our case is well illustrative of how detrimental to science dogmatism and dogmatic statements are. If Foucaux and Rockhill would have expressed themselves to the effect that "the first year of the first cycle, in the calculation of Csoma, is the year 1026", their statements would be formally correct, while the positive form of their sentences proves them to be in silent agreement with Csoma and makes them share in the responsibility for his material error.

2) Where personal inquiry among Tibetans was possible, correct cycle dates have usually been given in recent years Jäschke (Dictionary, p. 552) correctly says that 1874 was a *dog* year (but on the same page gives impossible identifications for *wood-dog, wood-pig, fire-rat* and *fire-ox* years), and Chandra Das (Dictionary, p. 1221) has it correctly that the year 1903 is called *c'u yos lo, water-hare* year. In Schlagintweit's and Rockhill's joined communication to the Dalai Lama translated into Tibetan under the auspices of Chandra Das, the year 1901 is justly rendered *iron-ox* (E. Schlagintweit, Bericht über eine Adresse an den Dalai Lama in Lhasa, *Abhandlungen der bayerischen Akademie*, 1904, p. 666, and plate). In the edition by Chandra Das of the Tibetan prose version of Avadānakalpalatā (*dPag bsam ak'ri šiṅ*, Bibl. ind) the year *iron-tiger* indicated on the Tibetan title-page adequately corresponds to the year 1890 on the English title-page. A good authentic example is furnished by the convention between Great Britain and Tibet signed at Lhasa "this 7th day of September in the year of our Lord 1904, corresponding with the Tibetan date, the 27th day of the seventh month of the Wood-Dragon year" (Parliamentary Blue-books: *Further Papers relating to Tibet*, N° III, p. 271). Vidyābhūṣana (*A Tibetan Almanac for 1906—1907, J. A. S. B.*, N. S, Vol II, 1906, p. 455) noted from the very title of this almanac that the year 1906 was *fire-horse*, and from another one for 1903 that that year was *water-hare*; nevertheless in his other publications (for example, Gyantse Rock Inscription, *ibid.*, p. 95) he adhered to the chronology of Csoma A recent publication of the same scholar, an edition of the seventh chapter of Mi-la-ras-pa's life (Darjeeling, 1912) bears on the Tibetan title-page the year *water-rat*. A new confusion was caused by G. Sandberg (*Hand-book of Colloquial Tibetan*, p. 159, Calcutta, 1894) who allowed "the cycle now in progress in Tibet to commence in the year 1863", and then gives a wrong table of years running from 1893 to 1906.

first year of the first cycle could not have been 1026, but 1027.
Thus, the fact crops out that such a table as now offered by M. Pelliot
has never before existed in this form. The tables made up from the
Chinese point of view do not contain the names of the animals,
though, of course, it would have been easy to supply them [1]). The
tables made up from the Mongol point of view, as, for example,
accompanying the Mongol Chrestomathy of Kovalevski, were rejected
by students of Tibetan, because the conviction gradually gained
ground that there was a divergence in the application of the cycle
between Mongols and Tibetans.

If M. Pelliot subjects the chronological table of Mr. WADDELL [2])
to a critical analysis, it would have been a matter of justice to
refer also to the table of historical dates appended by M. L. FEER
to his *opuscule de vulgarisation* which under the title "Le Tibet, le
pays, le peuple, la religion" appeared in Paris (Maisonneuve), 1886.
All dates there given (pp. 99, 100) down to 1650 are literally copied
from Csoma, and even the year 1025 is retained as that of the
first year of the cycle of sixty years. The fact that M. FEER made
this opinion his own is clearly proved by his statement in "La
Grande Encyclopédie" (Vol. VII, p. 604) to the effect that "c'est
de l'introduction parmi eux d'un des livres du Tantra, le Kâlatchakra
que les Tibétains font dater le commencement de leur ère (en 1025
de la nôtre)". In this opinion he was fully joined by M. ED. SPECHT
who in the same cyclopædia (Vol. XXXI, p. 63) states: "A cette
époque (1025), les Tibétains adoptèrent le cycle de soixante ans".
M. SPECHT evidently had an additional reason for this belief, for he
adds immediately: "La période Mekha gya tsho finit en 1024".

---

1) A comparative view of the twelve Chinese "branches" and the twelve Tibetan animals has been given by KLAPROTH (*Description du Tubet*, p. 56, Paris, 1831).

2) In justice to Mr. Waddell it should be mentioned also that in his book *Lhasa and
its Mysteries* (p. 450, London, 1906) he gives a correct table of the cycle from 1862
to 1927.

Farther above on the same page, M. Specht explains that this period begins in 622 A.D., and that it is perhaps the era of the hegira which the Tibetans adopted, "nous ne savons pas au juste à quelle époque" [1]). The date "1355 à 1417 environ" given for the lifetime of bTsoṅ-kʻa-pa by M. S. Lévi in his excellent work "Le Népal" (Vol. I, p. 169, Paris, 1905) testifies to the fact that also M. Lévi, following the traditions of Foucaux and Feer, sided with the computations of Csoma. Also M. L. de Milloué (Bod-youl ou Tibet, p. 185, *Annales du Musée Guimet*, Vol. XII, 1906) accepts the date 1355 for the birth of bTsoṅ-kʻa-pa, but on p. 188 sets the date of his death at 1417 or 1419 [2]) (as he states that the

---

1) Thus, M. Specht pinned his faith on the year 1025, in order to arrive at the year 622, the date of the hegira; but the calculation is wrong. True it is that the Tibetans are acquainted with the Mohammedan era; six practical examples of this kind are found in two Tibetan documents drafted at Tashilhunpo (bKra-šis lhun-po) in 1781 and translated in the appendix to S. Turner, *Account of an Embassy to the Court of the Teshoo Lama* (p. 449, London, 1800). True it is further that the Arabs (*Ma-kʻai kla-klo*, the Mleccha of Mecca) play an extensive rôle in the Tibetan speculations on chronology beginning with the Kālacakra system (see for the present E. Schlagintweit, Die Berechnung der Lehre, *Abhandlungen der bayerischen Akademie*, 1896, chiefly pp. 594, 609). The period *me kʻa rgya-mtsʻo* mentioned by Specht, as the very name implies, is a period of 403 years which, if subtracted from 1027 leads to the year 624 (according to Schlagintweit 623), which according to Tibetan tradition was a *wood-monkey* year.

2) This doubling of years shows the influence of Schlagintweit's "improved" system of chronology (compare Pelliot, pp. 647, 648). — The date of bTsoṅ-kʻa-pa's life-time has had many varying fortunes. Rhys Davids (*Encl. Brit.*, Vol. XVI, p. 99) adopted Klaproth's date 1357—1419; Yule (article *Lhasa, ibid.*, p. 530), however, dated him 1365—1418, again in his edition of Marco Polo (Vol. I, p. 315) 1357—1419. It would, of course, be preposterous to infer that those adopting the date of Klaproth were actuated by a deep insight into the matter. It is an entirely different question whether the date 1357—1419 is really correct. W. F. Mayers (*The Chinese Government*, 3rd ed., pp. 106, 107) set the date of bTsoṅ-kʻa-pa from 1417 to 1478, and in his essay *Illustrations of the Lamaist System in Tibet* (*J. R. A S.*, 1868, p. 303) where also Koeppen is quoted in the case more specifically referred to the *Shéng wu ki* 聖武記 (by Wei Yüan 魏源, 1842) as his source, without deciding the question of the striking diversity of the Tibetan and Chinese dates. It is evident that Hilarion, who likewise gives 1417 as the year of the birth of the reformer, drew from the same or a similar Chinese source, and that Koeppen's (*Die lamaische Hierarchie*, p. 108) charge of confusion between the years of birth and death should be directed toward the latter, not toward Hilarion. The *Shéng wu ki*, of

reformer died at the age of 63, he should have consistently assumed 1418). The remark in the foot-note that the date 1429 imparted by Sarat Chandra Das "paraît tardive" is proof for the fact that M. de Milloué, in like manner as the present writer, entertained serious doubts as to the correctness of the prevailing system of computation. For the rest also M. de Milloué could not get away from the firm grasp of traditional convention, and throughout acquiesced in the accepted dates. M. BONIN (*Les royaumes des neiges*, p. 273, Paris, 1911) derives from the tables of Csoma the date 1071 as that of the foundation of the monastery of Sa-skya.

M. PELLIOT laments that Chandra Das does not give the cyclical determination for 1747, the alleged date of the chronological table *Reu mig* translated by him. The question of the date of this work cannot be decided at a blow, as it is devoid of a colophon, and the colophon is lacking for the reason that the *Reu mig* is not an independent work of Sum-pa mk'an-po but incorporated in his great historical work *dPag bsam ljon bzaṅ*. For this reason I regret that M. Pelliot did not turn to the latter, as he evidently knows it from the edition of Chandra Das which, for the rest, is a very meritorious piece of work; M. Pelliot would have then discovered that the *Reu mig* is not contained in this edition (at least I cannot find there a trace of it), although the editor in the preface to the latter as well as in that of the former expressly assures us that *dPag bsam ljon bzaṅ* contains the *Reu mig*. The date of the completion of the latter spontaneously results from the last date given in the list of dates, which is 1746 indicated by *me stag*, *fire-tiger*, and as *dPag bsam ljon bzaṅ* was published in 1748 (*earth-dragon*), this year must hold good also for the publication of *Reu mig*. In restoring the dates of this work wrongly reduced by Chandra Das,

---

course, is a recent work and can hardly be looked upon as a pure source for the life of bTsoṅ-k'a-pa. Presumably, the *Ming shi* may contain the dates of his birth and death

who simply acted under the hypnotizing influence of Csoma and Schlagintweit, M. Pelliot mainly insists on the dating of bTsoṅ-k'a-pa. It is somewhat surprising that as a sinologue he did not notice the fact that *Reu mig* is replete with data of Chinese history: the dates of the Yüan, Ming and Ts'ing emperors are all completely given and in perfect harmony with the well-known dates of the Chinese, if M. Pelliot's correct point of view in the identification of the Tibetan cycle is adopted, while according to the calculation of Chandra Das the dates are one year behind the Chinese. This argument is very forcible, for we clearly recognize that the cyclical determinations were really understood by the Tibetans in exact agreement with the Chinese (and accordingly with the indications of M. Pelliot) as early as the Yüan and Ming periods, while the practical examples pointed out by M. Pelliot all relate to the age of the Manchu dynasty. It is thus further obvious that the Tibetans entertained correct chronological notions of Chinese events, and this fact must influence our judgment favorably on behalf of their datings of contemporaneous Tibetan events; if the former group of dates is correct, there is a fair chance that the same will be true of the latter. Some examples may illustrate this. In *Reu mig* (p. 63 of the translation of Chandra Das) we read: "Yunglo became emperor of China 1402." We know from the exact chronology of the Chinese that Yung-lo ascended the throne in 1403. The Tibetan text of *Reu mig* runs thus: *rgyai rgyal-sar gsum-pa Yoṅ-loi c'os rgyal ak'od .... c'u lug,* "the third (in the series of the emperors of the Ming dynasty), the king of the law (Skr. *dharmarāja*) Yuṅ-lo was installed on the throne of China .... *water-sheep.*" Consulting M. Pelliot's table we find that *water-sheep* fell indeed in 1403. On the same page of Chandra Das we read the following: "The second Miṅ emperor Huṅ-wu tsha ascended the throne of China .... 1398," a sentence which must cause every sinologue to shake his head. Everybody knows that Hung-wu was the first Ming emperor and reigned 1368--99, and that the second Ming emperor was his grandson Huei-ti who succeeded to his grandfather in 1399. What Chandra Das takes for a proper name, means in fact "the grandson ($ts'a = ts'a-bo$) of Hung-wu". The text reads: *rgya-nag raṅ-gi yig-ts'aṅ rñiṅ ltar-na gñis-pa Huṅ-wu ts'a rgyal-sar aḳ'od . . . . sa yos*, "according to China's own ancient records, the second (emperor of the Ming dynasty), namely, the grandson of Hung-wu, was installed . . . . *earth-hare*", a determination coinciding with 1399. The words omitted in the rendering of Chandra Das are important, for they clearly show that Sum-pa mk'an-po availed himself of a Chinese source or sources in establishing the dates of Chinese occurrences [1]). Of Mongol data, the

---

1) The romanizations of the names and Nien-hao of the Chinese emperors in Chandra Das are often inexact; he always neglects to indicate the Tibetan cerebral *ṭ* (transcribed by him with a dental *t*) which is the equivalent of Chinese palatal *č*, — thus *Tṅi-te = Chéng-tê* 正 德. *Bsson-te* on p. 65 rests on a misreading of his text which is *ḣia-pa zon-te*, the latter being equal to *Süan-té*. It is important to know the correct Tibetan transcriptions of Chinese Nien-hao and imperial names, especially those of the Yüan and Ming dynasties, as they are frequently made use of in Tibetan literature without any warning or any clear specification to the effect that they are so intended. Tibetan books, for example, printed in the monasteries of Sze-ch'uan and Kan-su at the time of the Ming dynasty, are usually dated in the colophon with the Chinese Nien-hao only, even without the addition of the convenient Ta Ming 大 明. A Tibetan version of Jātakamālā printed in the monastery Tai-luṅ-šen in Sze-ch'uan is dated *Zvon-te-i lo ḣia-pa t'un-moṅ lo*, "fifth year of the period Süan-tê (1430), the year *t'un-moṅ* (Skr. *sādhāraṇa*)." The latter is a year of the Indian Jovian cycle corresponding to the 44th year of the Tibetan, and the 47th year of the Chinese sexagenary cycle, and answering a *metal* (or *iron*)-*dog* year, and such was the year 1430. As regards the two inadvertences ascribed by M. PELLIOT (p. 652, note 1) to Chandra Das in the translation of *Reu-mig*, the text (at least in Schiefner's copy before me) indeed says that the fourteenth Kulika ascended the throne in 1227 (*me p'ag, fire-piṅ*), and the Kulika succeeding in 1527 was indeed the seventeenth (*bcu bdun-pa*). M Pelliot's emendations, therefore, hold good. The above omission is not the only one occurring in the translation of Chandra Das; there are others, too, noted by me, and perhaps others not yet noted. For all these reasons, and in view of the fundamental importance of *Reu mig*, the urgent demand must be made that the very text of this work should be critically edited. It is not long and will hardly occupy in print fifty pages of octavo size. Here is surely a worthy task for the *Bibliotheca Buddhica* of St. Petersburg.

death of Mangu (Tib. *Muṅ-k'e*) [1]) in *earth-sheep* (*sa lug*) year = 1259, and the death of Kubilai (Tib. *Se-c'en*, not as Chandra Das writes, *Sa-c'en*) in *fire-monkey* (*me sprel*) year = 1296, may be pointed out.

But it can even be demonstrated that any Chinese dates of whatever period have correctly been reduced by the Tibetans to the years of their cycle. Take, for example, the early Chinese dates occurring in the epilogue to the Sūtra of the Forty-Two Articles translated from Chinese into Tibetan, Mongol and Manchu by order of Emperor K'ien-lung in 1781 [2]). There we see on the same page in interlinear print the Chinese date "26th [3]) year of King Chao of the Chou dynasty with the cyclical signs *kia yin* 甲寅"

---

1) Compare the interesting study of M. PELLIOT, *Mungü et Mòngkü* (*Mönka) in *Journal asiatique, Mars-Avril*, 1913, pp. 451—459.

2) The edition referred to is the polyglot Peking print, the same as utilized by Huc and Feer. Compare L. FEER, *Le Sutra en 42 articles traduit du tibétain*, p. 45 (Paris, 1878). Feer has not converted the Chinese and Tibetan dates into their occidental equivalents.

3) The text has the error 24, adopted also by Feer, but the 24th year of Chao Wang is B. C. 1029 with the cyclical signs 壬子. The indication *kia yin* and the Tibetan conversion based on this plainly shows that B. C. 1027 is intended. The error, however, must be very old, for it occurs as early as in the *rGyal rabs* where the following is on record: "When the statues of the lord Çakya and of the sandalwood lord had reached the country of China, the annals of the dynasties in the great Chinese archives were opened with the intention of finding as to how the holy faith could be best diffused in the country. They discovered the fact that the former kings of China were the Chou dynasty which was coeval with King Yuddhishthira of India, that after four rulers King Chao Wang ascended the throne, and after twenty-four years of his reign, on the 8th day of the 4th month of the *wood male tiger* year (there is no agreement in the determination of the two years except that Buddha's lifetime appears as the same in both, but in that manner the date is given in the Chinese records) in the western region light, voices and many other wonderful signs arose which were interpreted by the astrologers of China on due calculation as indicating the birth of Bhagavat." This passage obviously shows that the Tibetans were smart enough to notice the deviation between the two years, which probably has its cause in a different calculation af Buddha's birth in China on the one hand and in Tibet on the other. The author of *Grub-mt'a šel-kyi me-loñ* (compare *J. A. S B.*, Vol. XLI, 1882, p. 88) who narrates the same event as *rGyal rabs* correctly imparts the date "26th year of Chao Wang", but adds that some authors believe that it was the 24th year of his reign. In regard to the Chinese date of Buddha's birth see EITEL, *Handbook of Chinese Buddhism*, p. 136.

(corresponding to B. C. 1027) = Tibetan *Ṭiu Ṭou wang-gi lo ñer drug-pa šiṅ p'o stag lo*, repeating the Chinese date and adding *wood male tiger* year, and such was B. C. 1027, the alleged date of Buddha's birth. Turning to the next page we find "Mu Wang 53d year 壬申"[1]) (B. C. 949) = Tibetan *Mu waṅ lo ṅa gsum-pa c'u p'o sprel lo, water male monkey* year, and such was B.C. 949. The next date given "7th year Yung-p'ing of the Han" has no cyclical determination in the Tibetan rendering.

M. PELLIOT deserves special thanks for indicating the means of restoring the correct dates in HUTH's translation of *Hor c'os byuṅ* which is a mine of precious information. But it is not correct to say that Huth, as imputed to him by M. PELLIOT, has never observed the divergence of a year which he regularly established between his translation and that of Sanang Setsen by Schmidt. HUTH indeed was fully conscious of this discrepancy, as plainly shown by his remark (*Z. D. M. G.*, Vol. XLIX, 1895, p. 281) that "Sanang Setsen (p. 53) states the year of the birth of Rin-c'en bzaṅ-po to be in the šim (wood)-dragon[2]) year corresponding to 992, or as his chronology is ahead of one year (*um ein Jahr voraneilt*), to the year 991 A.D." HUTH, quite consistently with the wrong chronology which he adopted from Schlagintweit, had formed the opinion that Sanang Setsen's system of computation was deficient by being in excess of one year. He who is acquainted with the opinions of Huth will not be surprised that in "Die Inschriften

---

1) The text has the misprint 庚申 which would correspond to the 41st year of Mu Wang or B. C. 961 and to a *metal (iron)-monkey* year. The very context shows that 壬 must be the correct reading.

2) This is certainly a gross misunderstanding of Sanang Setsen's word *šim* which does not mean "wood" but is a transcription of the Chinese cyclical character *jen* 壬 (Tibetan transcription: *žiṅ*). The *wood-dragon* year would be 944 or 1004. Sanang Setsen understands the *water-dragon* year. A sudden flash of a wrong association of ideas must have crossed Dr. Huth's mind and led him to link Mongol *šim* with the Tibetan word *šiṅ*, "wood".

von Tsaghan Baišiṅ" he gives three correct identifications of cyclical years (PELLIOT, p. 697, note 2). These dates occur in a Mongol inscription, and from his point of view, Huth was perfectly logical in applying to it the Mongol system of calculation, presumably by availing himself of Kovalevski's table, while in his study of Tibetan works he utilized what he believed to be the Tibetan system [1]).

In the face of all these authorities, what could the students of the present generation do? It is perfectly human that they should accept what they were taught in the classroom. Csoma, soon after his tragical death in the prime of life, was canonized and honored almost with the rites of an ancestral cult. The great Foucaux fully endorsed and upheld him in his chronology. Schlagintweit, by profession a jurist yet for the rest a good and honest man, was not a philologist but what is worse, a bad logician; it was certainly foolish to trust him for a moment. And then — GINZEL entered the arena. Well known is the witty saying of King Ludwig II of Bavaria, "a painter must be able also to paint". And we should justly expect that "a permanent member of the Royal Prussian Institute for Astronomical Calculation" should be able also to calculate. Csoma was not an astronomer and chronologist, but a scientist, about to issue an authoritative handbook on chronology as a safe guide to the historian, plainly had as such the duty of recalculating his precursor's computations and rendering to himself

---

1) There is no reason to assume with M. PELLIOT (p. 656) that ḥJigs-med nam-mk'a, the author of *Hor c'os byuṅ*, was a Mongol writing in Tibetan. He was a Tibetan by birth, born in a place near the monastery bĺa-braṅ bKra-šis ḥk'yil (HUTH, p. 357) in the province of Amdo (political territory of the Chinese province of Kan-su), and after completion of his studies, was called into Mongolia as preacher; later on, he was appointed at Yung ho kung in Peking and at Dalainōr (La-ma miao).

40

and to his readers an account of what the real foundation of this system is. Ginzel's book, with its sanctification of the year 1026, denotes the climax in the singular history of this *comedy of errors*, and by virtue of its highly authoritative character, indeed proved fatal. The higher must be estimated the merit of M. Pelliot who ultimately possessed enough pluck and wit to point to the very seat of the evil, and to eradicate it with a skilful operation.

I do not wish to be misunderstood. I merely intended on the preceding pages to contribute objectively and historically to the understanding of the development and diffusion of the error in question, as the matter now presents itself to one who for fifteen years has gathered documentary material for writing a history of Tibetan philology. I did not mean, however, to write an apology, or to whitewash anybody entangled in the case, — and least of all myself. Errors are errors, and no matter whether they are small or great, there is no excuse for them, and for myself I can only say *stultum me fatuor*. The importance of the present case must by no means be underrated. An outsider may easily jump at the conclusion that it makes little difference whether the date of a Tibetan book is accepted as 1818 or 1819. As a matter of principle, it makes a great difference which, if not in that example, yet in many others, may be of grave consequence. Above all it is the total assembly of wrong dates which is distressing, — distressing because it has bred the germs of reflections and conclusions which now turn out to be wholly imaginary, — conclusions which were inherited through three generations. We labored under the belief that the application of the Tibetan cycle differed from that of the Mongols and Chinese, a difference poorly enough explained, and this alleged diversity certainly gave rise to reflections on the trust-worthiness of Tibetan history. We were ridden by a veritable night-mare which rendered our lives miserable, we were haunted by a fox-spirit which has now been felicitously exorcised by the new Chang T'ien-shi.[1]) The path is free, the fox has fled, and with a feeling of relief and encouragement we may hope to cope anew with the fascinating problems of the history of Tibet.

In regard to the origin of the Tibetan cycle M. PELLIOT enter-tains some notions to which I am not ready to subscribe. "C'est de ce cycle chinois que le système tibétain par éléments et animaux s'est, de toute évidence, inspiré" (p. 660). This opinion is suggested by the manifest consonance of the two systems, but it is not sup-ported by M. PELLIOT with any evidence derived from a Chinese or a Tibetan source. On the contrary, all evidence, as far as we know it, speaks against the opinion that the Tibetan cycle is in-spired by that of China. Before presenting this evidence, it is justi-fiable to raise the question, — why, if the Tibetan cycle owes its impetus to China, does it appear so late as 1027, why does it not make its début in Tibet during the T'ang epoch when this cycle was perfectly known in China, and when both countries were in close mutual relations? There is no trace of the application of this cycle in the Tibetan inscriptions of the T'ang period nor in the colophons of the Kanjur and Tanjur. The only date thus far revealed

---

1) The future historian of science will assuredly remain mindful of the word of Mau-rice Maeterlinck (*Le temple enseveli*) that in each error of the past to which we clung tenaciously is usually hidden an excellent truth awaiting its hour of birth. All superstition is ancient science, and all science is modern superstition. Progress advances in zigzags, and error is a potent and necessary factor in the struggle for truth. The man who yielded to his successors the opportunity of revealing an error was also a combatant for the good cause.

in the colophon of a treatise of the Tanjur is worded in a Nepalese era. [1]) All the Tibetan historical works, as far as we know them at present, were composed after 1027, and the cyclical dates which we encounter there for the earlier periods certainly are the result of subsequent recalculations. [2]) The Genealogy of Tibetan Kings · (*rGyal rabs*, written 1328, not 1327, as formerly stated) has it that King Sroṅ-btsan sgam-po received books on time-reckoning from China and Mi-ñag, and if the *T'ang shu* (BUSHELL, The Early History of Tibet, p. 11) informs us that he invited learned scholars from China to compose his official reports to the emperor, this means to say that a Chinese chancery was attached to the government offices of Lhasa where naturally the system of Chinese Nienhao was employed, but apparently restricted to the official correspondence with China. Ecclesiastic literature marched along in its own way, and fed from the fountainhead of India drew its chronological inspiration from the same quarter. Buddha's Nirvāṇa was made the basis of time calculation, and as there was no consensus

---

1) HUTH, *Sitzungsberichte der preussischen Akademie*, 1895, pp. 276, 282.

2) But they are most certainly not the outcome of "the imagination of the historians", as intimated by A. H. FRANCKE (*Anthropos*, Vol VII, 1912, p. 264) whose remarks on the chronological question, in my opinion, are not at all to the point. The fact that "the dates in the sexagenary cycle do not come down from the first centuries of Tibetan historiography but from much later times" is as well known to me as to Mr. Francke. The contradictory dates given by the various Tibetan authors for events of earlier history have nothing whatever to do with the sexagenary cycle but have entirely different reasons. After the introduction of the sexagenary cycle in 1027 it was as easy as anything to recalculate any earlier dates, in whatever form they may have been handed down, on the basis of the new system, and as plainly proved by all facts, the Tibetans made these recalculations to perfect satisfaction. The hasty conclusion of Mr. FRANCKE that "the dates occurring in the *bTsun-mo bka-t'an* refer to the thirteenth [why the thirteenth, and not the eleventh?] century, and not to the eighth or ninth century" is entirely unwarranted. The dates most obviously relate to the time for which they are intended, and have been made by a simple process of correct arithmetical calculation. The imagination, in this case, is not on the part of the Tibetans but exclusively in the mind of Mr. Francke.

on this date, several theories being expounded, different computations of events are met with among Tibetan authors according to the standpoint which they took in that question. The great change came about when in 1027 the Kālacakra system was introduced. In that year the Kālacakra was translated into Tibetan by Ñi-ma ak'or-gyi Jo-bo ('the Lord of the Disk of the Sun'); in the next year, 1028, the great commentary to the Kālacakra was translated into Tibetan by Gyi Jo.[1] Now we know that 1027 is the first year of the sexagenary cycle, and the coincidence of this event with the introduction of the Kālacakra doctrine is not accidental. Indeed, Kālacakra, "the wheel of time," as already intimated by me in T'oung Pao, 1907, p. 403, is nothing but a designation of the sexagenary cycle, and the vast literature on Kālacakra is filled with expositions of this system. As correctly stated by Csoma (J. A. S. B., Vol. II, 1833, p. 57), the Kālacakra was developed in the country of Šambhala,[2] introduced into central India in the latter half of the tenth century, and then by way of Kashmir into Tibet. I do not wish to take up again the discussion of the location of Sambhala, which is to be sought in Central Asia. Divested of the later legendary accounts, that country is not at all so fabulous, and viewed in the light of the recent discoveries it is easily disclosed as a country where Iranian and Turkish Buddhism flour-

---

1) According to *Reu mig* in Schiefner's copy. Chandra Das attributes the former translation also to Gyi Jo; I am unable to say whether this is contained in the text from which he translated.

2) M. PELLIOT (p 652, note 1), on what authority is not known to me, writes the name Žambhala The Kālacakra texts embodied in the Tanjur (Palace edition) as well as the extensive later literature on the subject by Tibetan authors throughout follow the spelling Sambhala. and so do Csoma, Jäschke, Desgodins, Chandra Das, and the Petersburg Sanskrit Dictionary. The Tibetan gloss *bde byuñ* shows that the name was connected with Skr. *çambhu*.

ished. According to Tibetan tradition, the sexagenary cycle formed by means of the Twelve Animals penetrated into Tibet from a region of Central Asia, not from China.[1]) This is all that can be said for the present. The fundamental texts on Kālacakra which are of intense interest must be translated *in extenso* to reveal to us this chapter of history in detail;[2]) giving only a few extracts, though I could, seems to me to be of little avail. Better progress in the study of Central Asia would have been made if the suggestion made by me six years ago (*l. c.*, p. 407) had been carried out, for that literature contains the key to the understanding of many problems which now confront us in this new field. But workers in this line are few, and men possessed of the courage of initiative are rare. So we have to wait.

An important observation made by Mr. ROCKHILL (*J. R. A. S.*,

---

1) A distinction must be made between the mere knowledge of the series of the Twelve Animals and its utilization for chronological purposes. There are indications that the series of the Twelve Animals was known in Tibet before the year 1027, as shown by the symbolical interpretation of it in the legends of Padmasambhava (*T'oung Pao*, 1907, p. 400) and in other ancient writings centering around this personage. — Another side of this question is presented by the iconography of the Twelve Animals in Tibet and China which I hope to discuss on another occasion when the necessary illustrative material can be published. It seems to me that the iconographic representation of the Twelve Animals, as figured in the Tibetan works of chronology, is entirely distinct from that of China and decidedly points to another source.

2) The study of these texts will place on a solid basis our knowledge of Tibetan chronology which is now very scant. Then we may hope also to understand successfully the native works of chronology. SCHLAGINTWEIT (Die Berechnung der Lehre, *l. c*) has made a remarkable beginning along this line by editing and translating the work of Sureçamatibhadra of 1592. Though the translation is not entirely satisfactory, he has accomplished a great deal in elucidating the difficult terminology of the text, and this work is doubtless the best that the author has left to us. A standard book on astrology and chronology has been printed in Peking under the title *rTsis gžuń yań gsal sgron-me*, containing numerous tables, calculations, and illustrations. The collected works (*gsuń abum*) of the Lamas contain many treatises pertaining to this subject, even one dealing with Chinese chronology.

1891, p. 207, note 1) [1]) merits to be called to mind in this connection. "Tibet is the only dependency of China on which the imperial Chinese almanac has not been imposed as a proof of its
vassalage. The Chinese almanac is sent from Peking on the first
of the tenth month of each year to the various provinces and
tributary states. See e.g. *Peking Gazette*, Nov. 19, 1887". A special
edition of the calendar for the Mongols was yearly prepared, down
to the end of the Manchu dynasty, by the Calendar Section, *Shi
hien k'o* 時憲科, of the imperial Board of Astronomy in Peking
and sent from Peking into Mongolia. The Tibetan calendar, however, was not made in Peking but in Lhasa. The privilege reserved
by Tibet in this matter is a clear index of the fact that there is
some kind of a difference between the Chinese and Tibetan calendars;
if there were perfect agreement between the two, the request for,
and the grant of, such a privilege would be baseless. The existence
of a difference was the immediate cause of that privilege. Certainly,
this difference does not lie in the application of the cyclical years
where perfect harmony obtains. But it exists in the manner of
counting the months and days. The Central-Asiatic origin of the
Tibetan cycle accounts also for the fact briefly commented on by
M. PELLIOT (p. 661, note) that the Tibetan reckoning after months
and days does not tally with the Chinese system. This fact, M. Pelliot
could have easily ascertained from the *Wei Tsang t'u shi* (ROCKHILL's
translation, *J. R. A. S.*, 1891, p. 207, or KLAPROTH's *Description du
Tubet*, p. 57) where it is expressly recognized on the part of a Chinese
writer that the intercalation of months as well as days is different
in Tibet from Chinese practice. For this reason, Tibetan and Chinese
New Year do not necessarily fall on the same date, and Tibetan

---

1) See also his *The Land of the Lamas*, p. 241.

and Chinese datings of months and days cannot agree [1]). Father
A. Desgodins [2]), again, had a correct estimation of this matter
when he stated: "Ce que je sais de certain, c'est que tout leur
système d'astronomie est emprunté du Turkestan ou heur [intended
for the Tibetan word *Hor*], que les noms des jours de la semaine,
ceux des diverses constellations et des figures du zodiaque, etc., sont
ceux dont se servent les Turcs, et dont nous nous servons nous-
mêmes; c'est aussi d'après le comput du Turkestan que le calendrier
est publié chaque année". In the same manner I had pointed out
(*l. c.*, p. 407) in opposition to Schlagintweit and Ginzel that the
basis of the Tibetan calendar is neither Indian nor Chinese but
Turkish. This fact is most clearly evidenced by the term *Hor zla*,

---

1) This may be illustrated by a practical example. In 1906 the Tibetan New Year
fell on the 24th of February (Saturday), the Chinese New Year on the 23d of February
(Friday). The following Tibetan dates of that year are taken from VIDYĀBHŪṢAṆA's paper
A Tibetan Almanac (*J. A. S. B*, Vol. II, 1906, p 456) and given in comparison with the
Chinese dates after *Calendrier-annuaire pour* 1906 published by the Observatoire de Zi-ka-
wei. May 14, 1906 (Monday) = Tib. 21/III = Chin. 21/IV (this example plainly shows
that the Tibetan day and month cannot be identified with the same in Chinese, for 21/III
in China was April 14, Saturday); June 6 = Tib. 14/IV = Chin. 15/IV intercalary; June
28 = Tib. 6/V = Chin. 7/V; July 9 = Tib 17/V = Chin. 18/V; July 30 = Tib. 9/VI =
Chin. 10/VI; August 31 = Tib. 12/VII = Chin. 12/VII; September 23 = Tib. 5/VIII =
Chin. 6/VIII; October 15 = Tib. 27/VIII = Chin. 28/VIII, October 26 = Tib. 9/IX =
Chin. 9/IX; November 18 = Tib. 2/X = Chin 3/X; December 12 = Tib. 26/X = Chin.
27/X; January 15, 1907 = Tib. 1/XII = Chin. 2/XII; February 8 = Tib. 25/XII = Chin.
26/XII; on February 13, 1907 New Year tallied in Tibet and China, but again March 4,
1907 = Tib. 19/I = Chin. 20/I; April 7 = Tib. 24/II = Chin. 25/II, etc M. PELLIOT
certainly is correct in saying that among all peoples who have adopted the hebdomad the
same days of the week are in mutual correspondence; when it is Monday in Tibet, it is
on the same day Monday in China and throughout the world, but this very same Monday
is expressed by a different number in the lunar system of both countries. The tentative
experiment of calculation made by M PELLIOT, accordingly, is illusory, for 8/IV of *water-
dragon* in Tibetan need not agree (and most probably will not agree) with 8/IV in Chinese
of that year.

2) In the book published by his brother C H. DESGODINS, *Le Thibet d'après la cor-
respondance des missionnaires*, 2nd ed., p. 369 (Paris, 1885).

"Turkish month" [1]), advisedly used by the Tibetans with reference to their own months of Turkish origin in contradistinction to the Indian and Chinese months whose names are known to their scholars and those employed only in literature. The date of the completion of *Grub-mt'a šel-kyi mc-loñ* into which M. PELLIOT (p. 648) makes an inquiry is indicated in the colophon as the *water-dog* year of the thirteenth cycle (*rab byuñ bcu gsum-pai c'u k'yi*, consequently 1742), on the tenth day of the sixth Hor month. Sometimes three styles of a month are specified, thus in a work of the Fifth Dalai Lama written in 1658 the month is indicated 1. by the Sanskrit name Çravaṇa corresponding to the Tibetan rendering *bya sbo*, 2. by the Chinese *pi ts'a yol* (*yol = yüe* 月), and 3. by the Tibetan way *Hor zla bdun-pa*, „the seventh Hor month" [2]).

Those who desire to compute into our reckoning the day and month of a Tibetan date must therefore not fail to ascertain whether it is indicated in Chinese or Tibetan style. The rules to be observed are simple. Is the year expressed by a Nien-hao, month and day are naturally Chinese. For example, a Tibetan work dealing with the Sixteen Arhat, according to the colophon, was printed *Tai C'iñ*

---

1) The term *Hor zla* in this sense is already registered in CSOMA's *Dictionary of the Tibetan Language* (p. 333). KLAPROTH (in his edition of Della Penna's *Breve notizia del regno del Thibet*, p. 24, Paris 1834) remarks on this term: "Il ne peut être question ici des mois des Mongols, qui ont le même calendrier que les Tubétains, tandis que celui des Turcs, et des Mahométans en général, diffère du calendrier de ces derniers". The various meanings of the word *Hor* are well known (see *T'oung Pao*, 1907, p 404) From an interesting passage in the Tibetan Geography of the Minčul Hutuktu (VASILYEV's translation, p. 32, St Petersburg, 1895) it appears that the word is identified by the Tibetans with Chinese *Hu* 胡 ; but whether it is really derived from the latter, is another question. At any rate, it is not an ethnic but a geographical term. Different from this word *Hor* vaguely denoting any peoples living in the north of Tibet is *Hor* as a tribal name of Tibetan tribes in the Tsaidam and in eastern Tibet.

2) See *Z. D. M. G.*, Vol. LV, 1901, p. 124. The year is *earth-dog*, and as also M. Pelliot will admit, was correctly identified by me with the year 1658; this was facilitated by the addition of the cyclical signs *wu zui* = 戊 戌 . The year is further given with the designation of the Indian Jovian cycle *vilamba* = Tib *rnam-ap'yañ*

Yuṅ-ceṅ rgyal-po lo dgu zla-ba brgyad yar ts'es la, "in the first half of the 8th month of the ninth year of King Yung-chêng of the Great Ts'ing" (1731); on the margin of the page, the same is indicated in Chinese 大清雍正九年八月吉日 [1]).

In the Lamaist inscriptions of Peking and Jehol the days, as a rule, are not given but only the months, the Tibetan dates appearing as translations from Chinese, the year of the animal cycle being added to the Chinese Nien-hao. In the great inscription of *Yuṅ ho kuṅ* (plates 2 and 3 in the forthcoming publication of the Lamaist Inscriptions by FRANKE and LAUFER) Tib. *dgun zla daṅ-poi yar ts'es-la*, "in the first part of the first winter month", corresponds to Chin. 孟冬月之上澣; *ston zla ạbriṅ-poi ts'es-la =* 仲秋月 (plates 30, 31, 42. 43); *ston zla daṅ-poi ts'es bzaṅ-por =* 秋七月之吉 (plates 45, 47); *dgun zla ạbriṅ-poi ts'es bzaṅ-por =* 冬十一月吉日 (plates 22, 23).

Is the year indicated only in the Jovian cycle, so also the month is given with the Sanskrit term. For example, a work on the worship of the Twenty-One Forms of the Goddess Tārā written by the Second Dalai Lama dGe-ạdun rgya-mts'o dpal bzaṅ-po (1480—1542) is dated *rña c'en-gyi lo snron-gyi zla-bai yar-ṅoi ts'es brgyad-la*, "on the 8th day in the first half of the month *jyeshṭhā* (5th month) of the year *dundubhi*". This year is the 56th year of the Tibetan (59th of the Chinese) cycle answering to *water-dog* which during the lifetime of the author fell in 1502. Jovian and animal cycle are often combined, day and hour being given in Indian style. The colophon of the biography of Buddha epitomized by Schiefner runs thus: *dmar ser žes bya šiṅ p'o stag-gi lo, smin drug-can-gyi*

---

1) There is sometimes disagreement. A Mahāyānasūtra printed at Peking in the 8th year of Yung-chêng (1730) imparts in the Tibetan colophon "first part of the fourth month" (*zla-ba bži-pa-la yar ts'es-la*) where the corresponding date in Chinese offers "the 8th day of the 8th month".

zla-bai ñi-šu gñis, rgyal-ba lha-las byon-pai dus ts'igs-la, "in the hour when the Jina descended from Tushita heaven, on the 22nd day of the month *kārttika*" (the year has been discussed above p. 573).

Dates with the addition of month and day occasionally appear also in the recording of events of early history; thus, in *rGyal rabs*, mGar, the minister of King Sroñ-btsan sgam-po, set out on his mission to China "on the 8th day of the 4th month of the *fire male monkey* year" (636 A.D.) [1].

M. Pelliot points out that Tibetan chronology, in its principles, is very plain and easy. We do not doubt this for a moment. The principles of Tibetan grammar are still much easier, and yet they are violated every day by experienced Tibetan scholars in their work of translation. Theory and practice are antipodal, and whoever will dive into the study of Tibetan books on chronology and colophons with their often very complicated wordings of parallel dates in Sanskrit, Chinese and native styles, teeming with astrological determinations where the very terminology is still a mystery to us, will soon recognize that it is not exclusively aeroplanes in which it is hazardous to fly [2].

---

1) M. L. AUROUSSEAU (*B. E. F. E. O.*, 1910, p. 698) somewhat rashly accuses Mr. ROCKHILL, who on one occasion gave the date 635 for this mission of having confounded "la date de la demande en mariage (634) avec celle du mariage lui-même (641)". Mr. ROCKHILL is not guilty of any confusion in this case and is as familiar with the dates cited as M. Aurousseau who ought to have turned to his *The Life of the Buddha*, p. 213, where both dates are plainly given. The date 635 (to be converted into 636) which is entirely independent from the Chinese dates is simply that of Tibetan tradition. There are always many sides to every question.

2) To those who have the inclination to solve puzzles and can afford the time the following problem may be presented for solution. The *Lha-ldan dkar c'ag*, a work of the Fifth Dalai Lama, according to the colophon, was composed in 1645 (*sa skyoñ-gi lo*). The day is expressed in a double manner; first, it was the day of Chinese New Year, secondly it was in Tibetan *ñin byed dbañ-po gźu k'yim-du ñe-bar spyod-pai p'yogs sña-mai bzañ-po dañ-po | dbyañs ,i ac'ar-bai ts'es-la*. What is the Tibetan day, and how does it compare with the Chinese day?

Additional Note. In regard to the employment of Nien-hao on the part of the Tibetans in the T'ang period an example is given in *T'ang shu* (ch. 216 下, p. 6) in the case of King *K'o-li k'o-tsu* 可 黎 可 足 (= Tib. *K'ri gtsug*, usually styled *K'ri lde sroñ btsan*) who reigned under the Chinese title *Yi-t'ai* 彝 泰 (compare BUSHELL, The Early History of Tibet, p. 87, *J. R. A. S.*, 1880). The Nien-hao *Cheng-kuan* 貞 觀 and *King-lung* 景 龍 are utilized in the text of the Tibetan inscription of 783 published by Mr. WADDELL (*J. R. A. S.*, 1909, p. 932). —

The fact that Šambhala was a real country is evidenced by the colophon to Kanjur No. 458 (I. J. SCHMIDT, *Der Index des Kanjur*, p. 69), a text "collated with a book from Šambhala in the north". The spelling Žambhala is adopted by GRÜNWEDEL (*Mythologie des Buddhismus*, pp. 41, 42, 58, 244), which is not authorized by any Tibetan text known to me; but in *Die orientalischen Religionen*, p. 161, GRÜNWEDEL writes correctly Šambhala. What is more important, Grünwedel concurs with me in the opinion that the calendar of Tibet is derived from Šambhala, and more specifically refers to Atiça as having introduced the present form of the calendar and time-reckoning based on sexagenary cycles (Mythologie, p. 58). Grünwedel is likewise correct in stating (p. 205) that "the saints practising the cult of the sun-chariot" in Šambhala point to Iranian conditions, and this chimes in with his view that the country of this name should be located on the Yaxartes. —

After the above was written, I received, through the courtesy of the Oriental Institute of Vladivostok, Part II of Lama TSYBIKOV's *Lam-rim chen-po* containing the Russian translation of the Mongol text published in Part I and with a very interesting introduction. On p. XIII, Lama TSYBIKOV, in discussing the date of bTsoñ-k'a-pa, alludes to the year 1027 as that of the first year of the first cycle.

099

巴比伦尼亚的孔雀

Orientalistische Literaturzeitung

Monatsschrift

für die Wissenschaft vom vorderen Orient

und seine Beziehungen

zum Kulturkreise des Mittelmeers

Herausgegeben

von

Felix E. Peiser

Sechzehnter Jahrgang
1913

Leipzig
J. C. Hinrichs'sche Buchhandlung
ED

Eine Schwierigkeit macht nur das β, für das ein μ zu erwarten wäre und das nur im Griechischen auftritt, obgleich hier ein $\gamma\nu$ ungleich häufiger auftritt als ein $\gamma\beta$. Da nach Wilamowitz die Timotheos-Handschrift bereits das β enthält, ist natürlich nicht anzunehmen, dass erst aus einer Verdrehung Ex-$\beta\alpha\tau\alpha\nu\alpha$ mit Anlehnung an $\beta\alpha\iota\nu\omega$, das übrigens lautgesetzlich = iran. gam ist, das β an Stelle eines μ getreten wäre. Es wird also kaum etwas anderes übrigbleiben, als die Annahme einer mundartlichen Form, die schon iranisch das m durch b ersetzte, wie wir umgekehrt $M\alpha\varrho\delta o\varsigma$, $\Sigma\mu\varepsilon\varrho\delta\iota\varsigma$ für $Bardija$ finden.

Der Pfau in Babylonien.
Von B. Laufer.

Zu der Frage des Herrn Meissner (Sp. 292/3 dieser Zeitschrift) möchte ich bemerken, dass die Feststellung eines assyrisch-babylonischen Namens für den Pfau dem Kulturhistoriker von grossem Wert wäre. Haben wir doch von indischer Seite das Zeugnis des Bāveru-Jātaka (zuerst in Text und Uebersetzung herausgegeben von J. Minayeff, *Mélanges asiatiques*, Bd. VI, 1872, S. 577—599), worin der Transport eines Pfaus von Indien zur See nach Bāveru (Skr. *Babiru*, d. i. Babylon) erzählt wird (vgl. auch G. Bühler, On the Origin of the Indian Brāhma Alphabet, p. 84). I. Kennedy (The Early Commerce of Babylon with India, JRAS, 1898, p. 269) hat auf Grund dieser Tatsache und der griechischen Nachrichten berechnet, dass Pfau und Reis bereits im sechsten Jahrhundert v. Chr. von der Westküste Indiens im überseeischen Verkehr nach Babylon gelangten, und nach Bühler hätte dieser Handel wahrscheinlich schon weit früher bestanden. Die orientalischen Elemente, die im Pfauenkultus der Hera auf Samos mitspielten, werden in letzter Instanz wohl auch auf Babylonien zurückgehen. Dass der Pfau bereits 738 v. Chr. in einer Inschrift Tiglatpilesers III. genannt sein soll, erscheint durchaus plausibel, und es ist wohl zu hoffen, dass Herr Meissner noch weitere Entdeckungen in dieser Richtung machen wird.

Warum Herr Meissner an der üblichen Erklärung von hebr. תֻּכִּים aus Tamil *tokei, togei* (Caldwell, Comparative Grammar of the Dravidian Languages, p. 66), dazu Skr. *çikhin*, griech. $\tau\alpha\acute{\omega}\varsigma$ (*tahōs, tawōs*), pers. *tavus*, zweifelt, ist mir nicht recht verständlich. Mir erscheint diese Gleichung als durchaus gesichert und annehmbar. Wenn auch Indien nicht als Bestimmungsort in Kön. und Chron. genannt ist, so kann doch gar kein Zweifel daran bestehen, dass König Salomos Pfauen wirklich aus Indien nach Palästina gebracht worden sind, denn der Pfau kam damals nur in Indien vor; in Hinter-

indien, im südlichen China, und auf Java erscheint er erst als später Import in nachchristlicher Zeit.

Schliesslich möchte ich mir als Laie in assyrischen Dingen erlauben, an Herrn Meissner zwei Fragen zu richten. Warum ist in der achten Auflage von Hehns Werk, vermutlich auf seine Veranlassung (s. S. XXIV), da in der siebenten Auflage noch nicht vorhanden, auf S. 363 ein „assyr. *pa-'-ú* ein Vogelname?" hinzugekommen, was den Eindruck macht, als sei hiermit auf lat. *pavo* angespielt, während sowohl F. Delitzsch (Assyrische Thiernamen, S. 109) als auch W. Houghton (The Birds of the Assyrian Monuments, TSBA, Vol. VIII, 1884, p. 81) dieses Wort als „Krähe" deuten, und Meissner selbst Anm. 2 auf Sp. 293 die Bedeutung „Pfau" anzweifelt? Wie steht es mit den von Delitzsch (S. 105 und 106) und Houghton (p. 111) mit der Bedeutung „Pfau" bedachten assyrischen Vogelnamen? Sind diese Deutungen nicht mehr zulässig, und wie sind dann diese Wörter zu erklären? Neben Hehn ist jetzt für die Geschichte des Pfaus im Altertum O. Keller (Die antike Tierwelt, Bd. II, S. 148—154, Lpz., 1913) zu empfehlen, ein Werk, das jedem Kulturhistoriker die reichsten Anregungen geben wird.

Besprechungen.

Stephen Langdon: Babylonian Liturgies. Sumerian Texts from the Early Period and from the Library of Ashurbanipal, for the most part transliterated and translated, with introduction and index. With 75 plates. Gr. 8. LII, 151 pp.; LXXV pl. Paris, Paul Geuthner, 1913. Bespr. v. A. Ungnad. Jena.

Dieses neuste Werk Langdons ist vor allem eine Textpublikation sumerischer Liturgien. Sie umfasst etwa 200 Nummern, teils gut erhaltene grössere Stücke, teils kleine Fragmente, mit denen sich oft nicht viel anfangen lässt. Zeitlich kann man das publizierte Material in zwei Gruppen scheiden: der Hauptsache nach gehören die Texte der Kujundschik-Sammlung des Britischen Museums, also der Zeit Asurbanipals, an. Ausgenommen sind zunächst acht grösstenteils ziemlich gut erhaltene Texte des Royal Scottish Museums zu Edinburgh, die augenscheinlich aus dem gleichen Funde herrühren wie die kürzlich von Zimmern edierten Hymnen des Berliner Museums (VS. II): sie gehören etwa der zweiten Hälfte der Isin-Dynastie oder, was dasselbe ist, der ersten Hälfte der Hammurapi-Dynastie an.

Etwa derselben Zeit entstammen drei sumerische religiöse Texte des Ashmolean Museums zu Oxford, darunter ein ganz eigenartiger, den Langdon als Nr. 197 bringt: es ist ein vierseitiges leider nicht gerade gut erhaltenes Prisma,

100

史密斯夫人所藏中国古代鼻烟壶目录

Catalogue of a Collection of
Ancient Chinese Snuff-Bottles

In the possession of
Mrs. George T. Smith
Chicago

By
Berthold Laufer

Chicago
Privately Printed
1913

INTRODUCTION

IN the art and culture of China snuff-bottles occupy the same place as the engraved gems or intaglios in ancient Greece. In range of material and scope of artistic execution, however, the former cover a much larger ground than the latter. Not only are the finest stones capable of carving utilized for this class of work but also glass, strass, pottery, porcelain, enamel, carved lacquer, ivory, amber, coral, nutshells, mother-o'-pearl, are laid under contribution to the same end. In short, almost every available substance in the three kingdoms of nature is here represented, and geographical regions so widely apart as the Baltic coast of Prussia furnishing amber, the Arctic waters supplying the ivory tusk of the walrus, and the tropical jungles of Borneo and Sumatra yielding the wonderfully colored beak of the hornbill are closely united in the present collection. Owing to the wealth of design displayed on the exquisite pieces it becomes a microcosm of Chinese art with many marvelous examples of the skill and versatility of the lapidary and potter.

The origin of the manufacture of snuff-bottles in China may be traced back to the latter part of the seventeenth century. It is naturally an event contemporaneous with the introduction of snuff into the country. Tobacco is a plant of indigenous growth on this continent and became known in the Old World only after the discovery of America. It was first brought to Europe by Francisco Hernandez in 1560 and introduced into England by Capt. Ralph Lane in 1586. About the same time, as told in Chinese records, it made its appearance in China from the Philippine Islands where the Spaniards had planted it. The first place in China where tobacco was cultivated was at Amoy where many Portuguese had been settled since 1544. Another Chinese tradition ascribes the first importation of tobacco to the Japanese who became familiar with the seeds brought to them by Portuguese merchant-vessels around 1585. In all likelihood, several introductions of tobacco into China took place; it is certain that it was transplanted into Manchuria from Japan by way of Korea. Tobacco was first consumed only in the dry pipe and the water-pipe. Snuff came into vogue but a century later when, toward the close of the seventeenth century, it was brought over from Japan where it had been imported by the Dutch. From that time also snuff-bottles came into use. Taking snuff has never become a

[3]

very general or popular habit, but has remained a luxury confined to the well-to-do classes who pay fancy prices for foreign
brands. They attribute to snuff medicinal virtues and beneficial
effects, particularly after a heavy dinner, so that it is also taken
for curative purposes or made the vehicle of conveying other
medical agents into the system. The Chinese believe also in
ancient snuff, and the dealers in antiques sell snuff supposed to
be a century old or even older. A snuff-bottle is part and parcel
of a gentleman's dress; it holds about a tablespoonful of the dust,
which when taken is placed on the thumb-nail with the ladle
attached to the stopper.

Most forms of bottles which were employed for holding snuff
were not novelties invented at that time but were rather reproduced from much older types—small bottles containing medicines,
especially eye-medicines. Such a relic of the Sung period, a barrel-
shaped vessel of glazed pottery, is preserved in this collection
(No. 466); it was originally a medicine-bottle but in that period
when snuff appeared, since snuff was placed in the category of
medicines, was easily turned into a snuff-bottle by the addition of
a stopper with ladle. All the many cylindrical forms of bottles
in this collection are the direct and legitimate offspring of the drug-
phial of olden days. Also the shape of the calabash chosen with
so much predilection goes back to the apothecary; the druggists
still preserve their remedies in calabashes which are credited with
all sorts of marvelous properties.

Of the precious materials listed in the cause of snuff-bottles,
jade claims the first rank. It has been the most highly appreciated jewel of the Chinese from remote ages and idealized by their
poets and artists as the quintessence of nature, as the embodiment
of all virtues instilling virtue, sublime thoughts and good conduct
into the hearts of its wearers. Most of the jade utilized in this
collection is the variety known as jadeite quarried on the upper
Chindwin River in Burma and imported in large quantities to
Canton. Several bottles are carved from ancient jade, so-called
Han jade, because it was chiefly employed during the Han dynasty,
obtained from rolled pebbles fished from the bed of the Yellow
River and its affluents. No words can describe the endless variety
and beauty of color displayed by these carvings of nephrite, rock-
crystal, amethyst, agate, heliotrope and other stones, and the
ingenuity of the lapidary whose master-hand has brought out to
their fullest effect these wonders of nature. Their carvings of
glass are no less admirable than those of stone.

The early acquaintance of the Chinese with glass is due to their
intercourse trade with the emporiums of Syria and Asia Minor of
the Roman Empire, dating back to the second century before

[4]

Christ. They received from them, in exchange for their silk products, woolen weavings, drugs, precious stones, pearls, and glass "of ten different colors." A flourishing importation of this material continued until the beginning of the fifth century, when merchants from the northwestern borders of India arrived at the court of the Wei dynasty in A. D. 435, and taught the process. The plasticity and flexibility of glass softened by heat did not appeal to the Chinese, who cut and polished glass in its hard state, like semi-precious stones. Directions and models for this, as well as for cutting ornamental objects out of jade, rock-crystal, amethyst, and other quartzes, were contained in their ancient works of glyptic art, which was the preparatory schooling for glass-cutting. They understood also how to produce an astonishing wealth of color in glass in all the tinges imaginable, for which our weak nomenclature of color has no designation. They imitate in glass the color of jade, agate, malachite, lapis lazuli, amber, jet and coral. This is effected by means of metal oxides—iron pyrites, iron oxides, copper oxides, acetate of lead, and others. Sometimes the color of the glass is brought to its full perfection by a simple process of polishing; sometimes feet, handle, and neck are added to the body of the vessel made of glass of a different color. In many cases there are several layers of glass of various tints, placed side by side, or one above the other, out of which scenes or figures are cut in relief with great technical skill, that stand out from the body of the vessel as in cameo-work, or in the style of antique Roman glass.

The principal seat of the glass industry is the city of Poshan in Shantung, which is also famous for its pottery. A sort of grayish calcareous stone containing carbonate of lime, which occurs in the mountains of the neighborhood, is mixed with saltpetre for obtaining the raw material which is cast in pigs and exported to Peking and Canton. Opaque glass is made by the aid of fluorspar or fluorite. They are skilful also in fusing together glasses of different tinges, or in introducing into the mass spots, veins, or ribbons, always with the view of rivaling nature in the imitation of stones which are taken as models.

An industry peculiar to the Chinese is formed by snuff-bottles with decorations colored by hand. In this case, the paints are applied to the inside of the bottles, to preserve them from friction, which requires a large amount of skill and patience. The bottles used for this purpose are generally imported to Peking from Canton, and are made from the first quality of glass, called "rock-crystal." As the surface of the glass is too smooth to take the paints, the inside is specially prepared for this purpose with pulverized iron-oxydul mixed with water, which is shaken in the

[5]

bottle for about half a day.   A rough milk-white coating is thus formed, suitable for taking the colors.   In doing the work the painter lies down on his back, holding the bottle up to the light with the thumb and index-finger of his left hand, and a very fine brush in his right.   His eyes are constantly fixed on the outer surface of the glass, thus watching the gradual development of the picture as it emerges from under the glass.   He first outlines the picture with India ink, starting from below, and then passing to the middle and the sides, and finally inserting the paints.   Half a day is sufficient to complete an ordinary piece, while a whole day and more is spent on more elaborate work.   The subjects include landscapes, flower-pieces, and genre paintings.   The specimens in this collection are remarkable, because they are attributed to the Tao-kuang period (1821–1850).   Very little has survived from this time owing to the ravages of the T'ai-p'ing rebellion which then swept over the country.

Many of the soft-paste and blue and white porcelain bottles included in this collection belong to the finest of their type, and nowhere else may be seen such a magnificent variety of superb ambers, agates, and rock-crystals.   The majority of the specimens in this collection come down from the K'ien-lung period (1736–1795) with the exception of those marked otherwise.

BERTHOLD LAUFER.

[6]

Catalogue of a Collection of Ancient Chinese Snuff-Bottles

1. Rock-crystal without flaw and of perfect carving. Two lateral tiger-heads holding rings in their jaws are represented in relief. This is a motive of art going back into times of great antiquity when it appears on sacrificial bronze vessels and mortuary pottery. The rings were originally practicable and movable, serving the purpose of holding and carrying a vessel, while in the early pottery of the Han dynasty they began to be molded into reliefs, and to shrink into a conventional, artistic motive. The tiger was regarded by the ancient Chinese as a supernatural creature and solar animal capable of warding off demons and obnoxious influences.—Green jadeite stopper.

2. Rock-crystal, of extraordinary size and imposing shape. Pink crystal stopper mounted on plated silver; plated spoon. On one side is carved in high relief a group of five horses, each in different action and motion, a mountain scenery in the background. On the opposite side there is an etching in low relief of a flower and bamboo still-life.

3. Flawless rock-crystal, with stopper of pale green jadeite coming from Burma. One side displays a flat design of mountain scenery with brook and bridge. The other side displays a high-relief carving of the God of Longevity, a personification of the pole-star, represented as an old, long-bearded man with high forehead holding a sceptre of good augury which is believed to grant the fulfillment of every good wish. He is standing under a pine-tree, an emblem, from its evergreenness, of strength, endurance and long life, and accompanied by a deer, the emblem of good income. This scene implies for the recipient of the gift the wish, "May the God of Longevity, as you desire, confer upon you numerous years and good income!"

4. Milk-white, semi-transparent agate, with ruby-colored chrysoprase stopper, of unusual dimensions. Decorated with lateral tiger-heads holding rings in relief, otherwise plain.

5. Pudding-stone of gray background filled with yellow, brown and black speckles. Swelled body, with softly rounded outlines. Plain. Green jadeite stopper.

[7]

6. Chocolate-brown stone of lustrous polish, with green jadeite stopper, and ivory spoon. Magpie, a bird of lucky omen, and stag turning his head toward the bird are carved in relief on one side. On the other side is depicted a scene of a monkey trying to catch a butterfly. These designs present a readable rebus based on punning, with the meaning: "May happiness and good income await you in your old days!"

7. Chocolate-brown glass, overstrewn with brilliant gold dots, with bottle-green glass stopper mounted on a white marble ring. Plain.

8. Agate-like brown-red stone with vertical streaky zones, of high specific gravity. Silver stopper in shape of a rosette inlaid with a coral in the centre which is surrounded by three turquoises and three red stones bordered by a ring of small knobs. Decorated with genre-scenes in relief: on one side a Taoist genius called The Old Recluse is shown asleep leaning toward a big calabash known under the name "the great calabash of the male and female energy." It is supposed to be filled with a cosmic ether which has the property of hypnotizing and capturing any evil demon causing harm and disease to man. Healing power is therefore ascribed to the calabash, and the druggists preserve their medicaments in gourds or in bottles imitative of their shape. On the other side we notice a well-fed man and his wife conversing with a Taoist monk followed by a boy. The man and the monk point with their fingers toward the background from which it becomes evident that the Old Recluse and exorcist is the object of their confabulation; his efficient services are evidently recommended by the monk to his interlocutor.

9. Yellow fossil amber, veined like wood, and interspersed with whorls and black flecks and stripes. Of extraordinary size and elegance of shape; short neck, broad bulging shoulders, the sides gradually tapering downward, oval flat bottom. Pink crystal stopper. Unornamented except the two lateral lion-heads holding rings in their jaws, their manes being represented by rows of spirals.

10-11. Pair of bottles carved from transparent tea-leaf colored crystal.—Stoppers of green and white jade mounted on marble rings. Plain.

12-13. Pair of bottles carved from fine brown, transparent agate, plain, of elegantly rounded outlines. Stoppers of green jadeite mounted on white jade rings.

[8]

14. Whitish agate, of gray and brown background overstrewn with snow-white crystals looking like icicles. Of unusual size. Plain. Stopper of blue glass, carved with six spiral-like ornaments styled by the Chinese "reclining silkworm cocoons."

15. Brown, semi-transparent agate, of unusually large dimensions. Plain, except two lateral tiger-heads in relief, holding rings in their jaws. Stopper of green jadeite.

16. Bloodstone, of uniform peculiar olive hue with some sepia or purple flecks. Plain, extraordinary in size. Stopper of carnelian mounted on plated silver; spoon plated.

17. Glass carved in cameo style, with yellowish background. Decorated all over with various fruits, Buddha-hand citrus, peach and pomegranate in yellow, green and ruby colors. Stopper of green jadeite from Burma, mounted on marble ring.

18. Red clouded agate with green and black flecks. Undecorated, except the two lateral tiger-heads holding rings in their jaws, in relief. Stopper of green jadeite.

19. Green and brown agate. Plain. Stopper of lapis lazuli with a bead of pink crystal inlaid in the centre.

20. Semi-transparent violet amethyst. Extraordinary in dimensions. Decorated with lateral tiger-heads in relief, otherwise plain.—Stopper of pale-green jadeite.

21–22. Pair of bottles carved from semi-transparent pale agate. Stoppers of crystal. The one bottle shows a single horse of yellow tinge in relief, accompanied by a seal which designates it as "a breed of first quality." The other bottle displays a horse tethered to a pine-tree, and a monkey eating a peach. This is a famous mythological subject: the peaches conferring immortality upon man are believed to grow in the paradise located upon Mount Kun-lun over which the goddess Si Wang Mu presides; the monkey stealthily pilfers it to ensure rebirth in the body of a man.

23. Carved from a nut, also the stopper which is mounted on a silver ring and exhibits flowers in open-work carving. Silver spoon. The scene carved on the one side shows the celebrated art-motive of squirrels stealing and eating grapes; vine leaves and bunches of grapes are gracefully outlined and bordered above

[9]

by a wreath with a peach at the left and a calabash at the right, a squirrel rushing along and timidly looking backward. On the opposite side there is a pine-tree in the centre, a deer standing at its foot, a Buddha-hand citrus above to the left, a persimmon with leaf to the right, and two flower vases on the sides. These ornaments form a rebus reading: "May you have happiness, long life and good income, and peace in all your affairs!"

24. Crystal of violet tinge, with stopper of red crystal. On one side carvings in high relief showing vine leaves with grapes and a squirrel, in gray, white and brown. The other side displays a fine etching of pine-tree and bamboo which as a rebus means "we beg for long life," and five birds on the wing in sketchy outlines.

25. Violet transparent crystal with stopper of same substance and color. A pine-tree is engraved in low relief on one side; on the reverse eight horses, each showing a different action and motion, are carved in white relief. These are known as "the eight stallions of Mu Wang" (B. C. 947) who was the first sovereign of the Chou dynasty and undertook many campaigns and journeys toward the West when he was driven by his charioteer Tsao Fu with his eight famous steeds "wherever wheel-ruts ran and the hoofs of horses had trodden." These eight horses have been painted by numerous artists.

26. Brown, black and white clouded agate. Stopper of green glass. Plain.

27. Milk-white semi-transparent agate with relief carvings of ivory hue. A goat is feeding on a kind of agaric or fungus; on the reverse a crouching goat and another recumbent. Altogether three goats are represented which by a pun upon the words is symbolic of "threefold happiness." The fungus is the symbol of immortality, so that this design is intended to express for the recipient the wish: "Threefold happiness and old age!" Stopper of carnelian.

28. Light-yellow transparent glass cut in cameo style. Stopper of green jadeite. A white cat with black paws and tail is sitting in a tree and leering at a butterfly, while a widely opened peony and a bird perching on a tree are represented on the reverse. This is a famous subject of art represented by many painters. The Chinese have made the observation that the cat changes the form of her pupil in the course of a day, that it is round in

[10]

the morning, gradually growing longer and thinner during the day, and becomes a mere slit toward noon. A cat with such a pupil serves to illustrate a hot summer noon, with the addition of widely opened peonies. In the morning, the Chinese say, the blossom of the peony wet with dew is contracted, but displays its petals at noon. Aside from its naturalistic import, this picture has an allegoric significance suggested by a rebus and expresses the wish: "May you obtain wealth, honorable position and old age!"

29. Milk-white transparent agate, with stopper of green jadeite. The reliefs are carved in cameo style, black, brown and yellowish layers being skilfully utilized. The designs of the one side are a monkey squatting on a rock, another standing upright, a spotted stag looking backward at the magpie flying into the open from a pine-tree. The reverse exhibits a bat which is 'an emblem of luck, a leaved fungus of immortality, and another such fungus growing out of a rocky boulder. Thus, these ornaments imply good wishes for the owner of this bottle, happiness, good income and long life.

30. Glass of powdered white background carved in cameo style with reliefs of ruby color. Stopper of green jadeite. The subject is "cranes in a lotus-pond." Three cranes are wading in the water, one pecking at a leaf, the second drinking, the third with outspread wings and open bill turning its head backward toward its companion on the wing above in the air. The crane has the reputation of being a long-lived creature; of the four kinds assumed by the Chinese, viz., black, white, yellow and blue, the black is believed to be the longest lived. Cranes serve as aeroplanes to the Taoist gods, and saints who have led a life of purity and solitude are carried heavenward on their backs. At funerals paper images are cremated, in the belief that the departed spirit will soar with it into the regions of eternal bliss. The lotus is the emblem of purity of heart, because its petals and leaves are spotless, although its stalk takes root in the mire.

31. Reddish-brown agate carved in cameo style with reliefs of ivory color, cut out of the matrix. Stopper of green jadeite. On one side four lions are represented in the act of playing ball; the old lioness is watching her three cubs, each of which is standing on a ball; they are anxiously looking at one another, as if they apprehended that the ball might be taken away from them. The representation of this scene is remarkable for the unity of composition. On the reverse the happy Taoist genius, the Old

[11]

Recluse, with smiling face and straw sandals on his bare feet is cowering on a rock; in front of him there is a calabash emitting an ether on the top of which two bats, a symbolic expression for blessings, appear (compare No. 8). The calabash, in this case, takes the place of our horn of plenty.

32. Reddish-brown semi-transparent amber; broad-shouldered, unusually large piece. Stopper of green jadeite.

33. Fiery red and black semi-transparent amber, styled "gold amber" by the Chinese, with fine lustre. Lateral tiger-heads holding rings in their jaws carved in high relief. Stopper of green jadeite.

34. "Longevity bottle" of carved white porcelain of soft paste, with stopper of coral. The designs and characters are sharply cut out in strong relief. The background is formed by continuous rows of swastikas; the characters laid over them and enclosed in medallions, altogether seventy, are all ornamental variations of one and the same word *shou*, "longevity." The wish expressed by this design for the owner of the bottle is that he may live to the age of seventy. This is a work of exceptional beauty and quality.

35. Porcelain of deep black and lustrous glaze, painted with enamel colors. Decorated with sprays of chrysanthemums in white and pink. Barrel shape; extremely light and thin. Globular stopper of malachite.

36. Blue and white porcelain of soft paste, with gray crackles. Stopper of pale coral mounted on turquois ring. Underglaze painting of two five-clawed imperial dragons playing with the flamed pearl, simultaneously emblematic of the sun. The dragons' eagerness for reaching the pearl is compared with the human struggle for an ideal which may be approached but never seized. Along the lower border the sea is represented; one of the monsters is floating amidst the waves exhibiting only its massive head and three claws of a paw, the other soaring in the clouds is laid in full size around the middle and upper portion of the bottle. Date-mark in blue on bottom: "Made in the period Yung-chêng (1723–1735).

37. Lion-skin jade. The one side is milk-white and spotless, the other side is clouded with reddish-yellow speckles similar to those occurring in Han jade. This side is perfectly flat, and the

[12]

neck is not set off by an incision as on the opposite side. Stopper of pink crystal mounted on blue turquois plaque.

38.   Lion-skin jade, whitish with yellow veins and stripes, of irregular shape as found in nature. Stopper of pale green jade carved in the form of the stem of a fruit, as the shape of the bottle is suggestive of a fruit. Of considerable age.

39.   Milky agate carved with reliefs of deeper tinge. Stopper of coral on which a coiled hydra is cut out in relief, mounted on a green turquois ring. The subject of the scene is the abode of the blest according to the notions of the Taoists. A bearded saint holding a wine-cup is standing on a bridge, while a boy in attendance on him is heating water on a pottery stove; the scene is overshadowed by a pine-tree. On the other side a saint is rowing a boat in which a boy is seated, the pond being screened off by a wall. A camel-shaped bridge is on the left-hand side, and a double-roofed building is hovering on clouds in the air.

40.   Black and white agate in its natural shape, carved with reliefs. Stopper of green jade carved into the shape of a crouching hare hiding its head between its front-paws. This is a representation of the mythological hare believed to inhabit the surface of the moon and to pound in a mortar the drugs composing the elixir of life. This hare is called "jade hare," and for this reason is here carved from jade. The hare as an animal is reputed as deriving its origin from the vital essence of the moon and is subject to her influence. A uniform composition is laid around the bottle. The entire surface is taken up by a plum-tree in blossom, an emblem of beauty. The other elements of the composition are a butterfly sucking at a blossom, emblem of old age; a bat, emblem of luck; two magpies, birds of lucky omen, connected by a large lotus leaf which as a rebus is read in the sense of union, in this case indicating the happy union of the two birds; a spider-web with two spiders, emblems of joy (a spider is a good omen when seen in the morning); and a pheasant with long tail-feathers and open bill, looking backward.

41.   Bottle carved from the hornlike excrescence of the upper mandible of the helmeted hornbill (*Rhinoplax vigil*), a bird inhabiting Sumatra and Borneo. It is composed of a dense and solid mass of horn capable of being carved into plates, figures, charms, brooches and other ornaments. The natural colors of the bill, yellow and crimson, are skilfully utilized in the carving of this bottle in that the yellow portions form the two large surfaces, and

[13]

the crimson parts the two narrow sides. Stopper of green turquois. The large surfaces are plain. One of the lateral sides shows a coiled hydra surrounded by floral designs, in red relief; the other floral designs.

42. Glass of white ground, treated in cameo style. Stopper of green jade. The reliefs, of pink color, represent a phenix holding in his beak a fungus of immortality and looking backward at the sun-ball, and on the reverse a dragon soaring in clouds, both reliefs enclosed in a medallion formed by a pink ring. On the narrow sides the eight lucky Buddhist emblems introduced from India where they were carved on the sole of Buddha's foot are figured, viz., the wheel, the symbol of the king and the Buddha who turns the wheel of the law; a conch-shell, used as a trumpet at religious festivals, state-umbrella and canopy, attributes of royalty and high rank, the lotus flower, emblem of purity, the vase containing sacred relics, pair of fish, and the so-called çrivatsa (a Sanskrit word), an ornament known in China under the name "felicitous threads," originally a mystic sign on the body of the Hindu god Vishnu.

43. Opaque brown glass painted with enamel colors in green, yellow and red. Lotuses growing in a pond on the one side and orchids on the other; a band of floral designs is displayed around the neck. Stopper of coral mounted on turquois-blue glass ring.

44-45. Pair of bottles carved from clouded agate, dark and light green with red and purple flecks, of fine lustre and polish. Lateral tiger-heads holding rings in their jaws, carved in relief. Stoppers of pale pink crystal mounted on white marble rings.

46. Carved cinnabar-red lacquer. Stopper of white jade. The background is formed by continuous meander patterns combined with swastikas, and sixty ornamental forms of the character *shou*, "longevity," are represented on each side and enclosed in a medallion. "May you attain a hundred and twenty years!" is the wish expressed by this design for the owner of the bottle.

47. Gray agate, with stopper of green jadeite. The two circular sides are carved with the obverse and reverse of a Spanish Carolus dollar, the letters of the Latin legends having partially been misunderstood or inverted by the Chinese engraver. It is styled "pillar" dollar from the design of the two pillars of Hercules (the two rocks at the entrance to the Mediterranean Sea) joined by a scroll inscribed with the legend *Ne plus ultra*,

[14]

"nothing beyond," and supporting the arms and crown of Spain. This dollar was known to the Chinese under the name "devil's head" money, from the royal head on the obverse.

48. Amber-colored glass carved in cameo style: Stopper of dark-green jade. A pine-tree and a willow with bamboo-leaves growing from behind a rock are carved on the narrow sides. The large surfaces are each divided into an upper and lower panel, and decorated with figures of horses, each with a different action, running, grazing, looking upward or backward, and evincing friendly sentiments for one another.

49. Opaque glass colored red and black, with veins and clouds, in imitation of agate. Stopper of green glass. Plain.

50. Opaque glass colored brown with dark and red speckles bordered by yellow lines, in imitation of agate. Stopper of amethyst. Plain.

51. Glass of white background colored in brown horizontal zones. Decorated with lateral tiger-heads, holding rings in their jaws, in flat relief. Stopper of pink crystal.

52. Blood-red opaque glass with lighter bands and flecks, in imitation of the glaze of sang-de-boeuf porcelain. Stopper of green jadeite. Plain.

53. Pale-green opaque glass, in imitation of the sea-green glaze of celadon porcelain. Stopper of green glass.

54. Yellow opaque glass, in imitation of beeswax color. Stopper of green glass mounted on yellow glass ring. Plain.

55. Black opaque glass, in imitation of the color of jet or agate. Stopper of green jadeite. Plain.

56. Blue glass in imitation of blue crystal. Decorated in relief with lateral tiger-heads holding rings. Stopper of pink crystal. Plain.

57. Semi-transparent opalesque glass. Stopper of pale-green jadeite. Plain. The narrow sides have oval medallions in relief.

58. Dark-gray colored opaque glass. Stopper of green

[15]

jadeite. Circular medallions on the large surfaces and oval medallions on the narrow sides.

59. Gray opaque glass, in imitation of agate, of circular shape. Stopper of green jadeite. Plain.

60. Brown opaque glass of elegantly slender flask shape. Stopper of dark-green glass. Plain.

61. Glass colored in imitation of black lacquer interspersed with gold speckles. In shape of watch. Stopper of imitation coral. Plain.

62. Garnet-red opaque glass. Stopper of carnelian mounted on silver. Plain.

63. Brown and yellow glass, in imitation of agate. Stopper of green jadeite. Plain.

64. Chocolate colored opaque glass of deep, bright lustre. Stopper of green glass. Plain. In shape of a covered teacup.

65. Glass, white with yellowish tinge, in imitation of agate. Stopper of green jadeite. Plain.

66. Milk-white glass, in imitation of white jade. Rectangular shape imitative of a form of ancient pottery vessels. Stopper of pink glass. Plain.

67. Milk-white glass, in imitation of white jade. Octagonal shape formed by eight rectangular facets. Stopper of green jadeite. Plain.

68. Milk-white glass, in imitation of white jade. Shape of teacup with cover. Stopper of red crystal.

69. Grayish glass, in imitation of jade. Same shape as 68, but larger. Stopper of red crystal.

70–71. Pair of bottles of milk-white opaque glass, in imitation of jade. Stoppers of carnelian. Plain.

72–73. Pair of bottles carved from milky opaque glass, in imitation of jade. Octagonal shapes formed by eight oval facets. Stoppers of pink crystal.

[16]

74. Milky glass, in imitation of jade. In shape of teacup with cover. Stopper of carnelian of cherry color.

75. Milky glass, in imitation of jade, carved into the appearance of a wickerware basket. Stopper of brown agate set in gold.

76. Milky glass, in imitation of jade and porcelain. In shape of a porcelain teacup with cover. Stopper of green jadeite.

77. Milky glass, in imitation of jade. Stopper of green jadeite. Upper portion plain, lower portion carved and decorated in three zones, a conventionalized blossom being encircled by a band of lozenges and rows of petals enclosing an ornamental form of the character *shou*, "longevity."

78. Milky glass, in imitation of jade. In shape of watch, of very graceful forms. Relief carvings in blue of lateral lion-heads holding rings in their jaws, of fine workmanship. Stopper of coral.

79. Glass colored in imitation of the sea-green glaze of celadon porcelain. Trapezoidal shape. A blossom of four petals is carved out in relief in the lower portion from which the upper part of the bottle emerges. Stopper of carnelian mounted on white marble ring.

80. White glass inlaid with various colors and imitation white jade. Stopper of coral.

81. Semi-transparent lilac colored glass. In shape of watch. Stopper of carnelian. Plain.

82. Lapis-lazuli colored glass dotted with black spots. Stopper of dark-red crystal.

83. Dark-red colored glass, in imitation of agate. A single leaf is painted under the surface. Stopper of green jadeite.

84. Blue-colored opaque glass. Circular shape, with white neck. Stopper of carnelain. Plain.

85. Blue-colored semi-transparent glass. Circular shape. Stopper of crystal. Plain.

86. Blue-colored semi-transparent glass. Hydras holding

[17]

in their jaws the fungus of immortality are carved in relief on the lateral sides. Stopper of carnelian.

87. Glass of white background, carved in cameo style. The reliefs, of ultramarine blue, represent a dragon rising from the ocean on the one side, and on the reverse a rock merging from water, a striped tiger with a swastika, a design of happy augury, on its chest, and a bat above, the emblem of luck. The narrow sides are decorated with lion-heads holding rings in their jaws. Glass stopper of amber color.

88. Transparent green glass. Carved in shape of a gourd. Large leaves with tendrils and gourd fruits are spread all round in relief. Stopper of coral representing a stem.

89. Transparent blue glass. Carved in shape of gourd, with the same designs as on the preceding piece. Stopper of red and whitish coral carved à jour.

90–91. Transparent light-blue glass. Hydras with long manes are cut out in relief on the narrow sides; their tails are so intertwined as to form simultaneously the rim laid around the bottom of the bottle. This design is an emblem expressive of friendship. Stoppers of pink crystal.

92. Glass carved in cameo style. Transparent background, the characters standing out in high relief of lapis-lazuli color. They are enclosed in medallions and fashioned in the form of seals of very ancient style going back to the pre-Christian era (Han period), expressing the wish: "May you have joy without end!" Stopper of coral.

93. Purple-colored glass strewn over with black spots, in imitation of crystal. Surfaces and sides rounded. Stopper of green glass. Plain.

94. Opalesque glass strewn over with red patches. Stopper of dark-green jade. Plain.

95. "Longevity bottle," of light-yellow, semi-transparent glass carved with forty-eight ornamental forms of the character *shou*, "longevity." The bottle was evidently presented to a man on the occasion of his forty-eighth birthday. Stopper of carnelian.

[18]

96. Tea-leaf colored porcelain veined with fine yellow lines. In shape of watch. Stopper of carnelian. Plain.

97. Transparent green glass. Carved with a pair of hydras with double tails playing hide and seek and looking away from each other—an emblem of friendship. Stopper of pink crystal.

98. Bluish glass. Hydras with long manes are carved in relief on the narrow sides, their tails being intertwined and simultaneously forming the rim of the base of the bottle—emblem of friendship. Stopper of amethyst.

99. Glass carved in cameo style, the reliefs of pink hues standing out from a milk-white background. Scenery formed by rocky boulders emerging from water, pine-tree, two fungi of immortality and a flying bat—emblems of long life and happiness. On one of the narrow sides is displayed a large cloud in massive spirals with a constellation of three stars above. Stopper of green jadeite.

100. Green glass in imitation of the color of emerald. Each side exhibits a hydra in relief; the two creatures are playing hide and seek and form an emblem of friendship. Stopper of crystal mounted on ring of white jade.

101. Gray glass, in imitation of agate, interspersed with red spots. Stopper of green glass. Plain.

102. Colored glass in imitation of the color of ruby. Octagonal in shape. Each side is carved into a medallion of ellipsoidal form with four facets of triangular shape. Stopper of green glass mounted on milk-white glass ring.

103. Ruby-colored semi-transparent glass. Of trapezoidal shape. Narrow sides cut into oval facets. Stopper of turquois carved into a coiled hydra. Plain.

104. Ruby-colored semi-transparent glass, pear-shaped. Stopper of green glass. Plain.

105. Orange-colored transparent glass. Stopper of greenish glass mounted on turquois ring. Plain.

106. Milky clouded glass interspersed with pink and deep-red flecks, in imitation of the hues of a flower. Stopper of green jadeite. Plain.

[19]

107. Glass, in imitation of pink crystal. Stopper of green jadeite. Plain.

108. Milky glass with gray and deep-red speckles, in imitation of agate. Stopper of green glass. Plain.

109. Opaque glass covered with red, yellow and clouded flecks, probably in imitation of tortoise-shell. Stopper of green jadeite. Plain.

110. Transparent ultramarine glass. Of elegant shape and outlines. Stopper of green and white jade. Plain.

111. Colored glass, in imitation of the tinge of wax stone. Stopper of pale-green jade. Plain.

112. Coffee-colored glass. Of rectangular shape, the four sides being faceted into medallions. Globular stopper of green jade set in gold. Plain.

113. Sienna-colored transparent glass intermingled with yellow stripes and flecks. Stopper of green jade. Plain.

114. Glass carved in cameo style. Decorated with a still-life—lotuses on one side, and peach-blossoms on the other side, emblems of purity and long life, respectively. Stopper of pale-green jade.

115. Colored glass, in imitation of amber. Carved in shape of gourd with relief of bat, symbol of happiness. Stopper of green jade with reliefs of fruit and leaves.

116. Yellowish and purple transparent glass. Carved in shape of a bean, probably the soy-bean (*Soya hispida*) forming one of the staple foods of the Chinese from ancient times. The seams of the bean are indicated by deep grooves along the edges, colored a deep purple. Green jade stopper of trapezoidal shape carved into floral relief designs.

117. Green-colored glass, in imitation of jade. Carved in the shape of Buddha-hand citrus (*Citrus decumana*), the "fingers" of the fruit pointing downward; leaved branches with the same fruit are carved in white relief on the surface. It symbolizes happiness and long life. Stopper of coral carved into a magpie pecking at the same kind of citrus.

[20]

118. Manicolored glass. Carved in shape of gourd. Stopper of white and green jade.

119. Colored opaque glass, ruby-red interspersed with yellow flecks, neck and shoulders being chocolate-brown, the narrow sides being veined with brown longitudinal stripes,—the whole in imitation of a porcelain glaze. Stopper of bluish glass.

120. Opaque yellow glass, in imitation of agate. Stopper of green jadeite. Plain.

121. Dark-red opaque glass interspersed with yellow stripes and speckles. Stopper of green glass. Plain.

122. Opaque yellow glass. Designs carved in relief and enclosed in medallions. The one side shows a huge horned fish in a pond teeming with reed and aquatic plants, the reverse lotus flower and leaf. The neck is decorated with a design of pearl fringes. The narrow sides are adorned with the eight precious Buddhist emblems; viz., wheel, state-umbrella, felicitous threads, conch, canopy, vase (compare No. 42); the fish and lotus are not repeated in this series, as they are represented on the large surfaces. Stopper of green jadeite.

123. Opaque glass of yellow background clouded with red speckles and blue lines, presumably in imitation of tortoise-shell. Stopper of green jadeite. Plain.

124. Milky glass filled with pink and yellow stripes, in imitation of agate. Stopper of jadeite. Plain.

125. Crimson-colored glass, of almost grape color. Oblong flask shape. Stopper of green jadeite. Plain.

126. Red-colored glass interspersed with black speckles, in imitation of clouded agate. Stopper of green jadeite set in gold. Plain.

127. Dark-colored glass filled with brownish and purple clouds, in imitation of agate. Stopper of pink crystal. Plain.

128. Brown glass, in imitation of amber. Stopper of violet-colored glass. Plain.

129. Semi-transparent blue glass. In shape of watch.

[21]

Each side faceted into circular medallions. Stopper of crystal. Plain.

130. Mother-o'-pearl. Stopper of green jadeite.

131. Mottled red and yellow opaque glass. Stopper of green jadeite. Plain.

132. Colored glass coming near to the sang-de-bœuf porcelain glaze, with a yellow ring around the neck and yellow zones along the narrow sides. Stopper of green jadeite. Plain.

133. Ruby-colored glass. Carved in shape of a pilgrim's bottle, a type of vessel going back to the days of antiquity, and worn suspended from the girdle. This feature is imitated in this miniature bottle, a green silk cord being drawn through the lateral perforations. Stopper of yellow crystal.

134. Porcelain bottle painted with a representation of the Feast of Lanterns, celebrating boys carrying lanterns in the form of fishes, and others beating drums and gongs, or blowing trumpets. This feast takes place on the first five evenings of the new year and is the occasion for merry-making and jollification. Floral decorations in blue and gold are displayed around the neck and the narrow sides. Stopper of pink glass in shape of a cover for a teacup. Tao-kuang period (1821–50).

135. Canary-yellow porcelain with relief designs alike on the two sides—two butterflies with leaf patterns and drooping gourds, an allegory hinting at old age and numerous offspring. Stopper of imitation coral.

136. White porcelain painted in many colors. On one side is depicted a genre-scene showing a mother-in-law in yellow dress seated on a porcelain stool and brandishing a stick, while her daughter-in-law is kneeling before her, and a woman is standing in the door and listening. The other side represents in a garden a portrait of Li with the Iron Staff, a famous Taoist genius, one of the group of the Eight Immortals, figured as a beggar leaning on an iron staff. He is engaged in conversation with two men, one of whom is holding a fan; the other is carrying on his shoulder a bundle suspended from a pole. The narrow sides are painted with palmetto designs in blue and gold. Date-mark in vermilion on bottom: "Made in the years of the period Kia-k'ing" (1796–1820).

[22]

137. Blue and white porcelain, very thin and light, of flat shape. Decorated with seal characters of very ancient style, fifty on each side, in underglaze blue, believed to act as a charm in warding off calamity from the owner. Coral stopper in shape of cover for a teacup. Date-mark in blue on bottom: "Made in the years of the period K'ien-lung of the Great Ts'ing dynasty" (1736–1795).

138. White porcelain painted in colors with a hundred figures of boys celebrating the Feast of Lanterns (so-called Picture of Hundred Boys). Compare No. 134. Stopper of crystal. Date-mark in blue on bottom: "Made in the years of the period K'ien-lung" (1736–1795).

139. Glass carved in cameo style. Buff-colored background from which the reliefs stand out in green and red. They represent lions playing with a ball. A coiled hydra is carved in relief on the bottom. Of excellent workmanship. Stopper of dark crystal.

140. Creamy-white porcelain of soft paste with large fine crackles. Shaped as a lotus-leaf, the stem emerging from the centre. A butterfly is carved in relief on one side. The reverse displays five lotus stems, one with leaf, the second with closed blossom, the third with bud half-open, the fourth with blossom wide open, and the fifth with free seed-receptacles—the various stages in the growth of the lotus.—Stopper of pink coral. Period Yung-chêng (1723–1735).

141. Cream-white porcelain of soft paste with carvings in relief. On the one side, the goddess Ma-ku, one of the celebrities of Taoist fable, is represented as holding an oar and rowing in a hollowed-out tree-trunk over a lake. The porcelain vessel in front of her is believed to contain supernatural gifts and blessings which she bestows on mankind. She was an adept in the black art, and by her agency, a large area on the sea-coast of Kiangsu Province was reclaimed from the ocean and transformed into mulberry-orchards. Her brother was Wang Yüan shown on the reverse of the bottle in the act of floating over the surface of the water, while seated on a leaf. He was likewise a magician and a reputed astrologer.—Stopper of green jade carved in the shape of the cover for a teacup. Period Yung-chêng (1723–1735).

142. Creamy-white porcelain of soft paste, beautifully carved, in high relief, with the figures of the Eighteen Arhat (in Chinese

[23]

Lohan), the famous disciples of Buddha, who are believed to cross the sea on their voyage from India to China, and on this occasion perform magical tricks to show their supernatural power. The crossing of the ocean is a metaphorical expression of the Buddhist thought that they have safely crossed the ocean of existence and reached the eternal blessing of Nirvāna. They further symbolize the power which the Buddhist saint has attained over the forces of nature by safely overcoming the dangers of the sea and rendering the fiercest animals submissive to his will. The Arhat Bhadra (that is, the Good One) is riding on the back of a tiger raised on clouds; in pictorial art, the tiger is often seen peacefully crouching by his side. Another saint is carrying on his shoulder a young lion (compare saint Christopher carrying the lamb, across a river); another, to escape the noise of the roaring waves, is covering his ears with both his palms; another, with the same end in view, is closing his ears with his fingers. Others are characterized by their emblems, as a bundle of books, a pagoda, a walking-staff, a rosary, or a rattle-staff which the mendicant monks of India, while begging for alms, used to carry in order to attract the attention of charitable people. Again, another is folding his hands for prayer (a practice which originated in India and was subsequently adopted by Christians), another is joyfully tossing his hat into the air, and another is holding an almsbowl in his right hand and in his left a ball which is the plaything belonging to the dragon brought out on the neck of the bottle. He is able to conjure the dragon, and once in a while indulges with him in a harmless ball-match. This is one of the most admirable compositions ever brought out on a snuff-bottle, remarkable for its great scope concentrated on so small a space, for the spirituality in the treatment of the subject, and for the masterly technical execution and the delicate tracery of the lines. Every line is alive and secure, essential and eloquent; no stroke is superfluous, and everything is blended with perfection into an harmonious picture of fascinating beauty.

Stopper of green jade. Period Yung-chêng (1723–1735).

143. Creamy-white porcelain of soft paste, showing a representation in relief of the sea all round and seven dragons playfully sporting in it. Stopper of crystal.

144. Creamy-white porcelain of soft paste, the surface being filled with fine gray crackles (so-called *craquelé*). The bottle is fashioned into a large lotus leaf with its exquisite veins, the stalk being formed in relief. A crane standing on a single foot is added to one side, while the other side exhibits three blossoms of lotus,

[24]

two half-open buds and one wide open with visible seed-receptacle. The lotus is an emblem of purity; the crane a symbol of longevity, the popular belief being that it lives to a thousand years. Stopper of crystal, mounted on green jade ring. Yung-chêng period (1723–1735).

145. Creamy-white porcelain of soft paste, the surface being filled with large crackles. Decorated on one side with carved reliefs of four horses, trotting and grazing; on the other side with three goats, the sun-ball surrounded by clouds illuminating the scene. The goat, as the result of punning, becomes an emblem of felicity and prosperity, and the picture of three goats implies the wish of a threefold happiness for the owner of the bottle. Stopper of mottled dark-green jade. Period Yung-chêng (1723–1735).

146. White porcelain decorated with incised and relief designs. The background is formed by an etching of continuous square meander ornaments, a larger meander being laid around the neck and continued by a fringe of spirals. The relief designs are composed of the eight Buddhist emblems of happy augury: Canopy and double fish on the large surface; wheel and lotus on the one narrow side; state-umbrella and vase on the reverse; and conch and felicitous threads on the opposite narrow side (compare No. 42). Arabesques and floral designs are distributed between these emblems. Stopper of porcelain with coral red glaze.

147. White porcelain painted in colors with two scenes in the life of the demi-monde. Five women engaged in a meal are sitting around a table on yellow porcelain seats. The other side shows a girl in her bed-chamber, while a maid holding a burning candle escorts a guest into the room. The narrow sides are decorated with floral designs in pink on green background. Date-mark in vermilion on bottom: "Made in the years of the period Tao-kuang" (1821–1851).

148. Porcelain of coral red glaze, fashioned into the form of a lichee fruit. The stopper of enameled metal is intended for the stem of the fruit. Spoon of silver.

149. Painted enamel on foundation of copper. The same design is painted alike on both sides. The male and female phenix, famed for their inseparable fellowship, are surrounded by floral designs of Rococo style, an ornamental form of the character *shou*, "longevity," being in the centre. Enamel paintings

[25]

of this perfection are now very scarce. Stopper of embossed silver. This technique of enamel painting is originally foreign to Chinese art. It was highly developed at Limoges in France and introduced into China by the early Jesuit missionaries at the end of the seventeenth century when it became known there under the name "foreign porcelain." Limoges enamels were actually taken to China to be copied, and hence the influence of the epoch of Louis XIV is manifest in the pieces of Chinese workmanship.

150. Milk-white glass carved in cameo style. The palm-leaf designs stand out in ruby-red from the white background, except the stems which are likewise white. The narrow sides are decorated with lion-heads in relief, of peculiar design. A red rim borders the base. Stopper of green jade.

151. Carved from ivory. Decorated with lateral tiger-heads with projecting tusks and rings in their jaws, otherwise plain. Stopper of lapis lazuli.

152. Glass carved in cameo style, of pink background with reliefs in green and brown. The subjects represented are various stages in the life of the silkworm, a theme of inexhaustible interest to the people who conceived the idea of the silk industry. The cocoons are shown as feeding on the mulberry-leaves, and a butterfly has been evolved from a pupa.—Stopper of green jade.

153. Carved from the hornlike excrescence on the upper mandible of the hornbill (compare No. 41). Decorated with two lateral lion-heads in red with drooping ears, protruding tusks and rings in their jaws, otherwise plain.—Stopper of same substance in red.

154. Glass carved in cameo style, decorated in relief of various colors, with dragons and bats, emblem of happiness.— Stopper of green jade.

155. Carved transparent glass in cameo style, decorated with a so-called picture of antiques and the emblems of erudition. The one side displays a sacrificial tazza in yellow, a painted scroll in blue, three wrappers of books piled one above the other, and a conch-shell on top of them; to the right there is a brush-holder from which the tips of three writing-brushes project. On the opposite side is arranged a round black flower vase with slender neck standing on a draught-board for playing the famous Game of War, one of the national games far more complicated than

[26]

chess; beneath it is a picture-scroll, a round covered jar, a lute, an unrolled scroll in red, and another black vessel with lid. The vase, by means of punning, suggests the word "peace." The elements of this design express the wish: "May you have peace and become an accomplished scholar," lute, draughts, books and picture-scrolls being styled the four treasures of the scholar, and representing music, elegant pastimes, literature, and painting.— Stopper of green jade in shape of a bottle.

156. Glass carved in cameo style. The waves of the sea are displayed around the base, and a dragon, in plant-green relief,— the dragon is symbolic of prolific vegetation,—his body immersed in the watery element, is raising his head to look at his companion soaring high in the air. This lofty dragon is making for the flamed pearl.—Stopper of pink crystal, mounted on yellow glass ring.

157. Transparent glass carved in cameo style, on each side decorated with three hydras joining their heads in the centre, and a bat below—expressing the wish of a threefold happiness.— Stopper of green jade.

158. Glass carved in cameo style, with reliefs of two mandarin-ducks, the symbol of conjugal felicity, swimming in a lotus-pond, on the one side, and iris growing in a pond on the other side. Lateral lion-heads holding rings in their jaws, in relief.— Stopper of green jade set with ruby.

159. White porcelain, light in weight and fine in quality. Decorated on both sides alike with fifty-seven seal-characters of ancient style in vermilion over glaze. It is their object to act as a charm, and to safeguard the owner of the bottle from disease and evil influences. The narrow sides are painted with floral decorations in vermilion. Date-mark on the rectangular bottom: "Made in the years of the period K'ien-lung" (1736–1795).— Stopper of green jade.

160. White porcelain of Fukien with underglaze etching of three roosters and bamboo. Read as a rebus, this design means: "I pray that you may be promoted in office by three degrees!" Stopper of coral.

161. Glass carved in cameo style; pure white background with reliefs of ruby color. The subject is a so-called picture of antiques and the emblems of scholarship (compare No. 155).

[27]

The one side shows on the left-hand a flute enveloped by silk bands, a cylindrical jar posed on a wooden base, in the centre a cabinet of fantastic shape cut out into square spirals on which are placed a rectangular vessel holding two leaves and bowl containing a Buddha-hand citrus. On the opposite side are displayed a cabinet of similar style supporting a seal with handle in the shape of an animal and a tripod vessel holding a fungus of immortality; on the right-hand a slender vase upon a wood-carved pedestal, two books, and a bamboo receptacle holding a bunch of writing-brushes. The narrow sides are each adorned with the figure of a hydra.—Stopper of green jade.

162. Glass carved in cameo style, blue foundation with reliefs in green. An ornamental form of the character *shou*, "longevity," in the centre, and two hydras with manes and a single horn on their foreheads following each other. The wish symbolically expressed by this design on the part of the donor is "May our friendship be long-enduring!"—Stopper of crystal.

163. Glass carved in cameo style, with reliefs of ruby color. On each side is represented a basket united by a cord laid around the whole bottle and falling down the narrow sides. There are in each basket a branch with two peaches and lotus blossoms and leaves. The symbolic interpretation of this composition is: "May you have harmony, permanent union with your family, and long life!"—Stopper of green jade.

164. White lustrous porcelain painted with the figure of a Taoist recluse seated on a rock, mountains forming the background. His garb is decorated with ornaments in gold; he wears straw sandals and carries a staff. He is examining a scroll spread out on a boulder, bordered by gilt lines, on which the symbol of the male and female energy is outlined. The Chinese have a dualistic conception of nature reducing all phenomena to the interaction of a male and female principle forming the two primordial essences and conceived as heaven and earth, light and darkness, positive and negative, heat and cold, etc.—Stopper of green jade.

165. Glass carved in cameo style, of dark-green background with reliefs in white and pink, representing genre-scenes. A man and woman are having a match of draughts (compare No. 155) on a stone table standing on a rocky platform overshadowed by pine-tree and banana. A looker-on is standing behind the table, and a maid is serving a cup of tea. The other side shows a

[28]

garden of bamboo and trees with a double-roofed pavilion; a man is seated at a table, a woman holding a fly-whisk and followed by a servant is approaching him. The neck is adorned with relief-designs.—Stopper of green jade.

166. Glass carved in cameo style with reliefs of ruby color representing a uniform composition on both surfaces. Three wild geese are swimming in a lotus-pond.—Stopper of green jade.

167. Glass carved in cameo style with designs of gourds, fruits and leaves, in relief of ruby-red, standing out from a white background and laid around in a uniform composition.—Stopper of green jade.

168. Milk-white glass carved in cameo style, decorated with nine bronze vessels of ancient types in black and red reliefs. This motive is known as the nine tripod vessels of Emperor Yü. He is one of the legendary ancient culture-heroes who rescued his country from the disastrous effects of an enormous flood, and in the course of nine years, brought the waters under control. He divided the empire into nine provinces, and from the metal sent to him as tribute by the nine chiefs governing the provinces had nine tripod vessels cast on which all objects of nature were delineated for the instruction of his people that they might know the gods and evil spirits, and be no longer assailed by the terrors of the elements. For decorative purposes, the nine vessels are here figured in different outlines and decorated with various designs. A rectangular bowl, for instance, posed on four straight long feet, is inscribed in front with the character *fu*, "happiness"; a round tripod jar is decorated around the neck with a row of lotus petals and over the surface of the body with a net of hexagons. Other vases are adorned with the so-called Eight Trigrams, mystic symbols anciently serving as the basis of an elaborate system of philosophy and divination; others are decorated with spirals and that favorite pattern, the plum-blossom.—Stopper of coral.

169. Glass carved in cameo style with reliefs of ruby color, the prevailing motive of which is the strongly developed desire for happiness expressed by the bat. The bat is called *fu*, and another word *fu* means "happiness," hence the origin of this symbolism. Despite this highly abstract notion, the composition is vivid and artistic. On the one side we observe sea-waves spread around the base and the sun rising from beneath them; above, there are three bats beautifully connected by cloud designs encircling the symbol of the swastika. The latter is a felicitous emblem coming from

[29]

India where it is frequently employed on the palms and chest of Buddha; it is a symbol of solar heroes, of light, of royal power and victory. The Chinese take it as a rebus and read it in the sense of "ten thousand." The significance of the three bats combined with the swastika, therefore, is: "May you have a ten thousandfold enjoyment of the three kinds of felicity!" which are personal welfare, long life and numerous offspring. The same motive is brought out on the opposite side of the bottle, except that a rock is floating on the crest of the waves. While the sun on the obverse indicates the wish that every day when it rises may bring you new luck; this design on the reverse implies that your luck may be as stable and permanent as the rocky islands in the midst of the ocean. On the two narrow sides we see again a bat hovering over a rock; that is, happiness founded as safely as a rock, and a rock in the ocean surmounted by cloud bands, meanng, "May your luck be as vast as the ocean spanning the distance from the earth up to the sky!"—Stopper of dark-green jade.

170. Glass carved in cameo style with reliefs of ruby color. The designs represent two large pomegranates, on account of their numerous seeds the emblems of child-blessing, an open chrysanthemum used by means of punning for the word "long, enduring," a Buddha-hand citrus illustrating, owing to a pun upon the words, "wealth and honorable position," and two peaches, symbols of longevity. The good wishes indicated by these ornaments, accordingly, are: "May you enjoy long life, wealth, honorable position, and numerous offspring for all time to come!"—Stopper formed by a pearl.

171. Glass of milk-white background carved in cameo style, with reliefs in dark-blue. The designs show a lotus-pond and four goldfishes, which implies the wish: "May you receive official promotion and have abundance of gold!"—Stopper of coral.

172. Glass carved in cameo style, decorated with reliefs of ruby color representing four flowerpots posed on wooden stands. They are suggestive of the wish: "May you have peace!" The two pots on the main sides are each inscribed with four seal-characters of ancient style, meaning: "May your wealth, honorable position, honor and glory be long-enduring like the mountains and the oceans."—Stopper of green jade.

173. Glass carved in cameo style, of white background, with reliefs in pink representing three hydras. The male is carved on one side, the mother and her young on the other side. This

[30]

is a very ancient and famous motive of art appearing on the early jade buckles and known under the name "the hydra watching her cub," an emblem of motherly care and love.—Stopper of green jade.

174. Glass carved in cameo style, the relief designs in ruby color standing out from the white, powdered background. The composition consists of a lotus-pond with seven cranes, five of which are wading through the water, some pecking at a leaf, one biting its foot, and two are on the wing. The lotuses are displayed in various stages of growth, showing buds, open blossoms, and those with visible receptacle from which the petals have dropped. The crane, supposed to live a thousand years, is emblem of longevity. The wish expressed by this design is: "May you live seven times as long as a crane and receive promotion in your official career!"—Stopper of green jade.

175. Glass carved in cameo style, with reliefs in green, violet, white, red and amber-yellow. It is the celebrated motive "squirrels stealing vine-grapes," outlined with great delicacy and charm. The narrow sides are decorated with rocks of fantastic forms piled together, such as are met in Chinese parks and gardens.—Stopper of pink crystal.

176. Glass carved in cameo style, the reliefs in ruby color being set off from the white background. The one side shows a couple of mandarin-ducks, an emblem of conjugal felicity, swimming in a lotus-pond; the other side two leaved bamboo trunks, a bird perching on one of them. Lateral lion-heads holding rings in their jaws, in relief.—Stopper of green jade.

177. Glass carved in cameo style, decorated with reliefs of ruby color representing eight horses, three on each main surface and one on each narrow side, each designed in a different action or motion. These are the famed eight coursers of Emperor Mu Wang of the Chou dynasty, each of which bore a distinguishing name, and with which he was driven by his charioteer Tsao Fu on his journeys and campaigns through his empire (compare No. 25). —Stopper of green jade.

178. Glass carved in cameo style, with reliefs in ruby color set off from a white background. On one side is represented a frog cowering on a lotus leaf and overshadowed by another, the stems of the two leaves being joined in the centre. On the opposite side are carved a serpent, whose body forms the ring laid around

[31]

the rim of the bottom, and a tortoise turned away from, and looking back at, the serpent. Of the many long-lived creatures of
Chinese fable, the tortoise is the longest-lived, reaching an age of
three thousand years, and also the most revered; it is counted
among divine or supernatural beings. Its hibernating indicates
the arrrest of life in nature, and its awakening signals the beginning of the spring. Frog and serpent have the same function, and
this triad thus depicts the awakening of life during the spring.—
Stopper of green jade.

179. Glass carved in cameo style, with reliefs of ruby color.
On the one side is represented a greyhound with long muzzle
and neck-collar bound by means of an iron chain to a pillar. On
the other side is carved a parrot perching on a piece of knotty wood,
one of its feet fettered to a rope. Both pictures are symbolic of
all-potent fate which nobody may escape.—Stopper of coral
surmounted by an inlaid pearl and resting on a ring of lapis lazuli.

180. Glass carved in cameo style, with reliefs in lapis-lazuli
blue representing on each side a coiled hydra with long mane.—
Stopper of pink crystal.

181. Glass carved in cameo style, with reliefs in ruby color,
representing the emblems of scholarly attainment. A brush-
holder, a book and a bowl holding Buddha-hand citrus are arranged
on a table of fantastic form; above, there is a board for draughts.
The other side illustrates a similar table on which are displayed
a covered round jar, a tripod bronze vessel with two slender loop-
handles and lateral elephant-heads containing a sceptre of happy
augury, a cylindrical vase holding lotus leaves and adorned with
the Eight Trigrams, the ancient mystic symbols used in divination, and the emblem of the male and female principle. A dragon
head holding a basket with the fungus of immortality appears as
the bearer of good luck and long life.—Stopper of green stone.

182. Glass carved in cameo style, with reliefs in pink and
green. The one side illustrates a carp in curved outlines and
raising its head, a string of coins (so-called "cash") being placed
on its tail. Here we have again a readable rebus, as the name for
the carp can be understood also in the sense of "profit and abundance." This bottle was evidently intended as a gift for an assidu-
ous business man to whom the wish was expressed: "May you
always have good profit and abundance of ready money!" The
opposite side shows lotus leaves and blossoms and a green toad
resting on clouds. This is a mythical creature believed to dwell

[32]

in the moon which it swallows during an eclipse. It serves also as an emblem of the merchants.—Stopper of green glass mounted on yellow glass ring.

183. Milky glass carved in cameo style, with reliefs of ruby color. A fish-shaped dragon is emerging from the sea and lifting a six-sided pavilion surrounded by clouds. The opposite side represents a fungus of immortality in combination with a crab.

184. Transparent yellow glass carved in cameo style, with reliefs representing the motive "squirrels stealing grapes," the vine-leaves in green and blue, the grapes in purple, the squirrels in yellow. The base is bordered all round by rocks.—Stopper of pink crystal.

185. Glass carved in cameo style, with reliefs of flowers in ruby color,—lotus, chrysanthemums, plum blossoms, and peach.—Stopper of green jade.

186. Glass carved in cameo style, reliefs in blue of lotus flowers and leaves enclosed in round medallions being set off from the white background. A meander design is painted in red around the neck and bordered by blue lines. The narrow sides are decorated with pendants, of the style of textile designs, a rose blossom being combined with the ornament of felicitous threads, originally the mystic emblem of the god Vishnu.—Stopper of coral mounted on a ring of turquois.

187. Glass carved in cameo style, with reliefs of ruby color. The one side shows the gate of a city emerging from the waves; it is a solid crenelated brick wall surmounted by a two-storied watch-tower which is overshadowed by pine-trees. A crane on the wing, holding a letter in its bill, is flying down toward the city gate. Birds as messengers and carriers of news are frequently alluded to in Chinese stories. The reverse illustrates the mythical toad blowing forth from its mouth a two-storied palace-building soaring on clouds; it is raised on a high foundation of bricks, stone staircases leading up to a railed veranda. It is one of the magical palaces believed, as the toad itself, to exist in the moon. The narrow sides are decorated with pine leaves growing out of a rock and fungi of immortality sprouting on a rock, respectively.—Stopper of green jade.

188. Milky glass carved in cameo style, with reliefs in ruby red representing a creeping plant with blossoms, fruits and leaves.

[33]

A red ring is laid around the bottom. Oblong, slender shape with long neck, but without shoulders.—Stopper of green and white jade.

189. Glass carved in cameo style, decorated with a still-life of dark-ruby color,—branches of the plum-tree in blossom, bamboo and three magpies, expressing the thought: "We always pray for good luck."—Stopper of green jade mounted on a coral ring.

190. Glass carved in cameo style, decorated with reliefs of ruby color representing chrysanthemums growing from a rock and the leafed branch of a peach tree in blossom, expressing the wish: "May you live long and to a great age!"—Stopper of green jade.

191. Glass carved in cameo style, with reliefs of ruby color of exquisite carving. The one side shows a lion of Indian-Buddhist type surrounded by flames and placing a front-paw on the ball with which he is playing; he is just in the act of jumping downward, while the lion on the reverse is quietly standing on his ball. The lion is the emblem of the highest official rank of the empire which formerly was the office of the grand preceptor to the heir-apparent. This design conveys the wish: "May you become the first dignitary at the imperial Court!" The narrow sides are decorated with two conventional bats holding rings, symbolizing "all round" happiness.—Stopper of green jade.

192. Glass carved in cameo style, decorated with reliefs in ruby color representing a plant still-life,—water filled with fanciful rocks and boulders, leaved branches of cherry-tree in blossom, bamboo leaves, and chrysanthemum.—Stopper of green jade.

193. Glass carved in cameo style with high relief in dark-green of five-clawed imperial dragon soaring between clouds above (in red) and the sea below, from the midst of which an island emerges.—Stopper of amethyst.

194. Glass carved in cameo style, with reliefs of ruby color representing flower still-life,—chrysanthemum and leafed branch of peach in blossom.—Stopper of pink crystal.

195. Crimson-colored glass carved in cameo style with designs derived from an ancient bronze vessel—spiral patterns terminating in heads of hydras and encircling an ornamental form of the

[34]

character *shou*, "longevity,"—the same alike on either side.—Stopper of green jade.

196.   Clouded brown and yellow fossil amber, having the appearance of veined and stained wood.   Plain.—Stopper of green jade.

197.   Yellow amber interspersed with brown clouds.   Miniature size.   Plain.   Stopper of green jade.

198.   Clouded Burmese amber.   Decorated in relief with lateral lion-heads with manes, broad noses and rings in their noses. —Stopper of green and white jade.

199.   Glass carved in cameo style, with reliefs of lotus blossoms in pink and blue, leaf in green, and a bird on the wing in yellow.—Stopper of coral.

200.   Orange-colored clouded amber, of rounded outlines. Stopper of green jade set in gold.   Plain.

201.   Dark-red semi-transparent amber, carved into the appearance of a wickerware basket.—Stopper of green jade.

202.   Artificial amber, probably copal, shining in brown, red, green, yellow, and purple.   Miniature size.   Decorated in relief with two lateral lion-heads holding rings in their jaws, with elaborate manes composed of spirals.   Stopper of red globular coral.

203.   Similar to 202, with red, yellow and dark-blue spots. Stopper of red coral in shape of cover for a teacup.

204.   Glass carved in cameo style, with reliefs of ruby color representing fungi of immortality, chrysanthemums and a couple of butterflies.—Stopper of green jade.

205.   Yellow and brownish-red clouded amber.   Plain.—Stopper of green jade.

206.   Glass carved in cameo style with reliefs in green, set off from a mottled white background, of bamboo and plum-tree in blossom.—Stopper of pink crystal.

207.   Red transparent amber, of flat shape.   Plain.—Stopper of green glass, in shape of cover for teacup, mounted on coral ring.

[35]

208.   Brilliant deep-yellow clouded amber, carved in shape of a persimmon.   On one side is delineated a persimmon branch with leaves and fruit, and a bat and fungus of immortality below, implying the wish: "Good luck in all your enterprises and long life!"   The reverse illustrates a famed motive of pictorial art, a fox struggling with a hawk dashing on him from a height.—Stopper of green jade forming the stem of the fruit.

209.   Reddish amber with yellow clouds.   Plain.—Stopper of green jade.

210.   Honey-yellow amber.   Plain.—Stopper of green jade.

211.   Fossil clouded amber of yellow color with brown rings. Plain.   Stopper of coral carved in relief with a coiled hydra, inlaid with a pearl in the centre, and mounted on a ring of turquois.

212.   Chocolate-brown amber.   Plain.—Stopper of green jade.

213.   Clouded amber.   Plain.—Stopper of pink coral.

214.   Semi-transparent reddish amber, cylindrical in shape, carved with relief of continuous floral design of most elegant and graceful composition.—Stopper of coral.

215.   Dark-brown and yellow fossil amber.   Plain.—Stopper of green jadeite.

216.   Clouded   amber   with   deep-yellow   and   deep-brown flecks, of wonderful brilliancy.   Decorated in relief with lateral tiger-heads holding rings in their jaws, and an etching of bamboo and fungus of immortality on one of the large surfaces.   This design implies the wish: "We pray for long life."—Stopper of green jade.

217.   Walrus tusk, so stained as to imitate the appearance of jade.   The one side shows a peach-tree with fruit and leaves, a bat, a large spotted deer, standing on a sceptre of happy augury to which a fluttering band is tied, and bamboo-leaves between the feet of the deer.   This composition is expressive of the wish: "I pray that long life, happiness and every advantage may fall to your lot, as you wish."   The opposite side illustrates a pine-tree, a magpie, a bird of happy augury, perching on a branch and pecking at a cone, bamboo-leaves, and lion.   Decorated in relief with lateral lion-heads with drooping ears and holding rings in their jaws.—Stopper of pink crystal.

[36]

218–219. Pair of bottles carved from clouded amber. Plain.
—Stoppers of green jade.

220. Carved red lacquer decorated with genre-pictures in high relief. On one side is represented a woman kneeling in front of a basket and engaged in some work, while two men are chatting together, the background being formed by rocks and pine-tree. On the other side are carved three boys engaged in play, one of them turning a somersault.—Stopper of lacquer with a double row of lotus-petal designs.

221. Carved red lacquer exhibiting scenery with rocks and trees laid around in uniform composition. The celebrated poet Li T'ai-po (705–762) is twice represented, on the one side reclining on a boulder, while a boy, attendant on him, is bringing a box; on the other side, seated on a bench, while a boy is performing a dance before him. He was a poet of wine, woman and songs, leading a life of wild dissipation, and the story goes, was drowned from leaning one night over the edge of a boat, in a drunken effort to embrace the reflection of the moon in the water. The cover, of carved lacquer, is decorated with designs of meanders combined with swastikas, leaf patterns being laid around the knob in the centre.

222. Carved from a small coconut mounted on silver which is cut out in a row of leaves. Decorated in relief with floral designs.—Stopper of plated silver.

223. Chocolate-colored wax-stone decorated all round with bamboo in relief. Of trapezoidal shape.—Stopper of green jade.

224. Brownish wax-stone decorated on each side with the design as occurring on the projecting disks of the roofing-tiles of the Han dynasty. The four ornamental characters laid around the central knob mean: "Thousand and thousand of years without end!"—Stopper of green jade.

225. Semi-transparent gray agate. Plain.—Stopper of pink crystal.

226. Yellow wax-stone carved into rectangular facets on the main sides and into oval medallions on the narrow sides.—Stopper of green jadeite.

227. Wax-stone carved in shape of an egg, intended for some

[37]

kind of fruit. The coral stopper forms the stem of the fruit, and a leaf is carved on it in relief.

228. Wax-stone of amber-yellow color. Plain.—Stopper of green jade.

229. Brownish-yellow, lustrous wax-stone, with relief on one side of a European design of Renaissance style, two monster-heads facing each other. Reverse plain.—Stopper of green jadeite.

230. Olive-green pottery of soft clay, decorated with low reliefs of clouds, bamboo, and pine-trees, plum-blossom, and banana-leaves. Very flat.—Stopper of light coral.

231. Gray and brown wax-stone. Trapezoidal shape. Plain. Stopper of green jade.

232. Yellow-brown agate, decorated with floral designs in relief—plum-blossoms, butterfly, and swallow.—Stopper of coral set in brass.

233. Brown agate clouded with red flecks. In shape of watch. Plain.—Stopper of green jadeite.

234. Gray clouded agate. Trapezoidal shape. Plain.— Stopper of green jadeite set in gold.

235. Black-colored glass carved in cameo style, decorated on one side with reliefs in white of two squirrels rushing over vine-leaves and gnawing at grapes, on the other side with pine-tree and decayed tree mutually connected, in black relief.— Stopper of green jade.

236. Semi-transparent, striped brownish agate. The two main sufaces are cut and polished into facets of medallion form. The narrow sides are decorated in relief with lion-heads composed of spiral designs and holding oval rings in their jaws.—Stopper of green jade.

237. Yünnan marble with black and white horizontal stripes. This kind of marble is mined near Ta-li-fu, the capital of Yünnan Province. Its appreciation and esthetic effects rest on the black zones and clouds interrupting the white substance. The Chinese are fond of cutting this stone out in slabs to be used for screens, or to be inlaid in tables or chairs, and sawing it in such a way that,

[38]

with some stretch of imagination, the black masses form veritable scenery with streams, hills, clouds and animals. For this reason, this bottle is not carved, as it speaks for itself with the sole beauty of its natural color effects.—Stopper of green glass.

238. Hairy crystal, or sagenitic quartz. This is a form of quartz variously known as sagenite, *flèche d'amour* (love's arrow), hair stone, needle stone, and if the included mineral be rutile, rutilated quartz. These terms all refer to colorless crystallized quartz which is penetrated by hairlike crystals of other minerals. Of the minerals so included rutile is the most common, but tourmaline, hornblende, epidote, goethite, and others occur. The inclusions have been formed in the quartz by crystallizing at the same time with it, the quartz in this case being the "host." For cut stones, the Chinese prefer pieces in which a multitude of short individual crystals is scattered. The included crystals often cross each other nearly at right angles, thus giving the appearance of a network, hence the name sagenite (from Greek *sagene*, a net).— Stopper of red coral, mounted on ring of turquois.

239. Hairy crystal.—Stopper of green jade.

240. Same. The bundles of black included crystals are here denser and distributed more regularly. Oblong and flat shape.— Stopper of green jade.

241. Gray agate carved with reliefs in brown and ivory yellow, representing on one side a dragon soaring in clouds, on the other side three rampant leaping goats forming a rebus which reads "three kinds of blessing"; that is, happiness, long life, and numerous offspring. The narrow sides are decorated with the sun-ball surrounded by flames and clouds in the form of spirals.— Stopper of green glass mounted on white glass ring.

242. White jade, carved in the shape of a Buddha hand citrus with four fingers directed downward and forming the base. This fruit, on account of its name, is suggestive of felicity and longevity. Leaves of the plant are laid around the surface in high relief. The stopper of green jade forms the stem of the fruit.

243. White jade, carved in form of an oil-jar, and decorated in relief with genre-figures,—a boy holding flowers and a woman clad in a shawl which she snatches away from a dog that is leaping up at her attempting to seize her shawl.—Stopper of green jade.

[39]

244. Purple crystal carved with figures in light-blue relief.—Stopper of crystal.

245. Black Yünnan marble penetrated by white veins. Plain.—Stopper of green jadeite.

246. Ivory-colored stone, carved on one side with high relief in dark-brown, representing a cat standing on a rock and leering at two butterflies, expressing the wish: "May you live to the age of seventy and eighty years!" Reverse plain.—Stopper of green jade.

247. Pink crystal, colored artificially. Carved in form of a basket.—Stopper of green jade.

248. Lapis lazuli of azure color clouded with white and black spots. Plain.—Stopper of pink crystal mounted on marble ring.

249. Lapis lazuli of sky-blue color intermingled with gray clouds and white spots. With rounded shoulders and straight outlines. Plain.—Stopper of pink crystal mounted on ring of green jade.

250. Lapis lazuli spotted with gold dots, hence the ancient Chinese name "stone with golden stars," and Pliny's description of it as "refulgent with spots like gold." Small constricted neck with flaring rim, straight and bulging shoulders, then tapering downward in elegant curve. Plain.—Stopper of coral mounted on ring of turquois and inlaid in the centre with a green glass bead.

251–252. Pair of bottles carved from lapis lazuli speckled with gold dots and white flecks. Rectangular shape. Plain.—Stoppers of coral mounted on rings of amber-yellow glass.

253. Onyx intersected by white stripes. Rectangular shape. The two large surfaces carved into medallion facets. Plain.—Stopper of green jade.

254. Milk-white agate carved with reliefs of pine-trees growing on rock and fungus of immortality, both emblems of longevity, and on the opposite side with relief in brown of stag and doe facing each other, an allusion to good income.—Stopper of coral inlaid with pearl.

255–256. Pair of bottles carved from black and green mala-

[40]

chite. Plain.—Stoppers formed by beads of mother-o'-pearl set in gold.

257. Light malachite. Plain.—Stopper of coral.

258. Brown, yellow and red agate. Plain, of lustrous polish. —Stopper of green jade.

259. Perfect rock crystal, carved on both sides alike with an ornamental form of the character *shou*, "longevity," encircled by detached spiral designs and the ornament known as "reclining silkworm-cocoon," the entire composition being framed by a row of continuous meander (fret).—Stopper of green jade.

260. Powdered glass, of trapezoidal shape. Carved on the one side with relief of pine-tree growing on rock, a symbol of longevity, on the opposite side with a bat and a fungus of immortality, meaning "good luck and long life!" On the narrow side are incised bamboo and fungus.—Stopper of green glass.

261. Reddish stone, of rounded shape. Plain. Stopper of green jade, mounted on glass ring.

262. Brown agate, of trapezoidal shape; countersunk oval bottom. Plain.—Stopper of green jade.

263. Rare dark-blue stone. Plain, of rounded outlines.— Stopper formed by garnet.

264. Semi-transparent brown and white agate. Plain.— Stopper of green jade.

265. Brown clouded agate. Plain.—Stopper of dark green jade.

266. Onyx, brown with milky zones. Plain.—Stopper of green jade.

267. Mustard-yellow agate, of square shape. Plain, carved into facets.—Stopper of amethyst.

268. Semi-transparent gray agate overstrewn with numerous black spots, the natural irregular shape of the stone being retained and suggestive of a fruit.—The stopper of green jade forms the stem of the fruit and is carved à jour with leaves in relief.

[41]

269. Purple agate, of rounded shape. Plain.—Stopper of carnelian of cherry color.

270. Gray agate carved out of the matrix with reliefs in brown, showing on one side two lions in wild motion fighting for the ball with which they are playing, symbolic of the wish, "May you obtain the first dignity in the imperial Court!" and on the other side a Kilin running down a rock. This creature is one of the most renowned fabulous animals of China and generally translated by "unicorn." It is described in ancient texts as having the body of a deer, the tail of an ox, the scales of a fish, and one horn. The latter is covered with flesh, showing that while able for war, it desires peace. It does not tread on any living thing, not even on living grass. It is the emblem of perfect goodness and benevolence. It is looked upon as the noblest form of the animal creation, and said to attain the age of a thousand years. It appears as a happy portent only in an era when a virtuous sovereign is born or a good government is inaugurated.—Stopper of green jade.

271. Yellowish onyx intersected by milk-white zones. Plain. —Stopper of green jade.

272. Green moss agate. This stone is produced when chalcedony is penetrated by branching forms of manganese or iron oxide. The best occur as rolled pebbles in the beds of streams, and the finest specimens now known come from India. The sentiment attached to the stone is "a disperser of melancholy." Plain.—Stopper of red coral.

273. Same. Plain.—Stopper of green jade.

274. Same. Decorated in relief with lateral tiger-heads holding rings in their jaws.—Stopper of red agate.

275. Purple agate covered with brown, green and gray spots. Plain.—Stopper of green jade.

276. Agate of various colors, in the upper portion brown, in the lower portion green. Plain.—Stopper of pink crystal.

277. Brown and green moss agate. Plain.—Stopper of green jade.

278. Brown and green moss agate. Plain.—Stopper of carnelian set in silver.

[42]

279. Dark-green agate interspersed with iron-rust spots. Plain.—Stopper of pink crystal.

280. Carnelian. Plain, of circular shape.—Stopper of blue crystal set in gold.

281. Clouded red and whitish carnelian. Plain.—Stopper of green jade.

282. Carnelian, on one side almost milk-white, on the other side light-red intersected by circular bands. Plain.

283. Dark-red carnelian. Plain.

284. Carnelian of pure color, of oval shape. Plain.—Stopper of green jadeite.

285. Semi-transparent reddish and yellow carnelian. Flat and rounded shape.—Stopper of green jadeite.

286. Semi-transparent brownish stone intersected with black veins on one side and overstrewn with starlike white spots on the other side. The shape is suggestive of a fruit, and a single leafed twig of an orchid is engraved in the surface.—The stopper of jade, perforated at the top, forms an oblong stem carved into fruits.

287. Brown onyx intersected by white bands. Plain.—Stopper of crystal set in gold.

288. Brown onyx with black, milk-white and red zones. Plain.—Stopper of green jade.

289. Onyx intersected by brown, red and white horizontal zones. Decorated in relief with lateral tiger-heads holding rings in their jaws.—Stopper of bluish-green jade.

290. Brown, reddish and white-colored glass, decorated with a brickwall design.—Stopper of green glass.

291. Dark-brown stone, probably agate, with yellow veins, in its natural shape, suggestive of a fruit like a chestnut.—Stopper of white and greenish jade carved à jour in form of leaves.

292. Onyx with black spots scattered here and there and a milk-white zone running through the middle on one side. Oval facets are carved on the narrow sides.—Stopper of yellow crystal.

[43]

293. Brown agate carved on one side with the figure of a boy standing on a covered jar and holding a lotus-flower. He is turning toward a quail perching on a single foot. The quail, by means of punning, is suggestive of a wish for peace. A single lotus is represented on one of the narrow sides.—Stopper of mottled green and white jade.

294. Leaf-green agate, covered with green flecks standing out from a black background. Plain.—Stopper of green glass.

295. Milky and reddish-brown agate, the natural shape of the stone being retained, carved with reliefs in yellow, representing a lotus-pond with conventionalized waves. On one side is delineated a large lotus leaf with disclosed blossom, on the other side three leaves with open bud, the one leaf flat and spread out with edges slightly rolled, the second entirely rolled up, and the third with edges turned in. Such detail work reveals the keen observation of, and intimate acquaintance with, nature on the part of the Chinese.—Stopper of pink crystal.

296. Brown and gray clouded agate. Plain, the narrow sides being cut into oblong medallions in relief.—Stopper of white and green jadeite.

297. Brown and gray, semi-transparent agate. Plain.—Stopper of pink crystal.

298. Semi-transparent onyx or ribbon agate. Plain. The artistic effect is based on the lustrous white bands intersecting the middle of the bottle.—Stopper of carnelian.

299. Clouded onyx intersected by three brown horizontal zones. Plain.—Stopper of coral carved with coiled dragon in relief.

300. Brown onyx with horizontal white, yellow, and black stripes. Plain.—Stopper of green jade.

301. Onyx with two white zones running horizontally around the middle of the bottle. Plain.—Stopper of green jade.

302. Semi-transparent onyx with a narrow white zone running all round through the middle. Plain.—Stopper of green jade.

[44]

303. Onyx carved in shape of a calabash. The calabash is the emblem of abundance, fertility and felicity. At the marriage ceremony bride and groom drink wine alternately from two cups united by a thread of red silk; these bridal cups are now made from porcelain or jade, but formerly consisted of the half of a gourd or calabash. The physicians and apothecaries preserve their drugs in gourds or in flasks made of this shape, because the gourd is believed to exert a beneficial and conservative influence on any substance that it contains. For this reason it was chosen also as a suitable receptacle for snuff.—Stopper of green jade.

304. White onyx interspersed with red spots and intersected by brown and gray bands. Plain.—Stopper of pink crystal.

305. Agate carved in relief with a family of three monkeys, male and female crouching together, and the cub reclining on their shoulders. Their interest centres around a large peach which the monkey is believed to steal from the paradise of the goddess Si Wang Mu, the royal mother of the west.—Stopper of green jadeite.

306. Dark agate, black, filled with yellow speckles. Elegant shape. Stopper formed by ruby.

307. Milk-white agate clouded with black and brown flecks. Plain. Stopper of pink crystal.

308. Brown agalmatolite engraved with inscriptions in seal characters of very ancient style. Plated spoon. Stopper of green jadeite. The significance of this writing cannot be deciphered; these ancient characters play a purely magical rôle and are merely intended to act as a charm and to safeguard the owner from weird and wicked influences, from the malice of demons and the intrigues of his enemies. Similar writings put down on paper slips are pasted on the doors and walls of houses or carried around by persons as efficient talismans for the same purpose.

309. Whitish agate with black circular bands, the narrow sides carved into facets.—Stopper of green glass.

310. Gray agate filled with black and brown spots forming natural design. Oval, rounded shape. Plain.—Stopper of green jade.

311. Gray glass with brown wavy stripes, in imitation of agate. Plain.—Stopper of green glass.

[45]

312. Gray agate dotted with black and yellow spots. Plain. Stopper of green jade.

313. Same. Decorated in relief with lateral tiger-heads holding rings.—Stopper of green jade.

314. Same, dotted with yellow and brown spots. Decorated in relief with lateral tiger-heads.—Stopper of jadeite.

315. Brown agate filled with yellow flecks. Plain.—Stopper of green jade.

316. Red and brown agate. Of graceful miniature shape. Plain.—Stopper of blue glass.

317. Stone of various colors. Plain.—Stopper of green jade.

318. Heliotrope with dark-green and red spots and bands. This stone, popularly called bloodstone, is a variety of plasma containing spots of red jasper, looking like drops of blood. The ancients were acquainted with it under the name heliotrope, meaning "sun turning," which refers to the belief that the stone when immersed in water would change the image of the sun to blood-red. There is an ancient tradition that the stone originated at the crucifixion of Christ, from drops of blood drawn by the spear thrust in his side, falling on a dark green jasper. The supply is obtained almost wholly from India, especially from the Kathiawar Peninsula west of Cambay whence agate, carnelian, and chalcedony are also quarried.—Stopper of green jade.

319. Same.—Stopper of pink crystal.

320. Dark-green glass with reddish stripes, in imitation of agate. Plain.—Stopper of coral surmounted by green glass bead.

321. Brown agate intersected by yellow clouds and reddish zones. Plain.—Stopper of ruby-colored chrysoprase of irregular surface mounted on plated metal. Spoon plated.

322. Green, yellow, and red-spotted agate. Plain.—Stopper of green jade.

323. Pudding-stone, gray with large yellow and brown designs, a sandstone composed of siliceous pebbles, flint, etc., united by cement. Miniature shape. Plain.—Stopper of globular coral.

[46]

324. Same, with yellow, red, brown, gray and black flecks. Plain.—Stopper of red crystal set in gold.

325. Same, with red spots and blue veins. Plain.—Stopper of green jade.

326. Ancient jade of apple-green ground, carved with reliefs in brown, showing on the one side two large lotus leaves with blossom and a crane on the wing, and on the other side lotus with goldfish and bat, expressing the wish: "May you have luck and abundance of gold!"—Stopper of pale-green jade.

327. White flawless jade, with engraving of a small bamboo branch with leaves in the upper portion of one side, otherwise plain.—Stopper of green jade.

328. Semi-transparent gray agate, carved with reliefs in brown, representing a walking camel, a dog chasing a small mammal, and a crouching monkey.—Stopper of clouded jade.

329. Red agate. Plain. Rectangular, flat shape.—Stopper of green jade.

330. Green jade. Flat, rectangular shape with six facets in relief.—Stopper of amethyst.

331. Green Burmese jadeite, the one side being lighter than the other. Plain, with rounded shoulders and straight outlines.—Stopper of pink amethyst.

332. Milk-white, flawless jade, of perfect carving and polish. Plain, of straight outlines.—Stopper of green jade.

333. Gray jade of the variety styled "mutton-fat," carved into the appearance of a wickerware basket.—Stopper of dark-green jade.

334. Gray jade, carved in the shape of a calabash (compare No. 303). Simultaneously the lower portion is wrought into the figure of an elephant in kneeling posture, head and trunk being brought out on the narrow side, and the tusks being well formed. The saddle with which the animal is covered is carved in brown-red relief and adorned with a bat, the emblem of felicity, and a goose, the symbol of loyalty. The upper portion of the calabash is so represented as to form a vase called *tsun*, an ancient type of

[47]

sacrificial bronze vessel, and the elephant carrying this vase on its back is an ancient motive of the art of the bronze-founder. It served the purpose of holding wine in the ancient worship of the ancestors. In this case, the idea conveyed by this motive is simply that the elephant has many blessings in store which he carries in the vessel, and this thought is strengthened by the form of the calabash which is supposed to be a storehouse of blessings of all sorts.—Stopper of green jade carved into the figure of a frog.

335. White jade carved with relief scenes of Taoist lore. The one side pictures a Taoist sage stroking his beard and followed by a boy-attendant holding a quadrangular box, in scenery of rocks, pine-trees and clouds. The opposite side illustrates another sage reclining on a rugged tree-trunk near the sea. It is the belief of Taoism that man living in the solitude of the mountains and in permanent contact with the beauty of nature may finally attain to sainthood and immortal life.—Stopper of globular coral carved with a coiled hydra in relief.

336. Green glass. Plain. Stopper of pink glass mounted on plated ring. Spoon plated.

337. Milk-white agate clouded with green and rusty patches. Plain.—Stopper of red carnelian mounted on ring of lapis lazuli.

338. Leaf-green jade, with light-grayish shades. Carved in relief with five bats emblematic of the five kinds of happiness, which are long life, wealth, peace, love of virtue, and tranquil death.—Stopper of coral mounted on ring of gray jade.

339. Pale-gray agate, carved in crimson relief with pine-tree on which an eagle is perching, and a tiger turning its head around toward the eagle. Reverse plain.—Stopper of green jade.

340. Clouded green and white jadeite. Plain.—Stopper of red coral mounted on ring of lapis lazuli.

341. White jade, of perfect carving. Plain.—Stopper of globular pink crystal.

342. Gray and red clouded carnelian. Plain.—Stopper of green jade.

343. Chestnut-brown stone in its natural shape.—Stopper of green jade carved à jour in shape of peach.

[48]

344.   Reddish-gray stone overstrewn with white stars.   Decorated in relief with two lateral animal-heads holding rings.—Stopper of mottled green and white jade.

345.   Red jade.   Decorated in relief with lateral tiger-heads holding rings.—Stopper of green jade.

346.   Wax-stone, of trapezoidal shape, decorated on one side with a relief of a boy astride a water-buffalo, accompanied by an old man holding a sword, on the other side with engraving of a sceptre of happy augury with fluttering bands and a leafed branch of peaches.   This design expresses the wish: "May you have as many years as you desire!"—Stopper of clouded jade.

347.   Brown agate.   Plain.—Stopper of green jade.

348.   Brown agate with yellow spots.   Decorated in relief with lateral lion-heads holding rings.—Stopper of green jade.

349.   Agate clouded with yellow and reddish zones and spots. Plain.—Stopper of green jade.

350.   Green agate with purplish and gray spots.   Plain.—Stopper of coral mounted on lapis-lazuli ring.

351.   White jade, carved in the shape of a rounded wickerware basket, of fine workmanship.—Stopper of pink crystal.

352.   Gray agate, carved with flat reliefs in brown representing a large tree in which two birds are nesting; a fox and other animals are greedily leering at peaches.—Stopper of coral.

353.   Semi-transparent milky agate, with natural designs formed in the stone, appearing as if a woman seen from the back were meeting a tiger crouching on a rock.—Stopper of pink crystal mounted on ring of turquois.

354.   White jade.   Plain.—Stopper of green jade.

355.   Carnelian, of miniature shape, carved with two phenixes, their heads in all round carving standing out from the sides.   The style is derived from that of ancient bronzes.—Stopper of green jade.

356.   Semi-transparent agate, carved into four quadrangular medallions.—Stopper of red carnelian.

[49]

357. Yellow carnelian with a circular milk-white zone on one side. Rounded, flat shape. Plain.—Stopper of green jade.

358. Carnelian. Plain.—Stopper formed by a precious stone.

359. Same. Plain.—Stopper of green and white jade.

360. White jade, carved with the figures of the Twin Genii symbolizing harmony and union, one on each side, the two being connected by a lotus leaf which, as the result of punning, indicates union. The bottom is carved with cloud designs in relief.

361. Black and gray agate. Plain.—Stopper of green jade.

362. Rock-crystal, carved into the appearance of a wicker-ware basket.—Stopper of red coral.

363. Brown agate, carved in relief with figure of elephant with large tusks and covered with a saddle-cloth on which the ornament styled "reclining silkworm cocoon" is brought out.—Stopper of dark-green jade set in silver. Spoon of silver.

364. Clouded crystal, of rectangular shape. Plain.—Stopper of green and white jade.

365. Black crystal, of hexagonal shape, each of these six sides forming an ellipsoidal facet.—Stopper of pink crystal mounted on ring of blue turquois.

366. Brown and red agate, carved on one side with relief of running grasshopper. Reverse plain.—Stopper of green glass.

367. Clouded brown and gray agate. Plain.—Stopper of green jade.

368. Clouded white and brown agate. Plain.—Stopper of green jade.

369. Gray agate, carved with reliefs in brown representing a lotus-pond with lotus-leaf floating on water and duck swimming amidst rushes.—Stopper of crystal with violet tinge.

370. Brown, yellow and white agate with a vertical white zone running through the centre of each side. Plain.—Stopper of green jade.

[50]

371.   Variegated agate, similar to 370, of high specific gravity. Plain.—Stopper of green jade.

372.   Heliotrope (compare No. 318), dark-green overstrewn with red spots. Decorated in relief with lateral lion-heads.— Stopper of pink crystal.

373.   White jade-stone, carved in shape of a gourd, the upper green portion carved into leaves. Of very elegant workmanship. The stopper of dark-green jade forms the stem of the fruit.

374.   Milk-white semi-transparent agate, carved with reliefs in black, brown, and white. The one side shows two old gentle-men in mountain scenery walking along with their canes and just about to cross a bridge. On the other side is represented the celebrated poet Li T'ai-po of the eighth century seated under a pine and holding a wine-cup, from which he drew the inspiration for his songs. The legend here engraved says that a thousand days of anachreontic intoxication won him laurels for ten thousand years.—Stopper of coral mounted on amber-yellow glass ring and carved into a bat and peach, symbols of luck and longevity.

375.   Dark-green mottled jade, carved in shape of a bean, with relief of a smaller bean on the side and plum-blossoms on the main surface.—Stopper of coral.

376–377.   Pair of bottles carved from green turquois in the natural irregular form of the stone. The one bottle is decorated in relief with leaf designs laid around the neck and a separate bunch of palm-leaves near the base. The other bottle has its sur-face carved into peaches and a row of lotus petals around its neck. Stoppers of red coral mounted on silver. Turquois and coral form a favorite combination.

378.   Wax-stone, carved in the shape of a gourd. Leaved branches and a fruit, a bat, the emblem of happiness, and a running squirrel are laid around in relief.—Stopper of green jade carved into sprays of leaves.

379.   White jade, carved in the shape of a gourd and decorated with high undercut reliefs of twigs, plum-blossom, and a large phenix on the top.—Stopper of greenish stone, on which a bird perching on a rock is represented.

380.   Rock-crystal, carved with inscriptions in relief. Stop-

[51]

per of green jade. Each of the large sides contains a stanza. The one reads:

This is a queer product of distant lands which did not suffer from smoke and flame of a common workshop. Like red snow and dark hoar-frost is the flake of snuff enclosed in the crystal phial, bidding a welcome to every guest. From the song *I mu lan hua*.

The other is as follows:

Musk-scented tobacco is pounded into dust and conveys to you a fragrant message. When you take a spoonful after a wine-banquet, you will not long a moment for the crystal salt of the gods.

The legends on the lateral sides read, "Precious object of Siu P'ing" (being the name of the manufacturer), and "Property of Yen Pan-cho" (being the name of the former owner, for whom the bottle was made).

381. Pure-white jade, carved in relief with two genre-scenes in the life of boys; in the one, a mother is holding a child who is looking at two boys on the ground busy with lighting candles; this is a puerile pastime for ascertaining the future. In the other, a boy standing on a chair is leaning over a tub and turning water into it, while the water is pouring forth again from an aperture in the tub; two other boys are deeply interested in this pastime; pine and rock are represented on the right-hand side.— Stopper of pink crystal.

382. Semi-transparent, gray agate, carved with relief in brown of large grasshopper crawling over a reclining jar on which seven coins are represented. The grasshopper is regarded as a symbol of fecundity, and its image is expressive of the wish: "Numerous and remarkable sons and grandsons!" here with the addition of wealth (indicated by the coins) and peace (indicated by the jar). Stopper of mottled green jade.

383. Pale-blue transparent rock-crystal, of elegant shape, with softly rounded outlines. Plain.—Stopper of red carnelian.

384. Purple amethyst, carved in the shape of a gourd, decorated with leaves and tendrils and a bean-like fruit in relief.— Stopper of green jade.

385. Whitish crystal, carved in the shape of a Buddha-hand citrus (*Citrus decumana*), called in Chinese *fu shou*, Buddha's hand. As there is another expression *fu shou* with the meaning "happiness and longevity," the fruit is chosen to illustrate this thought. A leaved branch is carved in relief.—Stopper of green jade.

[52]

386. Agate intersected by white zones. Decorated with lateral tiger-heads holding rings.—Stopper of green jade.

387. Rock crystal, of excellent workmanship.—Stopper of green glass and pink amethyst.

388. Black crystal. Plain.—Stopper of pale coral.

389. Red agalmatolite, carved in the shape of the fruit lichee (*Nephelium lichi*). The symbolic interpretation of this name, based on punning, is "to add a new branch to the family pedigree." A bottle of this design is presented to a man, on the occasion of the birth of a son, the greatest event in a Chinese family, or if the donor is desirous of expressing the wish that he should be blessed with a son in the future.—Stopper of green jade.

390. Red agalmatolite with white zone. Plain.—Stopper of green jade mounted on marble ring.

391. Dark-red agalmatolite, carved in shape of a bean, the stopper of green jade forming the stem. Decorated in relief with large crawling hydra turning its head downward, surmounted by leaves and tendrils, and a butterfly surrounded by leaves.

392. Dark-red agalmatolite, carved in shape of a fruit, decorated in high relief with butterflies, dragon-fly, and peach.— Stopper of green jade.

393–394. Pair of bottles carved from coral, decorated with different designs in relief, and stoppers of dark-green jade. The one shows a medallion picture in which a Kilin (see No. 270) blowing up a cloud is enclosed, bamboo and blossoms on the narrow sides, and a stanza engraved on the reverse, reading: "A modest man dares not to write poetry, and to thrust it on his fellow-men, for fear that they might be scared, as if shooting-stars would fall down on them." The other bottle exhibits the famous Taoist recluse Liu Hai holding a fly-whisk and a play-ball with which he amuses his inseparable companion, the toad. He is a sort of Chinese Democritus noted for his jovial humor and serene laughter; he is the type of the happy-go-lucky, contented hermit who knows the art of living on nothing. At first he was a minister and student of Taoism when one day, as the story goes, a Taoist sage called on him and asked for ten eggs and ten pieces of gold. These the stranger piled one upon another in the form of a pagoda; whereupon Liu Hai cried out in fear lest the whole should topple over.

[53]

Then the sage said: "For him who dwells amid the pomps and vanities of the world, the danger is even greater!" With these words, he dashed the pagoda into two parts and bade his host farewell. Deeply impressed with this scene, he doffed his official garb and adopted the life of a recluse. On his death he was canonized and pronounced an immortal by the Taoist clergy.

395–396. Pair of bottles of red agalmatolite. Plain.—Stoppers of green jade.

397–398. Another pair of same substance.—Stoppers of green jade mounted on amber rings.

399. White and brown agate carved with reliefs, representing on one side old lion with cub, the latter coaxing for the ball which the parent is withholding from him, on the other side rock scenery with couple of spotted deer and bat. The meaning of this design is: "May you obtain the rank of the first dignitary at the imperial Court and enjoy good income and happiness!"—Stopper of carnelian.

400. Agate, carved with reliefs in white, representing old Chang Kuo, one of the Eight Immortals of the Taoists, riding on the back of his white mule and accompanied by a boy carrying large branch of plum-tree which is intended to ward off demons. This mule is a magical creature able to run thousands of miles in a day. When they halt, the sage folds him up and hides him away in his wallet. When his services are required again, he spurts water upon it, and the animal at once resumes its proper shape.—Stopper of white coral mounted on ring of turquois.

401. Milky agate, decorated on one side with flat relief in brown of two horses, the upper one turning its head toward the left, the lower one toward the right.—Stopper of white coral mounted on ring of turquois.

402. Canton glass bottle, with stopper of green and white jade. The interior is painted with a medley of scraps and papers, such as are often found as wall decorations in Chinese rooms. We notice the title of the Four Classical Books, the name-card of the famous Governor Tso Tsung-t'ang (1812–1885),—Chinese cards are always printed on a long rectangular sheet of red paper,—a public official notice; portion of a rubbing with characters standing out in white, being an extract from a well-known essay; a copy of the imperial almanac and calendar; a letter stating that a man

[54]

from Lo-yang in Honan Province is on his way to the province of Kuangtung to meet there a friend; the envelope of a letter mailed from Shanghai and bearing the address of a certain Mr. Wang; a fan with painting of roses, signed by Nan T'ien, the greatest flower-painter during the Manchu dynasty in the seventeenth century; a scroll containing an autograph in white on black, framed by blue borders; and a copy of the Peking Gazette.

On the reverse the following is inscribed: "During the middle summer-month of the sixth year of the period Chêng-kuan (that is, 632 A. D.) Emperor T'ai-tsung resided at his summer-resort, the palace styled Kiu-ch'êng ('Nine successes'), and subsequently in the Palace of Benevolence and Longevity rising on a lofty peak and surrounded by a deep ravine like a city by a moat. Written by Ma Shao-süan in the midsummer of the 18th year of the period Tao-kuang (1838)." Ma Shao-süan is the artisan responsible for the brush-work on this bottle, and the year 1838 yields the date for its manufacture. The above writing is a scrap from the History of the T'ang dynasty and certainly has no definite relation to this bottle; Mr. Ma is simply intent on showing off his erudition and knowledge of history.

403. Canton glass, with stopper of dark-green jade, painted on the inside by the same artisan as 402 with a still-life of flowers and insects, accompanied by the legend: "Delight fills heaven, and joy penetrates earth. Written on an autumnal day of the year yi-wei (that is, 1835) by Ma Shao-süan." On the reverse, a poem is written by the same hand, but dated 1822 (2d year of Tao-kuang). It is not very likely that the same bottle was inscribed at two such different dates as 1822 and 1835; it seems more probable that these alleged dates are fictitious and added par courtoisie to lend this piece another attractive feature. The poem reads as follows: "In the spring creation is animated like the world of the spirits. Spring's small garden is clean and dustless. The mysteries of creation are in full play, and as if the first creation would repeat itself, life grows anew. Vegetation, cultivated and wild, sprouts luxuriantly. I feel that our existence is like the image of the bright moon in her waxing and waning."

404. Pure transparent rock-crystal, carved in shape of a bell. Stopper of green jade, mounted on ring of agate.

405. Amethyst, a purple variety of quartz, carved in shape of a fruit, on the edge of which peanuts and leaves are carved in relief.—Stopper of green glass.

[55]

406.   Light-blue crystal, carved with relief of dragon immersed in the sea on which a rocky island is floating in the centre, and raising its head above the water to make for the pearl soaring in the air,—symbolizing the striving for an ideal.—Stopper of pink amethyst.

407.   Greenish transparent crystal, carved on the narrow sides with reliefs of two hydras hovering amidst the clouds, their heads being laid around over the large surfaces which are plain with the exception of some spiral designs representing clouds.   It is the picture of a serene sky over which only a few clouds are scattered.—Stopper of pink amethyst.

408.   Brown agate with white sections looking like icicles. Plain.—Stopper of yellow crystal.

409.   Milk-white agate, carved on one side with relief in yellow-brown of hawk dashing on fox from the height and making vehement attack upon him.   Reverse plain.—Stopper of green jade.

410.   Brown stone, carved with relief of horse.—Stopper of clouded green jade.

411.   Purplish-colored glass, with relief on one side in white, of two grazing horses.   Reverse plain.   Bell-shaped stopper of green glass mounted on yellowish glass ring and inlaid with a red glass bead.

412.   Black, brown, yellow and rusty clouded agate.   Plain. Stopper of yellow agate set in gold.

413.   Yellow wax-stone, carved in shape of a gourd, and decorated with reliefs of leaved branches and fruits, one fruit in high undercut relief.—Stopper of green jade carved into leaves.

414.   Celadon-green jade.—Plain stopper of pink crystal.

415.   Green old turquois, of the natural shape of the stone, suggestive of a fruit, with deep-black grooves.—Perforated stopper of coral with leaves carved around it.

416.   Agate of milky and yellowish color, carved with reliefs; on one side two lion-cubs wearing neck-collars with bells and running in a wild chase, on the other side a crouching cat leering at a butterfly, expressing a wish for old age.—Stopper of pink crystal.

[56]

417.   Grayish-blue jade, carved with reliefs in white, representing Liu Hai (see No. 393) standing on a rock and teasing with a string of coins his companion, the toad, enclosed by a stone rail. The copper coins in his possession indicate his contentment. The toad has become an emblem of money-making and a favorite sign of the merchants.   A few lines of poetry alluding to this subject and written in grass characters, a kind of shorthand, are carved into the reverse.—Stopper of pink coral, mounted on amber-yellow glass ring.

418.   Pale-green jade, carved in flat relief with an ornamentation derived from bronzes, a combination of bands and spirals.— Stopper of pink amethyst.

419.   Stone with red and yellow flecks.   Of cylindrical shape. Plain.—Stopper of pink coral carved into a coiled hydra.

420.   Porcelain of soft paste, decorated with etchings, and in reliefs in cream color of meander and lotus-petal designs.— Stopper of red coral.

421.   Violet-colored glass.   Plain.—Stopper of green jade.

422.   Yellow-colored glass.   Rectangular shape.   Decorated in relief with two lateral lion-heads holding rings in their jaws and a seal bordered by a rectangular frame, the same on each large surface.   The seal consists of four characters reading: "May you have luck and prosperity according to your wish!"—Stopper of coral carved with hydra in relief.

423.   Green glass imitating the color of jade.   Barrel shape, plain.—Stopper of brass ornamented with lion holding ball.

424.   Turquois-blue glass, painted on each side in enamel paints with still-life of flowers.—Stopper of pink crystal.

425.   Ordinary glass, the inside painted green and overstrewn with gold dots, carved with reliefs of dragon and bird-heads on one side, and two frogs on the other side; bands with conventionalized animal-heads on the narrow sides.—Stopper of globular coral.

426.   Transparent, blue glass, with gold spots inserted in the interior of the glass.   Barrel shape.   Cover of white jade on which a six-petaled lotus-flower is carved.

[57]

427. Transparent, ruby-red glass. Octagonal shape, four large medallion-like facets connected by four smaller chamfered facets. Metal cover with dragon à jour.

428. Soft-paste porcelain painted in five colors. Cylindrical shape, bottom unglazed. The picture represents Chung K'uei, the great magician believed to wield powers of exorcism over malignant demons, and the great destroyer of all malefactors. He is here portrayed with red hair and beard, wearing red costume and blue girdle, manipulating a fan in his left hand, while the right governs the bridles, accompanied by a devil-servant carrying an umbrella over him. Two other red-haired demons clad in aprons of tiger-skin and triangular hats crowned with plumes and carrying clubs open the procession. A girl is escorted by a devil in a push-cart. The picture of Chung K'uei is still hung up in doorways on New Year's Day, in order to keep off wicked spirits.—Stopper of carnelian.

429. White porcelain with open-work carvings of two layers in lapis-lazuli blue, representing dragon and phenix soaring in clouds. Meander designs around neck.—Stopper of porcelain.

430. Terra cotta, imitating bronze which is incrusted with gold, painted in gold with plum-tree in blossom. Shape of oil-bottle.—Stopper of coral.

431. Pale-green, transparent crystal. Carved with relief, of excellent workmanship, of hydra with head in the centre and body coiled around the rim, and iris-leaves on both sides.—Stopper of coral mounted on amber ring.

432. Transparent, flawless crystal. Carved with elephant heads holding movable rings in their trunks and hydras climbing up from below and eager to catch the rings.—Stopper of crystal, surmounted by crouching lion.

433. Transparent, flawless crystal.—Stopper of carnelian.

434. Pure-white crystal. Plain.—Stopper of coral mounted on turquois ring.

435. White jade interspersed with green patches. Plain.—Stopper of old coral.

436. "Mutton-fat" gray jade, of uniform color. Plain.—Stopper of green jade mounted on coral ring.

[58]

437. "Lion-skin" jade, in the irregular natural shape of the stone, one side spotless white, the other filled with iron-rust flecks. Stopper of rose crystal.

438. Han jade, white with red and rusty speckles, in the natural shape of the stone, triangular.—Stopper of coral branch.

439. "Mutton-fat" jade. Plain.—Stopper of coral and turquois.

440–441. Pair of bottles carved from light-red coral, carved with relief designs derived from the geometric style of the bronze vessels of the Chou dynasty, dragons facing each other with bodies formed by square spirals.—Stoppers of green jade.

442. Yellowish agate, decorated with relief carvings in black. The one side shows a Buddhist monk walking toward the gate of a temple, above which is visible a tiled roof covered by clouds. On the opposite side is represented a pine-tree rising from rock, its branches with cones reaching the neck of the bottle; a boy is seated on a boulder under the tree and raising his head to gaze at peaches above him. The beginning of a verse "there is a tree in the grove" is here engraved.—Stopper of green jade set in gold.

443. Greenish moss agate, decorated in relief with lateral tiger-heads holding rings.—Stopper of silver inlaid with corals arranged like the petals of a flower.

444. Ivory, carved in the shape of a cloth wrapper in which books are kept, enclosed in a black wooden case; the slip in green would contain the title of the book. The cover of the ivory box moves on a hinge. Inside there are two figures, each carved from ivory, a showman and a monkey. The former has the appearance of a dwarf and wears a pointed cap; he is a professional juggler. The monkey is standing beside him and forms the handle of the spoon with which to scoop out the snuff.

445. Dark-green stone. Plain.—Stopper of silver in form of rosette inlaid with turquois and coral.

446. Painted enamel over brass foundation (see No. 149), in shape of calabash, decorated with genre-picture: two cranes on the wing, a palace in clouds, boys and girls engaged in play, one holding a halberd, another lotus. The symbolism of this picture is: "May you be promoted a degree!" The cover, likewise

[59]

enameled, shows four bats, emblem of happiness, laid around a rosette.

447. White porcelain with five-color decoration of the eight auspicious emblems of Buddhism (see No. 42), flowers, fruits, butterflies, and lateral lion-heads in blue. Circular shape.— Stopper of imitation coral.

448. Pale-green porcelain carved with high reliefs in coral-red of nine lions playing ball, alluding to a wish of official promotion for the recipient. Neck painted with rows of small circles.— Stopper of silver in form of rosette, inlaid with a ruby-like stone in the centre surrounded by six corals.

449. Lustrous black glass, imitating the color of jet. Plain. Stopper of green glass.

450. Yi-hsing terra cotta, bottom inlaid with porcelain glaze, cylindrical shape. Decorated with two imperial, five-clawed dragons, one soaring in the clouds, the other rising from the sea, the space between them being filled by clouds.—Brass stopper in open work.

451. Blue and white porcelain of soft paste, *craquelé* decorated with imperial five-clawed dragon playing with flamed ball. Cylindrical shape.—Stopper of rose crystal. Period Yung-chêng (1723–1735).

452. Blue and white porcelain, in shape of long-necked vase, decorated with imperial five-clawed dragon rising above the water, the background being formed by chrysanthemums. Date-mark on bottom in blue: "Made in the years of the period Yung-chêng (1723–1735) of the Great Ts'ing dynasty."

453. Glass carved in cameo style with characters in blue standing out from the white background and enclosed in medallions. The meaning of this legend is that a stately tree spreads out its branches in all directions and cannot conceal them, alluding to the owner of a fine snuff-bottle who cannot keep it solely for himself but is proud of passing it around among his friends to share his admiration. The four characters on the reverse mean: "Precious trinket of the imperial Court!" a euphemistic polite phrase to convey the impression as if the owner had received the bottle as an act of imperial grace.—Stopper of coral.

[60]

454. Blue and white porcelain of soft paste, *craquelé*, decorated with a dragon floating in the sea and a phenix soaring in the clouds. Of perfect quality.—Stopper of coral. Date-mark in underglaze blue on bottom: "Made in the years of the period Yung-chêng (1723–1735) of the Great Ts'ing dynasty."

455. White porcelain of soft paste, *craquelé*, carved with the famous motive "squirrel stealing grapes"; the squirrel is squatting on the ground, holding a bunch of grapes between its forepaws.— Stopper of coral mounted on silver. Period Yung-chêng (1723– 1735).

456. White biscuit porcelain, carved in shape of a fig-leaf. Buddha obtained his enlightenment while meditating under a fig-tree, and clay plaques of fig-leaf form stamped with Buddhist pictures were given as souvenirs to the pious pilgrims visiting Buddhagayā. This motive spread to all Buddhist countries of the East. A medallion filled with square meanders styled "thunder pattern" is bordered by a row of leaves with drooping points. Stopper of amethyst carved in shape of the mythical toad inhabiting the moon. K'ang-hi period (1662–1722).

457. Semi-transparent pale-gray agate with natural design in brown relief suggestive of the figure of a lion.—Stopper of pink crystal mounted on turquois ring.

458. Carved cinnabar lacquer, engraved with two genre-pictures. A man is on a journey with his wife, and they are making a stop at an inn; a maid is serving tea to them; the background is formed by rocks, banana and pine. The other side shows us their traveling vehicle, a two-wheeled push-cart, and the coolie attending it clad in a straw-hat. He is squatting on the ground, in China a convenient position for taking a rest. He is chatting with another coolie who is carrying a pole with two baskets containing the baggage of the couple. On the right-hand side the good humor of the carver has brought to life a fantastic forest-sprite peeping from behind a rock, an animal with head of a stag seated on its hind-legs and stretching forth a forepaw.—Stopper of carved lacquer.

459. Clouded spinach-green jade. Plain.—Stopper of pink crystal.

460. Iron, of octagonal shape, each panel incrusted with silver wire forming various designs. The shoulders are formed by eight

[61]

small fields in which the Eight Trigrams, ancient mystic symbols used for purposes of divination, are represented. The eight panels display two bats, emblems of happiness, encircling the character *shou*, "longevity"; a butterfly surrounded by leaf patterns; a design of sea-waves; geometrical ornaments; rows of spirals; a row of continuous swastikas, an emblem of victory and felicity; a composition consisting of fish, bowl and the ornament of felicitous threads, the mystic emblem of Vishnu.—Stopper of green jade mounted on silver ring.

461. Yellow ivory, each side being inlaid with the brown shell of a chestnut, decorated with fine etchings of landscapes—mountain-scenery with pine and willow.—Stopper of coral inlaid with pearl.

462. White porcelain of soft paste, *craquelé*, carved in shape of a crouching lion holding a ball in its jaws, its eye-balls being glazed black.—Stopper of coral.

463. Ribbon agate, semi-transparent, with bluish-white zones. Plain.—Stopper of coral in shape of cover for teacup mounted on turquois-ring. Spoon of ivory.

464. Blue and white porcelain of soft paste, decorated with imperial five-clawed dragons facing each other and playing ball.—Stopper of coral inlaid with pearl. Yung-chêng period (1723–1735).

465. White porcelain, of cylindrical shape, etched under the glaze with an imperial five-clawed dragon soaring in clouds, its eyes being painted with cobalt blue.—Stopper of porcelain decorated with chrysanthemums. Bottom unglazed, laid out in concentric circles. Yung-chêng period (1723–1735).

466. Sung ivory-glazed porcelain, with leaf designs in black and brown spread over the whole surface. The bottle was originally used to hold medicines, and only subsequently converted into a snuff-bottle.—The stopper of coral inlaid with a pearl certainly is a later addition. A piece of great rarity. Sung period (960–1278).

467. Canary-yellow porcelain, *craquelé*, cylindrical, in shape of oil-bottle with long neck. Plain.—Stopper of carnelian mounted on a ring of blue glass.

[62]

468. Canton glass bottle, with green stopper, painted in the interior with a genre-picture: a man clad in a blue gown, black hood and straw hat, astride a donkey, is going to pass over a brook on a bridge; he is followed by a boy in pink suit walking with a long cane and shouldering a calabash. Rocky boulders with blossoming plum-trees form the scenery. The picture is signed by Ma Shao-süan, the same artisan who made the two bottles 402 and 403, and dated by him 1850. On the reverse he has added the following poem in graceful hand-writing:

"One night a north gale rose, and for many miles the sky was thickly enveloped by clouds. The air was filled with snowflakes tossed around in wild dance, and the mountains were coated with ice. Leaves of conifers and opening buds flew around throughout the world for some time. Astride my donkey I am crossing a small bridge, solitary, heaving a sigh that the plum-blossoms are so slender." This stanza of four verses with the same rhyme is signed again by Ma Shao-süan.

469. White porcelain bottle painted in enamel colors with portraits of a man holding pink umbrella and a European lady of the style of the Rococo. Stopper of pink crystal. Presumably end of seventeenth century.

470. Brick-red colored opaque glass, carved in shape of tea-cup with cover. Stopper of imitation coral mounted on stained glass ring. Spoon of bone.

471. Black-colored opaque glass overstrewn with gold speckles, of octagonal shape. Stopper of imitation jade mounted on brass.

[63]

PRINTED BY R. R. DONNELLEY
AND SONS COMPANY, AT THE
LAKESIDE PRESS, CHICAGO, ILL.

101

中国纸拓片的价值

ZWEITER JAHRGANG · HEFT 3 · OKTOBER/DEZEMBER 1913

OSTASIATISCHE ZEITSCHRIFT

BEITRÄGE ZUR KENNTNIS DER KUNST UND KULTUR DES FERNEN OSTENS

THE FAR EAST

AN ILLUSTRATED QUARTERLY REVIEW DEALING WITH THE ART AND CIVILISATION OF THE EASTERN COUNTRIES

L'EXTRÊME ORIENT

ÉTUDES ILLUSTRÉES TRIMESTRIELLES SUR L'ART ET LA CULTURE DE L'ASIE ORIENTALE

EDITED BY   HERAUSGEGEBEN VON   DIRIGÉES PAR

OTTO KÜMMEL und WILLIAM COHN

OESTERHELD & Co. VERLAG
BERLIN W. 15

VIERTELJÄHRLICH EIN HEFT. PREIS · JÄHRLICH 30 MK. EINZELHEFT 8 MK.

zeptionen", die jedenfalls erst dann besonders nachhaltige Umwälzungen auf dem Gebiet der Ornamentik herbeiführen, wenn die Volkspsyche der neuen, der zu rezipierenden Stufe entgegengereift ist.

Hat dann der Einbruch der neuen Formen stattgefunden — bei dem Kinde wirkt er geradezu revolutionierend: die rhythmische Anordnung wird für lange Zeit verlassen; das gegenständliche Interesse, die Bemühungen, die Dinge der Außenwelt zu erfassen und wiederzugeben, läßt gegen alle ornamentalen Tendenzen gleichgültig —, dann tritt bei Kind und Volk eine Zeit der Besinnung und Sammlung ein. In dieser Zeit geht nun die Entwicklung, von Ausnahmen abgesehen, nicht geradlinig in der Richtung fortschreitender Naturannäherung weiter, sondern es wird mit der neuen Form „gespielt", alle möglichen Versuche, sie zu verschönern, setzen ein[1]. Diese Tendenzen zur Verschönerung können eine Naturannäherung mit sich führen, sie können aber auch immer weiter von der Natur wegführen.

Nachdem dieses ornamentale „Spielen" eine Zeitlang gedauert hat, findet ein neuer Einbruch noch entwickelterer Formen statt, dessen Art und Verlauf um so modifizierter ist, je höher die Entwicklungsstufe ist, auf der er stattfindet.　　　　　　Friedrich Muth (Bensheim).

ÜBER DEN WERT DER CHINESISCHEN PAPIER-ABKLATSCHE.

Herrn Fischers Studie in dieser Zeitschrift (Bd. II, S. 52) verdient als erster Versuch in dieser Richtung alle Anerkennung. Bevor jedoch die Reliefs der Hanzeit zu kunstgeschichtlichen Zwecken benutzt werden, ist eine technische Vorfrage in ernstliche Erwägung zu ziehen. Über den Wert der chinesischen Papierabklatsche, auf denen Herrn Fischers Essay beruht, darf man sich keinen Illusionen hingeben. Für die Zwecke des Archäologen, der die Sujets und Motive festlegen will, mögen sie zur Not allenfalls ausreichen; für kunstwissenschaftliche und insbesondere stilistische Untersuchungen genügen sie indessen nicht, und zwar aus drei gewichtigen Gründen. Erstens läßt der Papierabklatsch nicht die plastische Form erkennen; zweitens sind zahlreiche feinere Linien in den neueren Abklatschen überhaupt nicht mehr wiedergegeben, aus Gründen, die von mir (Chinese Pottery, p. 70) dargelegt sind, wo auch der besondere Wert des *Kin shi so* für das Studium der Hanreliefs ins Licht gesetzt ist, was Herrn Fischer (S. 55) entgangen ist; drittens ist der Abklatsch nur ein Negativ, aus dem erst das Positiv mit der schwarzen Linie hergestellt werden muß, und auf die Beobachtung der Linie sollte doch für eine stilistische Studie alles ankommen. Nur auf einem in kritischer Hinsicht so verwahrlosten Gebiete, wie es die chinesische Forschung leider noch immer ist, konnte man überhaupt auf den ungeheuerlichen Gedanken verfallen, Skulpturen auf Grund luftiger Papierabklatsche abzuhandeln. Was würde man in der klassischen oder indischen Archäologie dazu sagen? Um die negative Zeichnung des Abklatsches in das Positiv zu verwandeln, lassen sich verschiedene Methoden einschlagen; die einfachste dürfte jedenfalls die der Doppelphotographie sein, wie sie diese Zeitschrift (Bd. I, S. 46) und Confucius and His Portraits (p. 5) von mir beschrieben und praktisch ausgeführt worden ist. Herr Fischer hätte jedenfalls der Sache besser gedient, wenn er sich dieses technische Verfahren angeeignet hätte, anstatt in manchen Fällen ohne sachliche Begründung subjektive Deutungen vorzutragen. Zu diesen unannehmbaren Deutungen gehört vor allem der Baum des süßen Taus (S. 55 und 61), den Herr Fischer in dem Stein von Tsi-ning chou (Chavannes, Mission, No. 182) erkennt. Der Baum des süßen Taus, inschriftlich als solcher bezeichnet (Mission, Nr. 167), ist aber erstens ganz anders gestaltet und stilisiert, und er ist zweitens durch einen an seinem Fuße stehenden Mann mit emporgehaltenen Handflächen, der den Tau auffängt, deutlich genug charakterisiert. Derselbe Gedanke wird aber in der Kunst der Han stets durch das gleiche Motiv ausgedrückt, und wo zwei verschiedene Motive vorliegen, sind auch verschiedene Handlungen und Intentionen zu suchen. Es ist auch ohne weiteres klar, daß man nicht den Regenschirm aufspannt, um den süßen Tau aufzufangen, und daß man an diesem Baum nicht so eilends vorbeimarschiert, wie die beiden Frauen auf dem Stein von Tsi-ning chou. Um eine „genremäßige

[1] Vgl. hierzu S. 100 u. ff. meiner Stilprinzipien und S. 266 u. ff. in meinem zweiten Aufsatze über Ornamentationsversuche Bd. 7, H. 2 u. 3 der Zeitschr. f. angew. Psychologie.

Darstellung", wie Herr Fischer meint, handelt es sich hier auch nicht, sondern wie in so vielen Reliefs, um die Illustration einer Szene aus dem *Ku lie nü chuan* 古烈女傳, der Sammlung von Biographien tugendhafter Frauen von Liu Hiang 劉向 (erstes Jahrhundert v. Chr.), einem Werke, das von dem Maler Ku K'ai-chi 顧愷之 illustriert worden ist (Faksimile Druck der Sung-Ausgabe in 4 Bänden, Yang-chou, 1825), und an dem sich bereits bekanntlich der Staatsmann Ts'ai Yung 柴邕 (133—192 n. Chr.) mit Skizzen versucht hatte (Hirth, Scraps, p. 49). Daraus geht hervor, daß die Steinschneider der Hanzeit aus denselben Quellen geschöpft haben wie die Maler. Nur Vertiefung in altchinesische Sage und Geschichte, das ist sicher, kann unser Verständnis der Hanskulptur fördern, wie das z. B. in der Juninummer des *Open Court* (p. 378) an der bisher unerklärten „Schlacht der Fische" von mir gezeigt ist. Die Außerachtlassung der chinesischen Tradition muß aber zu grotesken Auffassungen führen, wie Herrn Fischers Erklärung seiner Abbildung 6 (= Chavannes, Mission, Nr. 143) auf S. 57 zeigt: „Unter einem Baum spielt eine Szene zwischen einem Mann und einer am Boden liegenden Herrscherin. Ein Vogel im Baum und ein Bär vervollständigen das Bild." Ich möchte Herrn Chavannes, dem die Erklärung dieses Reliefs im zweiten Bande seines Werkes vorbehalten ist, nicht vorgreifen und daher nur in Kürze bemerken, daß es sich laut der chinesischen Beischrift um Liu-hia Hui 柳下惠, mit seinem eigentlichen Namen Chan Huo 展獲 (Giles, Biogr. Dict., Nr. 18) handelt, von dem Giles sagt: „He was a man of eminent virtue, and is said on one occasion to have held a lady in his lap without the slightest imputation on his moral character" (vgl. ferner Legge, Chinese Classics, Vol. I, p. 336, und Chavannes, Se-ma Ts'ien, Vol. V, p. 418), und das ist, was hier im Bilde vorgeführt wird. Auch diese Geschichte ist in dem erwähnten *Ku lie nü chuan* (Kap. 2, p. 10) enthalten. Die Weide (denn um eine solche handelt es sich) *liu* 柳 ist eine deutliche Anspielung auf den Distrikt Liu-hia, dessen Gouverneur Hui war, und daher ein integrierender Bestandteil der Darstellung selbst. Von dem Bären ist keine Spur; dieser Bär ist der bekannte Typus eines Luftgeistes oder Dämons, der so manche Szene auf den Hanreliefs begleitet, und von dem in Chinese Pottery (p. 157 et passim) vielfach die Rede war.

Schließlich möchte ich mir im allseitigen Interesse den Vorschlag erlauben, daß unsere Museen, für die der Papierabklatsch auch ein wenig sicherer und dauernder Besitz ist, sich gute Gipsabgüsse der Hanreliefs von Shantung verschaffen sollten, die in der ursprünglichen Anordnung der Grabkammern aufgestellt werden müßten. Dann könnten wir auch der noch ungelösten fundamentalen Frage der Gesamtkomposition der Reliefs nähertreten; denn es scheint Grund zu der Annahme zu sein, daß gewisse Gruppen von Reliefs zu rhapsodieartigen Kompositionen zusammengehören. Gemeinsame Aktion der interessierten Museen könnte die aus der Beschaffung der Abgüsse entstehenden Kosten auf ein Minimum reduzieren.   B. Laufer (Chicago).

ERWIN BÄLZ.
(† 31. August 1913.)

Die deutsche Wissenschaft hat im modernen Japan keinen zweiten so einflußreichen und allgemein bekannten Vertreter gehabt wie den jetzt heimgegangenen Professor emeritus der Medizin an der Kaiserlichen Universität Tokio Geheimen Hofrat Dr. Erwin Bälz. Zugleich hat er auch zur Kenntnis des japanischen Volkes und aller Ostasiaten die wichtigsten Bausteine geliefert, mit deren Zusammenfügung zu einem anthropologischen System er bis zuletzt unablässig beschäftigt war. Ihm gebührt deshalb auch an dieser Stelle ein Wort des Nachrufs und der persönlichen Würdigung.

Als ältester Sohn eines Architekten wurde Erwin Bälz am 13. Januar 1849 geboren, und er verdankt seinen Vornamen, den er später nebst der japanischen Übersetzung Toku seinem einzigen Sohne beilegte, der Verehrung für den Schöpfer des Straßburger Münsters, dem 1845 in Baden ein Denkmal gesetzt worden war. Aber aus innerem Drange wurde der lebhafte Knabe dem in der Familie erblichen Berufe untreu, um sich der Heilkunde zu widmen. Als junger Privatdozent an der Universität Leipzig hatte er sich bereits durch die Entdeckung und genaue Beschreibung einer seltenen, oft nach ihm benannten Krankheit bekannt gemacht, als ihn 1876 eine Berufung an die Kaiserliche Medizinschule nach Japan führte, das für ihn ein

考古学与民族学关系探讨

AMERICAN ANTHROPOLOGIST

NEW SERIES

ORGAN OF THE AMERICAN ANTHROPOLOGICAL ASSOCIATION,
THE ANTHROPOLOGICAL SOCIETY OF WASHINGTON,
AND THE AMERICAN ETHNOLOGICAL
SOCIETY OF NEW YORK

PUBLICATION COMMITTEE

VOLUME 15

1913

American Anthropologist

NEW SERIES

| VOL. 15 | OCTOBER–DECEMBER, 1913 | No. 4 |

SOME ASPECTS OF NORTH AMERICAN ARCHEOLOGY[1]

By ROLAND B. DIXON

ARCHEOLOGICAL investigations in North America may for convenience be divided into two classes—those, on the one hand, which are concerned mainly with the question of the existence of early man in the continent, and, on the other, those which relate to later prehistoric peoples, to the immediate predecessors of the historic Indians. With the former class I do not propose to deal here, but wish rather to confine my attention to certain aspects of the latter which have a more or less direct bearing on American ethnology and ethnography. A very considerable mass of archeological material and information of this type has been accumulated in the last half-century. It seems therefore not inappropriate to consider a few of the broader and more general results of this work, the character of some of the problems which it presents, and some of the lessons which we may draw from what has already been done that will help us to more efficient and productive work in the future.

Anyone who may make a general survey of the archeology of North America as it is known at present, cannot fail to be impressed, I think, by one broad and fundamental contrast which exists between the western portion of the continent and the eastern. The contrast lies in this, that in the former area the archeological

---

[1] Presidential address delivered at the annual meeting of the American Anthropological Association, New York, December, 1913.

record is, relatively speaking, simple and intelligible, whereas in the latter it is complex and to a large degree baffling. The fact of this contrast and the character of it lead to several interesting conclusions, but before considering these and their bearing on problems of American ethnology and ethnography, it will be well, even at the risk of stating facts which are familiar to all, to refer very briefly to a few concrete examples.

The shell-heaps and burial places along the southern California coast and on the adjacent islands, have, as is well known, furnished a large amount of archeological material. Many of these shell-heaps seem, by virtue of their relation to raised beaches, to be of very respectable antiquity. From some of them and from some of the graves, on the other hand, objects of European manufacture have been obtained, showing that a portion of the sites were occupied in historic times. The character of the objects as a whole, however, is quite uniform, and except for the things of European origin, there is little or no evidence in this region of any other type of culture from the earliest period down to that of the establishment of the missions.

The vicinity of San Francisco bay is characterized by abundant shell-heaps and shell-mounds. Investigation of a number of these has shown that the lower strata lie at present several feet below water-line. There is geological evidence that the shore-line has been slowly sinking, and while the rate of depression is not yet known with certainty, the conditions are such as to lead one to infer a very considerable age for the lower layers of these mounds. In the mounds themselves are found the remains of a culture which is on the whole uniform from the lower to the upper strata, and which merges directly into that of the historic tribes of the vicinity. The uniformity of the culture is paralleled by a similar uniformity of physical type, the crania from the shell-mounds being similar to those of the tribes in residence at the time of the first European settlement. In this region, as in that to the south, no remains indicating the presence of any other type of culture have been found.

Continuing farther to the north, abundant shell-heaps, frequently of large size, are found along the lower Fraser river and the coast of British Columbia. Here again evidence afforded by forest growth, and by the relation of the shell-heaps to the present shore-line, indicates that the lower layers of these heaps are of consider-able age. Careful investigation of these sites has shown that here also there is no sign of any noticeable change in culture from the lower to the upper layers, and that this culture as shown by its remains in these shell-heaps is substantially that of the historic Indians of the vicinity. Unlike the previous case, however, there seems to be indication of a rather radical change in physical type, dolichocephalic crania being present in the lower, but not in the upper layers. Although there would thus appear to be evidence of some considerable change in physical type, the culture has remained virtually constant.

The conditions still farther north, as shown by the shell-heaps of the Aleutian islands, are practically a repetition of those about San Francisco bay. From the lowermost layers containing objects of human manufacture to the uppermost there is revealed no important change in type, only an increasing perfection of the products of a uniform culture, accompanied by a change in the proportions of the food supply obtained from fish and from sea-mammals. Here, as in the other regions to the south, the culture of the shell-heaps is one with that of the historic tribes.

It would appear, therefore, that on the basis of the archeo-logical investigations so far made, we are justified in concluding that in each of the respective areas considered, one and only one type of culture is evident; that such differences as are found to exist between the lower or earlier and the upper or later strata are of such a character and degree as to be most probably ascribed to gradual and uninfluenced development; and that as these various prehistoric types of culture are similar to the cultures of the historic tribes in the respective regions, the various culture types have been in permanent and continuous occupancy from very early times to the present day. There is, in other words, no evidence of any succession of distinct cultures or of any noticeable influence on the local cultures exerted by those of other areas. This purely archeo-logical indication of permanence and stability is in large measure corroborated by the evidence of the historic tribes themselves, since they seem for the greater part to have been long resident in their present habitats, and to preserve no recollection of migration. Linguistic evidence, to be sure, indicates that some of the tribes are really immigrants, yet they seem to have brought with them little that is recognizable as exotic, and to have been so completely brought under the influence of the new environment that in some cases they have come to be taken as typical exponents of the culture of their respective areas.

If we turn now to the eastern portion of the continent the contrast is at once apparent, for instead of permanence and stability, we find relative impermanence and instability; in place of uniform, coherent archeological remains, we have varied and unrelated types; and compared with the relative absence of apparent relationship to other culture areas, we have clear if baffling similarities with other and widely separated types. Let me again illustrate by a few concrete examples.

Beginning in the northeast, with what is perhaps the simplest case, we find that in northern New England and the maritime provinces of Canada, there seem to be indications, from the archeological evidence, of two somewhat different types of culture. One of these, clearly revealed up to the present chiefly in Maine, is represented in the very old graves which are characterized in part by large deposits of red ochre, and in part by the frequency of the adze, the gouge, and especially the ground slate points, which are often of large size. Objects of other materials than stone do not occur in these graves, and as a rule the burials themselves have completely disappeared except for faint traces of teeth or a few particles of bone dust. In the shell-heaps, which are abundant in the region, no trace of the peculiar ground slate points occurs; the adze and gouge so typical of the old graves are either scarce or entirely lacking; whereas articles of bone and shell, which were absent in the graves, are here abundant, and pottery of a crude variety usually occurs. The two types of sites occur in close proximity, yet each is in the character of its artifacts quite distinct. It seems therefore most probable that we are justified in distinguishing in this region two different and presumably successive cultures.

Turning next to the region about lakes Erie and Ontario, occupied in historic times by tribes of the Iroquoian stock, a somewhat more complex situation presents itself.   Here it would seem that three varied types of culture are indicated by the archeological material at hand, although the evidence is as yet in some ways obscure and perhaps insufficient.   Most characteristic everywhere, and at least in the more fertile sections of this area predominant, are the remains typical of the Iroquoian tribes found in occupancy in the seventeenth century.   Objects of stone, shell, bone, and metal, together with abundant pottery, are found at a great number of sites, usually but not always further characterized by defensive works of a simple nature, many of which are quite accurately datable throughout the seventeenth and eighteenth centuries.   Others again are clearly prehistoric, but objects from all the sites show well-marked common features, and the changes and development in form and other respects can be traced from the earlier to the later times.   Scattered alike in the fertile region about the lakes, as well as in the more rugged uplands, are various locations from which implements of stone have been gathered, quite unlike any found on Iroquoian sites.   These are principally ground slate semilunar knives, short, ground slate points with notched bases, and gouges.   While none of these forms are very abundant, they occur in considerable numbers in the area north and south of the eastern end of Lake Ontario, in the St Lawrence valley, and about Lake Champlain, but are absent or scarce in southern and western New York and western Ontario. Rather more widely distributed, perhaps, is another class of objects, also largely foreign to sites of known Iroquoian occupancy.   This group comprises the stone tubes, the so-called banner-stones, and various types of gorgets, bird-stones, etc.   Technically as products of the stone-worker's art, many of these show a relatively high development both as compared to the known products of Iroquoian tribes and to the group of ground slate objects just mentioned.   So far as any evidence at present available goes, these two small groups of objects are quite distinct from each other, in both type and occurrence, as well as from the types of artifacts everywhere characteristic of the Iroquoian sites in this area.

The extreme southeastern corner of the continent also affords archeological indications of more than a single culture. Taking the area of the peninsula of Florida together with the immediately adjacent territory to the north, the remains of several types may be distinguished. The well-known investigations along the St John river have demonstrated that in the shell-heaps of this section we have traces of a very simple culture. The finds comprise a comparatively small variety of implements of shell and bone, stone objects being remarkably scarce. Pottery and metal objects are in many sites totally lacking, and in others are found only in the uppermost layers. Ornaments of any sort are rare, and evidences of the practice of agriculture comparatively meager, the people apparently living largely on fish and shell-fish. Interspersed with these shell-heaps and also widely distributed throughout the peninsula, particularly in its northern portion, are a large number of mounds, of both the domiciliary and burial types. Extended investigation of these has brought to light the remains of a different type of culture. While objects of shell and bone are still numerous, a much larger proportion of stone objects occurs and ornaments are quite abundant. Pottery, moreover, of several types appears to be generally present, and not a few ornaments and one or two implements of copper have been found. Pipes, which do not occur in the shell-heaps, are of not infrequent occurrence in the mounds. A further contrast with the shell-heaps is shown by the fact that whereas the few crania obtained from these are dolichocephalic, those from the mounds show a predominant brachycephaly.

While the remains as a whole in these mounds would seem to indicate a different culture from that of the shell-heaps, certain of the finds deserve special mention. I refer to the so-called "spade-shaped" objects and circular spool-like ear-ornaments of stone, to the copper plates with repoussé and excised decoration, the rectangular fluted copper ornaments, and copper spool-shaped ear-ornaments in one case overlaid with silver, in one with meteoric iron. With these may perhaps be included certain biconate earthen-ware tubes. These objects have been found, in the main, at two sites only, and are of types characteristic of the Ohio valley, Kentucky, Tennessee, and part of northern Alabama and Georgia. At first thought it would be natural to consider these exotic objects as brought to this remote point through the channels of aboriginal trade. It is however suggestive to note that in the two sites where the majority were found, burials at length were largely predominant, whereas the typical form of burial elsewhere in the region is in the flexed position.

In some respects distinct from either the culture of the shell-heaps or of the mounds, are the remarkable remains so far known only from Key Marco on the southwestern coast. I need not do more than refer to these well known and very interesting finds and to their curious apparent relationship alike to more northerly as well as to more southerly regions. Whatever may with fuller knowledge be the final verdict on the evidence which they supply, they clearly reveal a type or at least a stage of culture which differed from others in the area. Whether we are to regard the evidence of Antillean affinities derived from the study of the pottery designs of Florida and adjacent regions as indicating still another cultural stratum, or to consider it as merely a separate or closely related phase of the southern influence shown at Key Marco, is not wholly clear. Certain it is however, that, taken as a whole, the archeological record shows this southeastern corner of the continent to have had a far from simple history.

The last area the archeology of which I wish to consider briefly, is that of the Ohio valley. The richness and interest of this field is proverbial; the collections obtained from it have been large and varied; and the literature dealing with the region is abundant in quantity if at times disappointing in quality. It requires little acquaintance with the sites, the collections, or the literature to recognize that we have here the remains of more than a single culture, that indeed the problem is one of rather baffling complexity. A satisfactory classification even of the various types present is by no means easy, and I shall not therefore attempt to do more than refer briefly to some of the more important features.

Scattered rather widely, although nowhere very common, and more abundant in the northern than in the southern portion of the area, are groups of burials in gravel banks of glacial origin. Commonly placed in a flexed position, the bodies are either without accompanying artifacts or supplied with only a few chipped stone implements of a limited number of types. More abundant by far, and even more widely scattered, but predominant more in the south than in the north, are the so-called stone box-graves. These show a considerable number of variations from the typical cist form, and occur both in cemeteries of varying size and in mounds, the latter form being most characteristic of the Tennessee region. Some contain characteristic burials at length, others show flexed burials, while a few contain cremated remains. Some of these stone box-graves are associated apparently with defensive earthworks often of large size, others seem equally closely related to groups of mounds of complex sacrificial or ceremonial character. Some contain burials devoid of any associated artifacts or are supplied with simple objects of stone only, while from others objects and ornaments of stone, shell, and copper have been taken, showing a relatively high development of culture. In some the crania are apparently dolichocephalic and without any artificial deformation, in others the type is often strongly brachycephalic, and occipital deformation is present. In the great majority of cases nothing of European manufacture is found in these graves, but in some instances evidence of European contact is clear. From the wide variation in the details of this type of burial it would seem that we had here to deal with more than one group of people and more than one type of culture, or at least with one group at two different periods in its history.

A third type of remains in the region under consideration is that of the village sites. These again are of somewhat varied character. Some are clearly associated with large defensive works, or with small mounds of simple structure, whereas others occur quite independently. Many show traces of circular lodge sites and are characterized by extensive ash and cache pits. Burials in some cases were made in the stone box-graves, in others at length without the use of stone and in close proximity to the houses.   The people were dependent largely on agriculture, but also drew a large part of their food supply from hunting, although curiously they would seem not to have made any use of the buffalo.   The pottery which they made was of an inferior type, and they had little or no acquaintance with copper.

Still another and in many ways the most important type of remains is that limited largely to southwestern Ohio, and characterized by the well-known elaborate enclosures and complex ceremonial mounds.   Although in some instances associated with stone box-graves, the more typical method employed by the builders of these structures was cremation.   As evidenced by the elaborate structures they built, they must have developed a rather complex ceremonial life, and had attained considerable skill in the working of bone, stone, and metal, using copper, silver, gold, and meteoric iron.   Their pottery, on the other hand, was curiously crude, if we except the single case of the remarkable figurines found in the Turner group.

Whether or not the few cases of effigy mounds found in this area are to be regarded as representing a further distinct culture or are to be allied to one or another of those already referred to, the evidence at hand does not make clear.   The same is true in regard to the question of the large mounds of truncated pyramidal type which occur here in small numbers.   Without considering any further cases, however, it is clear enough that the history of this region is a more than ordinarily complicated one, and that we must admit here the presence of the remains of a number of different cultures.

This very hasty outline of some of the results of archeological investigation in the eastern part of the continent brings clearly into prominence the contrast referred to in the beginning.   On the Pacific coast we seem to have evidence of a number of local types of culture, each showing a continuity of development from the earliest times down to the present, and each being in its own area the only culture found; here in the eastern portion of the country, in each of the areas considered, two or more different types are revealed, some of which at least would seem to have been extinct or almost wholly superseded at the beginning of the historical period.

We have so far dealt with the archeological evidence only in and for itself, its bearing on ethnological or ethnographical questions not having been considered. This is, however, perhaps its most important side, for archeology is but prehistoric ethnology and ethnography—the incomplete and wasted record of cultures which, often in vain, we try to reconstruct and affiliate with their historic descendants. Looked at from this side, the broad contrast already pointed out is significant. The Pacific coast, as we have seen, has apparently been occupied from earliest times by peoples differing but little in their culture from the tribes found in occupancy in the sixteenth century. Cut off from the rest of the country by the great chain of the cordilleras and the inhospitable and arid interior plateaus, the tribes of this narrow coastal strip developed in comparative seclusion their various cultures, each adapted to the environment in which it was found. The immigrants who penetrated to this region from beyond its bounds, brought, it would seem, little with them which has left its mark, and have been so completely molded to their new environment that but for the test of language we should not suspect their distant origin. As is well known, this long strip of territory is conspicuous for its linguistic complexity, the causes of which have been not a little discussed. The long-continued seclusion, the permanence of occupancy, are in this respect therefore not without importance, for it is precisely under such conditions that wide differentiation and division into numerous dialects and languages might be expected. There would seem to be another inference which it would be justifiable to draw from these facts. In several of the ingenious theories relating to the development and origin of American cultures in general, it has been contended that considerable migrations both of peoples and of cultural elements passed along this coastal highway from north to south. If however the archeological evidence is to be depended on, such great movements, involving many elements of foreign culture, could hardly have taken place, for no trace of their passage or modifying effect is apparent. If from the general we turn to the particular, and consider the relations between the archeological material and the individual historic tribes, it appears that we can feel fairly sure that the prehistoric peoples of each area were in the main the direct ancestors of the local tribes of today, and that the culture of the former was the forerunner of the latter and can be explained by it—that, in short, we have here a developmental series, of which the middle and the end are known, although the beginning is yet to be discovered.

In comparison with the relative simplicity of the archeological record on the Pacific coast, that of the eastern portion of the continent is complex, and might indeed be best described as a palimpsest. This complexity leads inevitably to the conclusion that here there have been numerous and far-reaching ethnic movements, resulting in a stratification of cultures, such that later have dispossessed and overlain earlier.   These very natural inferences are indeed corroborated by the traditions of migration and conflict preserved by the historic tribes, whose culture in itself also bears witness to the discrete elements which have gone to its formation.   Antillean as well as Mexican and perhaps Central American influences have here been at work, and the possibility of others even cannot be neglected.   In the west it seemed possible to associate the archeological remains of each area with its historic tribes; in the east so soon as we attempt to go beyond the general evidence of mutual corroboration of archeological, ethnological, and traditional data, we meet with serious difficulties.   We are unable in many cases to affiliate with confidence the various types of prehistoric remains with particular historic tribes, so that as a result the archeological material remains in large part isolated and unexplained, as the modern representatives of these prehistoric peoples are unknown.

The shell-heaps, village-sites, and most of the burial places in northern New England can pretty confidently be ascribed to the Algonkian tribes of historic times, but where shall we look for the representatives or relatives of the so-called Red-paint People who seem to have preceded them?   There are, to be sure, various indications which point toward the now extinct Beothuk of Newfoundland, but clear evidence of the relationship is still lacking. The great mass of the remains in New York and Ontario can with certainty be attributed to the Iroquoian tribes in occupancy in the seventeenth century, but the archeological evidence itself shows them to have been comparatively recent comers, and it is not clear to whom we may ascribe either the simpler types of objects or those indicative of a higher and different culture, whose affiliations seem to run toward the region of the Ohio valley. In Florida we may recognize in the now extinct Timuqua the authors of the mounds of the northern part of the state, and with good reason suppose them to have succeeded in occupancy the builders of the shell-heaps of the St Johns. But whether these latter had formerly a greater extension or were related to any of the other tribes of the region, we do not know. Equally uncertain are the relations of the remarkable finds at Key Marco. Are they to be regarded as typical of the fierce, sea-roving piratical tribes of unknown linguistic affiliation who occupied the region in the sixteenth century? If so, how are we to account for the close relationship shown by many of the objects found to those typical of northern Alabama and Georgia and the country to the north?

Most difficult of all are the remains of the several cultures in the Ohio valley. In the extreme northeast the village sites and defensive works may reasonably be associated with the historic Erie, but it is quite uncertain how far southward and westward their remains extend. The Lenâpé, in their historic seats on the Atlantic coast, not infrequently, it would seem, constructed stone box-graves, and it is most probable that part at least of the numerous remains of this type in the Ohio valley (which area was by tradition their earlier home) are to be attributed to them. Graves of this type, however, containing typically undeformed dolichocephalic crania, are found clearly associated with the highest material culture of the valley. If we are to connect these, therefore, with the prehistoric Lenâpé, we must accept a radical change and considerable degeneration in culture coincident with their settlement on the Atlantic coast. We have again the problem of the typical stone box-graves of Tennessee, with their strongly deformed crania, absence of elaborate mounds and earthworks, and presence of types of pottery that are unknown in Ohio. The Cherokee traditionally occupied portions of the upper Ohio valley, and claim indeed to have constructed some of the larger elaborate burial mounds of the region.   The archeological material available, however, leaves something to be desired in substantiating this, and in determining the limits of their occupancy.

The earliest traditional home of a number of the western Siouan tribes lay in the lower Ohio valley, and the existence of a considerable body of tribes of the same stock in the middle Alleghanies has led to the belief that the Ohio valley must either itself have been the early habitat of both branches of the stock or that it served as a highway by which considerable portions migrated either east or west.   If this be true, we may ask which of the various types of remains in the region is to be attributed to this stock?   The association of the effigy mounds of Wisconsin and the adjacent area with the Winnebago or other Siouan tribes seems now reasonably certain, and one might therefore naturally regard the Serpent mound and the few others of this effigy type in the Ohio valley as due also to tribes of the same stock.   Yet these Ohio valley effigies are hardly to be considered as tentative and early forms, as they should be, if they are the first efforts in this direction in the prehistoric habitat of the stock.

Our difficulties are however by no means confined to this type, for how are the various types of remains, quite irrespective of their tribal affiliations, to be related to one another in time?   The builders of the stone box-graves would seem to have been at least in part contemporaneous with the builders of the elaborate mounds and earthworks, but they do not all show such evidence; and whether the beginning of the stone box-grave people overlapped the end of the period of construction of the ceremonial mounds and elaborate earthworks, or vice versa, is not wholly clear.   That the stone box-grave builders were themselves contemporaneous over the whole area would seem to be indicated by the close similarity, amounting in some cases to identity, between the finds made in the graves at points so far apart as Illinois and Alabama; they would seem, on the other hand, to have disappeared from some sections much earlier than from others.   The complete absence again from village sites such as that at Madisonville, of objects characteristic of the higher cultures, would indicate either that these sites completely antedated the higher culture of the Ohio valley or followed it only after it had entirely passed away. The absence of buffalo bones from such sites may be significant in this connection.

The archeological investigations in this eastern portion of the country present us with many other problems, such as those associated with the distribution of certain types of objects. Are we to regard this distribution as due to actual migration of tribal groups from one section to another, or to the results of aboriginal trade? Are the spool-shaped copper ear-ornaments, for example, found from Florida to Illinois, or the biconate tubes found from Florida to New York, so widely distributed merely as a result of trade? Were the pyramidal mounds with graded ways of the upper Ohio valley mere copies of those seen or heard of in the region farther south, or were they built by actual colonies or stray fragments of the builders of these southern mounds themselves? At present it is impossible to say.

Again, we have been able, on the basis of the material available, to determine a number of characteristic and more or less clearly defined types. We have, to take pottery as an example, a Middle Mississippi type, marked by certain peculiarities of form and ornament; and we have a southeastern type, characterized among other things by the use of stamped decoration, which same method is found employed again in the Northwest. We have, however, made little progress in correlating our different types: in indicating the relationship of the stamped decoration of the Northwest to that of the Southeast, or in tracing the origin and development either of this form of ornament or of the polychrome decoration and modeled type of pottery of the Middle Mississippi region.

It is unnecessary however to illustrate further the complexity of the problems or the difficulties surrounding any attempt to relate the archeology of much of the eastern portion of the continent to the historic tribes; to trace clearly the influences from distant cultures which have made themselves felt; to decide whether the wide distribution of certain implements and types is due to migration or trade; or to correlate the different types which we have defined, and follow out their development. The point which I want to make, however, and that to which much of what has been said, trite though it be, directly leads, is that to a large extent the difficulties and perplexities are of our own making. With honorable exceptions in more recent years, the archeological investigations so far made in this country have been woefully haphazard and un-coördinated, and the recorded data often sadly insufficient; the published reports have too frequently been unsystematic and incomplete; and there has been too little indication of a reasoned formulation of definite problems, with the attempt to solve them by logical and systematic methods. It is no doubt easier and perhaps pleasanter to skip about aimlessly in investigation, taking such opportunities as happen to present themselves; it makes a more attractive report to omit much uninteresting and supposedly unimportant detail, and to describe and illustrate by a few fine plates only the more striking objects, merely alluding to or passing over entirely the more common but often very important things; it requires considerable preliminary time and study to realize and define the real problems—all this is no doubt true, as well as that there are often practical difficulties in the way of carrying out a scheme that has been carefully considered. Nevertheless, these facts do not excuse us for the neglect of saner and more truly scientific methods.

A concrete example will make my meaning plainer. The separation of the Siouan stock into two main divisions, an eastern and a western, has already been referred to. These two groups, together with the other smaller fragments, must at some time in the past have occupied a single continuous area. The location of this early habitat, the order of separation of the various groups, their lines of migration, and the successive stages in the cultural modifications produced by new environment and association with other tribes and cultures—these and many other kindred questions are of much interest and importance not only for themselves, but in their bearing on the question of the growth of American culture as a whole, and on the still wider problems of the development of culture in general. We can trace historically the stages in this process as it

relates to one group at least of the stock, namely, in the movement
of some of the Sioux from the forested region out into the plains,
with the consequent transformation in the life and culture of the
people.   The facts in this case are historic, but a careful archeological investigation of successive sites from west to east in this
region would indicate the main features of these changes which in
this instance we happen to known from contemporary observation.
There is no reason to suppose that the earlier prehistoric movements
and changes among the other sections of the stock differed in character from those just referred to.   So that if the Quapaw formerly
lived on the Wabash and lower Ohio and were there ignorant of the
manufacture of polychrome pottery, they did not suddenly acquire
the art without some stimulus, nor at once attain to the highest
excellence in its practice.   There must have been stages between
the location on the Wabash without knowledge of this type of art
and the location in their historic sites, with the knowledge, and these
intermediate stages must lie somewhere between the two extremes.
It may well be replied that such a statement is puerile, that it is
self-evident and assumed as a matter of course; but if so, why have
not these self-evident principles been applied?   Why has no systematic attempt been made to trace back, let us say, the Quapaw to
their original or earlier home, to determine the stimulus which led
to this special development of art, and to follow out the line of its
growth?   We recognize, to be sure, a special Middle Mississippi
type of pottery, but so far as I know this group has not been
analyzed into its constituents, to trace the differences in detail due
to the practice of the same general form of art by several discrete
peoples, separating the various elements and influences which are
apparent, and following them wherever they may lead.   If there
are gaps in the evidence, why not make a systematic attempt to
fill them?   On the basis of evidence at hand a working hypothesis
or several alternative hypotheses may be framed, and material
sought which shall either prove or disprove them.

   Thus the eastern Siouan tribes have either been settled in their
historic habitat for a very long period, or have migrated thither
from elsewhere.   One hypothesis has already been framed according

to which they formerly lived in the Ohio valley, together with the majority of the remainder of the stock.   The Ohio valley contains, as already pointed out, archeological material of several different types, the authorship of which is still obscure.   If the Siouan tribes did formerly occupy the region, some of these remains must be attributed to them.   To settle this question and to determine which if any of these types is to be attributed to this stock, one would logically proceed to investigate a number of known Siouan sites, and work back from these toward the area in question.   It would be necessary to apologize for stating so simple a chain of reasoning, were it not for the fact that the puzzling problems of the archeology of the Ohio valley and of the origin and migrations of the Siouan stock have been before us for many years and are still unsolved, and so far as I am aware, no attempt has been made along such obvious lines to arrive at a definite or probable conclusion on this or on many other similar questions.

This is merely one out of many such examples which might be given of the probable advantage of carrying on our archeological investigations not only in a more systematic manner, but in one which rests firmly on an ethnological and ethnographical basis.   The time is past when our major interest was in the specimen, the collection, the site as a thing in itself; our museums are no longer cabinets of curiosities.   We are today concerned with the relations of things, with the whens and the whys and the hows; in finding the explanation of the arts and customs of historic times in the remnants which have been left us from the prehistoric; in tracing step by step the wanderings of tribes and peoples beyond history, beyond tradition; in attempting to reconstruct the life of the past from its all too scanty remains.   It is only through the known that we can comprehend the unknown, only from a study of the present that we can understand the past; and archeological investigations therefore must be largely barren if pursued in isolation and independent of ethnology.

This is all very well, all very true, one may say, but we live in a very practical world.   It is one thing to draw up an ideal plan of investigation, and evolve simple theories; it is another to apply

AM. ANTH., N. S., 15—38

the theory and to carry out the plan in practice.   Local and personal interests and prejudices in those carrying out or providing for archeological work must be reckoned with; important sites have either disappeared or been plundered or carelessly dug in earlier years, or are jealously guarded by unenlightened owners who refuse permission to excavate; the work really desirable is too costly, or not productive enough for the purposes of display—these and many other difficulties of course stand in the way of carrying out an ideal program.   Yet in spite of these facts is it not time that we made more of an effort than has yet been made to approach the subject from the ethnological point of view?   Is it not possible for us to carry through, before it is too late, even if not with ideal complete-ness, some of those investigations without the results of which we shall always be groping in the dark?   Is it not something of a reproach to American Archeology that it has so far failed to realize and appreciate, as fully as it ought, the need of applying to the solution of its problems the principles which have, in other lands, led to such substantial and magnificent results?

> HARVARD UNIVERSITY,
> CAMBRIDGE, MASSACHUSETTS

THE RELATION OF ARCHEOLOGY TO ETHNOLOGY

FOLLOWING the address of Professor Dixon at the New York meeting of the American Anthropological Association, the subject of the Relation of Archeology to Ethnology was discussed at length.   Of those who participated in the discus-sion, Mr W. H. Holmes, Dr George Grant MacCurdy and Dr Berthold Laufer have responded to the request to present their remarks, which follow.

REMARKS BY W. H. HOLMES

IT is natural that the ethnologist engaged in the study of the tribes and stocks and their culture should lay particular stress on the importance of the prehistory of these groups and seek to follow the various threads of their history far backward into the past   To him the chief value of archeology is that it may cast additional light on the particular subjects of his research.   To this attitude there can be no objection, and the archeologist stands ready to aid in this work; but he realizes his short-comings in this direction, having learned that traces of particular peoples fade out quickly into the generalized past.   A few generations, or at most a few centuries, close definite record of tribal history: beyond this the field of archeological research extends indefinitely and gleanings from this field are utilized in answering the greater problems of the history of the race as a whole. The field of the ethnologist has but a limited range when the entire history of man is considered, yet without the many hints which it furnishes for the interpretation of the past the archeologist would often find himself groping in the dark.

REMARKS BY GEORGE GRANT MacCURDY

On the Relation of Archeology to Ethnology from the Quaternary Standpoint

THE archeologist deals with the dry bones of ethnology. This is particularly true when it is a question of the same or of an adjacent geographic area. Under such circumstances the difficulties of bringing back to life the ethnology of the past and the liability to err in the drawing of conclusions are reduced to a minimum. As soon, however, as great distances are to be covered and great lapses of time are to be considered, the problem at once becomes vastly complex. Instead of dry bones we have to deal with fossil forms, some of which are wholly extinct.

The European prehistorians of the early days of the science were justified therefore in calling their special field paleoethnology. The term archeology covers a period that is in part historic and in part prehistoric. It has been so largely appropriated by the Egyptologist, and the student of Greek and Roman archeology, that a more definite terminology is needed for the remote past—prehistory for example, or prehistoric archeology.

After citing a few instances of the more or less near relationships between prehistoric archeology and ethnology I shall confine my remarks chiefly to the remoter relationships in time as well as space.

In the recent study of a series of ancient shell gorgets from graves in Perry county, Missouri, near Saint Marys,[1] I was very much impressed by the probability of a relationship between the symbolism on two of these gorgets and certain institutions that still persist among the Plains Indians. In the game of *itsẽ'wah* the Piegan Blackfeet make use of a metal ring wrapped with rawhide and cross-barred with sinew, on which beads of various colors are strung, and a wooden dart not unlike an arrow with its shaft. Before their acquaintance with the metals of the white man they employed flat stone disks of convenient size. A stone disk of this sort was given to Dr George Bird Grinnell in 1898 by the wife of Chief Three Suns. It had come down to Three Suns through

---

[1] *American Anthropologist*, July–September, 1913.

many generations. This stone disk, together with the wooden dart used by Three Suns and a modern metal ring disk wrapped with rawhide, were recently presented to Yale University by Dr Grinnell.

On one of the shell gorgets[1] from Saint Marys is represented a human figure evidently in ceremonial garb, and in the act of throwing a stone disk of approximately the same size and shape as the stone disk of Three Suns. Moreover in the left hand is held a wand that might well represent a variant of the Piegan wooden dart; for it is marked by an oblique band and the wooden dart is marked for nearly half its length by a painted spiral groove. Should a Piegan Blackfoot artist with the skill of the ancients wish to depict a player of the game itsë'wah he could hardly do better than copy the figure from this ancient shell gorget.[2]

Another shell gorget[3] from the same cemetery is likewise decorated with a human figure, but representing a very different scene. Each outstretched arm passes through the figure of a star. Below these and opposite the knees are two other larger stars, making four in all. The human figure is thus suspended, as it were, in the heavens from two stars through which the arms pass, while arrows are being shot at it from the east and the west—one at the forehead, one at the back of the head (in line with the ear ornament), one at the left side, and two at the feet. The portion of the shell broken away and lost probably carried with it a sixth arrow aimed at the right side. The designs above and overlapping the large lower stars are bilaterally symmetrical; their fragmentary condition leaves their meaning obscure.

This gorget is full of symbolic import. The stag horn, as suggested to me by Mr Stansbury Hagar, might be considered as an attribute of the sky-god, and the four stars as the four quarters of the sky. The arrows are suggestive of sacrifice and might point to some such ceremony as the Skidi rite of human sacrifice described by Dorsey.[4] The victim is a young woman taken from an enemy's camp and dedicated to the Morning Star. In the construction of the scaffold the four directions play an important part. The maiden's hands are tied to the upper cross-bar which points to the north and south; her feet to the topmost of four lower cross-bars. Her blanket is removed, and a man rushes up from a hollow in the east, bearing in his hand a blazing brand with which he touches her in the groins and armpits. Another man approaches and touches her

[1] Op. cit., fig. 70

[2] A shell gorget from Eddyville, Kentucky, depicts a like scene.

[3] American Anthropologist, op. cit., fig. 77.

[4] Congrès international des Américanistes, XV session, Québec, 1906.

gently with a war-club in the left groin; he is followed by three other men, the first touching her with a war-club in the other groin, and the other two in the arm-pits.   Then the man who captured the girl approaches from the east, bearing a bow and arrow which belong to what is known as the Skull bundle; he shouts a war-cry and shoots the maiden in the heart.   The chief priest opens the thoracic cavity of the maiden with the flint knife from the altar, and, thrusting his hand inside, besmears his face with blood.   All the men, women, and children press forward now and aim each to shoot an arrow into the body.

There is always danger of mistaking analogy for genealogy.   There is likewise danger of misconstruing the phenomena of parallelism and of convergence.   The pathway of the prehistorian who would delve into the remote past is beset by difficulties far greater than those in the way of proving a kinship between the culture of the modern Plains Indians and the ancient culture of the Mississippi valley.   His problem is bound up with the great, and as yet unsolved, problem of human origins.   He must take into consideration not only relationships but also beginnings; and the beginnings of things human, so far as we have been able to trace them, have their fullest exemplification in prehistoric Europe.   The cradle of the human race has not yet been definitely located.   When it is found it will prove to be at least within easy reach of Europe, which structurally is the keystone of the Old World arch—still firmly planted against Asia and once in more intimate contact with Africa than at present.

The Old World then is the ample stage on which the human drama has been played.   Here the cultural elements have had their exits and their entrances.   The character of a culture at a given time and place should be viewed in the light not only of the elements that were present, but also those that were manifestly lacking.   One can, for example, set about reconstructing the culture of *Homo heidelbergensis* or of Piltdown without danger of being misled by phenomena with which ethnologists have to reckon, namely, the disturbances resulting from a clash between cultures in almost totally different planes of development.   In those days there was no danger of being discovered by a Columbus or conquered by a Cortés.   Since the earliest times progress has been due in part to contact of one people with another and the resulting interchange of ideas.   Infiltrations and invasions, peaceable or otherwise, have also brought changes.   The evidence points to a diversity of human types as far back as the early Quaternary, but not to a corresponding cultural diversity.

Culture is a measure of man's power to control his environment. It depends largely on the inventive faculty and the facilities for transmitting racial experience. The dead level character of the so-called eolithic or pre-paleolithic industrial remains points to a long hand-to-mouth struggle for a racial bank account. Progress was slow even among the Chellean and Acheulian peoples. A rude Chellean industry was found associated with the Piltdown skull. Whether Mousterian culture was a direct outgrowth of the Chellean and Acheulian has not yet been determined. The human skeletal remains associated with Mousterian culture are of the Neanderthal type, representing a race of coarse mental and physical fiber, whose disappearance was coincident with the appearance of a new racial and cultural type. The ancestry of this new race, the Aurignacian, has not been definitely traced. The Aurignacians, represented by Cro-Magnon and Combe Capelle, were more nearly akin to the modern Europeans than to the archaic Mousterians. The cultural differences are at once so great as to make it difficult to conceive of the Aurignacian as having been an offshoot from the Mousterian age. The distribution of Aurignacian culture would in the opinion of Breuil seem to favor Africa rather than the east as a starting point.

The Aurignacians introduced the decorative as well as the fine arts: sculpture, bas relief, engraving, and painting. Through these we get a glimpse into their social and intellectual life. Some of their art works have been subjected to an interesting comparative study. For example, they left in a number of French and Spanish caves negative imprints of the human hand that manifestly point to phalangeal amputation, a practice that exists today among primitive peoples in widely separated parts of the earth. It was observed by Burchell among the Bushmen as early as 1812. It is also reported from Australia. According to Boas the Haida, Tlingit, and Tsimshian tribes of the Northwest Coast cut off a little finger on special occasions. Mindeleff reproduces a series of pictographs from the Cañon de Chelly, Arizona, in which representations of the human hand play an important rôle. He does not say however whether any of these show evidences of phalangeal amputation.

The Aurignacians likewise left us those perplexing female figures in the round from Brassempouy, Mentone, and Willendorf, as well as the bas reliefs from Laussel, all of which are reminiscent of the Bushman type of female beauty. The figures in question might however be explained on symbolic grounds rather than as realistic representations of a physical type.

If the Aurignacian culture came from the direction of the Mediterranean the same can hardly be said of the Solutrean which succeeded it and which seems to have come from the east. According to Breuil the early Solutrean is extensively developed in Hungary while the veritable Aurignacian is lacking there. It may be that the early Solutrean of the east is synchronous with advanced Aurignacian in France and that the Solutrean of the west was due to an invasion, which however did not remain long in the ascendency; for out of the contact between these two civilizations there arose the Magdalenian culture, to whose further development the east and not the Mediterranean contributed.

One encounters difficulties in comparing paleolithic art with any art period that has followed. It differs not only from neolithic art but also from the art of modern primitive races. The art of the untutored child is more like that of neolithic or modern primitive art than it is like paleolithic art. The child does not copy the thing itself so much as his ideas about the thing. Paleolithic art evinces a remarkable familiarity with the object combined with a skilled hand. The artists' models were almost without exception from the animal world, chiefly game animals. Conditions favoring progress in art are normally just the reverse of those that would make a hunter's paradise. With the increase in density of population there would be a corresponding decrease of game. The animal figures were no doubt in large measure votive offerings for the multiplication of game and success in the chase. The more realistic the figure the more potent its effect would be as a charm. The mural works of art—figures of male and female, scenes representing animals hunted or wounded—are generally tucked away in some hidden recess, which of itself is witness to their magic uses.

Mythical representations, so common to modern primitive art and to post-paleolithic art in general, are wholly foreign to paleolithic art. There were no gods, unless the human figures served also as such; and no figures with mixed attributes, as is so well typified in the gold figurines of ancient Chiriquian art of the Isthmus, or in the Hindu and the Egyptian pantheon. The paleolithic artist left frescoes, engravings, bas reliefs, and figures in the round of the horse, but there is not a single figure of a centaur.

The cave man's love for the real, the natural, as opposed to the mythical, the artificial, is also seen in his representations of the human form. A child will draw the figure of a man or a woman as clothed, but with the legs for example showing through the dress. The same thing was done by the artists of ancient Egypt. Not so with the cave artist.

That paleolithic man of the art period made use of clothing the numerous bone needles afford abundant testimony; but with a single possible exception (Cogul in southeastern Spain), and that, if an exception, dates from the very close of the paleolithic period, the human form was represented in the nude; some of the figures however suggest a more pronounced growth of hair over the body than would be common at the present time.

There is very little evidence that masks were used either ceremonially or for stalking purposes. An engraving of a male figure wearing a mask representing a horse's head has been noted from the Magdalenian deposits of the cave of Espelugues at Lourdes. Three engraved figures on a bâton de commandement from the rock shelter of Mège at Teyjat (Dordogne) have been reproduced by Breuil. A third example was found at Mas d'Azil—a man wearing a bear's head mask.

Art objects dating from the paleolithic period have every appearance of being originals and not copies. Earmarks of the copyist are singularly lacking. The work was done either in the presence of the model or with the image of the latter fresh in the memory.

Ethnology has done much toward illuminating some of the dark pages of European prehistory. But European ethnology is too far removed from paleolithic and pre-paleolithic Europe to be as good a guide there as the ethnology of the Indian is to prehistoric America. There are those who are inclined to criticize the temple of classification reared by the European prehistoric systematists. They call it too simple, too perfect, too academic—a system based on answers to the easy questions with all the puzzling problems left out of account, and therefore admirably calculated to attract the amateur. The critics however usually have very little first-hand knowledge of the European field. On the other hand those who have done most to develop the systematic side are the first to acknowledge not only the weaknesses of the classification, but also the complexity of the problems still confronting the prehistorian. No one who can speak with authority claims that the system can at present be applied anywhere except to central, southern, and western Europe. A certain definite succession of cultures already holds good over a large area. The horizon we call Solutrean, for example, need not however be synchronous in Hungary and southwestern France. When Asia and Africa shall have been studied with equal thoroughness there will be much to add and no doubt some to subtract. There can be a system of classification and still allow for all sorts of local rises and falls of the culture barometer as well as movements of peoples over large areas. All the people did not follow a retreating glacier to the north. But all who did follow were driven slowly back with the succeeding advance of the great continental ice sheet. And it is not likely that they recognized those whose ancestors had been left behind so many thousands of years before. Lapse of time and differences in the environment must have left their impress on both classes of culture, the contact between which would eventually result in a new phase of culture. The wonder is that any system could be discovered, and I say discovered rather than devised advisedly, which could long withstand so complex and heavy a strain. The system in its elemental outlines still survives; and where there is life there is hope, and the possibility of future growth.

REMARKS BY BERTHOLD LAUFER

THE value of a scientific method, in my estimation, cannot be determined by theoretical discussion. The academic exposition of a method may strike ear and mind favorably, and yet it may be unworkable if the practical issues of a science are at stake in broad daylight. The quality of a method is discernible only from the fruits which it yields. It remains a brutal fact that the worth of a man is estimated by the world at large from his outward success in life; in similar manner the merit and utility of a method are judged according to the degree of its success. It is sheer brutality and cold-hearted calculation if we are tempted to adopt the most successful method in the pursuit of our work. In matters of archeology it has always seemed to me that classical archeology, the oldest of the archeological sciences, has hitherto made the most successful advance; and for this reason it is deemed advisable to extend its methods, as far as feasible, to other fields of antiquarian exploration. But if a more effectual method should ever be contrived, I believe I should be inclined to abandon my own boat and embark on the new.

Archeology is largely a matter of practical experience; and, wide and unlimited as the range of experience is, the variability of methods applicable to specific cases is almost endless, and we may well say that each case must be judged by its own particular merits. Archeological problems may be likened to algebraic equations with one, two, or several unknowns: by starting from a given fact, we endeavor to unravel by it the one or more unknowns. If archeology is more than a mere description and classification of ancient remains left by past ages (and this could assuredly be only its technical foundation, which may be described under the term " museology "), but if it is the science of the ancient culture-phases of mankind illustrated by all accessible human monuments, it is needless to insist that archeological study cannot be separated from philology and ethnology.   It is a branch of historical research, a part of the history of human thought and culture; and as far as Asia, Africa, and Europe are concerned, it is obvious, without the shadow of a doubt, that only a combined knowledge of language, paleography, history, and culture will lead us to any positive and enduring result in archeological questions. Take, for example, the case of Egyptology.   The very word indicates the specific character of the science.   We do not speak of such divisions as Egyptian history, archeology, philology, and ethnology, but of *Egyptology* only, because a scholar desirous of promoting this research must be firm in every saddle.   The great architectural monuments of Egypt are covered with contemporaneous inscriptions revealing their significance; and well-trained familiarity with the script and language, with chronology and events, with religious and other ideas, becomes the indispensable equipment for any one serving the cause of the archeology of Egypt.   When we come to India, the situation is widely different.   India has no historical records, and lacks any sound chronology.   The accounts of the Greek, Chinese, and Arabic authors must partially supplement this deplorable gap.   Monuments are comparatively plentiful, some are also augmented by coeval inscriptions, but, on the whole, they are cut off from contemporaneous tradition.   The spirit of India is highly imaginative—essentially occupied with religious, mythological, and philosophical speculations, supported by an inexhaustible fund of good stories and legends.   The skilful interpreter of the monuments of Indian art must naturally have these at his fingers' ends, and, to make good for the lack of historical data, ought to have recourse also to the application of psychological methods.

In China we are confronted with a peculiar situation unparalleled in classical antiquity and elsewhere.   Here we face the unique fact that the Chinese themselves have created and highly developed a science of archeology beginning at a time when Europe still slumbered in the night of the middle ages.   The Chinese, indeed, were the first archeologists in the world: the first to explore the soil; the first to do field-work; the first to collect, arrange, catalogue, and illustrate antiquities; the first to study and describe their monuments—with most notable results.   This feature naturally offers to us many vantage points; and the study of Chinese archeology, accordingly, must begin with a study of the archeology of the Chinese.   The foreign student intent on the solution of a special problem will in this manner easily see a point of attack, and will find his path through the jungle cleared to some extent by the contributions offered by Chinese scholars. This state of affairs, however, has also grave drawbacks which must not be overlooked; and among these, two are important. The circumstantial evidence of Chinese antiquities, in general, is weak; the localities where they have been found are sometimes but vaguely known; the circumstances of the finds are seldom, and then but imperfectly, described to us. Again, the Chinese have their own peculiar theories, their point of view in looking at things, their peculiar logic and mode of argumentation, and have accumulated on top of their antiquities, and on the whole of their culture, huge strata of speculations and reflections which in most cases cannot withstand our sober criticism. It was a development easy enough to understand that until very recently our scholars meant to make Chinese archeology by merely reproducing the opinions of Chinese archeologists. This necessarily resulted in numerous errors, misconceptions, and wrong judgments, the effects of which are not yet overcome. These strictures being made, the outlook in this field is altogether hopeful. We have remains and antiquities in great plenty, and an overwhelming abundance of information accompanying them—often more, I should add, than we are able to digest. Above all, our conclusions can be built upon the firm basis of a secure and reliable chronology, and in the majority of cases we might say it is out of the question that a Chinese monument or object should not be datable within a certain period. The aim of Chinese archeology, as I understand it, should be the reconstruction of the origin and inward development of Chinese culture in its total range, as well as in its relation to other cultural provinces. A proper knowledge of China is bound up in this definition. We cannot comprehend any idea of modern China, or adequately treat any Chinese problem, without falling back on the past. The distinction between archeology and ethnology, consonant with the actual conditions in America, seems, at least to me, to be somewhat out of place in such fields as China, central Asia, and Siberia. The modern ethnographical conditions in these regions mean so little that they amount to almost nothing, being merely the result of events of the last two centuries or so. My conviction that there is in principle no essential difference between archeological and ethnological methods could not be better illustrated than by the fact that the method of Chinese archeology—at least, as I am inclined to look upon it—is in perfect harmony with the method of ethnology as conceived and established by Dr Boas. It is among the Chinese, even to a much higher degree than among primitive tribes, that we constantly have to reckon with such potent factors of mental development as recasting of old ideas into new forms; reinterpretation of ancient thoughts under the influence of new currents, theories, or dogmas; new associations, adaptations, combinations, amalgamations, and adjustments. The ideas expounded by Chinese scholars of the middle ages with reference to their classical antiquity one or two thousand years back are, in fact, nothing but subjective reconstructions of the past based largely on deficient associations of ideas. This feature is most striking, for instance, in decorative art. The Sung artists of the middle ages attempted to reconstruct all the primitive patterns on the ritual objects of the archaic period on the basis of the names of these patterns as handed down in the texts of the ancient rituals. All these names were derived from natural objects, but referred to geometrical designs. A combination of hexagons, for example, was styled a "rush" pattern, because it was suggestive of a mat plaited from rushes, and may indeed have been developed from a mat impression. In the Sung period, art was naturalistic, and these artists reconstructed the ancient geometric rush pattern in the new form of realistic rushes. In this manner a new grammar of ornaments was developed, purported to represent the real ornaments of the classical period, which, however, had never existed at that time. Cases like this may have happened a hundred or a thousand times among primitive tribes, not only in art, but in social and religious development as well.

The further advantage of this critical and reconstructive method is that it finally leads us to psychology, and allows us to recognize the laws working in the Chinese mind. And this, after all, must be the ultimate aim of all our research—the tracing and establishing of the mental development of a nation, the grasp of the national soul, the determination of its qualities, aspirations, and achievements. From this point of view, we may say paradoxically, and yet correctly, that all archeology should become ethnology, and all ethnology turn into archeology. The two, in fact, are inseparably one and the same—emanations of the same spirit, pursuing, as they do, the same ideal, and working to the same end.

Finally I may perhaps be allowed a word concerning the relation of American archeology to ethnology, although I must first apologize for talking of something about which I do not properly know. It is difficult for the present to bridge American archeology and ethnology; but it seems to me that this entire question has no concern whatever with methods, or that no alleged or real deficiency of methods could be made responsible for any disappointments in certain results that may have been expected. The drawback lies solely in the material conditions of the field, and prominent among these is the lack of a substantial chro-

nology.   Chronology is at the root of the matter, being the nerve elec-
trifying the dead body of history.   It should be incumbent upon the
American archeologist to establish a chronological basis of the pre-
columbian cultures, and the American ethnologist should make it a
point to bring chronology into the life and history of the postcolumbian
Indians.   This point of view, it seems to me, has been almost wholly
neglected by American philologists and ethnologists, and hardly any
attempt seems ever to have been made to fix accurately the time of traditions, mythologies, rituals, migrations, and other great culture movements.
This, however, must be accomplished, and I am hopeful enough to
cherish the belief that it *will* be accomplished.   When archeology and
ethnology have drawn up each its own chronology, then the two systems
may be pieced together and collated, and the result cannot fail to appear.
Whether we who are here assembled shall ever live up to that happy day,
is another question.   Meanwhile we ought not to be too pessimistic
about the outcome, or to worry too absorbingly about the issue of
methods.   We should all be more enthusiastic about new facts than
about methods; for the constant brooding over the applicability of
methods and the questioning of their correctness may lead one to a
Hamletic state of mind not wholesome in pushing on active research
work.   In this sense allow me to conclude with the words of Carlyle:
" Produce! Produce! Were it but the pitifullest infinitesimal fraction of
a product, produce it in God's name!   'Tis the utmost thou hast in thee:
out with it, then!"

103

西藏的鸟卜

T'OUNG PAO

通報

ou

ARCHIVES

CONCERNANT L'HISTOIRE, LES LANGUES,
LA GÉOGRAPHIE ET L'ETHNOGRAPHIE
DE
L'ASIE ORIENTALE

———

Revue dirigée par

Henri CORDIER
Membre de l'Institut
Professeur à l'Ecole spéciale des Langues orientales vivantes
ET
Edouard CHAVANNES
Membre de l'Institut, Professeur au Collège de France.

———

VOL. XV.

———

LIBRAIRIE ET IMPRIMERIE
CI-DEVANT
E. J. BRILL
LEIDE — 1914.

BIRD DIVINATION AMONG THE TIBETANS

(NOTES ON DOCUMENT PELLIOT No. 3530, WITH A STUDY OF TIBETAN PHONOLOGY OF THE NINTH CENTURY).

BY

BERTHOLD LAUFER.

Et illud quidam etiam his notum, avium voces volatusque interrogare.

TACITUS, *Germania* X.

Among the Tibetan manuscripts discovered by M. Paul Pelliot there is a roll of strong paper (provisional number 3530 of the *Bibliothèque Nationale*) measuring 0.85 ⸱ 0.31 m and containing a table of divination. This document has recently been published and translated by M. J. BACOT. [1]) This gentleman has furnished proof of possessing a good knowledge of Tibetan in a former publication, [2]) in which he gives a most useful list of 710 abbreviations occurring in the cursive style of writing (*dbu-med*) of the Tibetans, from a manuscript obtained by him on his journeys in eastern Tibet. It is gratifying to note that the tradition gloriously inaugurated in France by Abel-Rémusat, Burnouf and Foucaux, and worthily continued by L. Feer and S. Lévi, reincarnates itself in a young and fresh representative of the Tibetan field, who has enough

---

1) *La table des présages signifiés par l'éclair*. Texte tibétain, publié et traduit.. (*Journal asiatique*, Mars-Avril, 1913, pp. 445—449, with one plate)

2) *L'écriture cursive tibétaine* (*ibid*, Janvier-Février, 1912, pp. 1—78). M. BACOT is also the author of a pamphlet *L'art tibétain* (Châlon-sur-Saône, 1911), and of two interesting books of travel *Dans les marches tibétaines* (Paris, 1909) and *Le Tibet révolté* (Paris, 1912)

1

courage and initiative to attack original problems. It is likewise
matter of congratulation to us that the wonderful discoveries of
M. Pelliot will considerably enrich Tibetan research and reanimate
with new life this wofully neglected science. The volumes of the
ancient Kanjur edition discovered by him in the Cave of the Thousand Buddhas (*Ts'ien fu tung*) of Kan-su and dating at the latest
from the tenth, and more probably even from the ninth century,
together with many Tibetan book-rolls from the same place, [1]) are
materials bound to signal a new departure in the study of Tibetan
philology, hitherto depending exclusively on the recent prints of the
last centuries. We therefore feel justified in looking forward with
great expectations to the elaboration of these important sources.
The text published by M. BACOT is the first Tibetan document
of the *Mission Pelliot* made accessible to science, and there is every
reason to be grateful for this early publication and the pioneer
work conscientiously performed by M. BACOT. It is a document of
great interest, both from a philological and a religious point of view.
The merit of M. BACOT in the editing and rendering of this text
is considerable. First of all, he has honorably accomplished the
difficult task of transcribing the cursive form of the original into
the standard character (*dbu-can*), and, as far as can be judged by
one who has not had the opportunity of viewing the original,
generally in a convincing manner; he has recognized also some of
the archaic forms of spelling, and correctly identified them with
their modern equivalents; and above all, aside from minor details,
he has made a correct translation of the divination table proper.

There are, however, two points of prime importance on which
my opinion differs from the one expressed by M. BACOT. These
points are the interpretation of the meaning of the Table, and the

---

1) Compare P. PELLIOT, *La mission Pelliot en Asie centrale*, pp. 25, 26 (*Annales de
la société de géographie commerciale*, Fasc. 4, Hanoi, 1909) and *B. E F. E. O.*, Vol. VIII,
1908, p 507.

rendering of the introductory note prefacing the Table. In regard to the latter, M. BACOT is inclined to view it as a series of rebuses which seem to have the raven as their subject. He consequently takes every verse (the entire preface is composed of twenty-nine verses, each consisting of a dactyl and two trochees, — a metre peculiarly Tibetan and not based on any Sanskrit model) as a single unit; while in my opinion the verses are mutually connected, and their interrelation brings out a coherent account furnishing the explanation for the divination table. As indicated by the very title of his essay, M. BACOT regards the latter as a list of forebodings announced by lightning; and in column I of the Table worked up by him, we meet the translation *en cas d'éclair à l'est*, etc. The Tibetan equivalent for this rendering is *ñan zer na*, which literally means, "if there is evil speaking." No authority, native or foreign, is known to me which would justify the translation of this phrase by anything like "flash of lightning;" it simply means "to utter bad words," which may augur misfortune: hence *ñan*, as JÄSCHKE (*Dictionary*, p. 126) says, has the further meaning of "evil, imprecation." The phrase *ñan smras* is rendered in the dictionary *Zla-ḅai od-snañ* (fol. 29b, Peking, 1838) into Mongol *maghu käläksän*. In the present case, the term *ñan zer* refers to the unpleasant and unlucky sounds of the voice of the crow or raven, which indeed, as expressly stated in the prefatory note, is the subject of divination in this Table. Moreover, the preface leaves no doubt as to who the recipient of the offerings is. It is plainly told there in Verse 8 (4 in the numbering of M. BACOT): *gtor-ma ni bya-la gtor*, "the offering is made to the bird," and this bird certainly is the raven (*pʿo-rog*) [1]) spoken of in Verse 1, again mentioned in Verse 17, their various tones being described in V. 25—29.

In this Table, it is, accordingly, the question only of the raven,

---

1) The differentiation of the Tibetan words for "raven" and "crow" is explained below, in the first note relating to the translation of the preface.

not of lightning; no word for lightning (*glog* or *t'og*) occurs either
in the Table or in the preface. [1]) The fact that this interpretation

---

1) It must be said, in opposition to M. BACOT's explanation, also that neither the
Tibetans nor the Indians seem to have offerings to lightning, nor do I know that good or
bad predictions are inferred in Tibet from the manner in which a flash of lightning strikes.
M. BACOT assures us that analogous tables for divination from lightning are still in use
in Tibet and Mongolia. It would be interesting to see such a table referred to by M. BACOT.
In India, lightnings were classified according to color, a yellow lightning pointing to rain,
a white one to famine, etc. (A. HILLEBRANDT, *Ritual-Litteratur. Vedische Opfer und
Zauber*, p. 184, Strassburg, 1897) M. BLOOMFIELD (*The Atharvaveda*, p 80, Strassburg,
1899) speaks of a "goddess lightning" who is conciliated by charms to cause her to spare
the stores of grain; but then, again, he identifies the divine eagle with lightning. Among
the Romans, the lightning-flash was a solicited portent of great significance, not, however,
for the divination of the magistrates, but for certain priestly ceremonies of the augurs
(HASTINGS, *Encyclopaedia of Religion*, Vol. IV, p. 823). — In regard to thunder, a series
of omens regulated according to the quarters exists among the Mongols. P. S. PALLAS
(*Sammlungen historischer Nachrichten über die mongolischen Völkerschaften*, Vol. II, p. 318,
St. Petersburg, 1801) has extracted the following from a Mongol book styled by him
Jerrien-Gassool: "When in the spring it thunders in the south, this is a good sign for
every kind of cattle. When it thunders straight from an easterly direction, this signifies an
inundation threatening the crops. When it thunders from the north, this is a good sign
for all creatures. When it thunders in the north-west, this means much slush and wet
weather in the spring; and, moreover, many new and strange reports will be heard throughout the world. When it thunders from the west very early, a very dry spring will follow.
When it thunders early in the south-west, this means unclean diseases to men. When it
thunders early in the south-east, locusts will destroy the grass." In regard to auguries,
PALLAS states that the bird of augury among the Kalmuk is the whitish buzzard called
tsaghan chuldu; when it flies to the right of a tramping Kalmuk, he takes it to be a
happy omen, thanking it with bows, when, however, it flies to his left, he turns his eyes
away and dreads a disaster. They say that the right wing of this bird is directed by a
Burchan or good spirit, the left one by an aerial demon, and nobody dares shoot this bird.
According to Pallas, the flight of the eagle, the raven, and other birds, has no significance
among the Kalmuk. The white owl is much noted by them, and looked upon as a felicitous
bird. — Abou Bekr Abdesselam Ben Choaib (*La divination par le tonnerre d'après le
manuscrit marocain intitulé Er-Ra'adiya*, *Revue d'ethnographie et de sociologie*, 1913,
pp. 90—99) translates a Moroccan manuscript (date not given) treating of divination from
thunder-peals, according to their occurrence in the twelve months of the year. Also the
Malays draw omens from thunder (W. W. SKEAT, *Malay Magic*, p 561) and lightning
(p. 665). — The field of Tibetan divination and astrology is a subject as wide as ungrate-
ful and unpleasant for research. It has been slightly touched upon in the general books on
Tibetan Buddhism by E. SCHLAGINTWEIT and L A. WADDELL Some special contributions are
by A. WEBER, *Ueber eine magische Gebetsformel aus Tibet* (*Sitzungsberichte der preussischen Akademie*, 1884. pp. 77—83, 1 plate), and WADDELL, *Some Ancient Indian Charms*

is to the point, will be especially gleaned from the text of the *Kākajariti* given below. The first column of M. Bacot's Table finds its explanation in the last clause of this text, where it is said: "When an omen causing fear is observed, a strewing obla- tion must be offered to the crow" (*ajigs-pai rtags mt͡c'oṅ-na, bya- rog-la gtor-ma dbul-bar byao*), and the flesh of the frog is the most essential of these offerings. The crow does not receive offerings in each and every case when an oracle is desired from its sounds, but only when it emits disastrous notes pointing to some calamity, and the object of the offering is the prevention of the threatening disaster. It is therefore logical to find in the first column of our Table, headed "the method of offerings," and indicating the kind of offerings for the nine (out of the ten) points of the compass, the conditional restriction *ṅan ẕer na*, for example, "when in the east (the crow) should utter unlucky sounds, milk must be offered," etc. The crow is belived to fly up in one of the nine points of the compass, and exa ly the same situation is described in the beginning of the *Kākajariti*.

Among the offerings (*gtor-ma*, Skr. *bali*) enumerated in our Table, there are two distinctly revealing Indian influence, — the white mustard (Tib. *yuṅs-kaṛ*, Skr. *sarshapa*), and *guggula*, itself a Sanskrit word. [1]) The question must naturally be raised, Is this practice

from the Tibetan (*Journal Anthrop. Institute*, Vol. XXIV, 1895, pp. 41—44, 1 plate) The most common method of fortune-telling is practised by means of dice (*šo*) in connection with divinatory charts. Interesting remarks on this subject are found in the excellent works of Stewart Culin, *Chinese Games with Dice and Dominoes* (*Report of U. S. Nat. Mus. for* 1893, p. 536, Washington, 1895), and *Chess and Playing-Cards* (*ibid*, for 1896, pp 821—822, Wash., 1898). Also this practice doubtless originates in India, and should be studied some day with reference to the Indian dice games and oracles (compare A Weber, *Ueber ein indisches Würfel-Orakel*, *Monatsberichte Berl. Ak*, 1859, A. F. R Hoernle, *The Bower Manuscript*, pp. 09, 210, 214, J. E. Schröter, *Pāçakakevalī*, Ein indisches Wurfelorakel, Borna, 1900; and chiefly H Lüders, *Das Würfelspiel im alten Indien*, *Abhandl. der K. Ges der Wiss zu Göttingen*, Berlin, 1907). There are several Tibetan books treating especially of dice oracles (see also E. H. Walsh, *Tibetan Game of de šho*, *Proc. A. S. B.*, 1903, p 129)

1) Also rice and flowers are Indian offerings, the same as occur likewise in Burma

of divination from the notes of a crow of indigenous Tibetan origin, or
is it rather a loan received from India? The Tibetan Tanjur contains

---

among the offerings to the Nat (L. VOSSION, *Nat-worship among the Burmese*, p 4,
reprint from *Journal American Folk-Lore*, 1891), and the whole series of offerings may
confidently be stated to be derived from Indian practice "After bathing, with hands circled
by swaying bracelets, she herself gave to the birds an offering of curds and boiled rice
placed in a silver cup; ... she greatly honored the directions of fortune-tellers; she frequented all the soothsayers learned in signs; she showed all respect to those who understood the omens of birds" (*The Kādambarī of Bāṇa* translated by Miss C. M. RIDDING,
p. 56, London, 1896). — M. BACOT accepts the rendering *bois d'aigle* for *guggula* (Tibetanized *gu-gul*) given in the Tibetan Dictionary of the French Missionaries. But this is not
correct. *Guggula* or *guggulu* is not at all a wood but a gum resin obtained from a tree
(*Boswellia serrata*, sometimes called the Indian Olibanum tree) and utilized as incense
(W. ROXBOROUGH, *Flora Indica*, p. 365; G. WATT, *Dictionary of the Economic Products
of India*, Vol. I, p. 515). In more recent times this name has been extended also to the
produce of *Balsamodendron Mukul*, which became known to the Greeks under the name
βδέλλα (thus in *Periplus*, ed. FABRICIUS, pp. 76, 78, 90), then Grecized βδέλλιον (first
in DIOSCORIDES, Latinized BDELLIUM in PLINY, *Nat. Hist.* XII, 9, 19, ed. MAYHOFF, Vol.
II, p. 388, compare LASSEN, *Indische Altertumskunde*, Vol. I, p 290, and H. BRETZL,
Botanische Forschungen des Alexanderzuges, pp. 282—4, Leipzig, 1903) and to the Arabs
under the word *moql* مقل (L. LECLERC, *Traité des simples*, Vol. III, p. 331, Paris, 1883,
and J. LÖW, *Aramäische Pflanzennamen*, p. 359, Leipzig, 1881) The meaning 'bdellion' is
exclusively given for *guggula* in the Sanskrit dictionaries of St. Petersburg; this, however,
is not the original but merely a subsequent (and probably erroneous) application of the word,
nor is the identity of *bdellion* with *guggula*, as established by J. JOLLY (*Medicin*, p. 18,
Grundriss d. indo-ar Phil.), correct. WATT says advisedly, "Care must be taken not to
confuse this gum resin (*guggula*) with the olibanum or frankincense of commerce, or with
Mukul. The true Sanskrit name for this plant is most probably Sallaki." The Sanskrit
name which Watt has in mind is *çallakī* or *sillakī*, *Boswellia thurifera*, yielding frankincense which is called *silha* (Tib. *si-la*) The Greek words *bdella* and *bdellion* are derived
from Hebrew *bdolah*, *bĕdolah*; but "what it was remains very doubtful" (YULE and BUR-
NELL, *Hobson-Jobson*, pp. 76, 386). Regarding the Chinese names of *guggula* see PELLIOT,
T'oung Pao, 1912, p. 480. In his study of the names of perfumes occurring in Chao
Ju-kua, M. PELLIOT (*ibid.*, p. 474) alludes to the *Mahāvyutpatti* as one of the sources
to be utilized for such research; I may be allowed to point out that the Sanskrit and
Tibetan list of the thirteen names of perfumes contained in that dictionary was published
by me in *Zeitschrift für Ethnologie*, 1896, *Verhandlungen*, p. 397, in connection with the
Tibetan text and translation of the *Dhūpayogaratnamālā*; this certainly was *une œuvre de
jeunesse* on which I could now easily improve. The most important source for our purposes
doubtless is the *Hiang p'u* 香譜 by Hung Ch'u 洪芻 of the Sung period, reprinted
in *T'ang Sung ts'ung shu*. BRETSCHNEIDER (*Bot. Sin.*, pt. I, No 153) mentions a work
of the same title, but from the hand of Ye T'ing-kuei 葉廷珪 of the Sung.

a small treatise under the title *Kākajariti* indicated by G. Huth. [1]) The Indian method of divining from the calls of the crow is briefly expounded therein, and for this reason a literal translation of it may first be given. It will be recognized that the thoughts of this text move on the same line as the *document Pelliot*, and it will furnish to us the foundation for some further remarks on the latter. In order to facilitate immediate comparison of the two texts, I have numbered, in the Table published by M. Bacot, the series of the first vertical column with the Roman figures I—XI, and the nine series yielded by the nine quarters with the Arabic figures 1—9, so that by the combination of the two any of the ninety squares of the Table may be readily found. The references to the squares of this Table, placed in parentheses in the following text, indicate thought identity or analogy in the two documents. [2])

Translation of Kākajariti.

Tanjur, Section Sutra (*mdo*), Vol. 123, Fol. 221 (edition of Narthang).

---

1) *Sitzungsberichte der preussischen Akademie*, 1895, p. 275. Huth refers to "Schiefner in Weber's Indische Streifen I 275," which I have never seen, and which is not accessible to me.

2) After my translation was made from the Narthang edition of the Tanjur, I found that A. Schiefner (*Ueber ein indisches Krahenorakel, Mélanges asiatiques*, Vol. IV, St. Petersburg, 1863, pp 1—14) had already edited and translated the same work. In collating my rendering with that of Schiefner, it turned out that I differed from him in a number of points which are discussed in the footnotes Schiefner's text (apparently based on the Palace edition) and translation are generally good, though the mark is missed in several passages; I have to express my acknowledgment especially to his text edition, as my copy of the Narthang print, which is difficult to read, left several points obscure On the other hand, whoever will take the trouble to check my version with that of my predecessor, will doubtless recognize the independence of my work. As the principal point in the present case is to reveal the inward connection between the *Kākajariti* and the *document Pelliot*, it was, at any rate, necessary to place a complete version of that text before the reader, and not everybody may have access to the publication in which Schiefner's study is contained.

In Sanskrit: *Kākajariti* ("On the Sounds of the Crow"). [1]

In Tibetan: *Bya-rog-gi skad brtag-par bya-ba* ("Examination of the Sounds of the Crow").

This matter is as follows. The crows are divided into four castes; namely, Brāhmaṇa, Kshatriya, Vaiçya, and Çūdra. A crow of intelligent mind [2] belongs to the Brāhmaṇa caste, a red-eyed

---

1) The Sanskrit title is thought by SCHIEFNER to be corrupt. He made two conjectures, — first, in a communication to Weber, by restoring the title into *kākarutaṁ*, which he soon rejected. second, he accepted as foundation of the disfigured Sanskrit title the words *bya-rog-gi spyod-pa* occurring at the end of the treatise, which he took in the sense of *kākacaritra* or °*carita*, and he assumed that this title may have arisen through a retranslation from Tibetan into Sanskrit, at a time when the Sanskrit original no longer existed. Again, on p 14, he conjectures *spyod-pa* to be an error for *dpyod-pa* = Skr. *vicāraṇa*, "examination," and thus unconsciously contradicts his previous surmise on p. 1 I can see no valid reason for any of these conjectures The final words taken for the title do not in fact represent it, but only refer to the third and last part of the treatise, which is plainly divided into three sections: 1. Omens obtained from a combination of orientation and the time divisions of the day, 2. Omens to be heeded by a traveller, 3. Omens obtained from the orientation of the crow's nest. The *spyod-pa* of the crows refers to the peculiar activity or behavior of the birds in building their nests. Besides, the title of the work is simply enough indicated in its Tibetan translation, "Examination of the Sounds (or Cries) of the Crow (or Crows)," and the restoration of the Sanskrit title should be attempted only on this basis. It is evident that it is defective, and that a word corresponding to Tib *brtag-par bya-ba* is wanting, which, judging from analogies of titles in the Tanjur, it may be supposed, was *parīksha* The word *jarati*, corresponding to Tib. *skad*, seems to be a derivation from the root *jar, jarate*, "to call, to invoke."

2) Tib *zo-la rtsi-ba*. SCHIEFNER (p. 12) remarks on this passage which he renders *die in Karsha's rechnenden Brahmanen*: "The Tibetan text is not quite without blemish. Some passages of the original are wholly misunderstood; to these belongs the passage in question. I suspect a misunderstanding of *kārshṇya*, 'blackness.' As Weber observes, this supposition is confirmed by a classification of the Brahmans among the crows occurring elsewhere." This interpretation seems to me to be rather artificial; I think *zo* is a clerical error for *że*, and take *że-la rtsi-ba* in the sense of "to calculate in their minds." The crow is the object of divinatory calculation on the part of observing man, and the bird which, owing to its superior intelligence, easily adapts itself to this process, is considered to rank among the highest caste. The ability for calculation and divination is directly transferred to the bird The division into castes is found also among the Nāga and the spirits called *gñan* (see SCHIEFNER, *Ueber das Bonpo-Sūtra, Mém Acad de St Pét*, Vol. XXVIII, N° 1, 1880, pp. 3, 26 *et passim*; *Mém Soc finno-ougrienne*, Vol XI, 1898. p 105; *Denkschriften Wiener Akademie*, Vol XLVI. 1900, p 31).

one to the Kshatriya caste, one flapping its wings to the Vaiçya caste, one shaped like a fish to the Çūdra caste, one subsisting on filthy food and craving for flesh belongs likewise to the latter.

The following holds good for the different kinds of tones emitted by the crow. The layman must pronounce the affair the truth of which he wishes to ascertain simultaneously [with the flight of the crow]. [1]

I. When in the first watch (*t͏'un daṅ-po la*), [2] in the east, a crow sounds its notes, the wishes of men will be fulfilled.

When in the south-east it sounds its notes, an enemy will approach (Table II, 9, and V, 2). [3]

---

1) SCHIEFNER translates: „Die verschiedenen Arten ihres Geschreis sind folgende, (welche) der Hausherr einmal wahrgenommen verkunden muss.” But this mode of rendering the passage does not do justice to the text (*k'yim-bdag-gis cig-car bden-par agyur-ba ni b·rjod-par bya-ste*). Stress is laid on the phrase *cig-car*, alluding to the fact, which repeats itself in all systems of omens, that the wish must be uttered at the same moment when the phenomenon from which the oracle is taken occurs. SCHIEFNER overlooks the force of *bden par agyur-ba*, which is not *wahrgenommen*, but *was bewahrheitet werden soll*. Only he who seeks an oracle will naturally pay attention to the flight of the crow, and he must loudly proclaim his question, addressing the bird at the moment when it flies into the open.

2) SCHIEFNER takes the term *t͏'un* (Skr *yâma*) in the sense of night-watch This, in my opinion, is impossible. In this first section of the treatise, divination is detailed to five divisions of time, the fifth and last of which is designated as the sunset Consequently the four preceding divisions must refer to the time of the day, both *t͏'un* and *yâma* apply to the day as well as to the night, and simply signify a certain length of time (usually identified with a period of three hours in our mode of reckoning) of the twenty-four hour day The five watches named in our text would accordingly yield an average term of fifteen hours, the usual length of a day in India. It is also natural to watch crows in the daytime, and not at night, when, like others of their kind, they are asleep in their nests. The same division of the day into five parts, probably derived from India, exists also in Java (RAFFLES, *A History of Java*, Vol 1, p. 530, London, 1830).

3) The crow's prophecy of war is linked with the rapacious and bellicose character of the bird This notion appears as early as in the Assyrian inscriptions of Sennacherib, where we meet such comparisons as "like the coming of many ravens swiftly moving over the country to do him harm," and "like an invasion of many ravens on the face of the country forcibly they came to make battle" (F. DELITZSCH, *Assyrische Thiernamen*, p 102, Leipzig, 1874; and W. HOUGHTON, *The Birds of the Assyrian Monuments*, Trans. Soc. Bibl. Arch, Vol. VIII, 1884, p. 80) In Teutonic divination, the raven believed to possess wisdom and knowledge of events was especially connected with battle. should one be heard thrice screaming on the roof, it boded death to warriors, while the appearance of ravens

When in the south, etc., a friend will visit (Table VIII, 6; X, 3).

When in the south-west, etc., unexpected profit will accrue.

When in the west, etc., a great wind will rise (Table V, 4).

When in the north-west, etc., a stranger (guest) will appear. [1]

When in the north, etc., property scattered here and there (*nor ytor-ba*) will be found (Table X, 2).

When in the north-east, etc., a woman will come (Table VII, 8; IX, 5).

When in the abode of Brahma (zenith), [2] etc., a demon will

---

following a host or a single warrior would bring good luck in battle (HASTINGS, *Encyclo-paedia of Religion*, Vol. IV, p. 827).

1) In southern India, if a crow keeps on cawing incessantly in a house, it is believed to foretell the coming of a guest. The belief is so strong, that some women prepare more food than is required for the household (E. THURSTON, *Ethnographic Notes in Southern India*, p. 276, Madras, 1906). Among the Pārsī (J. J. MODI, *Omens among the Parsees*, in his *Anthropological Papers*, p. 4, Bombay, no year) the cawing of a crow portends good as well as evil. A peculiar sound called "a full noise" portends good. Such a noise is also considered to foretell the arrival of a guest or the receipt of a letter from a relative in some distant country. If a good event occurs after the peculiar cawing which portends good, they present some sweets to a crow. Another peculiar kind of cawing, especially that of the *kāgri*, the female crow, portends some evil. A crow making such a peculiar noise is generally driven away with the remark, "Go away, bring some good news'"

2) The four cardinal points (*p'yogs bži*) are expressed by the common words *šar, lho, nub, byaṅ*. The four intermediate points are designated *me* ("fire"), south-east, *bden bral*, south-west; *rluṅ* ("wind"), north-west; and *dbaṅ-lan*, north-east. These names are derived from those of the Ten Guardians of the World (see *Mahāvyutpatti*, ed of MINAYEV and MIRONOV, p 102; ed of CSOMA and ROSS, pt. I, p 57). The ninth point, Brāhmī, is there rendered by *steṅ-gi p'yogs*, the direction above, which is expressed in our text by *Ts'aṅs-pai gnas*, the place of Brahma. In the Table published by M. BACOT (II, 9) the term *ṅam-ka* (= *k'a, mk'a) ldin* is used in lieu of that one; this means literally "floating or soaring in the sky" (it occurs as a frequent name of the Garuḍa), and here "soaring in straight direction toward the sky," that is, the zenith. It will thus be seen that the nine points of the compass (out of the typical ten, *daṣadik*, which were assumed), as enumerated in the above text, are the same and occur in the same succession, as in M. BACOT's Table. The tenth point, naturally, is here out of the question, as crows cannot fly up in the nadir of a person. In the introductory to M Pelliot's roll the fact of nine cardinal points is distinctly alluded to in two verses (6 and 24), and M BACOT, quite correctly, has recognized there the eight quarters, making nine with the zenith. — The connection of crow auguries with the cardinal points may have arisen from the very ancient observation

come (Table X, 1). [1])

End of the cycle of the first watch.

II. When in the second watch (t'un gñis-pa-la), in the east, a crow sounds its notes, near relatives will come (Table VI, 4). [2])

of the crow's sense of locality, and its utilization in discovering land Indian navigators kept birds on board ship for the purpose of despatching them in search of land. In the *Bāveru-Jātaka* (No. 339 of the series) it is a crow, in the *Keraddhasutta* (in *Dīghanikāya*) it is a "land-spying bird." J. MINAYEV (*Mélanges asiatiques*, Vol. VI, 1872, p. 597), who was the first to edit the former text, explained the word for the crow *disākāka*, as it occurs there, as possibly meaning "a crow serving to direct navigators in the four quarters" (while the opinion of WEBER, added by him, that it might be an ordinary crow, as it occurs in all quarters, — seems forced). In my opinion, MINAYEV is correct: *disākāka* is the crow, whose flight is affiliated with the quarters, both in navigation and divination. GRUNWEDEL (*Veröff Mus. für Völkerkunde*, Vol. V, 1897, p 105) has published an allied text from the Biography of Padmasambhava, where the land-seeking bird of the navigators is designated "pigeon" (Tib. *p'ug-ron*). This will doubtless go back to some unknown Indian text where pigeons are mentioned in this capacity. PLINY (*Nat. Hist.* VI, 22, 83, ed. MAYHOFF, Vol. I, p. 465) relates that the seafarers of Taprobane (Ceylon) did not observe the stars for the purpose of navigation, but carried birds out to sea, which they sent off from time to time, and then followed the course of the birds flying in the direction of the land (siderum in navigando nulla observatio: septentrio non cernitur, volucres secum vehunt emittentes saepius meatumque earum terram petentium comitantur). The connection of this practice with that described in the Babylonian and Hebrew traditions of the Deluge was long ago recognized. In the Babylonian record (H. ZIMMERN, *Keilinschriften und Bibel*, p. 7) a pigeon, a swallow, and a raven are sent out successively to ascertain how far the waters have abated. When the people of Thera emigrated to Libya, ravens flew along with them ahead of the ships to show the way. The Viking, sailing from Norway in the ninth century, maintained birds on board, which were set free in the open sea from time to time, and discovered Iceland with their assistance (O. KELLER, *Die antike Tierwelt*, Vol II, p. 102) According to JUSTIN (XXIV. IV. 4), who says that the Celts were skilled beyond other peoples in the science of augury, it was by the flight of birds that the Gauls who invaded Illyricum were guided (DOTTIN in HASTINGS, *Encyclopaedia of Religion*, Vol IV, p. 787). In the *Ise-fudoki*, Emperor Jimmu engaged in a war expedition, and marched under the guidance of the gold-colored raven (K. FLORENZ, *Japanische Mythologie*, p 299). On the sending of pheasant and raven in ancient Japan see especially A. PFIZMAIER, *Zu der Sage von Owo-kuni-nushi* (*Sitzungsberichte Wiener Akademie*, Vol. LIV, 1866, pp. 50—52)

1) SCHIEFNER reads *agron-po*, and accordingly translates "guest" But it seems unlikely that the same should be repeated here that was said a few lines before in regard to the north-west. The Narthang print plainly has *agon-po*, which I think is mistaken for *agou-po*, "demon." The analogous case in Table X, 1, where the word *adre gdon* is used, confirms this supposition.

2) In the Kanjur, a little story is told of a crow uttering agreeable sounds auguring

[A reference in regard to the south-east is lacking in the text.]

When in the south it sounds its notes, you will obtain flowers and areca-nuts. [1]

When in the south-west, etc., there will be numerous offspring (*rgyud-pa ap^cel-bar agyur-ro*).

When in the west, etc., you will have to set out on a distant journey (*t^cag riñs-su agro-bar agyur-ro*; compare Table II, 2; IX, 3).

When in the north-west, etc., this is a prognostic of the king being replaced by another one (*rgyal-po gźan-du agyur-bai rtags*; compare Table VIII, 1). [2]

When in the north, etc., you will receive good news to hear (Table III, 8; VII, 7). [3]

---

for the safe return of a woman's absent husband, and being rewarded by her with a golden cap (A. SCHIEFNER, *Tibetan Tales*, English ed. by RALSTON, p. 355). J. J. MODI (*Anthropological Papers*, p 28) quotes the following lines, which he overheard a Hindu woman speak to a crow: "Oh crow, oh crow' (I will give thee) golden rings on thy feet, a ball prepared of curd and rice, a piece of silken cloth to cover thy loins, and pickles in thy mouth." A peculiar noise made by a crow, continues this author, is supposed to indicate the arrival of a dear relation or at least of a letter from him. When they hear a crow make that peculiar noise, they promise it all the above good things if its prediction turn out true In this case they fulfill their promise by serving it some sweets, but withhold the ornaments and clothes. — The following custom is observed in Cambodja. "Lorsque quelqu'un de la maison est en pays lointain, si le corbeau vient gazouiller dans le voisinage, la face tournée dans la direction de l'absent, il annonce son prompt retour Dans toute autre direction, il annonce un malheur" (É. AYMONIER, *Revue indochinoise*, 1883, p. 148).

1) Tib *me-tog dañ go-la t^cob-pa* SCHIEFNER renders *go-la* by "betel," but *go-la* is the areca-nut, which is chewed together with the leaf of betel, *piper betel* L. (see CHANDRA DAS, *Dictionary*, p. 227). We may justly raise the question whether anything so insipid was contained in the Sanskrit original, and whether the text is not rather corrupted here The Table contains nothing to this effect I venture to think that *go*, "rank, position," was intended. In Table I, 6, flowers are mentioned as offerings to the birds, and this may give a clew as to how the confusion came about.

2) In the text of the Table: *rgyal-po ajig-par ston*, "this indicates the overthrow or ruin of the king" (but not *indique un danger pour le roi*). I do not agree with SCHIEFNER's rendering: „Ein Zeichen, dass der König sich anderswohin wendet."

3) Tib *ap^crin-las legs-par t^cos-par agyur-ro*. P^crin, "news," will probably be the proper reading In the text of M Bacot *p^crin byañ* is printed, and translated *un courrier de nouvelles* M BACOT presumably had in mind the word *bya-ma-rta*, "a courier," but there is no word *byañ* with this meaning. We doubtless have to read *p^crin bzañ*, "good news, good message"

When in the north-east, etc., disorder [1]) will break out (Table V, 7).

When in the zenith, etc., you will obtain the fulfilment of your wishes. [2])

End of the cycle of the second watch.

III. When in the third watch, in the east, a crow sounds its notes, you will obtain property (Table X, 2).

When in the south-east a crow sounds its notes, a battle (*t'ab-mo*) will arise (Table V, 7).

When in the south, etc., a storm will come (Table V, 4).

When in the south-west, etc., an enemy will come (see above, I, south-east).

When in the west, etc., a woman will come (see above, I, north-east).

When in the north-west, etc., a relative will come (see above, II, east).

When in the north, etc., a good friend will come (Table VIII, 6; X, 3).

When in the north-east, etc., a conflagration will break out (*mes ats'ig-par agyur-ro*; Table VI, 7).

When in the zenith, etc., you will gain profit from being taken care of by the king. [3])

End of the cycle of the third watch.

---

1) Tib *ak'rug-pa* exactly corresponds in its various shades of meaning to Chinese *luan* 亂, "disorder, tumult, insurrections, war," etc This rendering is indeed given for the Tibetan word in the Tibetan-Chinese vocabulary of *Hua i yi yü* (Ch. 11, p. 33 b: Hirth's copy in Royal Library of Berlin). In the Table, the word *t'ab-mo*, "fight, battle," is used.

2) Tib. *adod-pai ajug-pa rñed-par agyur-ro.* SCHIEFNER translates: „Wird sich die gewünschte Gelegenheit finden."

3) SCHIEFNER's translation „wird der König den im Gemüth befestigten Gewinn finden" is unintelligible. The text reads: *rgyal-po t'ugs-la brtags-pai rñed-pa t'ob-par agyur-ro.* Schiefner's correction of *brtags* into *btags* is perfectly justifiable, indeed, the confusion of these two words is frequent. But *t'ugs-la adogs-pa* is a common phrase correctly explained by JASCHKE (*Dictionary*, p. 280) "to interest one's self in, to take care of." It should not be forgotten, of course, that, at the time when Schiefner wrote, this dictionary was not published.

IV. When in the fourth watch, in the east, a crow sounds its notes, it is a prognostic of great fear (*ujigs-pa c'e-bai rtags-so*; Table V, 6; IX, 1).

When in the south-east a crow sounds its notes, it is a prognostic of large gain.

When in the south, etc., a stranger (guest) will come (see above, I, north-west).

When in the south-west, etc., a storm will rise in seven days.

When in the west, etc., rain and wind will come (Table V, 4, 5). [1])

When in the north-west, etc., you will find property which is scattered here and there (*nor gtor-ba*).

When in the north, etc., a king will appear.

When in the north-east, etc., you will obtain rank. [2])

When in the zenith, etc., it is a prognostic of hunger.

End of the cycle of the three watches and a half.

V. When at the time of sunset (*ñi-ma nub-pai ts'e*; compare Table X), in the east, a crow sounds its notes, an enemy will appear on the road.

When in the south-east a crow sounds its notes, a treasure will come to you.

When in the south, etc., you will die of a disease (Table V, 8). [3])

---

1) The ability attributed to crow and raven of possessing a foreknowledge of coming rain has chiefly made them preeminently prophetic birds (*augur aquae* in Horace). The ancients observed that these birds used to caw with peculiar notes when rain was to fall, and that, if a storm was imminent, they were running to and fro on the beach with great restlessness, and bathing their heads (compare O KELLER, *Die antike Tierwelt*, Vol. II, p 98).

2) Tib. *go-la* (as above) *rñed-par agyur-ro*. The correction *go rñed-par* may here be allowed to pass, as the finding of areca-nuts seems such a gross stupidity.

3) In the story "The Death of the Magpie," translated from a manuscript of the India Office by A. SCHIEFNER (*Mélanges asiatiques*, Vol. VIII, p. 630), the raven has the attributes "the Uncle, the Judge of the Dead" (in Schiefner's rendering; the original is not known to me), and the following verses are addressed to it (p. 631): "Be kind to the nephews here, bestow fortune upon the children, direct the government of the country,

When in the south-west, [1]) etc., the wishes of one's heart will be fulfilled.

When in the west, etc., relatives will come.

When in the north-west, etc., it is a prognostic of obtaining property.

When in the north, etc, homage will be done to the king.

[A reference to the north-east is lacking in the text.]

When in the zenith, etc., you will obtain an advantage for which you had hoped.

End of the cycle of the fourth watch.

End of the description of such-like cries of the crow.

We shall now discuss the import of the crow's tones when one is travelling. When along dams and river-banks, on a tree, in a ravine, [2]) or on cross-roads, a crow sounds its voice on your right-hand side, you may know that this journey is good. When, at the time of wandering on the road, a crow sounds its voice behind your back, you will obtain the *siddhi*. When, during a journey, a crow flapping its wings [3]) sounds its voice, a great acci-

---

lend expression to good plans." In connection with these ideas of the raven as a bird of death, it is worthy of note that in two texts of the Tanjur, Mahākāla appears in the form of the Raven-faced one (Skr. *kākāsya*, Tib. *bya-rog gdon-can*), likewise the goddess Kālī (Tib. *k'va gdoṅ-ma*), see P. Cordier, *Cat du fonds tibétain de la Bibl. Nat.*, Vol. II, pp. 124, 127. The raven-faced Mahākāla is illustrated in the "Three Hundred Gods of Narthang" (section *Rin abyuṅ*, fol. 121) The raven as a bird announcing death is widely known in classical antiquity and mediæval Europe (O. Keller, *Die antike Tierwelt*, Vol. II, p. 97; E. A. Poe's poem *The Raven*). The imminent deaths of Tiberius, Gracchus, Cicero, and Sejan, were prophesied by ravens.

1) Is expressed in this passage by *srin-poi mts'ams*, "the intermediate space of the Rākshasa."

2) Tib. *grog stod*, as plainly written in the Narthang print Schiefner read *grog stoṅ*, and corrected *grog steṅ*, with the translation "on an ant-heap," regarding *grog* as *grog-ma*, *grog-mo*, "ant." I prefer to conceive *grog* as *grog-po* (related to *roṅ*), "ravine," which is more plausible in view of the other designations of localities which are here grouped together. Moreover, I do not believe that crows go near ant-hills or feed on ants. The reading *stod* is then perfectly good, the significance being "in the upper part of the ravine."

3) According to the introduction, one of the Kshatriya caste.

dent will befall one. When, during a journey, a crow pulling
human hair with its beak [1]) sounds its voice, it is an omen that
one will die at that time. When, during a journey, a crow eating
filthy food [2]) sounds its voice, it is an omen of food and drink
being about to come (Table VIII, 9).

When, during a journey, a crow perching on a thorn-bush
sounds its voice, it should be known that there is occasion to fear
an enemy. When, during a journey, a crow perching on a tree
with milky sap [3]) sounds its voice, milk-rice (o t'ug-gi bza-ba) will
fall to your lot at that time. When a crow perching on a withered
tree [4]) sounds its voice, it is a prognostication of the lack of food
and drink at that time. When a crow perching on a palace sounds
its voice, you will find an excellent halting-place. [5]) When a crow

---

1) Tib. skra mc'us gziṅs-śiṅ According to Jäschke (Dictionary, p 464) skra adsiṅs-pa
or gziṅ-ba is an adjective with the meaning "bristly, rugged, shaggy" (Dictionary of the
French Missionaries, p. 532: crines dispecti, cheveux épars). The verbal particle ciṅ and
the instrumentalis mc'u-s ("with the beak") indicate that gziṅs is a verbal form belonging
to a stem dsiṅs, adsiṅs, and means 'pulling about hair in such a way that it appears
rugged" Below, we find the same expression mc'us gos gziṅs-śiṅ, "pulling a dress with its
beak" The word adsiṅs-pa is used also of interlaced trees or thick-set vegetation, as indicated by the Polyglot Dictionary of K'ien-lung, according to which it is the equivalent of
ts'ao mu ts'ung tsa 草木叢雜, Manchu yabulehebi, Mongol kughunaldaji (s'entre-
lacer); we find there, further, the phrase sgro adsiṅs = liṅg ch'i ts'an k'ue 翎翅殘缺,
'with broken wings,' Mongol samtaran, se briser (the Tibetan equivalent in KOVALEVSKI
is a misprint). SCHIEFNER (p. 14) remarks that the form gziṅs is new to him, and questions its correctness, he takes it as identical with bzaṅ, and translates it by anfassen.
This derivation is not correct, it is merely surmised. The passage evidently means more than
that the crow simply seizes human hair; it is torn to pieces, and this destructive work
has a distinct relation to the foreboding of death

2) Tib mi gtsaṅ-ba za zur, the same expression as used in the introduction to denote
a crow of the Çūdra caste   Compare Subhashitaratnanidhi 37 (ed. CSOMA).

3) Tib o-ma-can-gyi śiṅ (Skr. kshīrikā, kshīraju). Indian medicine recognizes five trees
presumed to yield a milky sap. These are, according to HOERNLE (The Bower Manuscript,
p. 20), the nyagrodha (Ficus bengalensis), udumbara (Ficus glomerata), açvattha (Ficus
religiosa), plaksha (Ficus tjakela), and pārisha (Thespesia populnea)

4) As often in the Indian stories (SCHIEFNER, Mélanges asiatiques, Vol. VIII, 1877,
p. 96, or RALSTON, Tibetan Tales, p. 32).

5) SCHIEFNER translates erroneously, 'When you betook yourself to the royal palace,

perching on a divan sounds its note, an enemy will come. When a crow facing the door sounds its voice, it should be known that a peril will threaten from the frontier (*mtsʿams-kyi ajigs-par šes-par byao*). When a crow pulling a dress (*gos*) with its beak sounds its voice, you will find a dress (*gos*). When, during a journey, a crow perching on the cranium of a corpse [1]) sounds its notes, it is a prognostication of death. When a crow seizing a red thread and perching on the roof of a house sounds its notes, this house will be destroyed by fire (Table VI, 7). When, in the morning (*šňa-droi dus-su*, Table V), many crows assemble, a great storm will arise (Table V, 3). [2])

When, at the time of a journey, a crow seizing with its beak a piece of wood sounds its voice, some advantage will fall to your lot. When, at the time of a journey, at sunrise (*ňi-ma šar dus-su*, Table IV), a crow sounds its voice, you will obtain property. When, at the time of a journey, it sounds its voice, [3]) one's wishes will be fulfilled.

---

and when the crow then sounds its cries, you will receive a good seat." But it is the question of a traveller who, on his journey, happens to pass by a palace, and it is the crow which is sitting on the roof of the palace (the verb *gnas* means "to dwell, remain," but never expresses any act of motion): in the same manner as the crow has found a good resting-place, so the weary wanderer will find good quarters for the night. The text runs thus: *pʿo-braň-la gnas-nas gaň-gi tsʿe skad sgrogs-na, dei tsʿe sdod sa bzaň-po řñed-par agyar-ro* The word *sdod sa* does not mean "a seat," but a place where a traveller stops for the night, "halting-place." Likewise, in the two following sentences, SCHIEFNER refers the phrases *gdan-la gnas-nas* and *sgo lta ziň* to the man instead of to the crow

1) SCHIEFNER: „eine Krähe auf der Kopfbinde sich befindend " This is due to a confusion of the two words *tʿod* and *tʿod-pa*, the former means "turban;" but the text has *tʿod-pa* meaning "the skull of a dead person," and this only makes sense of the passage Crows congregate and feed on carrion, and are therefore conceived of as birds of death. The turban, for the rest, is out of the question in this text, as it was introduced into India only by the Mohammedans.

2) O. KELLER (*Die antike Tierwelt*, Vol. II, p 109, Leipzig, 1913), who concludes his interesting chapter on crow and raven in classical antiquity with an extract from Schiefner's translation, observes on this sentence that it is based on a fact, and that such grains of truth hidden among these superstitions account for the fact that they could survive for centuries

3) Apparently there is here a gap in the text, no definition of the activity of the crow being given

2

End of the signs of the journey (*lam-gyi mts'an-ńid*).

The symptoms (or omens) of the nest-building of the crow are as follows. [1]) When a crow has built its nest in a branch on the east side of a tree, a good year and rain will then be the result of it. When it has built its nest on a southern branch, the crops will then be bad. When it has built its nest on a branch in the middle of a tree, a great fright will then be the result of it (Table V, 6). When it makes its nest below, fear of the army of one's adversary will be the result of it. When it makes its nest on a wall, on the ground, or on a river, the king will be healed [from a disease]. [2])

Further, the following explanation is to be noted. When a crow sounds the tone *ka-ka*, you will obtain property. When a crow sounds the tone *da-da*, misery will befall you. When a crow sounds the tone *ta-ta*, you will find a dress. When a crow sounds the tone *gha-gha*, a state of happiness will be attained. [3]) When a crow sounds the tone *gha-ga*, a failure will be the result of it. [4])

---

1) In the first section of the treatise the crow is in motion, and the person demanding the oracle is stationary In the second section both the crow and the person are in motion. In this one, the third section, both the crow and the person are stationary; hence the text says: *qnas-pai bya-rog-gi ts'an-gi mts'an-ńid*, "the crows when they are settled . ''

2) Tib. *ats'o-bar agyur-ro*, translated by SCHIEFNER „so wird der König leben," which gives no sense Of course, the word *ats'o-ba* means "to live," but also "to recover from sickness." Here the Table (IX, 2) comes to our rescue. where we meet the plain wording *nad-pa sos-par ston*, "it indicates cure from disease." — Among the Greeks, the crow, owing to the belief in the long life of the bird, was an emblem of Asklepios (O. KELLER. *Die antike Tierwelt*, Vol. II, p. 105); compare Hesiod's famous riddle on the age of the crow and raven (W. SCHULTZ, *Rätsel aus dem hellenischen Kulturkreise*, p. 143, Leipzig, 1912; and K. OHLERT, *Rätsel und Rätselspiele der alten Griechen*, 2d ed., p. 146, Berlin, 1912). The idea of the longevity of the crow was entertained also in India (Skr. *dīrghāyus*, Tib *na-ts'od-can*, attribute of the crow given in the *Dictionary of the French Missionaries*, p. 86); it is striking that this quality of the crow is not alluded to in our text.

3) Tib. *don ṅgrub-par agyur-ro.* SCHIEFNER translates: „so geht die Sache in Erfüllung ''

4) Tib. *nor oń-bar agyur-ro.* SCHIEFNER, „so wird ein Schatz kommen," which is certainly correct, as far as the meaning of these words is concerned; but I doubt very much whether this is the true significance intended by the author, for what SCHIEFNER trans-

When an omen causing fear is observed, a strewing oblation must be offered to the crow. As the flesh of a frog pleases the crow, no accidents will occur when frog-flesh is offered. [1]

Oṁ mi-ri mi-ri vajra tudaṭe gilaṁ gṛihṇa gi svāhā!

End of the description of such-like behavior of the crow.

Translated by the Mahāpaṇḍita Dānaçīla in the monastery T'aṅ-po-c^ee of Yar-kluṅs in the province of dBus.

The translator Dānaçīla has been dated by HUTH in the ninth century, on the ground that he is made a contemporary of King K^cri-lde sroṅ-btsan of Tibet in the work *sGra sbyor* in Tanjur, Sūtra, Vol. 124. This fact is correct, as may be vouchsafed from a copy made by me of this work. Dānaçīla figures there, together with such well-known names as Jinamitra, Surendrabodhi, Çrīlendrabodhi, Bodhimitra, the Tibetan Ratnarakshita, Dharmatāçīla, Jñānasena, Jayarakshita, Mañjuçrīvarma and Ratnendraçīla. Dānaçīla is well known as translator of many works in the Kanjur [2] and Tanjur. From the colophon of a work in the latter collection it appears that he hailed from Varendrajigatāla. that is, Jigatāla

---

lates is exactly the same as what is said above in regard to the tone *ka-ka*. Further, the tone *gha-qa* stands in opposition to the preceding tone *gha-gha*; it thus becomes clear that *nor* stands for *nor-ba*, "to err, to fail," and is expressive of the contrary of *don agrub-pa*, "to reach one's aim, to obtain one's end, to attain to happiness." This case reminds one of the grammatical as well as other subtleties of the Indian mind. — Also the ancients seem to have distinguished between various kinds of raven's cries, judging from PLINY's words that they imply the worst omen when the birds swallow their voice, as if they were being choked (pessima eorum significatio, cum gluttiunt vocem velut strangulati. *Nat. Hist.*, X, 12, § 32; ed. MAYHOFF, Vol. II, p. 229). The crow, according to PLINY (*ibid.*, § 30), is a bird inauspicatae garrulitatis, a quibusdam tamen laudata.

1) In the belief of the Tibetans, the crow is fond of frogs; compare the jolly story "The Frog and the Crow" in W. F. O'CONNOR, *Folk Tales from Tibet*, p. 48 (London, 1906).

2) FEER, *Annales du Musée Guimet*, Vol. II, p 406.

(Jagaddala) in Varendra, in eastern India. [1] Then we meet him in Kāçmīra, where Tāranātha [2] knows him together with Jinamitra and Sarvajñadeva, in accordance with *dPag bsam ljon bzań* (ed. CHANDRA DAS, p. 115); while *rGyal rabs* has the triad Jinamitra. Çrīlendrabodhi, and Dānaçila. [3] It may therefore be granted that the *Kākajariti* [4]) was translated and known in Tibet in the first part of the ninth century. The original Sanskrit manuscript from which the Tibetan translation was made in all probability was defective. for three gaps in it could unmistakably be pointed out.

What is the position of *K.* in the history of Indian divination? H. JACOBI (in HASTINGS. *Encyclopaedia of Religion*, Vol. IV, p. 799) has formulated the result of his study of this subject in these words: "In India, divination has gone through two phases of development. Originally it seems to have been practised chiefly with the intention of obviating the evil consequences of omens and portents; in the later period, rather to ascertain the exact nature of the good or evil which those signs were supposed to indicate." In the Vedic Saṁhitās, birds are invoked to be auspicious, and certain birds, especially pigeons or owls, are said to be messengers

---

1) P CORDIER, *Cat. du fonds tibétain de la Bibl Nat.* II, pp 63, 122, 188 (Paris, 1909), and VIDYABHUSANA (the name of this author appears in his publications in four different ways of spelling, ₀bhusan, ₂bhusana, ₀bhusana, ₀bhusana: which is the bibliographer supposed to choose?) *Bauddha-Stotra-Saṁgrahah.* pp. XVIII, XIX (Calcutta, 1908). Mr V states that it is said at the end of the *Ekajaṭīsadhana* that the worship of Tārā originated from China, but that it is not clear whether this refers to Ekajaṭī Tārā alone or to Tārā of all classes. I fear that neither the one nor the other is the case. The Tibetan text plainly says, "The work *Tārāsadhana* which has come from China (scil. in a Chinese translation) is in a perfect condition." This implies that the Tibetan translator availed himself of a Chinese version The worship of Tārā most assuredly originated in India, not in China

2) SCHIEFNER's translation, p. 226.

3) SCHLAGINTWEIT, Konige von Tibet, p. 849. also ROCKHILL, *The Life of the Buddha*, p. 224.

4) Henceforth abbreviated *K*

of death (Nirṛti, Yama). [1]) But all these are no more than scant

1) The best investigation of the history of bird omens in India is found in the monograph of E. HULTZSCH (*Prolegomena zu des Vasantarāja çāku ia nebst Textproben*, Leipzig, 1879) The beginnings of bird augury in India may be traced back to the Vedic period. In the Ṛigveda occur the so-called çakuna, charms against pigeons, owls, and other black birds whose appearance or contact forebodes evil, or defiles (M. BLOOMFIELD, *The Atharvaveda*, p. 85, Strassburg, 1899) According to MACDONELL and KEITH (*Vedic Index of Names and Subjects*, Vol. II, p. 347, London, 1912) there are the two words, çakuna, usually denoting a large bird, or a bird which gives omens, and çakuni, used practically like the former, but with a much clearer reference to divination, giving signs and foretelling ill-luck; later the falcon is so called, but the raven may be intended; the commentator on the *Taittirīya Saṃhitā* thinks that it is the crow. Oracles obtained from an observation of crows seem to be contained particularly in the *Kauçika Sutra*. When the rite serving the purpose of securing a husband has been performed on behalf of a girl, the suitor is supposed to appear from the direction from which the crows come (H OLDENBERG, *Die Religion des Veda*, p 511, Berlin, 1894) Contact with a crow was regarded as unlucky and defiling He who was touched by a crow was thrice turned around himself, from the left to the right by the sorcerer holding a burning torch (V. HENRY, *La magie dans l'Inde antique*, p. 176, 2d ed., Paris, 1909; E. THURSTON, *Ethnographic Notes in Southern India*, p 277, Madras, 1906). A. HILLEBRANDT (*Ritual-Litteratur. Vedische Opfer und Zauber*, p. 183, Strassburg, 1897) believes he finds the explanation for this idea of bird omens in a passage of Baudhāyana, according to which the birds are the likenesses of the manes; but it seems rather doubtful whether the latter notion could receive such a generalized interpretation, and whether it is sufficient to account for the augural practice in its entire range The latter would naturally presuppose the idea of the bird being animated with a soul and being gifted with supernatural powers or instigated by some divine force, but Hillebrandt's opinion leaves the reason unexplained why the bird, even though it should represent a mane in every case, possesses the ability of divination. True it is, as shown by W. CALAND (*Die altindischen Todten- und Bestattungsgebrauche*, p. 78, Amsterdam, 1896), that especially the crows were conceived of as embodying the souls of the departed, as messengers of Yama, who, after the funerary repast (çrāddha), draw near, greedy for food (compare the Raven Spirit in the Lamaist mystery plays who attempts to filch the strewing oblation, and who is chased away by two stick-brandishing Atsara, the skeleton ghosts!), but plainly, in this case, no process of divination is in question CALAND, on this occasion, quotes DUBOIS on the modern practice that the chief of the funeral offers boiled rice and pease to the crows, — if they should refuse to eat, it is taken as an evil presage of the future state of the deceased, but this evidently is quite a different affair from that described in his above reference to Baudhāyana. Some authors allow the whole practice of auguries to go back into the prehistoric epoch of the Indo-European peoples (H. HIRT, *Die Indogermanen*, Vol. II, p 518, Strassburg, 1907; and S. FEIST, *Kultur*, etc., *der Indogermanen*, p. 326, Berlin, 1913), the latter even going so far as to speculate that the idea of a soul flying along in the shape of a bird was not foreign to the urvolk, since this augural divination is based on the transformation of the souls into birds. I am very skeptical regarding such conclusions and constructions, and must confess that

allusions; neither in the Vedic nor in the early Brahmanic epoch do we find anything like an elaborate augural system, as in *K.*, in which future events are predicted, — Jacobi's second stage. The same author tells us that the whole art of divination became independent of religion when Greek astronomy and astrology were introduced into India in the early centuries of our era; the Indian astrologer then took up divination, hitherto practised by the Atharva priest. It is of especial interest for our present case that in the *Brihat Samhitā* by Varāhamihira (505—587), written about the middle of the sixth century, in which a summary of the Indian arts of divination is given, the auspicious or unlucky movements of crows are mentioned. [1]) A work of the type of *K.*, ac-

I even belong to those heretics who are still far from being convinced of the existence of such a thing as the *indogermanische urzeit*, — at least in that purely mechanical and subjective formula in which it is generally conceived. The work of FEIST, however, is a laudable exception, perhaps the first sensible book written on this subject, and I read it from beginning to end with real pleasure — In regard to the crow or raven, we find also other ideas connected with them than those of a soul-bird, in India as well as among other Indo-European peoples. In a legend connected with Rāma, an Asura disguised as a crow appears to peck at Sitā's breast (E. THURSTON, *l c*, p. 276, and *Omens and Superstitions of Southern India*. p 87, London, 1912). Among the southern Slavs, the crows are believed to be transformed witches (F. S KRAUSS, *Slavische Volksforschungen*, pp. 57, 60, Leipzig, 1908), and in mediaeval legends, the devil occasionally assumes the shape of a raven. In Greek legend Apollo repeatedly appears in the disguise of a raven (O KELLER, *Die antike Tierwelt*, Vol II, p. 103) These various examples demonstrate that the raven as a divine bird cannot be solely explained as the embodiment of an ancestral soul. It seems to me that H. OLDENBERG (*Die Religion des Veda*, pp. 76, 510) is right in assuming that the animals sent by the gods were those of a weird, demoniacal nature, and were, for this reason, themselves deified, while at a later time they became mere stewards to divine mandators "The bird crying in the quarter of the fathers" (the south), mentioned in the Rigveda, according to OLDENBERG, should be understood as one being despatched by the fathers. The *document Pelliot* lends substantial force to this argument. It is there expressed in plain and unmistakable words that the raven is a divine bird of celestial origin and supernatural qualities, and the messenger who announces the will of a deity, the Venerable One of the Gods (*Lha btsun*); compare the Preface to the Table, translated below.

1) Ch XLV is taken up by the auguries obtained from the wagtails (see H. KERN's translation in his *Verspreide geschriften*, Vol I, p. 299, 's-Gravenhage, 1913, on crows, *ibid.*, pp. 130, 178) Regarding Varāhamihira's date of birth MUKERJI in *J A S B*, 1912, pp 275—8.

cordingly, must have been known at that time; but was it much earlier? I am under the impression that *K.* is hardly earlier than the sixth or seventh century, perhaps contemporaneous with the Çakuna of Vasantaraja, which, according to HULTZSCH (p. 27), is posterior to Varāhamihira; the striking lack of thought and imagination, and the somewhat flat treatment of the subject, plainly stamp *K.* as a late production. The absence of any mythological detail is a decided drawback; the religious function of the crow is not even set forth, and we remain entirely in the dark as to the religious concept of the bird in the India of that period. SCHIEFNER designated the little work a Buddhist retouch (*Überarbeitung*) of a Brahmanic text. It seems to me to be neither the one nor the other. It cannot be yoked to any definite religious system; it takes root in the domain of folk-lore, and closely affiliates with those manifold branches of divination which, independent of any particular form of religion, are widely diffused from the shores of the Mediterranean to almost the whole of continental Asia and the Malayan world. [1]) The tone and tenor of this text are not Buddhistic, nor

---

1) T. S. RAFFLES (*The History of Java*, Vol. II, p. 70, London, 1830) tells, in regard to the ancient Javanese, that when the crop was gathered and the accustomed devotions performed, the chief appointed the mode and time of the departure of the horde from one place to another. On these occasions, the horde, after offering their sacrifices and feasting in an open plain, left the remains of their repast to attract the bird *alunggiga* (supposed to have been a crow or raven), and the young men shook the *ánklung* (a rude instrument of music still in use), and set up a shout in imitation of its cry. If the bird did not eat of the meal offered to it, or if it afterwards remained hovering in the air, perched quietly on a tree, or in its flight took a course opposite to that which the horde wished to pursue, their departure was deferred, and their prayers and sacrifices renewed. But when the bird, having eaten of its meal, flew in the direction of their intended journey, the ceremony was concluded by slaying and burning a lamb, a kid, or the young of some other animal, as an offering of gratitude to the deity. RAFFLES adds that the Dayak of Borneo still hold particular kinds of birds in high veneration, and draw omens from their flight and the sounds which they utter. Before entering on a journey or engaging in war, head-hunting, or any matter of importance, they procure omens from a species of white-headed kite, and invite its approach by screaming songs, and scattering rice before it.

is there a particle of Buddhist color admixed with it. Nor is there
in it much that could be styled specifically Indian, with the exception, of course, of the outward garb in which it is clothed; but
most of the oracles could as well have been conceived in Greece
or Rome. [1])

We may justly assume that *K.* was not the only work of its
class, and that other Sanskrit books of an allied character may

---

If these birds take their flight in the direction they wish to go, it is regarded as a favorable omen; but if they take another direction, they consider it as unfavorable, and delay
the business until the omens are more suitable to their wishes. See now HOSE and
McDOUGALL, *The Pagan Tribes of Borneo*, Vol. 1, pp. 168—170, Vol II, p. 74 (London,
1912). Omens are taken either from the flight or the cries of certain birds, such as the
night-owl, the crow, etc. (W. W. SKEAT, *Malay Magic*, p. 535, London, 1900). Among
the tribes of the Philippines, bird omens play an extensive rôle My colleague F. C. COLE,
who has studied to a great extent their religious notions, kindly imparts the following
information on the subject: "With the Batak, a pigmy people living in northern Palawan,
the small sun bird known as *sagwaysagway* is considered the messenger of Diwata [evidently
Skr. *devatā*] Mendusa, the greatest of the nature spirits. Should this bird sing while they
are on the trail, the Batak will return home, for evil is sure to follow if they continue
their journey that day Should the bird enter a dwelling and sing, the place is deserted.
When a man desires to make a clearing in the jungle, he first addresses the sun bird,
asking it to sing and give him the sign if it is a bad place to plant, but to be silent if
it is a good plot for him to cultivate Similar beliefs are entertained by the Tagbanua
tribe which inhabits the greater part of Palawan" Further information will be found in
the publication of F. C. COLE, *The Wild Tribes of Davao District, Mindanao*, pp. 63,
108, 153, 173 (*Field Museum Anthr. Ser.*, Vol XII, 1913).

1) The Greeks distinguished five kinds of divination (οἰωνιστική) headed by auguration
(τὸ ὀρνεοσκοπικόν), Telegonos was the first to write on this subject (H. DIELS, *Beiträge
zur Zuckunyshteratur des Okzidents und Orients* 1, *Abhandl preuss Akad*, 1908, p. 4)
The typical Homeric method of foretelling the future was by the actions and cries of
omen-birds. In Homer, the omen-bird is generally an eagle, and is always sent by Zeus,
Apollo, or Athene. Its actions are symbolical, and need no complicated augury for their
interpretation (HASTINGS, *Encyclopaedia of Religion*, Vol. IV, p. 787) In Aristophanes'
Birds, Euelpides inquires what road is advised by a crow purchased at three obols. Ac-
cording to Virgil and Horace, a crow coming from the left-hand side is of ill omen. In
Works and Days by Hesiod it is said, "Do not let a house incomplete, otherwise a gar-
rulous crow will perch on it and caw" Even Epiktet believed in the correctness of the evil
prophecies of a raven (O. KELLER, *Die antike Tierwelt*, Vol. II, p 97). Compare L. HOPF,
Tierorakel und Orakelliere in alter und neuer Zeit (Stuttgart, 1888); and W R HALLIDAY,
Greek Divination, a Study of its Methods and Principles (London, 1913)

then have existed in Tibet; [1]) for, with all the coincidences prevailing between *K.* and the *document Pelliot*, there are, on the other hand, far-reaching deviations extant in the latter which cannot be explained from *K.* First of all, however, the interdependence of the two texts should be insisted upon. The main subject of the two is identical; it is the method of obtaining omens from crows which is treated in both on the same principle. This principle is based on a combination of two elements, — orientation of the augur and time-reckoning according to the hours of the day; divination is determined by space and time. In regard to the division of space, the coincidence in the two documents is perfect; the nine [2]) points of the compass forming the framework in both are one and the same. Time calculation is likewise the same in principle, except that *K.* follows the Indian, the Table the Tibetan method, — a point discussed farther on. The ideas expressed by the oracles show far-reaching agreements in both, and move within the narrow boundaries of a restricted area; no great imagination is displayed in them, they are rather commonplace and philistine, even puerile, but this is all that could be expected from this class of prophecy intended for the *profanum volgus.* Another feature which *K.* and the document of Pelliot have in common is the method of divining from the nature of the cries of the crow, independent of space and

---

1) Writings of similar contents are still extant in modern Tibetan literature. BRIAN H. HODGSON (*The Phoenix*, Vol. I, 1870, p. 94), in a notice on the *Literature of Tibet*, mentions a book "Ditakh, by Chopallah [C'os dpal?] Lama, at Urasikh, to interpret the ominous croaking of crows, and other inauspicious birds."

2) The number nine plays a great rôle in systems of divination In southern India, the belief prevails that ill luck will follow should an owl sit on the house-top, or perch on the bough of a tree near the house. One screech forebodes death; two screeches, success in any approaching undertaking; three, the addition by marriage of a girl to the family, four, a disturbance; five, that the hearer will travel. Six screeches foretell the coming of guests; seven, mental distress; eight, sudden death, and nine signify favorable results (E. THURSTON, *Ethnographic Notes in Southern India*, p. 281, Madras, 1906; and *Omens and Superstitions of Southern India*, p. 66, London, 1912).

time. The last six verses (24—29) of the prefatory note correspond in meaning to the conclusion of *K*.: "When a crow sounds the tone *ka-ka*," etc. It is a notable coincidence that in both texts five notes of the bird are enumerated in words imitative of its sounds, in *K*. conceived from an Indian point of view, in *document Pelliot* nationalized in a Tibetan garb.[1]) The character and quality of these tones, as well as the distinction between good and bad omens, necessarily lead to an effort toward reconciling the evil spirit which speaks through the organ of the bird. Offerings may counterbalance the mischievous effects of unlucky omens, — again a point on which the two texts are in harmony.

The differentiation of the two, in the first place, is due to a technical feature. The text of *K*. is a literary production and an analytic account. What is offered in the *document Pelliot* is an abstract of this divinatory wisdom worked up into convenient tabular form, manifestly with a view to handy and practical use. Any one who had encountered the necessary experience by observing a crow in a certain direction at a certain time of the day was enabled to

---

1) The number five is evidently suggested by the five elements, as shown by the five cries of the *piñjalī*, a kind of owl, distinguished according to the five elements in the *Çūkuna* of Vasantarāja (HULTZSCH, *Prolegomena*, p. 70). The beliefs in the omens of the owl in modern India are well set forth by E. THURSTON (*Omens and Superstitions of Southern India*, pp. 65—67). The enmity between crow and owl in Indian folk-lore deserves a word of comment in this connection. JÄSCHKE (*Dictionary*, p. 374) refers to *Suvarṇa-prabhāsasūtra* as describing the crow as an inveterate enemy of the owl. In the *Prajñādaṇḍa* ascribed to Nāgārjuna (ed. CHANDRA DAS, p. 9, Darjeeling, 1896) occurs the saying: "Those formerly vanquished by an enemy do not wish any longer for friendship. Look how the crows set fire to the cave filled with owls and burn them to death." In the same book (p. 8), the crows are credited with the killing of snakes. Compare also *Subhashita-ratnanidhi* 185 (ed. CSOMA). The animosity of the crow toward the owl seems to be based on the observation of a natural fact C. B CORY (*The Birds of Illinois and Wisconsin*, p. 548) has the following to say: "They seem to entertain an intense dislike to certain animals, especially an owl Often the peaceful quiet of the woods is suddenly broken by the harsh excited 'cawing' of a flock of crows, who have discovered a bird of that species quietly enjoying his diurnal siesta, and the din rarely ceases until the hated bird has been driven from his concealment and forced to seek other quarters"

read from this Table at a moment's notice what consequence this event would entail on his person. The subject-matter, therefore, was arranged here somewhat differently; the offerings placed at the very end of *K.* make here the very opening, and justly so, because, in accordance with the practical purpose of the Table, it was essential for the layman, or rather the priest acting on his behalf, to ascertain the kind of reconciliatory offering in case of threatening ill luck.

The greater fulness of the Table constitutes one of the principal divergences from *K.* In the latter, only five divisions of day-time are presented, while the Table offers double this number. This is infallible proof for the fact that the divination process revealed by the *document Pelliot* has been Tibetanized; it is by no means a translation from Sanskrit, but an adaptation based on some Sanskrit work or works of the type of *K.*, and freely assimilated to Tibetan thought. The Indian division of the day is abandoned; and the designations of the Tibetan colloquial language, as they are still partially in use, ¹) have been introduced into the Table. It is self-evident that these ten periods are not equivalents of the three-hour Indian *yāma*, but correspond to a double hour as found in China. In logical sequence these determinations run from about one o'clock at night to about nine o'clock in the evening. The plain Tibetan names for the points of the compass are all retained, while the fancy Indian names appearing in *K.* are all dropped. An attempt at adaptation to Tibetan taste has been made in the oracles. The killing of a yak and heavy snowfalls, for instance, are affairs peculiar to Tibet. It is manifest also that the prognostics given in

---

1) See G. SANDBERG, *Hand-book of Colloquial Tibetan*, p. 162 (Calcutta, 1894), and C. A. BELL, *Manual of Colloquial Tibetan*, p. 110 (Calcutta, 1905), where other terms also are included: also A. DESGODINS, *Essai de grammaire thibétaine*, pp. 90—91 (Hongkong, 1899).

the Table, in a number of cases, are more definite and specific than those of *K.*, which are rather monotonous and wearisome by frequent repetition of the same statement. Such repetitions, it is true, occur also in the Table (II, 2 = IX, 3; II, 4 = IV, 7 = VII, 4; V, 6 = IX, 1; VIII, 6 = X, 3), and there is certainly no waste of inventive power or exertion of ingenuity in this whole system. Apparently it appealed to the people of Tibet, where kindred ideas may have been in vogue in times prior to the infusion of Indian culture, [1]) and it is to this popularity that we owe the composition

1) For the inhabitants of the Western and Eastern Women Kingdom, the latter a branch of the K'iang, perhaps akin to the Tibetans, were in possession of a system of bird divination, *niao pu* 鳥卜 (*Sui shu*, Ch. 83, and *T'ang shu*, Ch. 122; the two passages are translated by ROCKHILL, *The Land of the Lamas*, pp. 339, 341, the former also by BUSHELL, *The Early History of Tibet*, p. 97, *J. R. A. S.*, 1880), which was based on the examination of a pheasant's crop, — a process of divination certainly differing from what is described in our Tibetan texts. Nevertheless we may infer that the shamans of those peoples, especially as the *T'ang shu* states that to divine they go in the tenth month into the mountains scattering grain about and calling a flock of birds, paid a great deal of attention to birds. (Whether the inhabitants of the two Women Kingdoms spoke a Tibetan language seems doubtful. The *T'ang shu* has preserved to us three words of the language of the Eastern one: *pin-tsiu* 賓就 "sovereign" 王, *kao-pa-li* 高霸黎 "minister" 宰相, and *su-yi* 蘇轝 "shoe" 履. None of these is traceable to a Tibetan word known to us. The vocabulary is so widely different in the present Tibetan dialects that this may have been the case even in ancient times; at any rate, these three examples are not sufficient evidence for pronouncing a verdict The word *su-yi* (not contained in GILES and PALLADIUS) is explained by the *Shi ming* as quoted in K'ang-hi's Dictionary 胡中所名也 "a word employed among the *Hu*". The *T'ang shu* (Ch. 216 下, p 6 a) relates that the great sorcerers *po ch'é pu* 鉢掣逋 (exactly corresponding to Tib. *nba č'e-po*, "great sorcerer"), taking their place on the right-hand side of the Tibetan king, wore, during their prayer ceremonies, head-dresses in the shape of birds and girdles of tiger-skin (巫祝鳥冠虎帶), while beating drums. They certainly were shamans, as indicated by the very Chinese word *wu* and the style of their costume, and it is difficult to see what made BUSHELL (*The Early History of Tibet*, p 101, note 81) think that the *po ch'é pu* would appear to have been a Buddhist. — Among the adherents of the Bon religion, transfiguration of saints into birds, and observation of and divination from birds' voices, are prominent (see *rGyal rabs bon-gyi abyun gnas*, pp. 12, 13; regarding this work compare *T'oung Pao*, 1901, p 24), there the verse occurs, "Omens are derived from birds, trees, the four elements, hills and rocks, from these the voices of the Bon doctrine have arisen."

of this divination table in the colloquial language. This point marks
the fundamental importance of the *document Pelliot*, which thus
becomes the earliest document of the Tibetan vernacular that we
have at present. And it is no small surprise to notice that the
style of this text is thoroughly identical with that of the living
language of the present day. Any one familiar with it will testify
to the fact that he can perfectly understand this Table through
the medium of his knowledge of colloquial Tibetan. The safest criterion for the correctness of this diagnosis is furnished by M. BACOT
himself, who had doubtless mastered Tibetan conversation during
his journeyings in the country, and, I venture to assume, was considerably aided by this knowledge in grasping correctly the meaning of the oracles in the Table. But let us not wholly rely on
such impressionistic opinions, when the text of *K.*, written in the
Tibetan *wén li*, the style of the early Buddhist translators, offers
such a tempting opportunity for comparing analogous sentences of
the two texts. In *T.* (Table) all oracles are concluded with the
plain verb *ston*; in *K.* *rtags-so* or the periphrastic future tense
with *ɑgyur-ro* are used, which do not occur in *T.* In *K.* we read
mes ɑtsᶜig-par ɑgyur-ro, "a conflagration will break out;" the same
is plainly expressed in *T.* by the words *mye ñan žig oñ-bar ston*.
In *K.* *rañ-gi ñc-bo oñ-bar ɑgyur-ro*; the same in *T.* *gñen žig oñ-
bar ston*. In *K.* *rluñ cᶜen-po ɑbyuñ-bar ɑgyur-ro*; the same in *T.*
rluñ ldañ-bar ston, etc. *T.* has the plain and popular words throughout, as *tᶜab-mo* for *ɑkᶜrug-pa*, *bza bca* ("food and drink") for *bza
dañ skom-pa* in *K.*, and, as shown, in the names of the quarters
and divisions of the day. Note that the termination *o* denoting the
stop, and restricted to the written language (discussed farther on), is
absent in *document Pelliot*; there is always *ston*, not *ston-no*, and
at the end of the preface *ston ñin*.

As to the time of the authorship of *document Pelliot*, there can
be no doubt that in the same manner as *K.* it is a production of
the ninth century. This is, first of all, proved by the date of *K.*,
which at the time of its introduction and translation was a live
source impressing the minds of the people, and hence gave the
impetus to further developments of the subject in a manner tangible
and palatable to the nation. Only at a time when the impression
of these things was deep, and the practice of such beliefs was still
fresh and vigorous, was the cast of these notions in the direct and
plastic language of the people possible. Secondly, the antiquity of
our document is evidenced by palaeographic and phonetic traits
(discussed hereafter) occurring in other writings of equal age; it
ranges in that period of language which is styled by the scholars
of Tibet "old language" (*brda riṅ*). Thirdly, there is the circumstantial evidence, the discovery of the document in the cave of
Tun-huang by M. Pelliot (see p. 2).

Let us note *en passant* that the Indian system of crow augury
has been transmitted also to China. H. DORÉ in his excellent book
"Recherches sur les superstitions en Chine" (pt. 1, Vol. II, p. 257,
Shanghai, 1912), has revealed a Chinese text on bird divination
which plainly betrays its connection with *K.* It is based in the
same manner on the division of the day into five parts and on
the local orientation of the cardinal points, eight of which are
given by DORÉ. The presages are identical in tone with those of
K. and *document Pelliot*; we meet predictions of wind and rain,
disputes, threatening of a disaster, reception of a visit, death of a
domestic animal, recovery of a lost object, malady, happy events,
growth of fortune, gifts, arrival of a friend or a stranger, etc.,
without reference to any specific Chinese traits. [1]

---

1) In regard to beliefs in crow and raven in China, the reader may be referred to
DE GROOT, *The Religious System of China*, Vol. V, pp. 638—640. J F DAVIS, *China*,

The Preface to the Table.

As M. BACOT's rendering of the preface accompanying the Table is in need of a revision, I take the liberty to offer a new translation of it, [1]) discussing in the notes the chief points in which my opinion deviates from that of M. BACOT. A Lama, bsTan-pa duldan by name, has been consulted by this gentleman, and has jotted down for him a number of notes, explaining certain phrases in the colloquial language. These notes are reproduced on pp. 447—448 of the essay of M. BACOT, but apparently have not been utilized. Most of the Lama's comments are correct, a few are wrong, and some, though wrong, are yet interesting. Anything of interest in his explanations is embodied in the notes which follow. It may not be amiss to give here a transliteration of the text, in order to enable the reader to compare my translation with it immediately. In M. BACOT's edition, the text (in Tibetan characters) appears as prose; but it is very essential to recognize its metrical composition. The metre is rigorously adhered to in the twenty-nine verses, and is $\perp \cup \cup \perp \cup \perp \cup$, a dactyl followed by two trochees (the signs $-$ and $\cup$ denote merely accentuated and unaccentuated, not long

---

Vol. II, p. 98 (London, 1857); J. DOOLITTLE, *Social Life of the Chinese*, p. 571 (London, 1868). The subject is still in need of special investigation. Crows and ravens are certainly very far from being exclusively birds of ill omen or productive of evil, as DE GROOT is inclined to think; on the contrary, the raven was even the emblem of filial piety, and the appearance of one of red color was a lucky augury, foreboding the success of the Chou dynasty (CHAVANNES, *Les mémoires historiques de Se-ma Ts'ien*, Vol. I, p. 226). Other augur birds, as the mainah (LEGGE, *The Chinese Classics*, Vol V, pt. II, p. 709; WATTERS, *Essays on the Chinese Language*, p. 444; and FORKE, *Lun-héng*, pt II, p. 3) and the magpie, who knows the future (FORKE, *l. c.*, pt. I, p. 358; pt II, p. 126), must be equally taken into consideration.

1) In a bibliographical notice of M. BACOT's study (*Revue de l'histoire des religions*, 1913, p. 122) it is remarked, "Un curieux préambule mériterait d'être tiré au clair; mais il ne semble plus compris aujourd'hui."

and short syllables). [1]) A. H. Francke [2]) observes that in Ladakhi poetry the dactyl is rather frequent, arising from a dissyllabic compound with a suffix. This certainly holds good of all Tibetan dialects and also of the written language. In this composition, all the dactyls are formed by the particle *ni* coupled with a trochaic element. It is curious that all verses are constructed in the same manner, having this *ni* in the third syllable (compare note to V. 19). At the same time, there is obviously a cesura after *ni*. [3])

Text of the Preface.

(The accents denote the metre.)

1 *p͡ó-rog ni myí-i mgón*

2 *dráń-sroń ni lhá-i bká*

3 *byáń ɋbroɋ ni ɋbróń ša-i rkyén*

4 *yúl-gi ni dbús mt͡ɕil dú*

5 *lhá btsun ni bdá (+ a) [4]) skad skyél*

6 *p͡yógs brgyad ni ltéń dań dgú*

7 *ɋáń toń ni t͡ɕábs gsum gsúńs*

8 *gtór-ma ni byá-la gtór*

---

1) On Tibetan metrics compare H. Beckh, *Beiträge zur tibetischen Grammatik, Lexikographie und Metrik (Anhang zu den Abhandl. der preussischen Akademie*, 1908, pp. 53—63) The author justly emphasizes that in the study of Tibetan works the metre is to be investigated in the first line, and that it should be kept in mind in all text-critical and grammatical questions, but he overlooks the fact that this principle had been fully brought into effect by the present writer in *Ein Sühngedicht der Bonpo (Denkschriften Wiener Akademie*, 1900), where textual criticism is fundamentally based on metrical considerations and statistical tables of the various metres.

2) *Sketch of Ladakhi Grammar*, p 7 (Calcutta, 1901).

3) My reading of the text is based only on the edition of M. Bacot, the general accuracy of which there is no reason to doubt. Not having had the privilege of checking it with the original, I do not hold myself responsible for eventual errors which may have crept in there. In V 20, *gsań*, printed in M Bacot's text, is apparently a misprint for *ysań. lhu* (V. 24), for *lteń* (as in V 6).

4) This graphic peculiarity is explained below, under the heading "Palaeographic Traits"

9 *ts⁽ó-ts⁽o ni yóns-su gyís*

10 *lhá-i ni p⁽yáŋ-du ʠbúl*

11 *gráɡs dɡu-r ni ltás myi bltá (+ a) ¹)*

12 *bzáṅ ṅan ni ltás-su ɡsúṅ*

13 *dráṅ-sroṅ ni lhá ʠdsin lá*

14 *lhá ston ni ɡṅén-bai byá (+ a) ¹)*

15 *mʲí sman ni ɡṅén-ɡis ɡsúṅs*

16 *dráṅ žiṅ ni brtán-por stén*

17 *p⁽ó-roɡ ni dɡúṅ-ɡi byá*

18 *ʠdáb druɡ ni ɡšóɡ druɡ pá (+ a) ¹)*

19 *lhá yul ni mt⁽ó-du p⁽yín*

20 *dmyíɡ rno ni sṅán ɡsan bás*

21 *lhá-i ni mán-ṅaɡ stón*

22 *myí rtoɡ ni ɡcíɡ-ma mc⁽ís*

23 *yíd c⁽es ni séms rton cíɡ*

24 *p⁽yóys brɡyad ni ltéṅ daṅ ̣dɡú*

25 *lhóṅ lhoṅ ni bzáṅ-por stón*

26 *t⁽áɡ t⁽aɡ ni ʠbríṅ-du stón*

27 *kráɡ kraɡ ni ríṅs-par stón*

28 *króɡ kroɡ ni ɡróɡ yoṅs smrá*

29 *,iṅ ,iu ni bár ston yín.*

Translation.

1 The Raven is the protector of men,

2 And the officiating priest (carries out) the order of the gods,

4 (Sending him, the Raven) into the middle of the country,

3 Where he has occasion for feeding on yak-flesh in the outlying pasture-lands,

5 The Venerable of the Gods conveys (his will) by means of the
   sound-language (of the Raven).

---

1) This graphic peculiarity is explained below under the heading "Palaeographic Traits"

3

6 When in the eight quarters, making nine with the addition of the zenith,

7 He (the Raven) sounds his notes, the three means (to be observed) are explained as follows:

8 The offering must be presented to the bird (the Raven),

9 And it should be a complete feeding in each instance.

10 (In this manner, the offering) is given into the hands of the god (or gods).

11 As to the omens, they are not drawn from the mere cries (of the Raven),

12 But in the announcement of the omens a distinction is made between good and evil cries.

13 The officiating priest is in possession of the knowledge of the gods,

14 He teaches (the orders of) the gods, and it is the bird who is his helpmate (in this task).

15 The remedies for warding off the demons are announced by the helpmate.

16 Truthful in his speech, he proves trustworthy,

17 For the Raven is a bird of Heaven;

18 He is possessed of six wings and six pinions.

19 Thanks to his visits above in the land of the gods,

20 His sense of sight is keen, and his hearing is sharp.

21 (Hence he is able) to teach (mankind) the directions of the gods.

22 There is for man but one method of examining (the sounds of the Raven),

23 And may you hence have faith and confidence (in his auguries)!

24 In the eight quarters, making nine with the addition of the zenith, (the following sounds of the Raven occur:)

25 The sound *lhon lhon* foretells a lucky omen.

26 The sound *t‘aq t‘aq* forebodes an omen of middle quality.

27 The sound *krag krag* foretells the coming of a person from a distance.

28 The sound *krog krog* announces the arrival of a friend.

29 The sound *,iu ,iu* is an augury of any future event (as indicated in the Table).

NOTES.

V. 1. The raven *p'o-rog* is still called *c'os skyoṅ* (Skr. *dharmapāla*), "protector of religion" (G. SANDBERG, *Hand-book of Colloquial Tibetan*, p. 170). The word *mgon* is employed in the sense of Sanskrit *nātha*. Our text gives the word only in the form *p'o-rog*, while in K. the form *bya-rog* is used exclusively. The latter, as shown by *Mahāvyutpatti*, seems to be the recognized form of the written language, while *p'o-rog* seems to be more popular; the latter occurs, for example, in the Tibetan prose version of the *Avadāna-kalpalatā*, which has been written for children. The distinction of *bya-rog* as "crow," and *p'o-rog* as "raven," is based on the Sanskrit-Tibetan dictionary *Amarakosha* (T. ZACHARIAE, *Die indischen Wörterbücher*, p. 48), where Tib. *bya-rog* is the equivalent of Skr. *vāyasa* ("crow"), and Tib. *p'o-rog* that of Skr. *droṇa* ("raven"), the two words being treated in different stanzas (ed. of Vidyābhūṣaṇa, *Bibl. ind.*, p. 134, Calcutta, 1911).

The word *bya-rog* appears twice in the *Mahāvyutpatti*, section on birds (Tanjur, Sūtra, Vol. 123, fols. 265b, 266a, Palace edition), — first, as translation of Skr. *dhvāṅksha*, "crow" (in *Amarakosha* rendered by *sgra ldan*), where the synonyms *spyi-brtol-can* (the Palace edition writes *sbyi-rtol-can*), "the impudent one," and *k'va*, are added; second, as rendering of Skr. *droṇakāka*, "raven," while the Skr. *kāka* and *vāyasa* are rendered by Tib. *wa* (not noted with this meaning in our dictionaries), evidently an imitative sound, in the same manner as Tib. *k'va*, *k'ra-ta*, and *k'a-ta*, "raven," and *ko-waŋ*, a word expressive of the voice of the raven. In *Se t'i ts'iṅ wên kien* 四體清文鑑 (Ch. 30, p. 25) the following distinctions are made: *k'a-ta* corresponds to *wu-ya* 烏鴉, Manchu *gaha*, Mongol *karya*: Tib. *bya-rog*, to *ts'e-ya* 慈鴉, Manchu *holon gaha*, Mongol *khoṅ karya*: Tib. *p'o-rog*, to *hua po ya* 花脖鴉 ("raven with colored neck"), Manchu *ayan gaha*, Mongol *torok karyā*. In the Appendix to this dictionary (Ch. 4, p. 12) we find Tib. *bya-rog* = *kuan* 鸛 (according to GILES a species of stork), Manchu *sungkeri gôwara* (according to SACHAROV a kind of large horned owl); and Tib. *ka-ka* = *hu k'ua ying* 呼哮鷹, Manchu *hurkun gôwara* In these two cases the Tibetan names seem to be artificial productions made *ad hoc* in order to translate the Manchu words. The *Polyglot List of Birds in Turki, Manchu and Chinese*, published by E. D. Ross (*Mem. A. S. B.*, Vol. II, No. 9, 1909), though in general a useful work, is incomplete in that the Appendix of the Polyglot Dictionary, containing about two hundred more names of birds, has not been utilized at all. For future work of this kind the following suggestions may be offered in regard to the methods of obtaining identifications of bird-names. In my opinion, it is an incorrect procedure, in most cases, to try to identify any Oriental bird-name with a *species* of our own ornithological nomenclature, because our scientific research has made out infinitely more species of birds than there are words for the species in any language: all we can hope for, at the best, is to establish the *genus*, and in many cases we have to be content to ascertain the family. Take, for example, the case of crow or raven, a popular name embracing a large family of birds, *Corvidae*. In 1877 A. David and M. E. Oustalet (*Les oiseaux de la Chine*, p. 366) stated that nearly two hundred species of it were known on the globe, and twenty-seven from China. At present we certainly know many more in addition. (A. Laubmann, *Wissenschaftliche Ergebnisse der Reise von G. Merzbacher, Abhandlungen der bayerischen Akademie*, 1913, pp. 37—42, enumerates ten genera of the family *Corvidae* from the region of the T'ien Shan.) Who can name those twenty-seven species in Chinese? Nobody. Our species are made from points of view which are entirely foreign to the minds of Oriental peoples. They see different "kinds," where our ornithologist may establish one species; and they may have one word, where we are forced to admit different species, and even *genera*; and they may even take the male and female of the same species for two distinct birds. It is further necessary to disillusion our minds regarding the production of the K'ien-lung lexicographers, which must be handled with great caution and pitiless criticism; it teems with artificial makeshifts in Manchu, Tibetan, and Mongol, which are not genuine constituents of these languages, and is vitiated by numerous blunders in spelling, which are to be corrected. The compilers were philologists, not zoologists; and their combinations of bird-names in the various languages offer no guaranty that these refer to really identical *genera*, not to speak of *species*, the greater probability in each case being that the species are entirely different (thus, for instance, as may be determined, in the majority of Tibetan and Chinese bird-names). — Tib. *bya rog* means "the black bird," and *p'o-rog* "the male black one." There is a dialectic form *,o-rog*, *,o-lag* (Walsh, *Vocabulary of the Tromowa Dialect of Tibetan*, pp. 11, 28, Calcutta, 1905), with the prefixed *,a* (here *,o* in consequence of vowel-attraction) forming nouns (Schiefner, *Mélanges asiatiques*, Vol. I, p 362; and Mainwaring, *Grammar of the Róng [Lepcha] Language*, p. 111). In meaning and grammatical formation this *,o-rog* corresponds to Lolo *a-nye*, "the black one," *i. e.* the raven (*T'oung Pao*, 1912, p. 13). The common raven, somewhat larger than the European species, is ubiquitous in

Tibet. Some remarks on it are made by P. LANDON (*Lhasa*, Vol. I, p. 404, London, 1905). According to H. v. SCHLAGINTWEIT (*J. R. A. S.*, 1863, p. 15), it occurs even in the ice-regions of the greatest elevation of the Himālaya: "some of the species of *corvus tibetanus* accompanied us during our ascent of the Ibi Gamin peak up to our highest encampment at 19,326 feet." Of especial interest with reference to the present case is the following observation of THOMAS MANNING, who travelled in Tibet 1811—12 (C. R. MARKHAM, *Narratives of the Mission of George Bogle to Tibet*, etc., p. 249, London, 1876): "Many of the ravens about this lake, and many in Lhasa, emit a peculiar and extraordinary sound, which I call metallic. It is as if their throat was a metal tube, with a stiff metal elastic musical spring fixed in it, which, pulled aside and let go, should give a vibrating note, sounding like the pronunciation of the word *poing*, or *scroong*, with the lips protruded, and with a certain musical accent. The other is similar to that of the ravens in Europe, yet still has something of the metallic sound in it. Whether there be two species of ravens here, or whether it be that the male and female of the same species have each their peculiar note, I cannot say."

V. 2. Who is the *draṅ-sroṅ* (corresponding to Skr. *ṛishi*? The Lama bsTan-pa du-ldan, whose explanatory notes in Tibetan have been published by M. BACOT, on p. 447 comments that the raven *pʻo-rog* is "the raven staying near the head of Vishṇu," and that Vishṇu should be understood by the term *ṛishi*. It is certainly the mythical bird Garuḍa, being the vehicle (*vāhana*) of Vishṇu, which crossed the Lama's mind, and it will be demonstrated farther on (V. 18) that an assimilation between Raven and Garuḍa has indeed taken place in Tibet (in the *Cakuna* of Vasantarāja the Garuḍa commands the *kāka* as an omen-bird: HULTZSCH, *Prolegomena*, p. 41). The beginnings of such an adjustment are visible even in our text when, in V. 17—18, it is said that the Raven is a bird of Heaven, and possessed of six wings and six pinions; he is, in a word, looked upon as a solar bird. Nevertheless, he is not identical with the Garuḍa, and I do not believe that the Lama's explanation is correct. Above all, *draṅ-sroṅ* cannot be identified with Vishṇu or any other god: for he is the person who executes the orders of the gods (V. 2; in this sense, at least, it seems to me, the passage should be understood), who has the knowledge of the gods (*lha gdsin*, V. 13), and who teaches the gods (*lha ston*, V. 14). The Raven is his helpmate (*gñen-pa*, V. 14), and he announces the will and the wishes of the gods transmitted by the divine bird. The *draṅ-sroṅ*, accordingly, is a person with a priestly function; and I should almost feel tempted to propose for the word, in this case, the translation "seer" or "augur." It is the *çakunika* of the Sanskrit texts who is designated also *guru* and *ācārya* (HULTZSCH, *Prolegomena*, p. 6). Moreover, we know that the word *draṅ-sroṅ* has obtained among the Lamas a meaning like "officiating priest, sacrificant,"

JASCHKE (*Dictionary*, p. 261) states *sub voce*. "At present the Lama that offers *sbyin-sreg* [a burnt-offering, Skr. *homa*] is stated to bear that name. and while he is attending to the sacred rites. he is not allowed to eat anything but *dkar-zas* [white food, like milk, curd. cheese. or butter]." Inevitably we must assume that our Table was not directly used by the laity, but that it was placed in charge of a priest who had due control over supernatural events. The layman who had encountered the vision of a raven applied to him for the proper oracle to be ascertained from the chart, and particularly, if necessary, for the making of the required offering, which was a ritual act along established rules. The Lama who fulfilled this function was called the *draṅ-sroṅ*. The origin of this word is explained in the work *sGra sbyor* (quoted above, p. 19; Tanjur. Sūtra, Vol. 124, fol. 6b) by the sentence *kāya-rāk-manobhir-ṛju-çete iti ṛishi*. rendered into Tibetan thus: *lus daṅ ṅag daṅ yid draṅ-por qnas-śiṅ sroṅ-bas-na draṅ-sroṅ c̀en-po źes blags*, "he who in regard to his body (actions), speech, and heart, remains straight and keeps them straight, is designated a great Ṛishi." Hence it follows that in the minds of the Tibetans the compound *draṅ-sroṅ* is formed of the words *draṅ-po* (Skr. *ṛju*. "straight," in the literal and moral sense) and the verb *sroṅ-ba*, "to straighten." and that the Tibetan interpretation is "one who is straight, upright in his conduct." Another definition given in the same work is "one who is possessed of knowledge" (*śes-pa-daṅ-ldan-pa*). The notion of "hermit" given in our Tibetan dictionaries is apparently not implied in the Tibetan definitions. It will thus be noticed that the literal interpretation of the word, "one who straightens out affairs in a straight manner." could result in the development of the notion "one who straightens out affairs relating to sacrifice, augury or divination."

V. 3. Tib. *byaṅ ḥbrog* is identified by M. BACOT with the well-known term *byaṅ t'aṅ*, "the northern table-lands." The two expressions are evidently synonymous (compare VASILYEV. *Geography of Tibet*, in Russian, p. 11, St. Pet., 1895). *Byaṅ ḥbrog* appears as one of the thirteen districts assigned by the Mongol emperors to the hierarchs of Sa-skya (*dPag bsam ljon bzaṅ*, p. 159, l. 1); but I do not believe that a definite locality in the geographical sense is here intended, any more than I believe that the word *dbus* ("centre") in the following verse need refer particularly to the Tibetan province of that name. The term *byaṅ t'aṅ* is also a general designation for uncultivated pastoral high lands (the proper meaning of *t'aṅ* is not "plain, steppe," as given in our dictionaries, but "plateau"), in opposition to *roṅ t'aṅ*, the low lands of the valleys. The former is the habitat of pastoral tribes; the latter. the seat of the agriculturists. The first element in *byaṅ t'aṅ*, in all likelihood, was not originally the word *byaṅ*, "north," but the word *ljaṅ*, "green" (*byaṅ* and *ljaṅ* are both sounded *jaṅ*: *ljaṅ t'aṅ*, "green plateau," is the name of a province in mNa-ris .jK'or-gsum, according to H. v. Schlagintweit. *Glossary of Tibetan Geographical Terms. J. R. A. S.*, Vol. XX, 1863, p. 13): for in Ladakh, for instance, the people apply the word *byañ t'añ* to the district of Ru-tog, situated on their eastern border, in the sense that it is more bleak and unreclaimed than their own sheltered and less elevated valleys (compare H. Strachey, *J. A. S. B.*, Vol. XVII, 1848, p. 331). The same evidently holds good for our text. for, in understanding *byañ abrog* literally, it would be unintelligible why the Raven despatched into the centre of the country should be supposed to gain his livelihood in the pastures of the north. The "centre." it should be understood, may be any settlement in Tibet with a sedentary farming population; and the term *byañ abrog* may refer to any nomadic district in its proximity where the Raven stands a better chance for his food than among the husbandmen. The word "centre" is probably chosen in view of the nine quarters which come into question for the Raven's flight; he has to start from a centre to make for the various directions. In regard to man, the cultivated land is conceived of as being centrally located, and surrounded on its outskirts by the wild mountains with their grassy plateaus suitable for cattle-raising. The tribal and social division of the Tibetan people into these two distinct groups of agriculturists and cattle-breeders meets its outward expression in the juxtaposition of the word-groups denoting „valley" and „mountain" („pasture," „plateau"), the one pertaining to cultivation, the other to everything uncultivated or of wild nature. The "valley pig" (*luñ p'ag*) is the domestic pig, a sedentary animal found only among the farmers, but never among the nomads; while the "mountain pig" (*ri p'ag*) is the wild boar: hence *ri* and abbreviated into the prefix *r-*, with predilection, enters into the names of wild animals (*W. Z. K. M.*, Vol. XIII, 1900, p. 206).

In regard to the yak-flesh we may remember the passage of the *T'ang shu* (Bushell, *The Early History of Tibet*, p. 7): "When they entertain envoys from foreign countries, they always bring out a yak for the guest himself to shoot, the flesh of which is afterwards served at the banquet." In the legends of the Buryat, the crow is invited by people to take part in a meal furnished by a slaughtered ox (Changalov and Zatopl'ayev, Бурятскія сказки и повѣрья, pp. 17, 21, Irkutsk, 1889).

V. 5. Tib. *lha btsun*, correctly translated by M. Bacot "le dieu vénérable," would correspond to Skr. *devabhadanta*. It is notable that the coming of *lha btsun* is the very first prediction appearing in the Table when the raven's voice sounds in the east during the first watch. His name appears again in Table VII, 6, where it is said that "the helper, or the assistance of the Venerable One (*btsun-pai-gñen*), will come." (I do not believe with M. Bacot that these words mean „un parent de distinction." In fact, M. Bacot sides with me in this opinion, for in Table V, 3, he very aptly and correctly renders the term *gñen lha* by „dieu protecteur"). The helper is referred to in V. 15 (*gñen*), and the expression *gñen-bai bya* ("the helping bird") in V. 14 leaves no doubt that the raven is meant. It seems futile for the present to speculate on the nature of this deity called *lha btsun*. All we may infer from this text is, that he seems to be a supreme god presiding over the *lha*, that he resides in the region of the gods (*lha yul*, V. 19), and that he reveals his will to mankind through the Raven, his messenger, whom he sends down on earth. On the whole, I am inclined to regard this deity as a native Tibetan concept, not as an adaptation to an Indian notion: possibly he is identical with the Spirit of Heaven 天神 invoked by the Tibetan shamans, according to *Kiu T̓ang shu* (Ch. 196 上. p. 1b). — As regards the name *lha btsun*, an analogous expression is met in Taoism in the name of the deity T̓ien tsun 天尊 (or Yüan shi T̓ien tsun, the first of the three divinities forming the trinity of the Three Pure Ones 三清): Tib. *lha* and Chin. *t̓ien* correspond in meaning, both serving for the translation of Skr. *deva*; and Tib. *btsun* and Chin. *tsun*, as already recognized by ABEL-RÉMUSAT and SCHIEFNER (*Mélanges asiatiques*, Vol. I, p. 340), are identical words.

M. BACOT translates, "Le dieu vénérable accompagne la parole qu'il prend avec lui," by taking *bda* for the verb *bda-ba*. Even granted that the latter could have this meaning, the construction of the sentence remains ungrammatical, and the rendering gives no sense. In these ancient texts we must be mindful of the fact that spellings at variance with modern usage occur, or, in other words, that different phonetic conditions are fixed in writing. There is no difficulty in seeing that *bda* here stands for the common mode of writing *brda*: and *brda skad* is a very frequent compound. which, as correctly interpreted by JÄSCHKE, means (1) language expressed by signs or gestures, (2) language expressed by words. Here it refers to the prophetic sounds or language of the Raven by means of which the Venerable One of the Gods conveys (*skyel*) his will and wishes.

V. 6. In the commentary of the Lama (p. 447), where the verses of the text. which are explained, are repeated in larger type, this verse terminates with the word *bcu*. so that the Lama brings out ten quarters, adding the nadir ("the region of the *klu*, the land below") as the tenth; but this is evidently a slip which occurred in the copy taken by or for the Lama.

V. 7. The expression *,an ton* presents some difficulties, as it is evidently an archaic and antiquated term not recorded in our dictionaries. The Lama maintains silence about it. M. BACOT has tentatively proposed to take it in the sense of *,an dan-po*, and renders the sentence, "Le meilleur est d'énoncer les trois moyens." But this is an entirely un-Tibetan way of speaking. and M BACOT's conception of the sentence contradicts the iron rules of Tibetan word-position. Such a translation would only be permissible if the reading were *t'abs gsum gsuñs ,añ dañ-po* (*red*). Aside from this, the identification of *,añ toñ* with *,añ dañ* is hardly acceptable: it is not supported by any native dictionary, nor can it be upheld by any phonetic law. Further, the Sanskrit-Tibetan hybrid, in the written language usually *,añ-gi dañ-po* (more rarely *,añ dañ*), has only the meaning of the ordinal numeral "the first" (in the enumeration of a series), while in the sense of "first quality, best," it is a very vulgar expression of the colloquial language, about the equivalent of Pidgin-English "number one." A few considerations may place us on the right track as to the meaning of the phrase. The preceding verse, "in the eight quarters etc.," demands a verb; in looking up the parallel passages of *K.*, we notice that each of the determinations of the quarters is followed there by the words *skad sgrogs na*, "if (the crow) sounds its voice," and this is what is apparently required and intended in this passage. In this case we recognize in *toñ* the verb *gtoñ* (compare *sod* for *gsod* in Table II. 8: VI, 2, and the phonetic remarks below). which, as shown by JÄSCHKE (*Dictionary*, pp 19a, 209a), is indeed used in this sense in Ladākhi: *skad tañ-ce*. "to utter sounds;" *ku-co. bó-ra tañ-ce*, "to raise, to set up a cry." But the phrase in question occurs also in writing, like many others given by JÄSCHKE as dialectic expressions: a number of those could be compiled from the prose version of *Avadānakalpalatā*. The word *,añ* (probably derived from the Sanskrit particle *añga*, pw. "anrufend oder auffordernd") means "cry, clamor." SARAT CHANDRA DAS (*Dictionary*, p. 1347) cites an example of this kind. without translating it, in the sentence *mi-yis bos kyañ ,añ mi k'ug*, which evidently means, "Although the man called, his cries did not draw any attention." GOLSTUNSKI, in his Монгольско-русский словарь (Vol. 1, p. 7b), assigns to Mongol *añ*, which has several other meanings, also the significance "shouting of fighters, cries of camels and donkeys." It is the same thing when JÄSCHKE quotes *,añ* as an interjection with the meaning "well, then! now, then! eh bien!" It is an exclamation Another use of *,añ* not noticed heretofore seems to be traceable to the same origin. *,Añ* appears as a particle joined to the imperative with or without *cig*, as well as to the prohibitive. In *Bya c'os* (see note to V. 28), p. 39, we meet five times with *šog ,añ*. In *sLob gñer byed ts'ul-gyi bslab bya le ts'an gñis*, a small work published by the monastery Kumbum (*sKu ḥbum*), we have *sgrims šig ,añ* (fol. 6), *gnas-par gyis šig ,añ* (fol. 7), *ma byed ,añ* (fol. 10), *ma rgyugs ,añ* (fol. 14), and many other examples. The meaning seems to correspond to French *donc* (German *doch*) in connection with an imperative, and this application seems to be derived from the original significance "cry, exclamation." In the case above, *,añ* is used as a noun synonymous with the word *skad* of *K.*, and refers to the cries of the raven which he emits (*gtoñ*) in his flight toward the various quarters. The phrase *,añ toñ* linked to the preceding verse is the psychological subject governed by *t'abs gsum gsuñs*: the augury derived from the sounds of the raven voiced in the eight quarters is explained as consisting of three means or modes of procedure. The explanation is inspired by the Venerable of the Gods. The three means are the offering (*gtor-ma*, Skr. *bali*), the discrimination between good and evil cries (and accordingly auguries), and the oracle proclaimed by the priest, with his superior knowledge of the supernatural.

V. 8. Tib. *gtor-ma gtor-ba* (as *ltas lta-ba* in V. 11) is a hendiadys favorite in Tibetan and other Indo-Chinese languages. A. CONRADY (*Eine indo-chinesische Causativ-Denominativ-Bildung*, p. 81, Leipzig. 1896) has given a number of good examples of this kind: others occur in *Ein Suhngedicht der Bonpo*, *l. c.*, p. 27. Compare the synonyms of the crow given in *Amarakosha* (*l. c.*), — *balipushṭa* and *balibhuj*, — and the Tibetan synonyms *gtor-mas rgyas* and *gtor za* in the "Dictionary of the French Missionaries," p. 86. Several others enumerated in the latter may be explained from *Amarakosha*: as *gci-med* = *arishṭa*; *gzan gso* = *parabhṛid*: *lan cig skyes* = *sakṛitpraja*, which accordingly does not mean "né une seule fois," but "one bearing young but once a year;" *bdag sgrog* (in the translation of *Amarakosha*, *sgrogs-pai bdag-ñid-can*) = *ātmaghosha*.

V. 9. M. BACOT translates, "Plus il y en a d'espèces, mieux cela vaut." He seems to have thought of *tso* ("number, host"), but, as already remarked by JÄSCHKE, this word hardly ever stands alone; in fact, it is only used as a suffix denoting a plural. As shown by the context, *tso* is written for *gtso* ("to feed, nourish"). and the duplication indicates the repeated action. Also the Lama, as shown by the wording of his comment, takes *tso* as a verb by saying that all birds *tso-nas* eat the offering; but, as he merely repeats *tso* in the same spelling as in the text, it is not clear in which sense he understands the verb. *Gyis* certainly is the imperative of *byyid-pa*. V. 8 and 10 have been correctly rendered by M. BACOT.

V. 10. The Lama understands this verse, "The raven is a bird soaring in the sky" (*nam ldiṅ-gi bya*), and possibly thinks again of the Garuḍa. It seems to me that the Raven as a bird of Heaven is understood to be the messenger sent down from heaven, as previously set forth, and it implies also that he is of celestial origin, as specified in V. 19.

V. 11. Tib. *grags* is not used here in the sense of „glory," but with the literal meaning "cry, outcry, clamor:" it is derived from the verb *s-grog-pa*, ("to call, to shout"), which is identical with Chinese *kiao* 叫 ("to call out: the cries of certain animals and birds"). in the same manner as Tib. *s-grog-pa* ("to bind") = Chin. *kiao* 絞 ("to bind"). and Tib. *ḥ-grogs-pa* (from *grogs*, "friend, to be associated") = Chin. *kiao* 変, "to be united, friendship, intercourse" (compare A. CONRADY, *Eine indochinesische Causativ-Denominativ-Bildung*, pp. VII, VIII, Leipzig, 1896). *Hua i yi yü* (Hirth's copy in the Royal Library of Berlin, Ch. 11, p. 67b) correctly renders Tibetan *gray* by *ming* 鳴. — Tib. *dgur* is not the word "crooked," as M. BACOT thinks, but is to be analyzed into *dgu-r*, terminative of *dgu* ("nine, many", and particle expressing the plural (FOUCAUX, *Grammaire de la langue tibétaine*, p. 27; A. SCHIEFNER, *Ueber Pluralbezeichnungen im Tibetischen*, § 23, in *Mém. Acad. de St.-Pétersbourg*, Vol. XXV, N°. 1, 1877). The question may be raised whether *grags-dgu* denotes the various kinds of cries of the raven, of an indefinite number, or whether exactly nine sounds are understood. It would be rather tempting to assume the latter possibility, and to set the nine sounds in relation with the nine quarters; but at the end of the Preface only five sounds of the raven are enumerated in accordance with *K*. Again, the fact that this section of the Preface is preceded by the verse, "In the eight quarters, making nine with the zenith," leads one to think that, besides the series of five, a series of nine sounds, corresponding to the nine quarters, may have simultaneously existed, and that the matter is confused in this text. A positive decision on this point, however, cannot be reached, and I prefer to regard *dgu* as a mere designation of the plural.

V. 12. As plainly stated in the first horizontal column of the Table, an offering is necessary whenever the voice of the Raven sounds ill luck. M. BACOT translates this verse, "Le bon et le mauvais, après qu'on l'a vu, qui en parle?" He accordingly accepts *su* as interrogative pronoun, while it is evidently the particle of the terminative belonging to *ltas*. Such slips are certainly excusable, and have been committed by other translators. Thus, for example, E. SCHLAGINTWEIT (*Die Lebensbeschreibung von Padma Sambhava* II, *Abhandl. der bayerischen Akad.*, 1903, p. 547) took the final *s-o*, denoting the stop, as the noun *so* ("tooth"), and translated the sentence *paṇḍita-rnams kun-gyis ma p'ub grags-so mts'ams ḫbyed-pas*, "All pandits praised him as the powerful one of the Abhidharma; if a tooth is hollow, its removal is desirable." There is nothing to this effect in the Tibetan words, which simply mean, "He is known under the name 'the One Unexcelled by all Pandits;' he began solitary meditation," etc. In the same author's *Die tibetischen Handschriften der k. Hof- und Staatsbibliothek zu München* (*Sitzungsberichte der bayerischen Akad.*, 1875, p. 73) occurs, in the title of a book, "the tooth of the fulfilment of the great Lama Rig-ḫdsin;" the Tibetan *bskaṅ-so*, of course, is a mere graphic variant of *bskaṅs-so*, and means "the fulfilment of vows."

V. 14. M. BACOT takes *gñen-bai bya* in the sense of "devoir des parents." It may be granted that these words could have such a meaning, though as a rule *byu-ba* retains its suffix, when it has the rôle of the word assigned to it by M. Bacot. But the point is that such a viewing of the matter has no sense in this context. I should think that *bya* is simply "bird," as it occurred in V. 8; while the suffix *bai* or *pai* sufficiently indicates the verbal character of *gñen*, "to help, assist" (in its sense somewhat synonymous with *myon*. V. 1). The whole term is to be construed like a Sanskrit Bahuvrihi: the *Dran-sroñ* is one having the bird as a helper. The fact that the helper refers to the Raven is manifest also from the following verse.

V. 15. M. Bacot translates, "remède de douleur, parole des parents." The meaning of *gñen* (V. 5) has been explained. The construction of the sentence is simple: in regard to the remedies, they are announced or explained by the helper (the Raven). The only difficulty is presented by the word *mu* preceding *sman*. Also M. Bacot has clearly seen that the word *mu* ("border, limit," etc.) cannot here come into question. In my opinion, we have to apply the rule laid down under V. 5. that a prefix has been dropped in *mu*: and I should like to propose to read *dmu* or *rmu* "evil demon," which befits the case very well: *dmu* is a demon causing blindness, dropsy, and other infirmities. In the Table (X, 4) the coming of demons is indicated as an oracle, and the augur is certainly obliged also to announce the means of escaping the evil effects or consequences of an oracle. In a wider sense, *mu sman*, accordingly, signifies the remedies releasing the person concerned from any threatening calamity in consequence of a prediction.

V. 16. This verse is explained by our Lama commentator (p. 442), "He who does not tell lies is reckoned as good by all men," which fairly reproduces the general sense, while the translation of M. Bacot is untenable. He takes *dran žiñ* in the sense of "en conduisant," and accordingly derives it from the verb *ḥdren-pa*: but "en conduisant" could be expressed only by *ḥdren žiñ*. The descriptive particle *ciñ* is hardly ever joined to a future tense (no example from literature is known to me), usually to a present tense, in the majority of cases to an adjective, rarely to a past tense (compare the examples in the grammars of Foucaux, p. 19, and Jäschke, p. 56). The chances, as a rule, are that the word preceding *ciñ* is an adjective with verbal force. As such it is used here, *dran* standing for *dran-po* (any suffixes may be dropped in verse), "honest, upright, truthful," and this attribute refers to the truthful sound-language of the raven. The phrase *brtan-por ston* cannot mean „on montre sa fermeté:" *ston-pa* with the terminative means "to show one's self as, to prove as, to furnish proof of being," etc. The word *brtan-po* or *brtan-pa* (also *rton-pa*, as in V. 23. *brton-pa*). with or without *yid*, means "to place confidence in a person" (Jäschke. *Dictionary*. p. 215a); *brtan-po*. more specifically, refers to a permanency of condition in which a person continues to enjoy the confidence once obtained, while *brtan-pa* signifies a temporary action. It occurs in *Saddharmapundarika*, where FOUCAUX (*Parabole de l'enfant égaré*, p. 54, Paris, 1854) renders it by "homme digne de confiance," and in *Bharatae responsa* (ed. SCHIEFNER, p. 46: *fidem habere*). The sense of this verse, accordingly, is, "(Le corbeau), en disant la vérité (ou, parce que ses augures sont véritables), se prouve digne de confiance."

V. 18. The two Tibetan expressions would theoretically correspond to Skr. *shatpaksha, shatparna*, but such Sanskrit terms do not exist. The whole idea apparently is not Indian. (M. BACOT's rendering, "six plumes devinrent six ailes," is not justified by the text, and yields no significance.) Here we must briefly touch on the religious ideas revealed by our text. Our knowledge of Tibetan folk-lore, and particularly of that of the past, is certainly still so scanty that for some time to come all speculations on such-like subjects must remain of a more or less tentative character. But with all their brevity, the twenty-nine verses of this Preface contain a good deal, and also, from the viewpoint of religious history, present a document of some importance. Above all, we notice that the ideas expressed by it are absent from the text of *Kākajariti*, and aptly fill the gap which we were obliged to point out there. It is the rôle of the Raven as a bird of divination which is here depicted. At first sight it is tempting to regard this description as breathing a certain Tibetan spirit. We know that the Raven plays a part in the sacred pantomimic dances of the Tibetan Lamas performed at the time of the New Year: he makes attempts at stealing the strewing oblation (*gtor-ma*), and is driven away with long sticks by two Atsara, skeleton ghouls, a skeleton being designed on their white cotton garbs, and their masks having the appearance of skulls. The mask of the Raven, though it is styled *bya-rog* by the Tibetans, has not at all the form of this bird, but that of the Indian Garuda, with big curved and hooked beak (while the raven's beak is straight). A specimen in the Field Museum, where are complete sets of Tibetan masks, shows the Raven's mask of dark-green color, with red bill, a blue eye of wisdom on his forehead, flamed eyebrows, and gold painted flames protruding from his jaws. The entire make-up is so unlike a raven, that the Chinese workman of Peking who manufactures the masks for the Lama temples of the capital styles it a parrot (*ying wu*). In the *Veda* the eagle carries off the *soma* or *amrita* for Indra, and in the *Kathaka* it is Indra himself who in the form of an eagle captures the beverage (A. A. MACDONELL, *Vedic Mythology*, p. 152; and H. OLDENBERG, *Die Religion des Veda*, p. 176). The *Mahābhārata* (*Āstikaparvan* XXXII) tells how Garuda, in order to take hold of the *amrita*, defeats the host of the Deva, kills the guardians, and extinguishes the fire surrounding the *amrita*. This Indian tradition seems to me in some way or other to be responsible for the cast of the Raven in the Tibetan sacred dances, and for certain elements of a sun-bird attached to the Raven in our text. The Indian source which has transmitted these ideas to Tibet certainly remains to be pointed out. If the raven was made the substitute of the Garuḍa in Tibet, this may be due to the world-wide reputation of that bird as a clever pilferer. The ancients regarded him as an all-round thief, particularly of sacrificial meat. In the sacred groves of Greece many ravens subsisted on the flesh which they seized from the altars and consumed in the trees (O. KELLER, *Die antike Tierwelt*, Vol. II, p. 93). The Kachin of Burma look upon the raven as the very first thief who sub-sequently was duly imitated by man (GILHODES, *Anthropos*, Vol. IV. 1909 p. 134).

On the other hand, the Tibetan mask of the Raven reminds us of the first of the seven degrees of initiation which the mystic successively assumed in the Mithraic cult. — the name of Raven (*corax*): the others being Occult, Soldier, Lion, Persian, Runner of the Sun, and Father (F. CUMONT, *The Mysteries of Mithra*, p. 152). CUMONT regards these as animal disguises going back to a prehistoric period when the deities themselves were represented under the forms of animals, and when the worshipper, in taking the name and semblance of his gods, believed that he identified himself with them. To the primitive titles of Raven and Lion others were afterward added for the purpose of attaining the sacred number seven, the seven degrees of initiation answering to the seven planetary spheres which the soul was forced to traverse in order to reach the abode of the blessed. It is in the Tibetan mystery-plays that we find the masks of the Raven and the Lion. In the belief of the Persians, the Raven was sacred to the God of Light and the Sun. On the Mithraic monuments he sits behind Mithras, sacrificing a bull, and, according to O. KELLER (*Die antike Tierwelt*, Vol. II, p. 104), the idea of the sacred Ravens assigned to Helios in Thessalia may have originated from Persia. The "six wings and six pinions" assigned in our text to the Raven in his quality as a bird of Heaven cannot be accounted for by any Indian notions, and it may well be doubted whether this feature is due to a creation of Tibetan mythology. It seems to me that also this trait savors of Mithraic elements, somehow inspired by the grotesque monsters of West-Asiatic imagination, par-ticularly the winged griffins (see, for example, PERROT and CHIPIEZ, *History of Art in Persia*, Figs. 71, 72, 158, also 187: another Tibeto-Mithraic parallel is pointed out by GRÜNWEDEL. *Baessler-Archiv*, Vol. III, 1912, p. 15). The Per-sian influence on Tibetan religion is established, though it remains for the future to work up the details of the problem (GRÜNWEDEL, *Mythologie des Buddhismus*, p. 205. note 38). The historical foundation of the Bon religion of Tibet, as shown by me (*T'oung Pao*, 1908, p. 13), is Persian The most significant feature revealed by this Preface, as already pointed out, is the Raven's function as the messenger of a god, so that his predictions appear as the expression of divine will. The Raven as a heavenly messenger is conscious of his presages The same idea is expressed by PLINY (*Nat. Hist*, X, 12. § 32:

ed. MAYHOFF, Vol. II, p. 229): corvi in auspiciis soli videntur intellectum habere significationum suarum.

V. 19. M. BACOT renders this verse, "La terre des dieux arrive au ciel." He has apparently been led into error (the same matter occurs in V. 3, 6, 7, 11, 12, 18) by assuming that the particle *ni* distinguishes the subject of the sentence. This was the erroneous view of I. J. SCHMIDT. which was refuted by SCHIEFNER (*Mélanges asiatiques*, Vol. I. p. 384). *Ni* is simply an emphatic particle added to any word or group of words in order to single them out (JÄSCHKE, *Tibetan Grammar*, p. 66). It may follow any adverb and any phrase expressing space or time, the genitive, dative, instrumentalis, or locative; and in metrical composition, it may take any place where a syllable is to be filled in (a peculiar case not discussed in our grammars is *na ni* forming the unreal conditional sentence). There are assuredly numerous cases where stress is laid upon the subject by the addition of this particle, then corresponding in meaning to Japanese *wa* and *ga:* but this rule must not be turned into the opposite, that wherever *ni* is employed, the subject is hinted at. Our text is very instructive as to the application of *ni*, since in each verse it occurs in the third syllable with intentional regularity, and lends to the style a somewhat oracular tinge. First of all, it is employed because of the metre to produce a dactyl in the first foot of each verse: simultaneously, certain words, as *p̓o-rog* and *dran̄-sron̄* in V. 1 and 2, are singled out with strong emphasis by its presence. In V. 4, 10, 11, 16. 21, 23, it is entirely superfluous and merely a rhythmic factor. As to V. 3 and 19, we should have *na* in its place in a prose text, in V. 9 *nas*, in V. 18 *dan̄*. If the author should have pinned his faith to a purely trochaic metre, which is the most frequent in Tibetan, he could easily have accomplished his purpose by dropping all the *ni*, and yet the sense of his words would have remained exactly the same.

V. 22. M. BACOT renders this verse, "Homme et raison ne font pas un." Whatever this may mean. it is evident that the Tibetan people do not indulge in metaphysical speculations of that sort, and that such a sentence has no *raison d'être* in this context. We notice that this text is a plain account of the Raven as a bird of augury, and that everything logically refers to it in a palpably concrete manner. For this reason we are justified in seeking the interpretation of the verb *rtog-pa* in the same direction. We met it in the Tibetan title of the *Kākajariti*, where it is used in regard to the "examination" of the sounds or cries of the crow, and I believe it is here used in exactly the same sense. The word *myi* preceding it is in parallel opposition to *lhai* of the previous verse, and, like the latter, may be construed as a genitive ("examination of the auguries on the part of man") or in the sense of a dative depending on *me̒is* ("to man .. there is"). The particle *ma* can, of course, be looked upon

as the negation, as M. BACOT considers it, but this does not make sense. I prefer
to read *gcig-ma*, "unity, oneness," (regarding *-ma* with words denoting space,
time, etc. see SCHIEFNER, *Mélanges asiatiques*, Vol. I, pp. 385, 386), and understand the verse to the effect that there is for man only one and the same
method of examining the forebodings of the Raven, that is, the method laid
down in the Table. This interpretation seems to be in keeping with the spirit
of the text. If the Raven is a heavenly bird, a messenger of the gods, and the
herald of their commands, if he is truthful and trustworthy, it is logical that
there should be but one way of studying and interpreting his notes. The comment
furnished by the Lama is quite in harmony with this point of view. He likewise understands the words *gcig ma mc'is* in a positive sense by transcribing
them *gcig gdra byed*, "make like one, might be one," and his note *mi t'ams-
cad rtog-pa ni* sufficiently indicates that these words mean an examination
referring to all men, and that *rtog-pa* is not intended for *rtogs-pa*, "knowledge, perception." The copula *mc'is* belongs to the *estilo culto*.

Analogous examples for the use of *gcig-ma* are *rkaṅ gcig-ma* "one-
footed," *rkaṅ gñis-ma* "two-footed" (SCHIEFNER, *Mélanges asiatiques*, Vol. III,
p. 12); *ral gcig-ma* = Skr. *ekajaṭā* (P. CORDIER, *l. c.*, pp. 122, 194, 195); *skad
cig-ma* "a moment," *skad gcig-ma* "instantaneousness" (in the philosophy
of the Sautrāntika: VASILYEV, *Der Buddhismus*, p. 305); and *skad cig-ma-
ñid*, "the short (instantaneous) duration of life" (in the commentary of *Suhṛil-
lekha*). The title of a small treatise describing the offerings to Vajrabhairava
is *drug bcu-pa-ma*. The title *ratnamālā* is once translated in the Tanjur *rin
c'en p'reṅ-ba-ma* (usually *p'reṅ-ba*), where *ma* is to express the feminine gender
of Sanskrit: and so it may be concluded that the influence of Sanskrit is
responsible also for the other cases of this kind.

V. 23. M. BACOT translates, "Croyance et confiance de l'esprit font un."
This is in contradiction to an elementary rule of Tibetan grammar. The
final *cig* does not mean "one," but is the well-known sign of the imperative :
besides, the form *rton* is an imperative in itself (from *rton-pa*), and also the
Lama has plainly indicated another imperative form, *t'ob cig*. The phrase *sems
rton (rton)* in this passage corroborates the interpretation given for *brtan-po*
in V. 16. *Yid c'es* may be taken as *adverbialis* ("with faith, faithfully"), or
as a verb to be supplemented by the following *cig* ("have faith and" ...). The
Lama explains this faith as "prayer to the gods" (*lha-la gsol*), which is hardly
necessary. Both faith and confidence, first of all, refer to the Raven and his
auguries, as presented in the Table: and faith in him naturally implies faith
in the gods who sent him.

V. 27. In Table IV, 1, M. BACOT translates the sentence *riṅs-pa źig oṅ-
bar ston* by "indique qu'une personne vient en hâte." But *riṅs-pa źig* is the

subject of the sentence, and means "a distant one, a person coming from a distance." True it is, *riṅs-pa* means also "swift, speedy." The spelling, however, must never lead us astray: it is here intended for *riṅ-ba*, meaning "distant" as to space and time, hence "long" (the Kien-lung Polyglot Dictionary confronts it with *yuan* 遠 and Manchu *goro*). The word *riṅs-par* in V. 27, in my opinion, contains an allusion to the passage of the Table quoted. M. Bacot's translation, "est signe de rapidité," has no meaning. Also the Lama is on my side when he interprets *mi yoṅ*, "a man will come." — Compare *Subhāshitaratnanidhi* 66 (ed. Csoma. *J. A. S. B.*, Vol. VII, 1912, Extra No., p. 116): *rin c̆en gliṅ-du riṅ-nas ḥdu*, "they flock from a distance to the Island of Jewels."

V. 28. The foretelling of the arrival of a friend, in all likelihood, is fraught with a deeper significance than may appear on the surface. In the Table (VIII, 6; and X, 3) we find twice the prophecy of a meeting with a great friend. The word used in each case is *grog*, which is pronounced and written also *rog*, *rogs*. Now, the Tibetans, for this reason, pun the word (*bya-*)*rog*, "raven" with *rog*, *grog*, "friend." An excellent example of this fact is furnished by the interesting little work *Bya c̆os rin c̆en ap̆reṅ-ba*, "The Precious Wreath (*ratnamālā*) of the Teachings of Birds," the text of which has been edited by S. Chandra Vidyābhusan under the title *Bya-Chos or the Religion of Birds: being an Old Tibetan Story.* Calcutta, 1903 (40 p.). Jäschke (*Dictionary*, p. 372) mentions this graceful work, styling it also *Bya skad*, "Bird Voices," or *Bya sgruṅs*, "Bird Stories," and characterizing it as a book of satirical fables, in which birds are introduced as speaking. I am under the impression that no satire is veiled under this text, at least not in the edition quoted, and that it belongs to the class of Nītiçāstra, as indicated by its very title. In order to teach the birds the tenets of the Buddhist doctrine, Avalokiteçvara transforms himself into the king of the birds, the large cuckoo (*kokila*), and finally attracts the attention of the other birds by his meditation carried on for many years in a sandal-tree. The birds congregate around him, and each recites in its language a number of stanzas in praise or support of Buddhist ethical teachings (compare *Mantic Uttair ou le langage des oiseaux*, poème de philosophie religieuse traduit du persan de Farid Uddin Attar par M. Garcin de Tassy, Paris, 1863, and the same author's *La poésie philosophique et religieuse chez les Persans d'après le Mantic Uttair*, Paris, 1864; this Persian work has doubtless received its impetus from that genre of Buddhist literature, as I hope to demonstrate in a future translation of the Tibetan book). The *Bya c̆os* is not a translation from Sanskrit, but a witty Tibetan production, though fundamentally based on Indian thought; it is full of fun and pun. The verses recited by the birds terminate in a refrain, and this refrain consists of a catchword forming a pun upon the name of the par-

4

ticular bird. The snipe (*tiṅ-tiṅ-ma*), for instance, puns upon *gtiṅ riṅ*, "a deep abyss," in this style: "The ocean of the misery of Saṁsāra is a deep abyss, the hell of Māra is a deep abyss," etc. Or the jack-daw (*skyuṅ-ka*) puns upon the verb *skyuṅ-ba*, "to leave behind:" the owl (*ug-pa*), on *u-sdug* (= *u-t'ug*), "destitute:" the ptarmigan (*goṅ-mo*), on *go-dka*, "difficult to understand." And the watchword of the raven (*p'o-rog*) is *grogs yoṅ grogs yoṅ*, "a friend will come, a friend will come," exactly as in the above verse of the *document Pelliot*. In this case, the coming of the friend is interpreted in the figurative sense of Buddhist blessings. The Raven speaks thus:

"When moral obligations have been fulfilled, happiness will come as a friend.

"When alms have been distributed, wealth will come in the future as a friend.

"When religious functions have been performed, thy tutelary deity will come as a friend.

"When the vows are pure, the delight of heaven will come as a friend.

"When the sacrificial feast was vigorous, the Protector of Religion (*dhar mapāla*) will come as a friend.

"When thy achievements correspond to the length of thy life, Buddha, in the future, will come as a friend.

"This *siddhi* of 'the friend who will come' take to heart and keep in mind!"

The coming of the friend appears also in *K*. (I, south; III, north), and from the viewpoint of Sanskrit, a play upon words can hardly be intended. We might therefore infer that simply the transmission of this Indian idea gave rise in Tibet to the formation of the quibble "raven — friend," which is apparent in *Bya c'os* (compare also the identical formations ,*a-rog*, "friend," and ,*o-rog*, "raven"). The date of this work is unfortunately unknown; the mention of the Siddha Saraha in the introduction, in a measure, may yield a *terminus a quo*. At any rate, *Bya c'os* is far posterior to *K*. and *document Pelliot*. Does the prophecy *grog yoṅ* in the latter imply an allusion to the name of the raven? The case would be interesting from a philological point of view; if the allusion could be established as a positive fact, it would prove that the word *grog* was sounded *rog* as early as the ninth century, for only under this condition is the *bon mot* possible; or another possibility would be that the two forms *grog* and *rog* co-existed at that time. At any rate, there is in our text an obvious relation between the sound *krog krog* and the word *grog*, accordingly a divination founded on punning (*krog krog* is a recognized word of the language and recorded as such in *Za-ma-tog: Studien zur Sprachwissenschaft der Tibeter*, p. 574). This etymological kind of augury finds an interesting analogy among the Arabs, among whom the appearance of a raven indicates parting or pilgrimage, as the word for raven comes from a root meaning "to be a stranger;" the name for the hoopoe suggests "guidance," whence its appearance is of good omen to the wanderer (HASTINGS, *Encyclopædia of Religion*, Vol. IV, p. 846). Among birds, the ancient Arabic poets most frequently mention a black and white spotted species of crow and a black one which it is disastrous to scare, and whose croaking signifies separation from a mistress (G. JACOB, *Altarabisches Beduinenleben*, p. 22, Berlin, 1897). Another explanation than the above is given by D. C. PHILLOTT (*Note on the Common Raven*, J. A. S. B., N. S., Vol. III, 1908, p. 115); the Arabs, according to him, call the raven "raven of separation," because it separated itself from Noah and failed to return. This bird of ill omen alights on the deserted habitations of men; it mourns like one afflicted: when it sees friends together, it croaks, and its croaking foretells "separation;" and when it sees well-peopled habitations, it announces their ruin and desolation. If it croaks thrice, the omen is evil; but if twice, it is good. Possibly the two explanations exist side by side. — Similar etymological punning in augury takes place in Annam with reference to the bird *khéc*. "Le mot *khách*, étranger, devient par corruption patoise, *khéc*, comme le nom de l'oiseau. De là un jeu de mots sur le nom de l'oiseau : Si le *khéc* crie à la porte d'entrée, c'est signe de l'arrivée de visiteurs venant de loin: s'il crie derrière la maison, ce sont des parents qui vont arriver" (L. CADIÈRE, B. E. F. E. O., Vol. I, 1901, p. 196).

V. 29. M. BACOT translates "est signe d'intermédiaire." I do not believe that this is the sense intended, as omens of middle quality (*ạbriṅ*) are referred to in V. 26. The Lama understands that "the sound ,*iu* ,*iu* is continually his (the raven's) note." It is not intelligible to me how he arrives at this view of the matter. The phrase *bar ston* is somewhat embarrassing. I should be inclined to construe *bar* as an abbreviation of *bar-c̣ad*, "accident, calamity," and as referring to the prophecy of calamities given in K., where this word is used; but the fact remains that it does not occur in our Table, and it is certainly to this our Table that we have to look for the interpretation of the term, as in the two preceding verses. There we observe that the greater number of oracles close with the words *oṅ bar ston*, and that in fact each of the ninety oracles ends in the two syllables *bar ston*, or, what is practically the same, *par ston*. This typical formula, I believe, should be recognized in the *bar ston* of V. 29, which accordingly means that the sound ,*iu* ,*iu* points to any of the ninety oracles enumerated in the Table, and therewith the Preface is happily closed with a direct appeal to the latter. This conception of the matter is satisfactory also from a grammatical point of view; for *bar* in this case is *ba + r*, and the terminative is required in connection with *ston*, as shown by V. 25—27 and the ninety examples of the Table, while *bar* taken in the sense of "intermediate, middle," would be the formless *casus indefinitus*, and decidedly present a grammatical anomaly.

Palæographic Traits.

The plain consonant, according to the rules of Tibeto-Indian writing, implies the vowel *a*. In seven cases we find an additional letter *a* following a consonant in this document, where no *a* is admissible in modern writing. The word *dgra* is four times written this way (Table II, 9; IV, 4; V, 2; VIII, 8); further, the suffix *pa* in V. 18, *blta* in V. 11, and *bya* in V. 14. Mr. BARNETT (in A. Stein, *Ancient Khotan*, Vol. I, p. 549) has made a similar observation in the fragments of the *Çālistambasūtra*. He says that before a short pause a final *a* sometimes appears to be lengthened to *ā*, the letter *a* being added on the line; and on p. 500 he adds in a note that this lengthening seems due to the short pause following. I regret being unable to share this opinion; I can see no reason (and Mr. BARNETT gives none) why this addition of *a* should indicate a lengthening of the vowel. True it is, a subjoined *a* (the so-called *a ṇdogs*) denotes *ā* in the Tibetan transcription of Sanskrit words; and it may even be granted with reserve that in the word *gso* (p. 553, note 6), as Mr. BARNETT is inclined to think, the subjoined letter *a* may be intended to give the phonetic value of long *ō*. [1]) But there must be some difference between *a* written beneath and *a* written alongside a consonant. Why, if the lengthening of the vowel is intended, is the letter *a* not subscribed too in the other

---

1) An analogous case is known to me in the Tibetan version of the *Jātakamālā*, a print of 1430, where (vol. II. fol. 9) the word *rgya-mts'o* is equipped with an additional letter *a* under the letter *ts'*. — The subscribed letter *a* occurs also in Tibetan transcriptions of Chinese words, and it would be wrong to conclude, that, because it denotes length in Sanskrit words, it does so also in the case of Chinese, which has no long vowels In the Tibetan inscription of 822, line 15 (see plate in BUSHELL, *The Early History of Tibet*), we have Tib. *bun bu* (each with subjoined *a*) as transcriptions of Chin. 文 武 *wên wu* (Japanese *bun bu*). Most certainly, the additional *a* was not intended by the Tibetans to express a Chinese *a*, but a peculiar Chinese timbre of *u*, which was not sufficiently reproduced by the plain Tibetan *u*.

cases mentioned? The further question arises, If the ancient Tibetan language should have made a clear distinction between short and long a, and if an attempt at discrimination between the two in writing should have been contemplated, why is this distinction not carried through with regular and convincing persistency? Why does it only appear in a few isolated cases? And if this project were once set on foot, how could it happen that it was dropped so soon, as not a trace of it has survived in later literature? Considerations like these should render us cautious in accepting the view of Mr. BARNETT. It is highly improbable that long a (and in general long vowels) existed in Tibetan. It seems to me that long vowels are in Tibetan merely of secondary origin, being the outcome of a fusion of two joining vowels, or arising from the elision of final consonants. ') In our text we notice that the word *bya*,

---

1) JASCHKE (*Tibetan Grammar*, p. 4), who assuredly possessed a good ear, expressly states, "It ought to be specially remarked that all vowels, including e and o (unlike the Sanskrit vowels from which they have taken their signs) are short, since no long vowels at all occur in the Tibetan language, except under particular circumstances mentioned below." Compare the same author's *Ueber die Phonetik der tibetischen Sprache* (*Monatsberichte Berliner Akademie*, 1866, p. 152). For the same reason I am unable to share the opinion of Mr. WADDELL (*J. R. A. S.*, 1909, p. 945) when he tries to make out short and long i in the Tibetan inscription of A. D. 783   The short i following its Indian Devanāgarī prototype, according to Mr. WADDELL, is represented there by a reversion of the tail of the superposed sign to the left, which is not found in modern Tibetan manuscripts. But what evidence is there that the letter i with tail to the left should denote in Tibetan a short, and i with tail to the right a long vowel? This is an arbitrary and unfounded opinion. Why should — taking the examples from the text of the inscription as transcribed by Mr Waddell — *gyi, kyi, srid, myi, ńi, yin, rin, k'rims, adi*, etc., have a short i, but *bris, šuń, gcig* (*gtsig* in line 2 is a misprint), *dgyis, žiń, bkris, bži, ciń, ži-ba, k'rim, drin, p'yin, p'rin, rńiń, lci*, etc , have a long i, — words which at present are all pronounced with the vowel short? There are, further, several inconsistencies due either to the original or to Mr. Waddell's transcript. The interrogative pronoun *ci* has the long vowel in line 3, the short vowel in line 45; the particle of the genitive *kyi*, otherwise short, becomes long in line 68; *rńiń* is long in line 55, but short in line 66; *-i*, the sign of the genitive, is usually long, but short in line 60. The author remarks that the distinction of the short i by reversal of the superscribed limb has not been noted in every instance. On p 1276, where two other inscriptions are transcribed, he says, "In this copy

"bird," is followed by the letter *a* in but a single case (V. 14), while in two other cases (V. 8 and 17) it is written without it. Why should it be *byā* in the one, and *byă* in the two other cases? In fact, however, the vowel of *bya* is not long, but short or quite indeterminate in regard to length. Nor can it be argued with Mr. BARNETT that the juxtaposition of *a* and the alleged vocalic lengthening are due to the pause, for we have *bya +- a* at the close of V. 14, and *bya* without *a* at the close of V. 17. Now, what is

---

the distinction between the long and short *i* has not been recorded." An important palaeographic and phonetic fact is revealed by these inscriptions: in the one case it is dealt with in a perfectly arbitrary manner, as suits the author's convenience; in the other case it is simply suppressed. This is a singular method of editing texts. The student who is desirous of investigating this phenomenon will therefore turn away from these artifacts and for the time being have recourse to the facsimile reproduction of the Tibeto-Chinese inscription of A. D. 822 appended to Dr. BUSHELL's *Early History of Tibet*, where the same distinction of the two *i*'s occurs. The inscriptions published by Mr. WADDELL, for this and several other reasons, will have to be studied anew in the future, on the basis of facsimile rubbings actually taken from the stones. In regard to this peculiar form of *i*, Mr. WADDELL is wrong in asserting that it is not found in modern Tibetan manuscripts. It occurs in all good manuscripts and prints denoting the vocalic *r* and *l* of Sanskrit words, as may be seen, for example, in pl. I of CHANDRA DAS, *The Sacred and Ornamental Characters of Tibet* (*J. A. S. B.*, Vol. LVII, pt. 1, 1888); and this is the only positive fact which we thus far know about the meaning of this sign in Tibetan. It is frequently employed in *P'yi rabs mi-la bslab bya*, a manuscript of the India Office Library alluded to by SCHIEFNER (*Mélanges asiatiques*, Vol. VIII, p. 624), in words as *mi, yin, p'yis, k'ri, odi*, and in the particles of the genitive *kyi* and *-i*, but with no apparent regularity. The sign, further, occurs in the rock-carved inscriptions of Ladakh published by A. H. FRANCKE (*Indian Antiquary*, Vol. XXXII, 1903, pp. 361—363, pl. VIII); there we meet it in the endings of the genitive, *gi* and *-i*, which proves how unfounded Waddell's opinion is, for the supposition that the genitive sign *-i* should be short in Ladakh and long in Central Tibet would be absurd. The distinction of the two *i*'s, in my opinion, does not relate to quantity, which did not exist, but was made to express two different phonetic values or timbres of *i*, which are determined farther on. The vowel system of Tibetan, also at the time of the introduction of writing, was far richer than it appears from the five main vowels *a, e, i, o, u*, the only ones expressed in writing; and for a certain length of time an attempt at discriminating between two values of *i* seems to have been made. — The inverted sign *i* is still employed also, for typographical reasons, in cases where there is no space for the ordinary vowel-sign; as occurs, for instance, when in the line above a word with the vowel-sign *u* (especially the combinations *-yu, -ru* hanging beneath the line proper) is printed.

the rule? Our material is certainly still too scanty to admit of positive conclusions. We have to wait till more ancient documents turn up. Meanwhile it is incumbent upon us to record all peculiarities *le cas échéant,* and to beware of premature and generalized judgments, which will do more harm than good to the future student, and which may be exploded at any moment by the reading of a new document. A conclusion as to the existence of long and short vowels in ancient Tibetan is certainly a case of importance, not only for Tibetan but also for Indo-Chinese philology, as the latter is vitally affected by the former; but such a case must be founded on facts, not on guesswork. Basing my opinion on the *document Pelliot,* I am under the impression that the addition of the letter *a* is not charged with a phonetic value, but has a mere graphic function. The writing of such words as *dgra* and *blta* with an additional *a* moves along the same line as words like *dya, bka, mkᶜa, dma,* etc., where the vowel *a* is still expressed by the presence of the letter *a* to avoid ambiguity, as without it the readings *dag, bak, dam,* would be possible (Csoma, *Grammar of the Tibetan Language,* p. 17). Writing was then in its initial stage; and the rule as to when the letter *a* was a necessity, and when it could be dispensed with, was not yet clearly developed. To all appearances it was then granted a wider latitude; and for the sake of greater distinctness, the *a* was rather added than omitted. In other cases it is neglected where it is demanded by modern rule: thus, in the *Çalistambasūtra,* the word *mkᶜa* is once expressed by the two letters *mkᶜ* (*Ancient Khotan,* p. 552, D 9). One point is clear, that at the time when, and in those localities where, the *da drag* was still in vogue, the rule necessarily had to meet a more extensive application; for there the word *brda,* for instance, if unaccompanied by the letter *a,* could have as well been read *bard.* As this word is written *bda* in our text, it was certainly necessary to add the letter *a*; but it is just this word *brda* which even in modern prints is spelled with *a* as well as without it; the spelling with *a* is, for example, the rule in K'ien-lung's *Dictionary in Four Languages*. If it should turn out through further investigations that this *a* occurs with special predilection in the suffixes *pa*, *ba*, etc., at the end of a sentence, it may very well be that it is a graphic sign employed to mark a certain stress or emphasis, or to denote a stop.

Our text is characterized by two negative features, — the absence of the final *o*, which may be explained by the fact that this text is written in colloquial style, whereas the final *o* is restricted to the written language; [1]) and the lack of the so-called *da drag*.

---

1) It is in full swing in the Stein fragments of the *Çālistambasūtra* and in the sgraffiti of Endere, as well as in the ancient inscriptions of Lhasa, — all documents of the written language. The origin and meaning of this final *o* have not yet been explained. A. Csoma (*Grammar of the Tibetan Language*, p. 84) has merely noticed the fact. When Foucaux (*Grammaire de la langue tibétaine*, p. 17) observes that the particle *o* has the signification of the verbs "to be, to have, to make," this is only to the point in that the sentence, in some instances, may thus be translated by us, but it is not correct from a Tibetan view-point. From Jäschke (*Tibetan Grammar*, p 45) it only appears that the principal verb of a sentence closing it receives in written Tibetan in most cases the mark *o*, by which the end of a period may be known. This *o*, in my opinion, is identical with the now anti-quated demonstrative pronoun *o* (compare Lepcha *o-re*) which, according to Schiefner (*Ergänzungen*, etc., p. 49), very rarely occurs He points out *padma o-ni*, "this lotus," in the Kanjur (Vol. 74, fol. 46), and *groṅ-k'yer o-nir agro*, "to go into that town," in *aDsaṅs-blun* (compare also *Mélanges asiatiques*, Vol. I, p. 385; and *Ueber Pluralbezeich-nungen*, *l. c*, §§ 21, 22). In the Tibetan prose version of *Avadānakalpalatā* (p. 262, line 20) we find, *k'yed ni . . lus so šiṅ bžin skam-pa aṅ srid*, "this your body seems to be dried up like wood;" and (p. 134, line 19), *o ri-dvags gser-logs adi-o žes*, "this one here is that gazelle gSer-logs by name" The latter example is very instructive in showing the pronoun *o* preceding a noun, and again at the end of the sentence linked to the related pronoun *adi*, *adi-o* apparently meaning "this is" The frequent phrase *o-na*, abbreviated into *on*, embodies a survival of this pronoun, the literal meaning being "if this is so." The pronoun *o* itself represents the remains of the entire vowel series which must have originally had pronominal significance. In Ladākhi (A. H. Francke, *Sketch of Ladakhi Grammar*, p. 23, Calcutta, 1901) we have *i* or *i-bo*, "this," and *a* or *a-bo*, "that." In eastern Tibet we have *e*, for example *e-de mi*, "that man" (beside *o-de*; A. Desgodins, *Essai de grammaire thibétaine*, p 39, Hongkong, 1899), and in Tsang and Sikkim *u-di* (Jäschke, *Dictionary*, p. 499, and G. Sandberg, p 85; also according to the writer's own observation), with the survival *u-nir*, *o-nir*, "hither," in the written language. Also the

This term means "strong *d*" or "strengthening *d*." A. Csoma was already acquainted with the occurrence of this phenomenon in ancient orthography, as shown by the spellings *stond-ka, dbyard-ka, rgyald-ka* (*Grammar of the Tibetan Language,* p. 28); *gsand-tam, k^cyerd-tam, gsold-tam* (p. 29); *gsand-to, gyurd-to, gsold-to* (p. 30), and his note on p. 11. Foucaux (*Grammaire de la langue tibétaine,* p. 14), in accordance with Csoma, speaks of three ancient double affixes, — *nd* or *nt, rd* or *rt, ld* or *lt* (the *d* was evidently pronounced with *auslautschärfung,* as the final media in many modern dialects), — and adds that this *d* is now omitted, and that probably, under the influence of this ancient spelling, *gyur-to, gyur-tam, zin-to,* are still written. The terminations *to* and *tam* cannot be considered as survivals; for the dental is nothing but the very *da drag* itself, the terminations proper being *o* (see the note below) and *am*. It is therefore wrong to say that the *dra drag* is obsolete: it is obsolete only as a graphic element, in that it is no longer actually written;

---

personal pronouns *u-cag, u-bu-cag, o-cag, o-skol,* etc. must be explained from this demonstrative pronoun. In the same manner, there was extant in a primeval period of the language a complete vowel series in the *d* group of the demonstrative pronoun, of which only *adi* and *de* have survived. But we have such remnants as *da nań* and *da rańs,* "this morning;" *da lo,* "this year;" *do nub,* "this evening;" *do ydoń,* "to-night;" *do żag* or *do mod,* "to-day," — examples in which *da* and *do* doubtless have the function of a demonstrative pronoun. — The Tibetan verb is, strictly speaking, a verbal noun, which for this reason could easily be connected with a demonstrative pronoun: the sentence *ńas mt^coń-ńo* literally means "by me this seeing (is done)" The fact that this final *o* is not a verbal particle proper follows from its association with any word category; it may be joined to a noun, an adjective, a pronoun, a numeral, the original function of the demonstrative pronoun still being in prominence, with the significance of a completed action or description (hence the Tibetan name for this final is *rdsogs ts'ig,* "word of completion," while its other designation, *slar bsdu-ba,* refers to its position at the end of the sentence). There is, for instance, *bstan bcos agyur-r-o-coų* (Laufer, *Dokumente,* 1, p. 49), and such combinations appear as subject or object within a sentence; compare *gsol-l-o mc^cod-d-o sruń skyobs mdsod* (A. H. Francke, *Der Wintermythus der Kesarsage,* p 9), "guard these prayers and these offerings!" (where Francke p. 66, comments that "the termination *o* is here inexplicable, unless it may have arisen from the emphatic articles *bo, po*") — It is noteworthy that at the conclusion of the Preface we find, not *ston-no,* but the popular *ston yin.*

but it is fully alive phonetically, as soon as certain affixes, to which also *ciṅ*, *ces*, and *cig* belong (*Studien zur Sprachwissenschaft der Tibeter, Sitzungsberichte der bayerischen Akad.*, 1898, p. 584), are joined to the word. We are easily deceived by the appearance of writing. In the Tibetan alphabet is developed the principle of writing separately each syllable of a word and of any composite formation; this, however, does not mean at all that what is separated by the use of the syllabic dot in writing presents also an independent part phonetically. If dissyllabic words, as *me-tog*, *me-loṅ*, *mu-ge*, *pᶜo-ṅa*, *tᶜa-ga(-pa)*, are written in two syllables for the mere reason that the monosyllable is the basic principle of Tibetan writing, it does not follow that these words are compounds; on the contrary, they are stem words consisting of two syllables, and should phonetically be written *metog*, *meloṅ*, *muge*, *pᶜoṅa*, *tᶜaga* (from *tᶜag*, "to weave"). In the same manner we find *rdsogs-so* written in two syllables, and *rdsoɡso* written in one graphic syllable; the pronunciation is not *rdsogs so*, but *rdsogs-o*. In other words, this is not a case of phonetic, but merely of graphic reduplication, caused by the principle of writing. Likewise it does not make any difference from a phonetic viewpoint whether the Tibetan spells *gyurd-to* or *gyur-to*; phonetically it is neither the one nor the other, but *gyurt-o*. Consequently the rule as expressed by JÄSCHKE (*Tibetan Grammar*, p. 45, and *Dictionary*, p. 246) — "*da drag* is a term used by grammarians for the now obsolete *d* as second final, after *n*, *r*, *l*, *e. g.* in *kund*, changing the termination *du* into *tu*; *no*, *ro*, *lo* into *to*; *nam*, *ram*, *lam* into *tam*" — is, from a scientific standpoint, wrong. The rule ought to be formulated that a number of stems at present terminating in *n*, *r*, *l*, were formerly capable of assuming a final *d* sharpened into *t*, and quite regularly assumed the terminations -*u*, -*o*, and -*am*; of course, the proper form of the particle denoting the terminative is -*u*, and not *ru, tu, du, su,* as our grammars merely state for practical purposes, the consonants *r, t,* and *d* being inserted for euphonic reasons, and *su* joined to a word with final *s* being solely a graphic picture of no phonetic value (*e. g., nags-su* of writing = *nags-u* phonetically). The presence of the *da drag* was known to us for a long time only through the medium of the native grammarians, till Mr. BARNETT (*J. R. A. S.,* 1903, p. 110, and *Ancient Khotan,* Vol. I, p. 549) found it written in a large number of cases in the Stein fragments of *Çalistambasūtra.* But, Mr. BARNETT observes, "in isolated instances it is omitted in our MS. from roots that elsewhere have it, a fact indicating that it was already beginning to be dropped in actual speech." This is a point which I venture to challenge. Spelling and speech are in Tibetan two matters distinct; and, as shown above, spelling is not a true mirror of the phonetic state in the present case. The vacillating spelling in the *Çalistambasūtra* simply proves that there was no hard and fast rule for the application of this *d* in writing; but it does not at all prove that if or because it was not written, it was not sounded, at least in many cases.[1]) In other cases when it was omitted, there was surely no necessity for it; and the problem, after all, amounts to this, — What is the significance of this additional *d?* This question is raised neither by Mr. BARNETT, nor by Mr. A. H. FRANCKE (*Ancient Khotan,* p. 564), nor by Mr. WADDELL (*J. R. A. S.,* 1909,

---

1) There is a practical example in our Preface from which it may be demonstrated that the *da drag,* though not fixed in writing, nevertheless may have been sounded (see note on p. 61). Further, Mr. Barnett may be refuted with examples furnished by his own text. In D 3 (p. 551) occurs the writing *rkyen adi,* and in the next line *rkyend adi.* Now, should this indicate two different pronunciations co-existing at that time? Certainly not. The pronunciation simply was *rkyendi* in either case. The two spellings solely indicate two modes of writing these words in that period; they could be written either way, say, for instance, in the same manner as we have the two systems of Webster and Worcester in English spelling, and the latter days' questionable boon of simplified spelling.

pp. 942, 1250), who notes the absence of *da drag* in the inscription of A. D. 783 and its occurrence in another inscription from the first part of the ninth century. The latter document, according to Mr. WADDELL, retained the old popular [why popular?] style of orthography, while it is lacking in the older inscription, because it was revised by the staff of scholarly Indian and Tibetan monks working under the orders of King K'ri-sroṅ lde-btsan [there is no evidence for such a statement]. The *document Pelliot* is highly popular and even written in the language of the people, and shows no trace of the writing of a *da drag*. The whole argumentation of Mr. Waddell, owing to its subjective character, is not convincing; [1]) and it is difficult to see how anybody could argue out this case with any chance of success, without previously examining what a *da drag* is.

First, we have to note that the application of this sign is not quite so obsolete as heretofore stated. It is upheld, no doubt under the force of tradition, in many manuscripts; I observed it repeatedly, for instance, in eighteenth century gold and silver written manuscripts of the *Ashṭasāhasrikāprajñāpāramitā* with the Tibetan title *šes-rab-kyi p'a rold tu p'yin-pa*. The mere occurrence of a *da drag* is therefore no absolute valid proof for the antiquity of a

---

1) On this occasion Mr. WADDELL remarks that the *drag* "has always [°] been recognized by the English lexicographers of Tibetan as a genuine archaism." The English lexicographers of Tibetan! — I regret that they are unknown to me. The first Tibetan dictionary edited by SCHRÖTER (Serampore, 1826) is based on the materials of a Roman Catholic missionary, Father Juvenal (see *The Academy*, 1893, pp. 465, 590; Father FELIX, *J. A. S. B.*, Vol. VIII, 1912, p. 385, without knowledge of this article, attributes the materials of this dictionary to Orazio delle Penna) Csoma, as known to everybody, was a Hungarian. I. J. Schmidt, A. Schiefner, H. A. Jäschke, were Germans. Vasilyev, to whom also Tibetan lexicography owes much, was a Russian "Les missionnaires catholiques du Thibet," figuring as the authors on the title-page of the Tibetan-Latin-French Dictionary published at Hongkong in 1899, were assuredly not Englishmen; and Sarat Chandra Das is a Bengali. Or does Mr. Waddell's philosophy include every English-speaking or English-writing person in the category of Englishmen°

manuscript; nor does its suppression constitute evidence against antiquity, as demonstrated by the *document Pelliot* and the inscription of 783. Secondly, we have to consult the Tibetan grammarians, and to study what they know anent the subject. The most complete native grammar is *Si-tui sum-rtags,* edited in 1743 by gTsug-lag c͑os-kyi snaṅ-ba of Si-tu in the province of K͑ams, and reprinted by the Bengal Secretariat Press in 1895.[1]) In this work, grammatical rules are illustrated by numerous examples, and the *da drag,* wherever applicable, is strictly maintained. Thus we meet on p. 19 the forms *kund-tu, p͑a-rold-tu, mts͑ard-tu, ḩdsind-la,*[2]) *ḩdsind-na, ḩdserd-la, ḩdserd-na, stsald-la, stsald-na;* on p. 24, *ḩbreld;* on p. 30, *bstand kyaṅ, ḩbyord kyaṅ, stsald kyaṅ;* on p. 33, *gyurd tam, ḩts͑ald tam;* on p. 102, *bstand, bkand, bkard, bstard, bcald, mnand, bgard, bsald, mk͑yend, mts͑ard, ḩk͑ruld, ḩdund byed, ḩdserd byed, gsold byed, mt͑ard byed, ḩp͑end byed, bstund bžin-pa, gsold bžin-pa,* etc., but *gnon bžin-pa, gtor bžin-pa;* on p. 103, *rtsald, rold, sbrand, zind, smind, byind, p͑yind, t͑ard, ts͑ard,* but *dul, šar, bor, ts͑or, t͑al,* further *stond, stend, rtend, sbyind, skurd, spruld, speld, lend, smond, seld, ṅand,* but *sgrun, snron, sgyur, k͑ur;* on p. 108, *stond-ka* ('autumn'), *berd-ka* ('staff'), *mk͑yend-pa, p͑and-pa, p͑yind-pa, stond-pa;* and on p. 110, *dkond-cog, rind-c͑en, lhand cig.* On pp. 15 and 16 the part played by this *d* is explained

---

1) This work is mentioned by A. Csoma, *Enumeration of Historical and Grammatical Works to be met with in Tibet* (*J. A. S. B.,* Vol. VII, 1838, p. 152); but Situ or lDom-bu-pa are not the names of the author, as stated by Csoma, but merely titles. He is styled "the great Paṇḍita of Situ" (compare *Si-tui sum rtags,* p. 137, and Chandra Das, *Dictionary,* pp. XXXI and 1272).

2) While the preface of *document Pelliot* (V. 13) has *adsin-la.* In V. 3 *rkyen,* while *rkyend* is repeatedly found in the fragments of *Çalistambusūtra;* in V. 14 *ston ṅi* instead of *stond ṅi;* in V. 23 *rton cig* instead of *rtond cig.* But in the latter example, *cig* in the place of *žig,* as required by the present rule, is testimony of the effect of a *da drag;* the palatal *c* or *č* is certainly a composite sound of the value of *tš,* and, though not actually written, the *da drag* may have nevertheless been actually sounded — *rtont-tšig.*

as purely euphonic (*brjod bde-ba*), and there is surely much in favor of such a view, at least in the final stage of the development of the matter, though this does not exclude the idea that in a former period of the language a more specific function of a formative character may have been attached to it. When in the fragments of the *Çālistambasūtra* the adverb *on kyań* is written *ond kyań*, we doubtless have here a wholly secondary application suggested by analogy where no other than a euphonic reason for the presence of *d* can be given; for the element *on* has arisen from *o-na* ("if this is so"), hence the *d* cannot have originally inhered in it, but must be a later addition to facilitate pronunciation (comparable to the French euphonic *t* in *a-t-il*, etc.). The euphonic character of *da drag* is visible also in its restriction to stems terminating in *n*, *r*, *l*; and even in these limited groups a certain selection seems to take place, in that certain stems are not capable of receiving it, as evidenced by the examples quoted, and many others occurring in literature. Thus, *t'ar-ba* forms only *t'ar-ro*, never *t'ar-to*, while *skul-ba* always forms *bskul-to*. An interesting case is presented by the verb *skur-ba*, which in the sense "to abuse" forms *skur-ro*, but in the sense "to send" *skur-to*. Here we almost gain the impression that the additional *d* was resorted to in order to discriminate between two different homophonous words.

In questioning the formative elements of the language, we observe that there is an affix *-d* forming transitive verbs from intransitive or nominal roots: for example, *skye-ba*, "to be born," — *skye-d-pa*, "to beget;" *nu-ma*, "breast," — *nu-d-pa*, "to suckle;" *ąbye-ba*, "to open" (intr.), *ąbye-d-pa*, "to open" (tr.); *ądu-ba*, "to assemble" (intr.) — *sdu-d-pa*, "to assemble, gather" (tr.); *ąbu-ba*, "to be lighted, kindled," — *ąbu-d-pa*, "to blow;" *dma*, "low," — *smo-d* (*dmo-d*)-*pa*, "to blame, contempt." [1]) Also *byed-pa*, "to do," compared with *bya*, "to be done, action," belongs here; and I am inclined to think that *byed* (phonetically *byŏd* or *b'ŏd*) has arisen from a contraction of *bya* + *yod*, *lit.* "he is doing." It is conceivable that this final -*d* may in general be a remnant of the copula *yod*: as, for instance, *sgo ḥbye*, "the door is open;" *sgo ḥbyed* (= *ḥbye* + *yod*, *ḥbyŏd*), "(I am) opening the door." This possible origin of the transitive -*d* would account also for the fact that formations with -*d* denote a state or condition, as there are *rga-d-pa*, "old man," from *rga-ba*, "to be old;" *na-d*, "disease," from *na-ba*, "to be sick." If this -*d* is a survival of a former *yod*, then *nad* formed of *na* + *yod* is "the state of being ill;" *rgad* formed of *rga* + *yod* is literally "one being old." Likewise we have *ḥgro-ba* and *ḥgrod-pa* (also *bgrod-pa*), "to go, travel," without apparent distinction of meaning at present, while the latter originally meant "to be on a journey."

The conclusions to be derived from these considerations may be summed up as follows. It is probable that the so-called *da drag*, in the beginning, was a formative element of grammatical character, or at least derived from such an element. In the earliest period of literature, this significance had entirely vanished from the consciousness of the speakers; and we then find the *d* applied in the *n*, *r*, and *l* stems inserted between stem and suffix for purely euphonic reasons. The degree to which the euphonic *d* was culti-

---

1) Compare SHTSHERBATSKOI in *Collection of Articles in Honor of Lamanski* (Vol. I, p. 646, St. Petersburg, 1907). The author who abstains from indicating what he owes to his predecessors is neither the discoverer of this law nor others propounded by him. The case under consideration has already been treated by A. CONRADY (*Eine indochinesische Causativ-Denominativ-Bildung*, p. 45); before the time when Professor Conrady published his fundamental book, I enjoyed the privilege, in the course of over a year, of being engaged with him in so many discussions of the Tibetan verb, that I am no longer conscious of what is originally due to him or to me

vated must have varied in different localities, or, what amounts to the same, dialects; it was not a stable or an indispensable constituent of the language, but could be used with a certain amount of freedom. This accounts for its uncertainty in writing, being omitted in some ancient documents, and being fixed in others, and even in these not consistently. The state of writing, in this case, does not allow of any safe inferences as to phonetic facts. In the spellings *t-o*, *t-am*, *t-u*, still in vogue in the modern written language, the *da drag* is practically preserved, the alteration inspired by simplification being of a graphic, not phonetic nature. For this reason it is justifiable to conclude that also in other cases the *da drag*, without its specification in writing, may have continued to be articulated.

Phonology of the Tibetan Language of the Ninth Century.

The Tibetan scholars distinguish two main periods in the development of their language, which they designate as "old language" (*brda rñiṅ*) and "new language" (*brda gsar*).[1] The difference between the two is largely lexicographical and phonetical, the latter distinction being reflected in the mode of spelling; the grammatical differences are but slight, while stylistic variation commands a wide latitude. The existence of a large number of archaic terms in the older writings, no longer understood at present, has led the Tibetans to prepare extensive glossaries, in which those words and

---

1) The translations "old and new orthography" proposed by JÄSCHKE (*Dictionary*, p. 298) take the meaning of these terms in too narrow a sense. Questions of spelling in Tibetan are at the same time those of phonetics and grammar, and in the native glossaries the two terms strictly refer to old and new words. They consequently bear on grammar and lexicography, and comprise the language in its total range. For the distinctions made by Mr. WADDELL (*J. R. A. S.*, 1909, pp. 1269, 1275) of pre-classic and classic periods (even "fully-fledged classical style," and semi-classic, p. 945) I see no necessity; the Tibetan division is clear and to the point, and is quite sufficient.

phrases are defined in modern language. The most useful of these works is the *Li-šii gur kᶜaṅ*.[1]) The well-known dictionary *rTogs-par sla-ba*[2]) contains a long list of such words in verses; and the lCaṅ-skya Hutuktu of Peking, Rol-pai rdo-rje (Lalitavajra), a voluminous writer, who has composed a number of special glossaries for various departments of literature, offers in this series a "List of ancient compared with the modern words" (*brda gsar rñiṅ-gi skor*).[3]) There is, further, a work under the title *Bod yul-gyi skad gsar rñiṅ-gi rnam-par dbye-ba rta bdun snaṅ-ba*, which has been carefully utilized in the "Dictionnaire thibétain-latin-français par les Missionnaires catholiques du Thibet" (Hongkong, 1899).[4]) It is a particular merit of this dictionary that the words and phrases of the ancient style are clearly indicated as such, and identified with the corresponding terms of the modern style (by the reference $A = R$, *ancien = récent*). This as well as another feature, the treatment of synonyms, constitutes a point in which the French work is superior to Jäschke. JÄSCHKE, it is true, includes a goodly number of archaisms (though far from being complete), but in most cases does not indicate them as such. As regards spelling, the

---

1) SCHMIDT and BOEHTLINGK's *Verzeichnis*, p. 64; SCHIEFNER, *Mélanges asiatiques*, Vol I, p 3. There is a good Peking edition (26 fols) with interlinear Mongol version, printed in 1741.

2) *Keleti szemle*, 1907, p. 181

3) It is published in Vol. 7 of his Collected Works (*gsuṅ abum*) printed in Peking (compare *Mélanges asiatiques*, Vol. I, p. 411).

4) According to kind information given by Father A. DESGODINS in a letter dated from Hongkong, October 7, 1901. Father Desgodins, with whom I was in correspondence on Tibetan subjects from 1897 to 1901, and whose memory is very dear to me, was good enough to furnish me with a list of the seven Tibetan dictionaries compiled for his great enterprise. It was at my instigation that Father Desgodins consented to send to Europe the single sheets of his Dictionary as they left the press, so that I was in a position to make practical use of his material in my work as early as 1897 and 1898. It seems singular that, perhaps with the sole exception of Mr v. Zach, I have thus far remained alone in recognizing the special importance of this dictionary and the way of using it.

5

system now generally adopted is traced by Tibetan tradition to
the reform of two scholars, dPal-brtsegs (Çrīkūṭa) from sKa-ba, [1])
and kLui rgyal-mts'an (Nāgadhvaja) from Cog-ro, [2]) assisted by a
staff of scholars, at the time of King K'ri-lde sroṅ-btsan (first part
of the ninth century; according to *T'ang shu*, his reign began in
816).[3]) Prior to this time, as we are informed by Rin-c'en c'os
skyoṅ bzaṅ-po (1440—1526) in his remarkable work *Za-ma-tog*,
there were different systems of spelling in vogue, but all traceable

---

1) dPal-brtsegs took part in the redaction of the first catalogue of the Tibetan
Tripiṭaka (*Dokumente*, I, pp. 50—51), was familiar with the Chinese language (*Roman*,
p. 4), and figures as translator in the Kanjur (*Annales du Musée Guimet*, Vol. II, pp. 182,
233, 337). In the Tanjur, for instance, he cooperated with Sarvajñadeva in the translation
of Nāgārjuna's Suhrillekha (translated by H. WENZEL, p. 32), and in that of Candrago-
min's Çikshalekha (ed. by A. IVANOVSKI, *Zap.*, Vol. IV, pp. 53—81). His portrait is in
GRÜNWEDEL, *Mythologie des Buddhismus*, p. 49.

2) This name occurs in the list of names of the Tibetan ministers in the Lhasa
inscription of 822 reproduced by BUSHELL (*The Early History of Tibet*, J. R. A. S.,
1880); he belonged to the Board of Ministers of Foreign Affairs (*p'yi blon bka-la gtogs-
pa*). The name *Cog* (or *Čog*) *-ro* is transcribed in Chinese *Shu-lu* 屬盧, which indicates that the former character was sounded in the T'ang period *čuk* (compare Hakka *chuk*,
Yang-chou *tsuk*, Hokk. *čiuk*, and CONRADY, *Eine indochinesische Causativ-Denominativ-
Bildung*, p. 165). An analogous case occurs in *Yüan shi*: 搠思 = Tib. *c'os*, indicated
by PELLIOT (*Journal asiatique*, Mars-Avril, 1913, p. 456), and formerly by E. v. ZACH
(*China Review*, Vol. XXIV, 1900, p. 256b). Compare p. 75, No. 14.

3) This king was honored with the epithet *Ral-pa-can* (Skr. *kesarin*), "wearing long
hair," because he wore his hair in long flowing locks. F. KÖPPEN (*Die lamaische Hierarchie und Kirche*, p. 72), with his sarcastic humor, has described how the weak and bigot
monarch became a plaything in the hands of the clergy and allowed the Lamas to sit on
the ribbons fastened to his locks, he intended, of course, to imbibe the strength and holi-
ness of the clergy. Mr. WADDELL (*J. R. A. S.*, 1909, p. 1253) tries to establish two new
facts, — first that the king wore a cue, and secondly that the cue is a Chinese custom
introduced by the king into Tibet (the undignified vernacular word "pigtail" used by
Mr. Waddell, in my opinion, is out of place in an historical treatise). The attribution of
a cue to the king is a rather inconsiderate invention. No Tibetan tradition ascribes to him
a cue or its introduction from China; on the contrary, it is expressly related that the
ribbons mentioned above were fastened to the hair of his head (*dbu skra*, see *dPag bsam
ljon bzaṅ*, p 175, line 14). The difference between wearing long hair and a cue is self-
evident. Neither could the king have introduced any cue from China, since in the age of
the T'ang dynasty, as known to every one, the Chinese did not wear cues; nor is the cue
a Chinese invention at all.

to the teachings of T͑on-mi Saṁbhoṭa, who, during the reign of King Sroṅ-btsan sgam-po (seventh century), introduced writing from India to Tibet.[1])   That reform of the language is expressly recorded in Tibetan history.   I. J. SCHMIDT [2]) has already pointed out this fact from the *Bodhi-mör*, the Kalmuk version of the Tibetan *rGyal rabs*, where it is said that at the time of King K͑ri-lde sroṅ-btsan (the name as given by SCHMIDT is erroneous), besides the new translations, also all previous translations were "recast and rendered clearer according to a more recent and corrected language."   In *dPag bsam ljon bzaṅ* (p. 175, line 12) the same is told still more distinctly in the words that the translations were made afresh (*gsar-du aṅ*) in a newly cast language.   The reflex of this tradition is conspicuous in the colophons of numerous treatises of the Kanjur translated at that period, where we meet the same phrase, *skad gsar c͑ad kyis kyaṅ bcos-nas gtan-la p͑ab-pa.*

In order to study successfully the phonology of a Tibetan text of the ninth century, it is an essential point to form a correct idea of the condition of the language in that period.   This task has not yet been attempted.   The material for the solution of this

---

1) It is known to what fanciful conclusions Messrs. BARNETT (*J. R A. S.*, 1903, p. 112) and FRANCKE (*Ancient Khotan*, p. 565; *Indian Antiquary*, 1903, p 363; *Mem. A. S. B.*, Vol. I, 1905, p. 44) have been driven in regard to the introduction of Tibetan writing.   Mr. BARNETT, sensibly enough, later withdrew his former view; while Mr. FRANCKE, who stamps as a myth, without any historical criticism, every Tibetan account not suiting his fancy, continues to create his own mythology.   There is no reason to dwell on these fantasies, or to waste time in their discussion   Mr. WADDELL (*J. R A. S.*, 1909, pp. 945—947) has already risen against these views with what seems to me to be perfect justice, and it gives me pleasure to acknowledge that I fully concur in Mr. WADDELL's opinion on this point.

2) *Geschichte der Ost-Mongolen*, p. 358.   The passage of *rGyal rabs* (fol. 89) runs thus: *c͑os t͑ams-cad skad gsar bcad-kyis gtan-la p͑ab*, "all religious treatises were cast into a new language and re-edited."   JASCHKE translates the phrase *gsar gcod-pa* by "to inquire into, investigate, examine;" but the literal significance is "to cut anew, to do something from a fresh start, to recast."   An examination of the language of the texts would have sense only if alterations in the language, its style, phonology, and spelling, were to be made.

problem is deposited in the Tibeto-Chinese inscriptions of the T͏ᶜang period and in the Chinese transcriptions of Tibetan words embodied in the Chinese Annals of the T͏ᶜang Dynasty. The bilingual epigraphical material in which Tibetan words are recorded, in comparison with their renderings in Chinese characters reproducing the contemporaneous Tibetan pronunciation of the language of Lhasa, is of primary importance; for it enables us to frame certain conclusions as to the Chinese method of transcribing Tibetan sounds, and to restore the Tibetan pronunciation of the ninth century on the basis of the ancient Chinese sounds. Thus equipped with a certain fund of laws, we may hope to attack the Tibetan words in the T͏ᶜang Annals. The most important document for our purpose is the sworn treaty concluded between Tibet and China in 821, and commemorated on stone in 822, known to the Chinese archæologists under the name *T͏ᶜang T͏ᶜu-po hui mêng pei* 唐吐 蕃會盟碑. This inscription has been made the object of a remarkable study by the eminent scholar Lo Chên-yü 羅振玉 in No. 7 of the journal *Shên chou kuo kuang tsi* (Shanghai, 1909).[1] This article is accompanied by two half-tone plates reproducing the four sides of the stone monument erected in Lhasa, which is 14 feet 7 inches (Chinese) high and 3 feet $1\frac{1}{2}$ inches wide. The recto contains a parallel Tibetan and Chinese text; the verso, a Tibetan text exclusively. The lateral surfaces are covered with the names of the ministers who swore to the treaty. There were seventeen Tibetan and seventeen Chinese officials participating in the ratification. The names of the Tibetan officials are grouped on one of the small sides; those of the Chinese, on the other. Both series of names are given in interlinear versions, — the Tibetan names being transcribed in Chinese, the Chinese names in Tibetan. It is obvious that from

---

1) Compare P. PELLIOT, *B. E. F E O.*, Vol. IX, 1909, p. 578.

a philological point of view, material of the first order is here offered
to us. From the reproductions of Lo Chên-yü it follows that
BUSHELL, [1] who has given a translation of the Chinese text, [2] merely
reproduced half of the stone. The first plate attached to his paper
contains the list of the Tibetan ministers, which is, accordingly,
one of the small sides of the stone; this part is not translated by
Bushell or referred to in his text; his second plate gives the
recto of the stone, while the verso and the other small side with
the names of the Chinese ministers are wanting. Bushell's photo-
lithographic reproduction is very readable, and my reading of the
Tibetan names is based on his Plate I. The Chinese reproduction
is too much reduced, and the glossy paper on which it is printed
considerably enhances the difficulty of reading. But Lo Chên-yü
deserves our thanks for having added in print a transcript of the
entire Chinese portion of the monument, inclusive of the thirty-
four names as far as decipherable; this part of his work proved to
me of great utility, as Bushell's small scale reproduction, in many

---

1) *The Early History of Tibet* (*J. R. A. S.*, 1880).

2) A drawback to BUSHELL's translation is that it appears as a solid coherent account,
without indication of the many gaps in the text. Bushell filled these from the text as
published in the *Ta Ts'ing i t'ung chi*. As the notes of Lo Chên-yü rectify and supplement this edition of the text on several points, a new translation of this important monument would not be a futile task, if made on the basis of Lo Chên-yü's transcript, in which
the lacunes are exactly indicated. — A. H. FRANCKE (*Epigraphia Indica*, Vol. X, 1909—
10, pp. 89—93) has given, after BUSHELL's rubbing (Pl. 11), a transcript of the Tibetan
version, and what, from a Tibetan point of view, he believes to be a translation of it.
BUSHELL's Plate I, the list of the Tibetan officials, is not mentioned by Francke. It goes
without saying that this Tibetan text, as well as the other Tibetan epigraphical documents
of the T'ang period, cannot be translated merely by the aid of our imperfect Tibetan dictionaries; sinology is somewhat needed to do them These documents were drafted in the
Tibeto-Chinese government chancery of Lhasa; and the Tibetan phraseology is to some
extent modelled after the Chinese documentary style, and must be carefully studied in the
light of the latter. BUSHELL (p. 102), it seems to me, is not correct in stating that the
Chinese text of the monument is a translation of the Tibetan original; the question as to
which of the two is the original is immaterial. Both express the same sense, and were
drafted simultaneously by the Tibeto-Chinese clerical staff of Lhasa

passages, left me in the lurch. The account of the erection of the monument as given in the Tibetan annals (rGyal rabs, fol. 92) may be of some interest. "During the reign of King Ral-pa-can, the son-in-law and father-in-law [the sovereigns of Tibet and China] were still in a state of war, and the Tibetan army, several tens of thousands, conquered all fortified places of China. The Ho-shang of China and the clergy of Tibet intervened and concluded a sworn pact. The son-in-law despatched pleasing gifts, and an honest agreement was reached. In the frontier-post rMe-ru in China, the two sovereigns each erected a temple and had a design of sun and moon engraved on a bowlder, which was to symbolize that, as sun and moon form a pair in the sky, so the sovereign son-in-law and father-in-law are on earth. It was agreed that the Tibetan army should not advance below rMe-ru in China, or the Chinese army above this place. In order to preserve the boundary-line, they erected visible landmarks in the shape of earth-mounds where earth was available, or stone-heaps where stone was available. Then they fixed regulations vouching for the prosperity of Tibet and China, and invoking as witnesses the Triratna, Sun and Moon, Stars and Planets, and the gods of vengeance,[1])

---

1) This passage occurs in the inscription 三寶及諸賢聖日月星辰請爲知 (BUSHELL: 和) 證. Tib. (line 62) dkon mc'og gsum dan ap'ags-pai dam-pa-rnams gñi zla dan gza skar-lu yan dpan-du ysol-te, "the Three Precious Ones (Skr. triratna), the Venerable Saints, Sun and Moon, Planets and Stars they invoked as witnesses." Mr. FRANCKE (l. c., p. 93) translates, "The three gods(!), the august heaven, etc., are asked to witness it." He has the wrong reading ap'ags-pai nam-k'a where dampa, "holy," is clearly in the text; the plural suffix rnams is inferred by me from the context (the stone is mutilated in this spot). The Tibetan phrase, as read by me, exactly corresponds in meaning to the Chinese chu hien shéng, "the holy sages." There is no word for "heaven" in the Chinese text, nor a Tibetan word for "heaven" in the above corresponding passage in rGyal rabs; consequently nam-k'a cannot be sought in the Tibetan version of the inscription, either. The gods of vengeance (lha gñan rnams) are omitted in the inscription, presumably for the reason that no exact Chinese equivalent for this Tibetan term could be found. The interpretation as above given is derived from JASCHKE (Dictionary, p. 192), with whom I. J. SCHMIDT (Geschichte der Ost-Mongolen, p. 361), translating from the Bodhi-mor ("die rachenden Tenggeri"), agrees. The gñan are a class

the two sovereigns swore a solemn oath by their heads.[1]) The text of

---

of demons whose specific nature is still somewhat uncertain; in the Bon religion they form a triad with the *klu* and *sa bdag* (see the writer's *Ein Sühngedicht der Bonpo*). The word *gñan* means also a species of wild sheep, argali (*Ovis ammon* L. or *Ovis Hodgsoni* Blyth., see M. DAUVERGNE, *Bull. Musée d'hist. nat.*, Vol. IV, 1898, p. 216; the definition of CHANDRA DAS *Dictionary*, p. 490 — 'not the *Ovis ammon* but the *Ovis Hodgsoni*" — is wrong, as both names, in fact, refer to the same species). Now, we read in *Kiu T'ang shu* (Ch. 196 上, p. 1b), in regard to the ancient T'u-po, 事源祇之神, "they serve the spirits of *nguan ti*;" *nguan* (this reading is given in the Glossary of *T'ang shu*, Ch. 23, by the characters 吾官 *ngu kuan*; Tib. *gñan* and Chin. *nuan* are perhaps allied words; *Erh ya* reads *yüan* 元) likewise refers to a species of wild sheep or argali, and *ti* is a ram. We know nothing to the effect that the Tibetans ever worshipped argali, nor can the Chinese words be explained as the transcription of a Tibetan word. It seems to me that Chin. *nguan ti* is a literal translation of a Tib. *gñan-p'o* (or -*p'a*, "male of an animal") caused by the double significance of the Tibetan word *gñan*, and that the Chinese annalist means to convey the idea that the Tibetans worship a class of spirits styled *gñan*. On two former occasions it was pointed out by me that the word *gñan*, presumably for euphemistic reasons, is frequently written *gñen* ("friend, helper") In the Table of *document Pelliot* (V, 3) we meet the oracle, *gñen lha skyes-po-la ats'e-ba-żig oṅ-bar ston*, where I am under the impression that *gñen lha* should be taken in the sense of *gñan lha*, and accordingly be translated, "It indicates that a terrific spirit doing harm to men will come" (the injury is not done to the god, as M. BACOT translates).

1) Tib. *dbu bsñuṅ daṅ bro bor-ro*. JÄSCHKE (*Dictionary*, p. 382a) has already given the correct translation of this phrase. Mr. WADDELL (*J. R. A. S.*, 1909, p. 1270) has misunderstood it by translating *dbu sñuṅ gnaṅ-ste* "(the king) was sick with his head." The word *sñuṅ* in this passage has nothing to do with the word *sñuṅ*, "disease," but is the verb *sñun-ba* (causative from *ñuṅ-ba*, "small"), "to make small, diminish, reduce." The phrase *dbu sñuṅ* is a form of adjuration corresponding to our "I will lose my head, if ..." The beginning of the inscription therefore is, "Land was granted (*sa gnaṅ*, which does not mean 'honor be given')... The father, the sovereign K'ri-sroṅ lde-btsan [the translation "the king's father's father" is wrong: the father, *yab*, is a well-known attribute of King K'ri-sroṅ] formerly made the grant under his oath." On this mistranslation the following speculation is based (p. 1268): "King K'ri-sroṅ lde-btsan is stigmatized as being of unsound mind — a condition regarding which there never has been the slightest hint in the national histories — and the rule of the kings generally is declared to have caused a cycle of misfortunes to the country." The entire "historical" interpretation of this inscription is unfortunately not based on the national histories, but is a dream of the author. There is nothing in the text of "the Sacred Cross of the Bon," which is plainly a Svastika designed on the silver patent (*dṅul-gyi yi-ge*, translation of *yin p'ai* 銀牌), nor is there "the P'an country of the Secret Presence of the Bon deity," which simply means "the district of aP'an in sKu sruṅs" (name of a locality). Neither the translation nor the explanation of this inscription can be accepted

the treaty was inscribed on three stone tablets. On the two large surfaces was written the text containing the sworn treaty concluded between the two sovereigns; on the two small sides of the stone was written the list of the names [1]) of the Tibetan and Chinese officials who were accredited as ministers of state. One of these stone monuments was erected at Lhasa, another in front of the palace of the Chinese emperor, another at rMe-ru on the frontier of China and Tibet. 'If regardless of the text of this treaty, the Tibetans should march their army into China, the Chinese should read three times the text of the inscription in front of the palace of the emperor of China, — then the Tibetans will all be vanquished. On the other hand, if the Chinese should march their army into Tibet, all Chinese will be vanquished in case the text of the inscription of Lhasa should three times be read,' — this oath was stipulated between the state ministers of Tibet and China and sealed with the signets of the two sovereigns."

The purpose of the following study is purely philological, not epigraphical or historical, though it simultaneously furnishes a not unimportant contribution to the then existing offices in Tibet; the latter subject, however, calls for a special investigation, for which also the numerous references in the Tibetan annals must be utilized, and it is therefore here discarded for the time being. The inquiry is restricted to the Chinese transcriptions of Tibetan words; their pronunciation is ascertained by restoring, as far as possible, the Chinese sounds, such as were in vogue during the T'ang period. It will be recognized that the Chinese applied a rigorous and logical method to their transcriptions of Tibetan words, and that in this manner a solid basis is obtained for framing a number of

---

1) Tib. *miṅ rus*. The same expression written *myiṅ rus* occurs likewise in the inscription of 822 (compare No. 12, p. 74), where it corresponds to Chin. *miṅg wei* 名位.

important conclusions as to the state of Tibetan phonology in the ninth century, with entirely convincing results, which are fully confirmed by the conditions of the ancient Tibetan documents. First the material itself is reviewed, to place everybody in a position to form his own opinion, then the conclusions to be drawn from it are discussed. The single items are numbered in the same manner as has been done by Lo Chên-yü. Nos. 1—3 contain no transcriptions, and are therefore of no avail for our purpose; in Nos. 4—8, the Tibetan text, with the exception of a few words, is hopelessly destroyed. Nos. 9—20 run as follows:

9. $C^cab\text{-}srid\text{-}kyi$ [1]) $blon\text{-}po$ $c^cen\text{-}po$ $\check{z}a\check{n}$ k^cri $btsan$ [2]) k^cod ne $sta\check{n}$ = 宰相同平章事尚綺立贊窟寧思當 ts^cai $siang$ t^cung p^cing $chang$ shi $shang$ k^ci li $tsan$ $k^cu(t)$ $ning$ se $tang$. The name of this minister, accordingly, was sounded k^cri $tsan$ $k^cod(t)$ $n\underline{e}$ [3]) $sta\check{n}$. His Tibetan title means "great minister of state," rendered into Chinese "minister and superintendent of affairs." [4])

10. $C^cab\text{-}srid\text{-}kyi$ $blon\text{-}po$ $c^cen\text{-}po$ $\check{z}a\check{n}$ k^cri $b\check{z}er$ lta $mt^co\check{n}$ = 宰相同平章事尚綺立熱 [5]) 貪通 ts^cai $siang$ t^cung p^cing $chang$ shi $shang$ k^ci li $\check{z}e(je)$ t^cam (t^can) t^cung. The Tibetan name of this minister, accordingly, was articulated k^cri $\check{z}e(r)$ $tam\text{-}t^co\check{n}$ (for explanation see farther on).

---

1) By the transcription i the inverted vowel sign i commented on p. 53 should be understood. Its phonetic value will be discussed hereafter.

2) The two words k^cri $btsan$ are destroyed on the stone, but can be correctly restored on the basis of the Chinese equivalents k^ci li $tsan$; Chin. k^ci li corresponds to Tib. $k\underline{,i}$ in No. 10, and Chin. $tsan$ is the frequent and regular transcription of Tib. $btsan$.

3) As indicated by Chin. $ning$, the vowel of Tib. ne was nasalized (pronounced like French $nain$).

4) See GILES, *Dictionary*, 2d ed , p. 1132b.

5) Lo Chên-yü transcribes this character 慹, but this is an error. The reproduction of BUSHELL shows that the character is as given above, and this is the one required for the rendering of the Tibetan sounds. This reading, moreover, is confirmed by *Kiu T'ang shu* (Ch. 196 下, p. 11b), where exactly the same personage is mentioned 尚綺立熱 who in 825 was sent on a friendly mission to the Chinese Court.

11. *Cab-srid-kyi blon-po ćen-po blon rgyal bzañ ḥdus kuñ* [1]) = 宰相同平章事論頬藏弩思恭 *tsʿai siang tʿung pʿing chang shi lun kia(p) (ɣʾap) tsang* [2]) *nu* [3]) *se kuñ*. The name of this minister was pronounced *gʾal* (or *ɣʾal*) *zañ dus kuñ*.

12. *Bod ćen-poi blon-po tsʿal-gyi tʿabs dañ myiñ rus* = 大蕃 諸寮案登壇者名位 *Ta Po chu liao ngan tëng tʿan che ming wei*. The Tibetan is a free translation from Chinese, the phrase *tëng tʿan*, "those who ascended the altar" (in order to swear to the treaty) being omitted. Note that *Bod ćen-po*, "Great *Bod*," does not occur in Tibetan records, but is only a stock phrase modelled in the Tibeto-Chinese chancery of Lhasa after the Great Tʿang Dynasty 大唐.

13. *nañ blon mćims žañ rgyal bžer kʿon ne btsan* = 曩論琛 尚頬熱窟寧賛 *nang lun chʿëm (chʿén) shang kia(p) (ɣʾap) že(je) kʿu(t) ning tsan*. In the name of the Minister of the Interior we note the pronunciations *ćim* (or *ćʿim*) for *mćims*, *že* for *bžer*, and again the nasalized vowel in *ne*.

14. *pʿyi blon bka-la gtogs-pa Cog-ro | blon btsan bžer lto goñ* =

---

1) In Bushell's reproduction, *kañ*. But the rubbing was sharply cut off around these last two words, so that the sign *u* may have been lost during this process. The Chinese transcription *kung* calls for a Tibetan *kong* or *kung*.

2) It doubtless represents an ancient *zang ('dzang); compare the Japanese reading *zō*. Also in *Yuan shi* Tib. *bzañ-po* is transcribed 藏卜 and Tib. *blo bzañ* 羅藏 (E. v. ZACH, *Tibetica, China Review*, Vol. XXIV, 1900, p. 256a). The character 臧 *tsang* serves in *Tʿang shu* (Ch. 216 下, p. 6a) to render Tib. *ytsan*, the name of the main river of Central Tibet.

3) *Nu* 弩 seems to have had the phonetic value *du* (Japanese *do*), and *du se* is intended for Tib. *ḥdus*. An analogous example occurs in *Kiu Tʿang shu* in the name of the Tibetan king *Kʿi nu si lung* 器弩悉弄 answering to Tib. *Kʿri du sron* (usually styled *Du sron mañ-po*). Compare *lo* 羅 transcribing Turkish *du* (CHAVANNES and PELLIOT, *Journal asiatique*, 1913, No. 1, p. 175). The character *lie* 獵 rendering Tib. *lde* (pronounced *de* in the ninth century) in the name of King *Kʿri sron lde btsan* 乞黎蘇籠 | 贊 (*Kiu Tʿang shu*, Ch. 196 上, p. 8b), offers another instance of Chinese initial *l* corresponding to *d* in a foreign language.

紕論伽羅篤波屬盧論贊熱土公 *pʻi lun kia lo*[1]) *tu(k) po šu (čuk) lu lun tsan že (je) tʻu kung.* The Tibetan words were accordingly articulated at that time, *pʻi lon ka-la tog-pa* (the Minister of Foreign Affairs) *čog-ro lon tsan že(r) to goṅ.*

15. *snam pʻyi-pa mcʻims žaṅ brtan bžer snag*[2]) *cig* = 思南紕波琛尙旦熱思諾市 *se nam (nan) pʻi po chʻêm (chʻên) shang tan že (je) se nak (no) shi.* Tibetan pronunciation, *snam pʻi-pa cʻim žaṅ tan že(r) snag(k) ciʔ.*

16. *mṅan pon baṅ-so o-cog gi ḥbal blon kru bzaṅ gyes rma* = 岸奔猛蘇戶屬勃羅末論矩立藏〇摩[3]) *ngan pên (pön, pun) mong* (Cantonese and Hakka *mang*, Japanese *bō*) *su hu* (Cantonese *u*, Ningpo *wu*, Japanese *o*)[4]) *šu* (**čuk*) *pu lo mo* (Hakka

---

1) Sounded *la*; see VOLPICELLI, *Prononciation ancienne du chinois*, pp. 161, 181, 183 (*Actes XIe Congrès Or.*, Paris, 1898).

2) Written as if it were *stag*, but the seeming *t* may have been intended for *n* which is required by the Chinese transcript; likewise in No. 17. The palaeographic features of Tibetan epigraphy of the Tʻang period remain to be studied in detail. — The character 諾 is sounded *nak* in Korean, *naku* in Japanese. The phonetic element 若 has the value *nik*; in the Manichean treatise translated by M. CHAVANNES and M. PELLIOT (*Journal asiatique*, 1911, No. 3, p. 538) it is combined with the radical 口 into a character which otherwise does not occur; but as the Pahlavī equivalent rendered by it is *nag*, this artificial character must have had also the sound *nak*, in the same manner as 諾.

3) Lo Chên-yu transcribes the last two characters 名〇. The first of these does not seem to be 名, though I cannot make it out in the reproduction of BUSHELL, which is too much reduced, but 名 cannot be the correct reading, as the sound *ming* is incapable of reproducing anything like Tib. *gyes*. The second character left a blank by Lo, I distinctly read *mo* (anciently *ma*), as above, in BUSHELL's plate, and this very well answers as transcription of Tib. *rma* (sounded *ma*).

4) The equation 戶 = Tib. *o* allows us to restore theoretically the name (姓) of King *Kʻri sroṅ lde btsan* given in *Tʻang shu* (Ch. 216 下, p. 1b) in the form *Hu lu tʻi* 戶盧提 into Tib. *O ro lde*. Chin. *lu* = Tib. *ro* we had in No. 14. The ancient sounds of *tʻi* were **te, de* (Japanese *tei, dei*), hence Tib. *de* or *lde* frequently occurring in the names of the kings may be inferred (it occurs likewise in the name of the ancestor 祖 of the Tibetans, *Hu tʻi pu si ye* 鶻提勃悉野 where *tʻi pu* corresponds to Tib. *de-po* or *lde-po*; the other elements of this name are treated farther on). A name of the form *O ro lde*, however, does not occur in Tibetan records; but in

mat, Korean *mal;* ancient sounds *mwat and *mwar* [1])) *lun kü li tsang*
◯ *mo (ma).* The sign of the genitive, *gi,* is not transcribed in
Chinese. Tib. *mñan,* accordingly, was sounded *ñan; blo* was sounded
blo (Chin. *pu-lo*), not *lo,* as at present; *ḥbal* was sounded *bal,* or
possibly *mbal* or *mwal; kru* was sounded *kru* (Chin. *kü-li*), not as
now *tru* or *ṭu; rma* was sounded *ma.* Tib. *mñan pon* must be a
compound written for *mña dpon* ("rulers and lords"), the prefix *d*
being altered into *n* under the influence of the initial guttural nasal
ñ and then pronounced and written *ñan pon.* The meaning of the
above passage is, "The minister *Kru bzañ gyes rma,* who was in
charge of the sepulchres of the sovereigns and lords." It was hitherto
unknown that such an office existed in Tibet, and this fact is of
great culture-historical interest. We know that the ancient kings
of Tibet were buried under elevated tumuli, and the *rGyal rabs*
has carefully recorded the exact locality and its name where each
king was interred.[2]) The *T*°*ang shu* (Ch. 216 下, p. 6) imparts a

---

the inscription of 783 edited and translated by Mr. WADDELL (*J. R. A. S.,* 1909, p. 931)
the name of a primeval king *O lde spu rgyal* is mentioned. I am therefore inclined to
regard the Chinese transcription *Hu lu t'i* as a reproduction of Tib. *O lde,* the Chinese
syllable *lu* rendering the prefix *l* in *lde,* which was sounded on account of the preceding
vowel, as still at present the prefix is articulated in the second element of a compound
when the first terminates in a vowel. The name *O lde* has not yet been pointed out as
a name or title of King *K'ri-sron* in any Tibetan document; it remains to be seen
whether it will be confirmed. The comment made by Mr. WADDELL (p. 933) on the king
named *O lde spu rgyal* is erroneous: he does not follow the Seven Celestial Rulers in
Tibetan tradition. This king whom Mr. WADDELL has in mind is styled in *rGyal rabs*
"*Spu de guñ rgyal*" (mentioned also by ROCKHILL, *The Life of the Buddha,* p. 209, but
the name does not mean "the tiger-haired king"), but there is no reason to assume that
he is identical with *O lde spu rgyal.* Although Mr. WADDELL (p. 949, note 3) expressly
states that there seemed no trace of a final *d* in the word *o,* Mr. A. H. FRANCKE
(*J. A. S. B.,* Vol. VI, 1910, p. 94) boldly and arbitrarily alters this name into *Od lde
spu rgyal,* and translates this *Od lde* by "beautiful light," which is pure fancy, as is the
whole article in which Mr. FRANCKE, to his great satisfaction, shifts the theatre of
action of Tibetan tradition connected with King *gÑa k'ri btsan-po* from central to
western Tibet.

1) CHAVANNES and PELLIOT, *Journal asiatique,* 1911, No. 3, p. 519.
2) The interment of King Sroñ-btsan sgam-po is thus described in *rGyal rabs* (Ch.

vivid description of the sepulchral mounds 丘墓 of the Tibetan nobles scattered along the upper course of the Huang-ho, white tigers being painted on the red-plastered walls of the buildings belonging to the tombs; when alive, they donned a tigers-kin in battle, so the tiger was the emblem of their valor after death.

17. *bkai p‘rin blon c‘en* [1]) *ka* [2]) *blon snag b̌er ha ñen* = 給事 中勃 ○ 伽論思諾熱合軋 *ki shi chung p‘o (pu)* ○ *kia lun se nak (no) še (je) ha (ho) yen.*

18. *rtsis-pa c‘en-po* ○ [3]) *blon stag zigs rgan k‘od* = 貧思 波折遍額論思 ○ [4]) 昔幹竄 *tse-se po ch‘e pu ngo(k) lun*

18, fol. 76): "His sepulchre (*bañ-so*) was erected at ạC‘oñ-po (in Yar-lun), being a mile all around. It was quadrangular in shape, and there was a vault made in the centre. The body of the great king of the law (Skr. *dharmarāja*) was laid in a composition of loam, silk and paper, placed on a chariot, and to the accompaniment of music interred in the sepulchre. The vault in the interior was entirely filled with treasures, hence the sepulchre became known under the name *Nañ brgyan* ('Having ornaments in the interior'). Five chapels were set up in the interior, and the erection of quadrangular sepulchres took its origin from that time. They are styled *sKu-ri smug-po* ('red grave-mounds')." I. J Schmidt (*Geschichte der Ost-Mongolen*, p. 347), translating from *Bodhi-mor*, the Kalmuk version of *rGyal rabs*, erroneously writes the latter name *sMuri*, and makes an image of the king fashioned from clay and buried in the tomb, while the burial of the body is not mentioned. The Kalmuk version is not accessible to me; the Tibetan text is clearly worded as translated above. The same work (fol. 87) imparts the following information on the tomb of King K‘ri-sroñ lde-btsan: "His sepulchre was erected on Mu-ra mountain, in the rear, and to the right, of that of his father. The king had it built during his lifetime. The posthumous name *aP‘rul ri gtsug snan* was conferred upon him. At the foot of his sepulchre there is a memorial inscription in stone. The sepulchre became known by the name *P‘yi rgyan can* ('Ornamented in the exterior')."

1) See *dPag bsam ljon bzan*, p. 151, l. 25 This term is not explained in our Tibetan dictionaries. The Chinese rendering shows that it is the question of supervising censors.

2) For *bka*.

3) This word is badly mutilated in the stone. The Chinese parallel is *ngo(k)*, so that I infer Tib. *rñog*, a well-known clan name. The Tibetans have no family names but clan names (Tib *rus*, Chin. *tsu* 族; compare the account on the Tang-hiang in *T‘ang shu*, Rockhill's translation in *The Land of the Lamas*, p. 338) named for the localities from which the clans originated.

4) This lacune corresponds to Tib. *stag*. The character 答 *ta* may be inferred from the name *Lun si ta je* 論悉答熱 (Tib. *Blon stag rje*) in *T‘ang shu* (Ch. 216 下, p. 6a).

se ◯ *si(k) han*[1]) *kᶜu(t)*. The word *rtsis-pa* was accordingly sounded
tsis-pa. The Chinese transcription of this ministry (instead of translation as in the preceding cases) indicates that there was no correlate institution for it in China. In the modern administration of
Tibet, the *rtsis dpon* had charge of the accounts. [2]) from which it
may be inferred that the *rtsis-pa cᶜen-po* of the Tᶜang period had
a similar function.

19. *pᶜyi blon ḡbro žan* (the remainder is almost destroyed and
cannot be positively deciphered) = 紕論沒盧尚 *pᶜi lun mu-
lu shang*. The transcription *mu* (compare Japanese *botsu*)-*lu* hints
at a pronunciation *bro* for Tib. *ḡbro*.

20. *žal-ce-ba*[3]) *cᶜen-po žal-ce* ◯ ◯ *god* (?) *blon rgyud ñan li
btsan* = 刑部尚書◯論結硏歷贊 *hing pu shang shu*
◯ *lun kie (γ'et) ngan (yen) li tsan*. The transcription of *rgyud* is
of importance; it was sounded *g'ut* or *γ'ut*, the prefix *r* being silent.

---

1) Chin. *han*, accordingly, renders Tib. *rgan*, which, after the elimination of the prefix
r, was presumably sounded χ*an* In a passage of *Ŷan shi*, the same Tibetan word is
transcribed *han* 罕 (E. v. Zach, *l. c.*, p 255). Chin *h*, therefore, in transcriptions, does
not usually correspond to Tib. *h*, but to Tib. *g* with or without prefix. The following
case is of especial interest. Tib. *la pᶜug*, "radish," is a Chinese loan word derived from
lo pᶜo 蘿蔔 (see Bretschneider, *Bot. Sin.*, pt. 2, No. 39); consequently also Tib.
guñ la pᶜug, "carrot," must be the equivalent of Chin. *hu lo pᶜo* 胡蘿蔔 of the
same meaning: so that we obtain the equation Chin. *hu* 胡 (Japanese *ko*) = Tib. *guñ*.
For this reason we are justified in identifying also the name *Hu* 鶻 with Tib. *Guñ*
in the name of the ancestor of the Tibetans mentioned on p. 75, note 4; and *Guñ rgyal*,
as correctly stated by Chandra Das (*Dictionary*, p. 221), according to Tibetan tradition,
is the name of one of the early kings of Tibet (the same name occurs also in *Guñ ri gun
btsan*, son and successor of King *Kᶜri-sron*, and in *Spu de guñ btsan*).

2) Rockhill, *J. R. A. S.*, 1891, p. 220.

3) Jäschke writes this word *žal cᶜe*, which is a secondary development; it is properly *žal lce* ("mouth and tongue"), thus written, for instance, *Avadānakalpalatā* (Tibetan
prose ed., p. 71, 7) and Chandra Das (*Dictionary*, p. 1068). The Table (II, 6) offers the
spelling *ža-lce*, which, together with the spelling of the inscription, shows that the word
was pronounced *žal-ce* in the ninth century. As proved by the Chinese translation 刑,
it had, besides the meanings "lawsuit, litigation, judgment," also the significance of
"punishment." Tib. *cᶜen-po*, "the great one," appears as rendering of Chin. *shang shu*.

There are, further, in the inscription, two interesting parallels of geographical names. In line 44 we meet Tib. *stse šuṅ čʿeg* (or *tsʿeg*) transcribing Chin. *tsiang kün ku* 將軍谷 ("Valley of the General"), and in line 46 Tib. *čeṅ šu hyvan* transcribing Chin. *tsʿing šui hien* 清水縣. The Tibetan word *stse* was pronounced *tsǫ* (the sign *e* including also nasalized *ō*). The addition of the prefixed sibilant *s-* does not prove that this *s* was sounded, but, as in so many other cases, it owes its existence only to the tendency of preserving the high tone which indeed is inherent in the Chinese word *tsiang*. The Tibetan word *tse* without the prefix would have the deep tone, while the prefix indicates that it is to be read in the high tone; the Chinese equivalent *tsiang* (Cantonese *tsōng*, Hakka *tsiong*) undeniably proves that the palatal sibilant was also the initial intended in the Tibetan word. It is entirely out of the question to regard the *s* in *stse* as the articulated initial consonant, and only the desire for regulating the tone can be made responsible for the presence of the prefixed *s*.[1]) We have here, accordingly, unassailable evidence for the fact that the tone system existed in the language of Lhasa at least as early as the first

---

1) An analogous example is presented by Tib. *spar kʿa* being a transcription of Chin. *pa kua* 八卦. Chin. **pat, par* (compare Tib. *pir* = Chin. *pit* 筆) never had an initial *s*, and there is no reason whatever why the Tibetans should articulate *spar* a Chinese *par*, of course, they did not, nor do they do so, but say *par*; the unprotected *par*, however, has with them the deep tone, while, if the prefix *s* is superscribed, it receives the high tone, and the high tone is required by the Chinese word; the letter *s* is simply a graphic index of the high tone. Also the high-toned aspirate *kʿa* instead of *ka*, which we should expect, seems to be somehow conditioned by the tone of Chin. *kua*. Vice versa, Chin. *mo-mo* 饃 | with the even lower tone is written in Tibetan *mog-mog* ("steamed meat-balls"), having likewise the low tone, but not *smog*, which would indicate *mog* in the high tone. — Another interesting loan-word is *lcog-tse* (*rtse*), "table," derived from Chin. *cho(k)-tse* 桌子; the final *g* indicates that the loan is old. The prefix *l* merely has the function of expressing the high tone of the Chinese word; the Tibetans certainly pronounce only *čog-tse* (later spellings are *cog-tsʿe* and *cog-tsʿo*, the latter in *Li-šii gur-kʿaṅ*, fol. 23).

part of the ninth century, and the reason for its coming into existence will immediately be recognized from our general discussion of the phonetic condition of the language in that period. Another interesting example of the presence and effect of tone at that time will be given hereafter in dealing with the word *žan*. Tib. *žun* as equivalent for *kün* 軍 is conceivable only when the Tibetans heard or understood the latter word as *ćun* or *šun* with a similar pronunciation, as still existing in the dialects of Wên-chou, Ning-po, and Yang-chou (compare W. *ciung*, N. *cüing*, Y. *chüng*, given in GILES's *Dictionary*); for Tib. *ž* and *j* are regular equivalents of the Chinese palatals *ć* and *š* (compare Tib. *kong jo* = Chin. *kung ću* 公主, Tib. *žo* transcribed in Chin. *šo*).[1]

The word *ćeg* (or *tsćeg*) is a Tibetan word, and has nothing to do with Chinese *ku*. The Tibetan transcription *ćen* for Chin. *tsćing* is striking; it is not known to me whether the latter word may have had an initial tenuis in the Tćang period. *Shui* 水 was then doubtless sounded *šu* or *žu*; we shall have to come back to the question why the Tibetan transcription is *šu*. The Tibetan *hyvan* [2]) consists in writing of initial *h* with subscribed *y* (*ya btags*) and following *va zur* which is the semi-vowel *ṿ*; phonetically, the word is *h'ṿan*, so that the pronunciation of 縣 must then have been something like the Korean reading *hiön*, or like *hiuan*. [3])

---

1) The case is fully discussed farther on, where more examples will be found.

2) BUSHELL (*l. c.*, p. 105, note *f*) has wrongly printed it *hrun*.

3) It has been asserted that Chin. *Lo sie* 邏些 (*Kiu Tang shu*, Ch. 196 a, p. 1 b) and *Lo so* 邏娑 (*Tʻong shu*, Ch. 216 a, p. 1) are intended to render *Lha-sa*, the capital of Tibet (BUSHELL, *l. c.*, p. 93, note 6; ROCKHILL, *J. R. A. S.*, 1891, p. 190; and CHAVANNES, *Documents*, p. 178). This identification seems to me rather improbable. The Tibetan word *lha* is phonetically *χla*; the initial *χ* is not a prefix which could be dropped, but an integral part of the stem, which is still preserved in all dialects. It is not likely that the form *χla* would be rendered in Chinese exclusively by the one syllable *lo* (formerly *la*, *ra*). The strict reconstruction of *Lo sie* and *Lo so* is *Ra sa*; and *Ra sa* ("Goat's Land"), as is well known, is the ancient name of the city of Lhasa, before it

In connection with this list of Tibetan offices and officials it may be appropriate to examine the designations of the Tibetan Boards of Ministry, as handed down in *T͑ang shu* (Ch. 216 上, p. 1). Not only are the Tibetan names here transcribed, but also their meaning is added in Chinese, so that for the restoration of the Tibetan originals a double test is afforded, — phonetic and semasiological. Nine ministries are distinguished:

1. *lun ch͑i* 論苴, styled also *ta lun* 大論 (that is, "great *lun*," Tib. *blon c͑en*) with the meaning 大相, "great minister." BUSHELL (*l. c.*, p. 6) transcribes the title *lunch͑ai*, although the Glossary of the *T͑ang shu* (Ch. 23) indicates the reading of the character 苴 as *ch͑i* (昌止). From the double interpretation of the term *lun ch͑i* it follows that it represents Tib. *blon c͑e*, "great minister."

2. *lun ch͑i hu mang* 丨丨扈莽, styled also *siao lun* 小論 (that is, "small *lun*," Tib. *blon c͑uñ*) with the meaning 副相, "assistant minister." Chin. *mang* strictly corresponds to Tib. *mañ*, "many." Chin. initial *h*, as noticed above under No. 18, represents Tib. *g* with or without prefix, and Chin. *u* represents Tib. *o*, so that Chin. *hu*, I am inclined to think, is the equivalent of Tib. *mgo*, "head." In this manner we obtain Tib. *blon c͑e mgo mañ*, "the many heads (assistants) of the great minister." I have not yet been able to trace this expression in any Tibetan record, but it may turn up some day.

---

received the latter name (CHANDRA DAS, *Dictionary*, p. 1161). The Chinese, as shown by their mode of transcription, were acquainted with the name *Ra-sa*, and perpetuated it even after the change of the name in Tibet. KOEPPEN (*Die lamaische Hierarchie*, p. 332) indicates *Julsung* as a designation of the city after VIGNE, and explains this *yul gsuñ*, "land of the teaching." This, of course, is impossible: those words could mean only "teaching, or words of the land." But the reconstruction is erroneous: VIGNE's transcription is intended for *yul gźuñ*, "centre, capital of the land."

6

3. *si(t) pien cʰê pu* 悉編掣 [1]) 逋 with the significance *tu hu* 都護, "commander-in-chief," corresponding to Tib. *srid* [2]) *dpon cʰe-po* (*srid*, "government, ruler, commander;" *dpon*, "master, lord;" *cʰe-po*, "the great one"), "the great commander."

4. *nang lun cʰê pu* 囊論掣逋 with the meaning *nei ta siang* 內大相, "chief minister of the interior," corresponding to Tib. *naṅ blon* (exactly so in the inscription No. 13) *cʰe-po*, "great minister of the interior." [3])

---

1) In the inscription 折.

2) Another explanation is possible. Chin. *si* 悉 is also capable of rendering a Tibetan initial *s-*, when followed by a consonant, as shown by *si lung* 悉弄 = Tib. *sroṅ* in the name of *Kʰri du sroṅ* mentioned on p. 74. Theoretically we should thus arrive at a Tibetan word *spon* (= Chin. *si pien*), which would represent the equivalent of *dpon*. While this alternation between prefixed *d* and *s* is possible, there is as yet no evidence that *dpon* was also anciently sounded *spon*; but the case deserves consideration, if such a reading should ever occur in an ancient text Provisionally I therefore prefer to adhere to the restitution *srid dpon*.

3) He is styled also *lun maṅ je* 論莽熱. The latter word is repeatedly utilized in the inscription to render Tib. *bžer*, which I think is an ancient form of *rje*, "lord." The Tib. *blon maṅ bžer* or *rje*, accordingly, would mean "the first among the many ministers." This expression appears also as the title of military officers, as in *Tʻang shu* (Ch. 216 下, p. 4 b): 南道元帥論莽熱沒籠乞悉篦, "the commander-in-chief of the Southern Circuit *Mo lung kʻi si pi* (probably Tib. *Mod sroṅ kʻri spyi*), with the title *blon maṅ rje*." *Kiu Tang shu* imparts only his title without his name. In this respect great caution is necessary, in that the Tʻang Annals frequently designate Tibetan officials merely by their titles, not by their names. The commander in question was captured in 802 by Wei Kao, and sent on to the Chinese emperor, who gave him a house to live in. On this occasion it is repeated in *Kiu Tʻang shu* (Ch. 196 下, p. 8 b) that *mang je* denotes with the Tibetans the great minister of the interior. The title *maṅ rje*, indeed, occurs in Tibetan: a contemporary of King *Kʻri sroṅ* was *Sva maṅ rje gsal* (*dPag bsam ljon bzan*, p. 171), and the son of King *Maṅ sroṅ* was *ɑDus sroṅ maṅ rje* (ibid., p. 150). Analogous titles are *maṅ sroṅ*, *maṅ btsun*, *maṅ bza* (title of a consort of King *Sroṅ btsan*). — In the following passage a gloss is imparted for the word *je*. In *Tʻang shu* (Ch. 216 下, p. 7 a) mention is made of a general *Shang kʻung je* 尚恐熱, military governor of *Lo mén chʻuan* 落門川, with the family name *Mo* 末, and the name (名) *Nung li je* 農力熱, "which is like the Chinese title *lang* ('gentleman') 猶中國號郞." Chin. *mo* (ancient sounds *mwat* and *mwar*), I am inclined to think, is intended for the Tibetan local and clan name *Mar* or

5. *nang lun mi ling pu* 囊論覔零逋 with the meaning 副相, "assistant minister" (that is, of No. 4). The sound *mi* was anciently *bi* (compare the Japanese reading *beki*). Since the ministers of the interior are divided into three classes, the first and the third of which are designated as "great" and "small," the Chinese transcription *bi-ling-pu* naturally refers to the Tibetan word *ɑbriṅ-po*, "the middle one of three." We arrive at the result: Tib. *naṅ blon ɑbriṅ-po*, "the middle minister of the interior," or "the minister of middle rank."

6. *nang lun chʿung* 囊論充 with the meaning 小相, "small minister," corresponding to Tib. *naṅ blon cʿuṅ*, "small minister of the interior." [1])

7. *yü χan (han) po chʿê pu* 喩寒波掣逋 meaning *chêng shi ta siang* 整事大相 (translated by BUSHELL [*l. c.*, p. 6] "chief

---

ɑBal (Inscription Nᵒ. 16); the words *nung li je* seem to represent Tib. *luṅ ri rje*, "the lord of valleys and mountains," and it is this Tibetan word *rje* to which the Chinese gloss *lang* refers. The words *shang kʿung je* (Tib. *žaṅ kʿoṅ* [?] *rje*) are certainly not part of the name, but a title. In *Sung shi* (Ch. 492, p. 1) we meet under the year 1029 the title of a Tibetan minister *Lun kʿung je* 論恐熱 (Tib. *blon kʿoṅ* [?] *rje*).

1) It is notable that both Tib. *cʿuṅ* and Chin. 充 agree in tone, which is the high tone. The importance of the tone for Tibeto-Chinese transcriptions is discussed on pp. 79 and 105. — In 751 and 754 the Chinese vanquished Ko-lo-fêng, king of Nan-chao, who took refuge with the Tibetans. These conferred upon him the title *tsan pʿu chung* 贊普鍾, that is, "younger brother of the *btsan-pʿo*" (not *po*, as is always wrongly restored; see the note on this subject farther on), *chung* in the language of the "barbarians" signifying "younger brother." M. PELLIOT (*B. E. F. E. O.*, Vol. IV, 1904, p. 153), who has translated this passage, observes, "C'est probablement le *čung* tibétain." This is not quite exact. The Tibetan word here intended is *gcuṅ* (*gčuṅ*, pronounced *čuṅ* in the high tone), the respectful word (*že-sai skad*) for a younger brother (otherwise *nu-bo*), with which Chin. 鍾 exactly harmonizes in sound and tone; this equation (as many other examples in the inscription) proves that the prefixed *g* was not then articulated. The Tibetan word *cʿuṅ* (*čʿuṅ*), "small, young," may denote the younger of two brothers, but cannot be rendered by the Chinese palatal tenuis, only by the aspirate, as proved by the above case Tib. *cʿuṅ*, "small," = Chin. 充 *chʿung*. A Tibetan initial aspirate is regularly reproduced by the corresponding Chinese aspirate.

consulting minister"), corresponding to Tib. *yul* [1]) *rgan-po c͑e-po.*
Chin. *han* answers to Tib. *rgan*, as we saw in the inscription
No. 18; *rgan-po* is still the elder or head man of a village, and
the Tibetan term relates to local (*yul*) administration.

8 and 9 do not require any further discussion. They are Tib.
yul rgan ᶜbriṅ-po (Chin. *yü han mi liṅ pu*), "the middle minister
of local administration," and Tib. *yul rgan-po c͑uṅ* (Chin. *yü han
po ch͑ung*), "the small minister of local administration."

These nine Boards are styled collectively *shang lun ch͑ê pu t͑u kiü*
尚 論 掣 逋 突 瞿, which is considered by me as a transcription
of Tib. *žaṅ blon c͑e-po dgu*, "the Nine Great Ministers." The word
žaṅ is fully discussed on p. 104. The word *t͑u* 突 formerly had the
initial *d* (Japanese reading *dochi*, Annamese *dout*), the word *kiü*
瞿 had the initial *g* (Japanese *gu*). [2])

The phonetic phenomena to be inferred from the Chinese transcriptions of Tibetan words may be summed up as follows.

We gain an important clew as to the determination of the two
vowel signs for *i*, the graphic differentiation of which in the ancient texts has been discussed above (p. 53). The inverted *i*, transcribed by me $i̤$, occurs in four examples: *myi̤ṅ* (= modern *miṅ*) =
名, *p͑yi̤* = 紕, *k͑ri̤* = 綺立, *zi̤gs* = 昔 *sik*. [3]) Hence it fol-

---

1) Chin. *yü* 喻 = Tib. *yul* occurs likewise in proper names. The *Sung shi* (Ch.
492, p. 2) mentions under the year 991 a governor (折 逋 = Tib. *c͑e-po*, "great")
of Si Liang-chou 西涼州, by name *Ngo yü tan* 阿喻丹, corresponding to Tib.
mṄa (compare 阿里 = Tib. *mṄa-ri(s)*) *yul brtan*; and under 994 a governor *Yü lung
po* 喻龍波, being Tib. *Yul sroṅ-po.*

2) It renders the syllable *go* in *Gotama* (T. WATTERS, *Essays on the Chinese Language*, p. 388), in *Gopāla* (*Life of Hüan Tsang*) and *Suvarṇagotra* (*Memoirs of Hüan Tsang*).

3) A fifth example is afforded by 悉 *sit* transcribing Tib. *srid* in the third Minis-
terial Board mentioned in *T͑ang shu*, and *srid* is written with inverted *i̤* in the sworn
treaty of 822 (9—11).

lows that the ancient Tibetan sound i exactly corresponded to the plain, short Chinese i. For the vowel i written in the regular modern form we have three examples; namely, mc^{c}ims = 琛 c^{c}öm, rtsis = 資思 tse (tsi)-se, and cig = 巿 ši. These varying Chinese transcriptions prove that this Tibetan vowel did not sound to the Chinese ear like a definite i, but must have been of somewhat indistinct value, something between i, $\ddot{\imath}$, [1]) and $\ddot{o}$.

The comparison of allied words which Tibetan and Chinese have in common is apt to confirm this result. There are Chinese sę 四 ("four") corresponding to Tibetan (b)ži, Chinese sę 死 ("to die") corresponding to Tibetan ši, indicating that Tibetan i was an equivalent of this indistinct Chinese vowel ę. The two Tibetan signs for i, therefore, have great significance in the comparative study of Indo-Chinese languages; and their distinction in the ancient monuments must be conscientiously noted and registered, instead of being neglected,[2]) as was done by Mr. WADDELL. The inscription of 822 indicates that the two timbres of i were still fairly discriminated, but that they were already on the verge of a mutual fusion, as shown by a certain wavering in the employment of the two signs. Thus we find in line 43 gñis, but in line 50 gñis; in line 43 kyi, in line 50 kyi; and other inconsistencies. Perhaps the phonetic differentiation was already wiped out at that period, and only the graphic distinction upheld on traditional grounds.

---

1) Compare SCHAANK, *Ancient Chinese Phonetics* (*T'oung Pao*, Vol. VIII, 1897, p. 369). — On the other hand, Chin. i is rendered by Tib. e in the nien-hao *King lung* 景龍 transcribed Tib. *Keń luń* (in the inscription of 783), probably sounded *Köń* (compare *čöń kuan* 貞觀 = Tib. *čeń kvan* [*ibid.*; accordingly, Tib. e = Chin. *o*]). For this reason it is possible that Chin. *king*, as heard at that time by the Tibetans, was sounded *k'oń* (compare Korean *kyòng*). Chin. *ti* 帝 (in *huang ti*) is transcribed by Tib. *te* (compare Jap. *tei*, Annamese *de*). *Vice versa*, Tib. *ne* in the inscription (above, Nos. 9 and 13) is rendered by Chin. *ning* (but Hakka *leń*, Korean *yòng*), which, in my opinion, goes to show that Tib. *ne* was nasalized: *nę(nē)* or *nẹ*.

2) The hypothesis of the two i's serving for the distinction of short and long i is herewith exploded once for all.

The most signal fact to be gleaned from the Tibeto-Chinese concordances is that phonetic decomposition, which was hitherto regarded as a comparatively recent process of the language, was in full swing as early as the first half of the ninth century. The superscribed and prefixed letters were already mute at that time in the dialect of Lhasa: *blon* was articulated *lon*, *btsan* was *tsan*, *bzań* was *zań*, *bžer* was *žer*, *bka* was *ka*, *lta* was *ta*, *lto* was *to*, *gtoys* was *tog*, *rqyal* was *gyal*, *rgan* was *gan* (probably *χan*); *brtan* was even sounded *tan* 且. Superscribed *s*, however, seems to have been preserved throughout: the pronunciation of *stang* and *snam* is indicated as *stang* and *snam*, that of *snag* and *stag* as *snag* and *stag*. *P^cyi* was sounded *p^c'i*; the alteration of the palatalized (*mouillé*) labials into palatal *ć* and *ć^c* had apparently not yet taken effect. In the combination of two monosyllables into a unit, the prefix of the second element, when the first terminates in a vowel, was articulated and connected into a syllable with the first element, exactly in the same manner as at present. This is exemplified by the interesting transcription *t^cam-t^cung* for Tib. *lta mt^coń* (No. 10), which simultaneously proves that the word *mt^coń* when isolated was pronounced *t^coń*, and by the transcription *ngan pén* for Tib. *mña dpon* (No. 16).[1]) Compare in recent times the name of the monastery *dGa-ldan*, pronounced *Gan-dan*, hence Chin. *Kan-tan* 甘丹; and Tib. *skye dman* ("woman"), pronounced *kyen* (or *kyer*) *män*, hence transcribed *king mien* 京面 in the Tibetan vocabulary inserted in *T^cao-chou t^cing chi* 洮州廳志, 1907 (Ch. 16, p. 48).

Of final consonants, *d*,[2]) *g*, *n*, and *ń* were sounded. Final *s* was

---

1) Compare also the above *žań blon c'e dqu*, which, judging from the Chinese mode of transcription, must have been articulated *c'et-gu*.

2) Final *d* was pronounced in *Bod*, as indicated by the transcription 蕃 **pat, pot, pön*. It is incorrect, as Mr. ROCKHILL (*J. R. A. S.*, Vol XXIII, 1901, p. 5) asserts, to say that "the word *Bod* is now, and probably always has been, pronounced like the French *peu*."

sounded when it followed a vowel (ʻʻdus), but it was eliminated when following a consonant (mcʻims was sounded cʻim, zịgs as zik).[1]

In regard to final *l*, I feel somewhat doubtful. If my identification of 喩 *yū*, which had no final consonant, with Tib. *yul*, holds good, this would rather indicate that final Tib. *l* was not sounded, or but indistinctly. The transcription 頰 *kiap* (γʻap) for *rgyal* in the inscription No. 11, however, may point to a pronunciation *gyal* (gʻal, γʻal). On the other hand, in the list of royal names in *Tʻang shu* (Ch. 216 上, p. 2a) we find the word *rgyal* rendered by 瘕 *kia* (BUSHELL [l. c., p. 9] transcribes *hsia*; Glossary of *Tʻang shu* 古牙 *ku ya*) in the first of King *Sroṅ btsan's* an-

---

JÄSCHKE, in the Phonetic Tables of his *Dictionary* (p. XVI), indicates the pronunciation *bhod* for Spiti, *ʻod* for Kʻams, *bhoʻ* for Tsang and Ü. In the latter the initial is an aspirate media, and, besides, the word has the deep tone; it has accordingly nothing in common with French *peu*. Mr. ROCKHILL himself (p. 6) indicates that in the tenth and eleventh centuries the sound *peu* was transcribed 不德 *pu-té* and 孛 (or 伯) 特 *po tʻé*; but surely it was not the sound *peu*, but the sound *bod*, which is clearly enough indicated by these transcriptions. If *bod* was thus sounded in the tenth and eleventh centuries, we are bound to presume that this pronunciation held its ground also in the preceding Tʻang period. Skr. *Bhoṭa* and Ptolemy's Βχʋται afford additional evidence for an ancient indigenous *Bod* sounded *bot*.

1) In final *s* a distinction must be drawn between the suffix -*s* (called Tib. *yaṅ ʻjug*) and radical *s* inhering in the stem. The latter seems to have survived until comparatively recent times, if we may rely upon the transcription 烏思藏 *Wu-se tsang* of the *Ming shi* for Tib. *dBus gTsaṅ* (the two large provinces of Central Tibet); the Chinese equivalent must be based on a Tibetan pronunciation *vus tsaṅ* during the Ming period, while the new transcription 衞 *Wei*, rendering the word *dBus* in the age of the Manchu, clearly indicates that the final phonetic decay resulting in the modern *vui, vü, ʻü*, is an after-Ming event. On the other hand, the name of the temple *bSam-yas* is transcribed *Sam-ye* 三耶 by the Chinese pilgrim Ki-ye in the latter part of the tenth century (CHAVANNES, *B. E. F. E. O.*, Vol. IV, 1904, p. 81, who did not identify this locality; this implies that Ki-ye made his return from India to China by way of Nepal and Tibet). Tib. *yas* is *ya* + *s* of the instrumental case (the temple was fine "beyond imagination," *bsam-yas*); *sam-ya* is still the current pronunciation in Central Tibet (JASCHKE, *Tibetan Grammar*, p. 6); but as the ancient pronunciation of 耶 was *ya* (compare 耶婆 *Yava*), it is necessary to assume that Ki-ye, at the time of his sojourn in the famous monastery, heard the pronunciation *Sam-ya*. If he had heard *yas*, he could easily have expressed it by the addition of 悉, as it occurs in 耶悉茗 *yasmin*, "jessamine."

cestors, 痕悉董摩 *Kia si tung mo*, which I provisionally take as reproducing Tib. *rgyal ston-mo*; further, 夜 *ye* in 弗夜 corresponding to Tib. *Bod rgyal*, "king of Tibet," as title of King *Sron btsan*, and 野 *ye* in 宰勃野 *Su pʿo ye* = *Su pʿo rgyal*, the Tibetan name and title of Fan-ni, and in 鶻提勃悉野 *Hu* (Tib. *Gun*) *tʿi* (*dɛ* = Tib. *lde*) *pʿo si ye* (= Tib. *rgyal*), the ancestor of the Tibetans. The Chinese symbols employed in these cases, *kia* and *ye*, correspond to an ancient pronunciation *gia* (γʹa) (Annamese *gia, ja*), without a final consonant, so that they seem to be indicative of a Tibetan sound *gyaʾ* (*gʹaʾ, γʹa*). Final *l* was articulated in the tribal name *Bal-ti* (*rGyal rabs: sbal-ti*), as shown by the Chinese rendering *Pu-lü* 勃律 (CHAVANNES, *Documents*, p. 149), the ancient sounds of this *pu* being *ba* and *bʿa* (Ningpo *ba*, Japanese *botsu*, Korean *pal*; it renders the syllable *bha* in Skr. *Bhamātra*), so that *Pu-lü* appears as a reproduction of Tib. *Bal*.[1])

An interesting example of the treatment of Tib. final *l* in Chinese is afforded by the Chinese word *pʿêng sha*, "carbonate of soda, natron" (*natrium carbonicum*), which has not yet been explained. Li Shi-chên (*Pên tsʿao kang mu*, 石部, Ch. 11, p. 12) confesses his ignorance in the matter (名義未解); and WATTERS (*Essays on the Chinese Language*, p. 378) is wrong in deriving the Chinese word from Tib. *ba tsa* (to which it has not the slightest similarity), "called also *pen-cha*," which is certainly nothing but the Chinese, and not a Tibetan word. The first and oldest mention of the term, as far as I know, is made in *Kiu Wu Tai shi* (Ch. 138, p. 1b), where *ta pʿêng sha* 大鵬砂 ("sand of the great rukh") is enumerated among the products of the Tʿu-po. This very name is suggestive of being the transcription of a foreign word (the

---

1) In *Tʿoung Pao*, 1908, p 3, *Po-lu* was connected by me with *Bolor*, the ancient name of Baltistān; but *Bolor* seems to be derived from *Bal*.

character 硼 certainly is an artificial formation, the two other characters given by WATTERS are taken from the *Pên ts͘ao*). The ancient sounds of the phonetic element *p͘êng* 朋 are *bung, and the Tibetan word answering in sense to the Chinese is *bul* (JÄSCHKE, *Dictionary*, p. 370), so that Chin. *p͘êng* (*bung*) appears as a reproduction of Tib. *bul*, [1]) simultaneously proving that the final *l* in *bul* was sounded; both words agree also in the low tone. [2])

---

1) Also in the ancient allied words of the two languages, Tib. final *l* corresponds to a final nasal in Chinese: for instance, *dṅul*, "silver" = Hakka *ngyin*, Fukien *ngüng* 銀 (*yin*); Tib. (*s*)*brul*, "snake" = Cantonese and Hakka *mong* 蟒 (*mang*, Jap. *bō*). In other cases Tib. final *ṅ* is the equivalent of Chin. final nasal, as Tib. (*ṅb*)*rug*, "dragon" = Chin. *lung* (Jap. *riū*) 龍. But Tib. *buṅ*(-*ba*), "bee" = Chin. *fung* (Korean *pong*) 蜂; Tib. *rṅa-boṅ* (*moṅ*), "camel" = Chin. *pong, fong* 峯, "hump of a camel" (Tib. *rṅa* is related to *rṅog*, "hump"); Tib. *maṅ*, "many" = Chin. *mang* 茫 and 厖; Tib. *spyaṅ*(-*ku*), "wolf" = Chin. *mang* 尨 (Korean *pang*, Jap. *bō*), "Tibetan mastiff."

2) On *p͘eng sha* see P. CIBOT (*Mém. conc. les Chinois*, Vol. XI, pp. 343—346); KLAP-ROTH (*Asiat. Magazin*, Vol. II, pp. 256—261, Weimar, 1802); SOUBEIRAN, *Etudes sur la matière médicale chinoise (minéraux)*, p. 13 (Paris, 1866); F. DE MÉLY, *Les lapidaires chinois*, p. 141; H. H. HAYDEN, *Geology of the Provinces of Tsang and Ü in Central Tibet* (*Memoirs Geological Survey of India*, Vol. XXXVI, pt. 2, 1907, p. 65). — The Chinese loan-words in Tibetan have not yet been studied, and are hardly indicated in our Tibetan dictionaries. Some of them are even passed off as Sanskrit: for instance, *pi-waṅ* or *pi-baṅ*, "guitar," is said to be derived from Skr. *vīṇā*, which is impossible; in fact, it is to be connected with Chin. 琵琶 *p͘i-p͘a*, ancient sounds *bi-ba* (Japanese *bi-wa*, Mongol *biba*). The nasalization of the final vowel *wa* or *ba* is a peculiarity of Tibetan sometimes practised in foreign words (compare *pi-pi-liṅ*, "pepper" = Skr. *pippalī*). The *Tang hiang* 党項, a Tibetan tribe in the region of the Kukunor, according to *Sui shu* (Ch. 83, p. 3), were in possession of *p͘i-p͘a*; according to Chinese tradition, the instrument originated among the *Hu* 胡, a vague expression generally referring to peoples of Central Asia, Iranians and Turks. GILES (*Biographical Dictionary*, p. 889) ascribes its introduction into China to the Princess of Wu-sun. The Djagatai word for it is *pišik* (*Keleti Szemle*, 1902, p. 161). The fact that the Tibetan and Chinese words refer to the same object is evidenced by the Polyglot Dictionary of K͘ien-lung. In the latter we meet also Tib. *coṅ*, "bell" = Chin. *chung* 鐘. There are, further, Tib. *p͘iṅ*, "pitcher, cup" = Chin. *p͘ing* 瓶; Tib. *la-c͘a*, "sealing-wax," from Chin. *la* 蠟, "wax;" Tib. *mog* (-*ša*), "mushroom" = Chin. *mo-ku* 蘑菇; Tib. *ts͘uu* (the double *u* indicates the fourth tone of Chinese), "vinegar" = Chin. *ts͘u* 醋; Tib. *giu* (*gi*)-*waṅ* (*bam*), "bezoar" = Chin. *niu huang* 牛黃 (Jap. *giu-kwō*); Tib. *kau*, "watermelon" = Chin. *kua* 瓜; Tib. *sraṅ*, "ounce" = Chin. *liang* 兩 (Korean *riang*, Jap. *riō*). Tib. *pi-pi*, "flute," and *bid-bid*, "hautboy

On the whole, the probability is greater that the final *l* was

reed," must be connected with *pi* 觱 (**bi, bit*; Korean *pʿil*), originally a horn used by the Kʿiang to frighten horses (definition of *Shuo wén*), but then in the compound *pi-li* 觱篥 a pipe (A. C. MOULE, *Chinese Musical Instruments, J. Ch. Br. R. A. S.*, 1908, p. 84), in *Huang chʿao li kʿi tʿu shi* (Ch. 9, p. 53) figured and described as a reed flute with three holes, metal mouthpiece and broadening funnel, 5.37 inches long, used for dance music by the Turkish tribe *Warka* 瓦爾喀. The word, therefore, is presumably of Turkish origin, but it is much older than the eighteenth century. We meet it in the transcription *pei-li* 貝蠡 in the chapter on music in *Kiu Tʿang shu* (Ch. 29, p. 8 b), where it is defined as a copper horn 銅角, two feet long, of the shape of an ox-horn, in use among the Western Jung 西戎. According to another tradition, it originated in Kucha, Turkistan (*Koʿ chi king yuan*, Ch. 47, p. 6 b). The original Turkish form seems to have been *beri* or *böri* (H. VÁMBÉRY, *Die primitive Cultur des turko-tatarischen Volkes*, p. 145, notes a word *boru*, "trumpet," properly "reed"); and we find this word in Mongol *boriya*, "trumpet," from which Manchu *buren* and *buleri* seem to be derived. The latter corresponds in the Polyglot Dictionary to Chin. *la-pa* 喇叭, Mongol *ghōlin boriyā*, "brass trumpet," and Tib. *zaṅs duṅ*. The Mongol word *rapal* given in the first edition of GILES, and repeated by MOULE, does not exist (Mongol has neither initial *r* nor a *p*); nor can Chin. *la-pa* be derived from Manchu *laba*, as stated in the second edition, the latter being merely a transcript of Chinese, as already pointed out by SACHAROV. *La-pa* is neither Mongol nor Tibetan; it is listed among the musical instruments of Turkistan in *Hui kiang chi* 回疆誌 (Ch. 2, p. 8), published 1772 (WYLIE, Notes, p. 64). The musical instrument *kan tung* 干動, left unexplained by MOULE (*l. c.*, p. 103), is Tib. *rkaṅ duṅ*, the well-known trumpet made from a human thigh-bone; I met also the transcription 剛洞. Among the interesting loan-words of cultivated plants, we have Tib. *se-ḥbru* (pronounced *se-ru*), "pomegranate" (*punica granatum* L.), derived from Chin. 石榴 *shi-liu*, anciently *se(shi)-ru* (Japanese -*ro*). The pomegranate does not thrive in Tibet, and, as is well known, was introduced into China by General Chang Kʿien (BRET-SCHNEIDER, *Bot. sin.*, pt. 1, p. 25; pt. 3, No. 280; HIRTH, *Tʿoung Pao*, Vol. VI, 1895, p. 439; *Pén tsʿao kang mu* 果部, Ch. 30, p. 8) Whether Chin. *ru, ro*, is connected with Greek ῥόα or Arabic *rummān*, Amharic *rūmān* (SCHRADER in HEHN, Kulturpflanzen und Haustiere, 8th ed., p. 247), I do not venture to decide. The Tibetan word must be regarded as a loan from Chinese, and not as indigenous, as W. SCHOTT (*Entwurf einer Beschreibung der chinesischen Litteratur*, p. 123, note, Berlin, 1854) was inclined to believe, who explained the word as being composed of Tib. *se*, "rose," and *ḥbru*, "grain, seed." These Tibetan words (the meanings "pomegranate" and "rosebush" interchange in South-Slavic) were doubtless chosen as elements of the transcription, because they conveyed to the national mind some tangible significance with reference to the object (in the same manner as there are numerous analogous cases in the Chinese transcriptions of foreign words). The Central-Tibetan pronunciation *sen-ḍu* and Ladākhi *sem-ru* represent secondary developments suggested by the mode of spelling, and application of phonetic laws based thereon (nasalization of the prefix *ḥ*, transcribed *nen* 恩 in *Hua i yi yü*).

articulated than that it was suppressed, and the same remark holds good of final *r*. For the latter we have the only example in the word *bžer*, transcribed by Chin. *že* (Nos. 10, 13—15, above). In this case the Chinese transcription certainly is not conclusive, since Chinese lacks final *r*, and, taking into consideration that the other finals were heard, there seems good reason to assume that *bžer* was pronounced *žer* at that period.

The subjoined *r* was still clearly sounded in the guttural and labial series. The word *kʿri*, as evidenced by the Chinese transcription *kʿi-li*, [1]) was actually heard as *kʿri* (not as at present, *ṭ ʿi*);

Lolo *sebuma* (P. VIAL, *Dict. français-lolo*, p. 176, Hongkong, 1909) possibly points to a former Tibetan articulation *seb-ru*.

1) The Tʿang Annals employ various methods of transcribing the word *kʿri* ("throne") in the beginning of the names of the Tibetan kings, 乞黎, 乞立, 棄隸 (*kʿi li*), and also only 棄 and 器 (*kʿi*). Probably also *kʿo* (*ka*) *li* 可黎 in the name *Kʿo li kʿo tsu* 可黎可足 (being identical with the Tibetan king Kʿri-lde sron-btsan) is the equivalent of Tib. *kʿri*. The Chinese rendering of his name has not yet been explained. The elements *lde sron btsan*, the Chinese equivalents of which are well known to us, cannot be made responsible for Chin. *kʿo tsu* (ancient sounds *ka tsuk*). In *rGyal rabs* this king is designated also *Kʿri gtsug lde btsan Ral-pa-can*; so that we are bound to assume that the Chinese name *Kʿo-li ka-tsuk* is intended for the first two elements of this Tibetan name, *Kʿri gtsug*. It is singular, however, at first sight, that in this case the prefixed *g* is expressed by the Chinese syllable *kʿo* (*ka*), while in another royal name Tib. *gtsug* is transcribed in Chinese regardless of the prefix (see p. 92, note 2). The Tibetan prefix is often preserved in the second element of a compound if the first word terminates in a vowel; the words *kʿri gtsug* could be sounded *kʿrik-tsug*, and hence the Chinese mode of transcription. The case is analogous to that of *lta mtʿon* pointed out on p. 86. An interesting Chinese transcription of a Tibetan word showing the preservation of *r* is the word *pʿu-lu* 氆氌, "woollen cloth," a reproduction of Tib. *pʿrug*. As far as I know, the Chinese term does not occur in the Tʿang period, but only from under the Yüan. The mode of writing (Manchu *pʿuru*) presupposes a Tibetan pronunciation *pʿru*', for the phonetic element *lu* 魯 is devoid of a final consonant. In the age of the Tʿang, when the word sounded *pʿrug* also in the dialect of Lhasa, a complement sounding *luk*, for instance 祿, would have doubtless been chosen in forming the second character in the word. The very mode of transcription thus betrays a post-Tʿang origin, but it must result from a time when the initials *pʿr* were still in full swing and had not yet undergone the *lautverschiebung* into the cerebrals *ṭ ʿr*, *ṭ ʿ* (see also KLAPROTH, *Description du Tubet*, p. 50, Paris, 1831; T. WATTERS, *Essays on the Chinese Language*, p. 378).

kru (*kü-li*) was sounded *kru*; the word *ḁbro* (No. 19) was sounded *bro*, and *ḁbriñ*, as shown by the Chinese transcription *bi-liñ*, was articulated *briñ*, [1]) and *blo* was *blo*. In the combination *sr*, the *r* seems to have been dropped, if the identification of 悉 *sit* in *T°ang shu* with Tib. *srid* holds good.[2])

---

1) Another good example of the initials *br* being sounded with perfect clearness is presented by the word 拂廬 *fu-lu* (*bu-ro*), imparted in the T'ang Annals as a gloss for the Tibetan word meaning "a felt tent." The word intended apparently is Tib. *sbra*, "felt tent," still sounded *bra* in western Tibet and so likewise in the T'ang period. The Chinese syllable *bu* reproduces the initial *b*, and the syllable *lu* the Tib. *ra*. It is strange, however, that the Chinese did not choose in this case an element *ra*, *la*; but this may be easily accounted for by the fact that the above Chinese word *lu* means "a hut, a hovel," and also the tent erected for the wedding ceremony. As in so many other cases, the Chinese selected a word approximately imitative of the foreign sound, and simultaneously indicative of the significance of the foreign word. The Tibetan word *gur*, "tent," can certainly not be sought in the Chinese transcription, as 拂 never had the sounds *gu* or *ku*. A good modern example of Chinese rendering of Tib. *br* is 老木郎 = Tib. *Lha-braṅ*; in this compound the second element is still pronounced *braṅ* (but never *daṅ*) throughout Tibet, while *p°o-braṅ*, "palace," is always *p°o-daṅ*. These two elements *braṅ*, therefore, seem to be two words of different origin.

2) But the word *sroṅ* in the names of several kings was doubtless articulated *sroṅ*, as evidenced by the transcriptions in the T'ang Annals 宗 (*tsung*; Japanese *sō, su*) 弄 (*lung*), *si lung* 悉弄, *su lung* 蘇籠 and *si lung* 悉籠. Mr. ROCKHILL (*The Life of the Buddha*, p. 211) is inclined to think that Chin. *K°i tsung lung tsan* renders Tib. *K°ri ldan sroṅ btsan*; but Chin. *tsung* cannot reproduce Tib. *ldan* (pronounced *dan*). In my opinion, the Chinese words are intended only for *K°ri sroṅ btsan*. In regard to the name of King *K°i li so tsan* 棄隷蹜贊, Mr. ROCKHILL (p. 217) takes it as "giving a quite correct pronunciation of the four first syllables of his Tibetan name," that is, *K°ri lde gtsug btsan*. But Chin. *li* cannot represent an equivalent of Tib. *(l)de*, which, as pointed out on p. 74, is rendered by Chin. *lie* 獵. The Chinese words exactly reproduce the Tibetan words *K°ri* (*g*)*tsug* (*b*)*tsan*. The character 蹜 is sounded in Cantonese *shuk*, Korean *suk*, *č°uk*, Japanese *shuku*, and seems to have had in the T'ang period the value of *tsuk*, *dzuk*. BUSHELL (*The Early History of Tibet*) unfortunately availed himself of the Wade system in the transcription of Tibetan names, so that they are useless for the purpose of identification, and wrote names sometimes consisting of five and six syllables into one solid word without divisions, which led his successors into error; for instance, HERBERT MUELLER (*Tibet in seiner geschichtlichen Entwicklung, Z. f. vergl. Rechtswissenschaft*, Vol. XX, p. 325), who transcribes *Ch'in-u-hsi-lung* instead of *K°i nu si lung*. An error of transcription was committed by BUSHELL (pp. 5, 39) in the name written by him after *Kiu T'ang shu Sohsilungliehtsan* (and so repeated by ROCKHILL, p. 219, and MUELLER, *l.c.*), where *P°o* (婆, confounded with

Initial and final consonants, in general, were still intact, but prefixed consonants were doomed to being silent. It it natural that tones began to be developed in consequence of this phonetic disintegration (p. 79); for we know, particularly from CONRADY's researches,

---

婆) si lung lie tsan (corresponding to Tib. *P̊o sroñ lde btsan*) must be read. The *T̓ang shu* (Ch. 216 上, p. 8a) writes the same name 挲悉籠臘贊 *So* (ancient sound *sa*) *si lung la tsan*; nevertheless BUSHELL's *So si lung lie tsan* remains inexact, as we have either *P̊o si lung lie tsan* of the *Kiu T̓ang shu* or *So si lung la tsan* of the *T̓ang shu*. The latter spelling, however, is erroneous. The historical observation inserted by Mr. ROCKHILL shows that this is a case of importance, as, according to him, this name has not yet been traced in Tibetan history. But if names are wrongly transcribed and inexactly restored, any attempt at identification is naturally hopeless at the outset. All the Tibetan words and names encountered in the T̓ang Annals are capable of rigorous philological research; and when this is properly carried through, much of the alleged diversity between Chinese and Tibetan traditions (BUSHELL, p. 4) will be blown up into the air. Mr. ROCKHILL's conclusion that in the *T̓ang shu* the king *So si lung lie tsan* is inserted between *K̓i li so tsan* and *K̓i li tsan*, whereas all Tibetan histories are unanimous in affirming that *K̊ri sroñ* succeeded his father on the throne, is not at all to the point; likewise BUSHELL (p. 5) is wrong in making *So si lung lie tsan* and *K̓i li tsan* two individuals and two different kings. They designate, indeed, one and the same personage, who is none other than the Tibetan king *K̊ri sroñ lde btsan*. This name appears in both *T̓ang shu* as that of the king who died in 755 (BUSHELL, p. 39), but this is the same king previously styled *K̓i li so tsan* (*K̊ri gtsug btsan*), so that it is evident beyond cavil that it is simply a clerical error which here crept in when the annalist copied from his state documents. It was *K̊ri gtsug btsan* who died in that year; and it was his son *K̊ri sroñ lde btsan* who succeeded to him, and who was styled — the annalist meant or ought to say — also *P̊o sroñ*. This reading of *Kiu T̓ang shu* is doubtless correct, whereas the *so* of the New Annals must be a clerical error. Tib. *p̊o*, "the male," is an ancient title occurring in the names of the Tibetan kings, as will be seen below in a discussion of the word *btsan-po*, which had originally the form *btsan p̊o*, "the warlike one, the male." Likewise *rgyal-po*, "the king," was originally *rgyal p̊o*, "the victorious male" (compare WADDELL, *J. R. A. S.*, 1909, p. 1268, whose explanation is certainly a fantasy; the title *p̊o* implies nothing derogatory). It is worthy of note that also the chief consort of the king, *P̊o yoñ* (or *γyoñ*) *bza*, bore the title *p̊o* in her name, whereas his other wives were not entitled to this privilege. In the transcription 勃弄若 *P̊o lung* (BUSHELL, p. 9, *mung*) *jo* (= Tib. *žo*) the same title *P̊o sroñ* appears in the name of the fifth of King *Sroñ btsan*'s ancestors (*T̓ang shu*, Ch. 216 上, p. 2 a). The title *P̊o rgyal* occurs in the name 宰勃野 *Su p̊o ye* (*gia*), adopted by Fan Ni 樊尼 on his election as king of the T̓u-fa (BUSHELL, p. 6), and in the name of the ancestor 祖 of the T̓u-po, 鶻提勃悉野 *Hu* (Tib. *guñ*) *t̓i* (Tib. *lde*) *p̓u* (Tib. *(p̊o)*) *si* (possibly Tib. *srid*) *ye* (Tib. *rgyal*) = Tib. *Guñ* (see p. 78) *lde p̊o srid rgyal*.

that tones are the substitutes of eliminated consonants. Presuming that writing, when introduced in the first half of the seventh century, rather faithfully fixed the condition of the language as then spoken, we are confronted by the fact that the first stage in the process inaugurating the remarkable phonetic decomposition of the Tibetan language took place within a period of hardly a century and a half. In the first part of the ninth century a deep gulf was yawning between the methods of writing and speaking, and due regard must be taken of this fact in our studies of the manuscripts of that epoch. The natural tendency of writing words in the same manner as they came from the lips of the speakers was then steadily growing. The inscription of 822 (above, No. 17) furnishes a curious example in writing the word *bka* with the single letter *k*, which, even more than the Chinese transcription *kia*, is undubitable proof that it was sounded simply *ka*.[1]

---

1) For the present I refrain from a discussion of the laws underlying the Chinese method of transcribing Tibetan words, as several intricate points remain to be cleared up. It will be observed that this method in some respects differs from what we are wont to have in the case of Sanskrit, Turkish, and Persian transcriptions, and that in the face of Tibetan the Chinese were compelled to struggle with difficulties which they did not encounter in other foreign languages. It is manifest that the Chinese transcriptions, as we have them now, were recorded at the time when the decomposition of the Tibetan prefixes and initials had set in, and when the tone system sprang into existence. The tones could not escape the Chinese ear, and were bound to influence their manner of transcribing. The fact that the new initials were affected by the eliminations of the prefixed consonants, most of which were grammatical elements of formative functions, is evident from what we observe in the modern dialects; thus far, however, we are not in a position to frame any definite conclusions in regard to such changes during the ninth century. Nevertheless they must have taken place, as we see from several parallels in the inscription of 822. Whereas all the Tibetan true initial aspirates are exactly reproduced by the corresponding Chinese aspirate, we notice that Chinese has an aspirate where Tibetan offers a tenuis + silent prefix; for instance, Tib. (*l*)*ta* = Chin. *t'an* 貪 (No. 10), and Tib. (*l*)*to* = Chin. *t'u* 土 (No. 14). Whether Tib. *t* was really aspirated or changed into the aspirate media *d'*, I do not venture to decide; but the Chinese transcriptions are a clear index of the fact that the tenuis had undergone some sort of revolution prompted by the elision of the prefixed *l*. In other instances, judging from

Phonology of Document Pelliot.

We now enter into a discussion of the phonology of the text of *document Pelliot*. M. BACOT himself has noted the addition of

the Chinese transcriptions, the tenuis remained unaffected; as, *gtogs = tuk* 篤, *brtan = tan* 旦. This case is of importance when we meet Tibetan names in the Chinese annals and are intent on restoring them to their original forms. Take, for instance, the name of the king 陀土度 *Tʿo tʿu tu(k)* (*Tʿang shu*, Ch. 216 上, p. 2 a), the second in the series of the ancestors of King *Sroṅ btsan*. At first sight, I felt much tempted to recognize in the first two elements the Tib. *tʿo tʿo* occurring in the name of King *Lha tʿo tʿo ri* of Tibetan tradition, but due regard paid to the case just cited makes me skeptical: the Chinese dental aspirate may correspond to this sound in Tibetan, but it may express also Tib. *lt* (hence also *rt*, and probably *st*). Since 土 in the inscription is the equivalent of *lto*, it may very well be that this is the case also in the above name, which may be restored *Tʿo lto bdag* (度 **dak*; Japanese *taku, do*; Korean *tʿak*). This consideration has a bearing also on the interpretation of the tribal name 吐蕃 *Tʿu po (fan)*, the second element of which has correctly been identified with Tib. *bod*; for the first element, Tib. *stod*, "upper," and *mtʿo*, "high," have been proposed (the various theories are clearly set forth by L. FEER, *Etymologie, histoire, orthographe du mot Tibet, Verh. VII. Or.-Congr.*, pp. 63—81; and YULE and BURNELL, *Hobson-Jobson*, p. 917). The first objection to be raised to these identifications is that they are merely based on guesswork, and not on any actual name of Tibet found in Tibetan records. Neither in *rGyal rabs* nor in any other Tibetan history did I ever come across such a name as *stod bod* or *mtʿo bod*, but Tibet and Tibetans are simply called *Bod*, with or without the usual suffixes. It is true, Mr. ROCKHILL (*J. R. A. S.*, 1891, p. 5) is very positive in his assertion that "Tibetans from Central Tibet have at all times spoken of that portion of the country as *Teu-Peu* (*stod bod*) or 'Upper Tibet,' it being along the upper courses of the principal rivers which flow eastward into China or the Indian Ocean" (in his *The Life of the Buddha*, p. 216, he still adhered to the fanciful *tʿub-pʿod* etymology of SCHIEFNER), but no documentary evidence for this statement is presented; and, as long as such is not forthcoming, I decline to believe in such invented geographical names as *stod bod* and *mtʿo bod*, alleged to have resulted in the Chinese word *Tʿu-po* of the Tʿang period. From a philological point of view, it is entirely impossible to restore Chin. *tʿu* 吐 to Tib. *stod*, for in the same manner as its phonetic element 土, it was never provided with a final consonant; it may be restored to a Tib. *tʿo, lto* or *sto* (*mtʿo* seems very doubtful). The Tʿang Annals impart an alleged older name 禿髮 *Tʿu-fa*, which was subsequently corrupted 語訛 into *Tʿu-po*. Mr. ROCKHILL (*l.c.*, p. 190) comments on this name that "the old sound of *fa* in *Tʿu-fa* was *bat* or *pat*; consequently *Tʿu-fa* represents *Teu-peu* (*stod bod*), our Tibet." I regret being unable to follow this demonstration; *tʿu* cannot represent *tö*, and *pat* does not represent *bod*. The word *tʿu* 禿 was anciently possessed of a final *k*, so that we have *tʿuk pat*, which certainly has nothing to do with *stod bod* or *mtʿo bod*, or anything like it. It is clearly indicated in the Tʿang Annals that the word *Tʿu-fa*

the subscribed letter *y* after *m* when followed by the vowels *e* or
i. We find here *myed = med*, "there is not;" *bud-myed = bud-
med*, "woman;" [1]) *mye = me*, "fire;" *myi = mi*, "man;" *myi =*

(apparently a nickname) was not of Tibetan origin, but derived from Li Lu-ku 利鹿孤
of the Southern Liang dynasty and carried over to the K'iang tribes by his son Fan Ni
樊尼. The name *T'u-fa*, accordingly, is not capable of restitution into Tibetan, and
the alleged change of the tribal name from *T'uk-pat* into *T'u-pot* is merely inspired by
a certain resemblance of these names. Nor can the Arabic designation تبت of Iṣṭakhri,
Khordādba, etc, which has been variously spelled Tobbat, Tibbat, etc., be set in relation
with this alleged *T'u(k)-pat*, as only the consonants are expressed by Arabic writing, and
the vowels are optional; it offers no valid proof for the attempt at restoring the
original Tibetan form, but it shows in the case of Iṣṭakhri that a name for Tibet with
the consonants *Tbt* existed toward the end of the sixth century. *T'u-po* must be regarded
as the correct and original tribal designation; but as to the proper Tibetan equivalent of
吐, we have to await thorough evidence. It is hoped that a Tibetan gloss for it will
turn up in some *document Pelliot.* — The identification of Tibetan proper names in the
T'ang annals with those of the Tibetan annals is beset with difficulties, as many names
of the Chinese annals are not mentioned by the Tibetans or given by them in a form
not identical with the Chinese. The famous minister *mGar*, as already recognized by
Rockhill (*The Life of the Buddha*, p. 216), is identical with *Lu tung tsan* 祿東贊
with the name *Ku* 薛氏 in *Kiu T'ang shu* (Bushell, *l.c.*, p. 12). Theoretically I
should restore *Lu tung tsan* to Tib. *Lug ston btsan*, but *rGyal rabs* has preserved to us
this name in the form *Se le ston btsan* (Schmidt, *Geschichte der Ost-Mongolen*, p. 359,
transcribes according to *Bodhi-mör: Ssele sDong bDsan*); *Se le*, nevertheless, cannot be
the model of Chin. *lu(k)*. *Lu tung tsan* had five sons, — *Tsan si jo* 贊悉若, *K'in ling*
欽陵 (perhaps Tib. *dKon glin*), *Tsan p'o* 贊婆 (Tib. *bTsan-p'o*), *Si to kan* 悉
多干, *Po lun* 勃論 (Tib. *Po-blon*). The third and fifth are not names, but mere
titles In *rGyal rabs* (fol. 77) I find only two sons of the minister mentioned, — *gÑa* (in
another passage *sÑan*) *btsan ldem-bu* and *sTag-ra k'on lod.* Except the element *btsan*,
there is nothing in these names that could be identified with any part of the Chinese
transcriptions.

1) The word *bud-med* has been interpreted by A Schiefner (*Mélanges asiatiques*,
Vol. I, p. 358) as meaning "the powerless one" (*die kraftlose*) on the mere assumption
that the element *bud* has developed from *bod*, and that *bod* is a *verdünnung* of the verb
p'od, "to be able, capable," which, according to him, holds good also for the word *Bod*,
"Tibet." These far-fetched etymologies are based on a now outgrown view of things phonetic The vowel *u* has not arisen from *o* owing to *trübung*, as assumed by Schiefner,
nor is there anything like a *schwachung* of an aspirate sound to a media. *Bud, bod,* and
p'od are three co-existing, distinct matters of independent valuation, and without mutual
phonetic relationship. There is no phonetic law to connect them. The whole explanation
is not prompted by any rigorous application of phonology, but doubtless inspired by the

mi, "not;" *dmyig = mig,* "eye." The same phenomenon has been observed in the fragments of the *Çālistambasūtra* found by A. Stein (*Ancient Khotan,* Vol. I, pp. 549, 564; observations of BARNETT and FRANCKE) and in the inscription of King K^cri-sroṅ lde-btsan of the year A. D. 783 (WADDELL, *J. R. A. S.,* 1909, p. 945). [1]) These authors merely point out this case as an instance of archaic orthography, as also M. BACOT speaks of "certains archaïsmes de graphie et d'orthographe." But it should be understood that this peculiar way of writing naturally corresponds to a phonetic phenomenon; the subjoined letter *y* (called in Tibetan *ya btags*) indicates the palatalization of the consonant to which it is attached. How this process came about is easily to be seen in the case of the negative copula *myed,* formed of the negation *ma* + the copula *yod,* yielding *myŏd,* in phonetic writing *m'ŏd.* The letter *e* covers

---

Sanskrit word *abalā* given as a synonym of the word "woman" in *Amarakosha* (ed. *Bibl. ind.,* p. 140). But we only have to cast our eyes on the Tibetan version to see that *abalā* corresponds, not to Tib. *bud-med,* but to Tib. *stobs-med,* while Tib. *bud-med* appears as equivalent of Skr. *strī.* Consequently Skr. *abalā* cannot be made responsible for Tib. *bud-med;* there is no relation between the two; Tib. *stobs-med* is an artificial rendering of Skr. *abalā.* The main objection to be raised to SCHIEFNER's etymology, however, is that it flatly contradicts the natural facts. The Tibetan woman is very far from being weak or without strength, but is physically well developed, — an observation made by all travellers, nor did it escape the Chinese writers on Tibet. "Tibetan women are robust and the men weak, and one may frequently see women performing in the place of their husbands the socage services which the people owe" (ROCKHILL, *J. R. A. S.,* 1891, p. 230). It is not necessary to expand on this subject, but "the weak sex" would be applicable in Tibet only to man. A more plausible explanation of the word may now be offered. It was, of course, doubtful whether the second element *med* was really identical with the negative copula *med;* it may have been, after all, a different word. But the old form *bud-myed* confirms the opinion that this *med* has arisen from *myŏd, ma yod.* In the first element the word *bu* ("child, son;" *bu-mo,* "girl, daughter") may clearly be recognized, and *bud* (as other monosyllables terminating in *d*) is a contraction of *bu + yod,* "the condition of being a child or girl." *Bud-med,* accordingly, means "one who is no longer a girl, an adult woman," and in this sense the word is indeed utilized.

1) It occurs likewise in the inscription of 822, presenting the interesting example *myi̯ṅ rus.* As has been pointed out, this expression is employed on the same occasion in *rGyal rabs* in the form *miṅ rus,* so that the identification of *myi̯ṅ* with *miṅ* is absolutely certain.

also the vocalic timbre ŏ.[1]) The word *myi* accords in sound with Russian мн. [2])

This alternation between hard and palatalized consonants, restricted to the guttural and labial series and to dental *n*, is still conspicuous in the modern language, and has already been noted by A. SCHIEFNER in his "Tibetische Studien."[3]) As to *m*, SCHIEFNER refers to the pairs *miṅ — myiṅ*, "name;" *mid — myad*, "gullet;" *smig — smyig*, "reed." He correctly compares Tib. *mig*, "eye," with Burmese *myak*, and he also knows that the older forms *myed* and *myin* have been preserved along with *med* and *min*; there are such alternations as *kᶜem — kᶜyem*, *kᶜab — kᶜyab*, *gon — gyon*, *ɑbo — ɑbyo*, *pᶜe — pᶜye*, *nag — ñag*, *rnil — rṅil*, and many others. In Ladāk and Lahūl we find the labial tenues, aspirates and mediae, where the written language offers the corresponding palatalized sounds, as may be gleaned from the Phonetic Table preceding JÄSCHKE's *Dictionary* (p. XVIII) and F. B. SHAWE. [4])

The verb *ɑsod*, "to kill," appears as *sod* without the prefix twice (Table II, 8; VI, 2) and with it once (XI, 3), which indicates that the spelling was as vacillating at that time as it is now. [5]) The stem of the verb is *sad* (Ladākhi *sat*), as shown also by Burmese *sat* and Chinese *ŝat* 殺. Likewise we have *toṅ* in lieu of *gtoṅ* in V. 7. Also in this case the stem is *taṅ* or *toṅ*. [6])

---

1) This is best attested by the Tibetan transcription *ceṅ* (*ċeṅ*) of Chin. *čüṅ* 貞 (in the nien-hao *Chêng-kuan*) in the inscription of 783 (WADDELL, *J. R. A. S.*, 1909, p. 950, l. 29; the writing *ceṅ ṅa kvan* must be due to a slip in copying the text of the inscription).

2) The Chinese transcriptions assist us again. Compare above under No. 20 Tib. *rgyud* = Chin. *g'ut*, *γ'ut*.

3) *Mélanges asiatiques*, Vol. I, pp. 370—371.

4) *J. A. S B.*, Vol. LXIII, pt. 1, 1894, p. 12.

5) LAUFER, *Ein Suhngedicht der Bonpo, l. c.*, p. 21.

6) Compare such cases as occurring in the inscription, *mtᶜoṅ* sounded *t'oṅ*, *gtogs* sounded *tog*, etc.

These spellings cannot therefore be explained as irregularities or negligence on the part of the writer. From a grammatical standpoint they are perfectly legitimate, for the prefixes *g* and *b* are purely formative elements indicating tenses of the verb. The Tibetan grammarians are fully conscious óf this process, as shown by me on a former occasion;[1]) the prefix *b* denotes the past and the active, the prefixes *g* and *d* the present, the prefix *ɑ* the passive and future, the prefix *m* an invariable state.

The prefix *r* is omitted in *bda* = *brda* (V. 5), the prefix *ɑ* in *ts͡ᶜo ts͡ᶜo* (V. 9), the prefix *d* (or *r*) in *mu* (V. 15). We accordingly· meet symptoms of simplified spelling prompted, as we saw above, by the phonetic conditions prevailing at that time.

The prefix *l* appears in *lteñ* (V. 6, 24) in the place of *s* (*steñ*); compare *ldib-pa* and *sdib-pa*, *lña*, "five," in Ladākhi *šña*, *lga* and *sga*, "ginger," *lbu-ba* and *sbu-ba*, "bubble."

The sound *n* in lieu of *l* appears in *nam nañs*, "daybreak," for the normal *nam lañs*. SCHIEFNER[2]) has pointed out the same form in the *ɑDsañs-blun* (where also *lañs* occurs), and considers both forms as equally legitimate.

In Table I, 6, we meet the word *me-tog*, "flower," in the form *men-tog*, which, according to JÄSCHKE, still occurs in the West Tibetan dialects; but it is heard also in eastern Tibet. Mr. BARNETT[3]) has pointed out the form *me-tᶜog* in the fragments of the *Çālistambasūtra*, and, as the *m* is not palatalized, arrays it as an exception among the palatalized *m*. The assumption that *men* presents the older form may account for the preservation of the hard *m*.

Of great interest is the form *nam-ka*, "heaven" (Table I, 9),

---

1) *Studien zur Sprachwissenschaft der Tibeter*, pp. 529, 543.

2) *Ergänzungen und Berichtigungen zu Schmidt's Ausgabe des Dsanglun*, p. 9, St. Pet., 1852.

3) *Ancient Khotan*, p. 549.

which occurs also once in the fragments of the *Çālistambasūtra* found by A. Stein (*Ancient Khotan*, p. 555), while in other cases it is written *nam-mk^c a*. This case is of importance, because the word has been looked upon as a loan from Sanskrit. O. Boehtlingk [1] was the first to entertain this opinion. W. Schott [2] explained *namk^c a* as developed from *nabk^c a*, "since evidently it has arisen from the combination of two Sanskrit synonyms for 'air' and 'heaven,' *nab^c as* and *k^c a*." In a rather dogmatic form the same question is taken up again by Mr. Waddell, [3] who makes the statement, "For the conception of heaven in the Indian and Western sense the Tibetans use the word *mk^c a*, which they clearly borrowed from the Sanskrit *k^c a*, as they evidently had no indigenous word of their own to express it." The somewhat generous application of "clearly" and "evidently" does not appeal to everybody; what is evident to one is not always so to another, as opinions largely vary on the nature and quality of evidence. The *Kiu T^c ang shu* (Ch. 196 上, p. 1) informs us that the shamans of the Tibetans invoke the gods of Heaven and Earth (令巫者告于天地), and that in the prayer during sacrifice the spirit of Heaven 天神 is implored. [4] If the Tibetan shamans invoked the deity of Heaven, they must "evidently" have possessed a word by which to call it; and that

---

1) In his article *Ueber eine tibetische Uebersetzung des Amarakosha* (*Bull. de l'Acad. de St. Pétersbourg*, Vol. III, No. 14, pp. 209—219).

2) *Altaische Studien* I (*Abhandlungen Berliner Akademie*, 1860, p. 614, note 2). The occasion for this observation is afforded by the Manchu word *abka*, which Schott, on hardly plausible grounds, considers as a corruption of Tib. *namk^c a*.

3) *J. R. A. S.*, 1909, p. 931, note 3.

4) Compare Bushell, *The Early History of Tibet*, p. 7; and F. Grenard, *Mission scientifique dans la haute Asie*, Vol. II, p. 404 (Paris, 1898). Also the *Tang hiang* 党項, a Tibetan tribe inhabiting the southwestern part of Kan-su and the region of the Kuku-nōr, worshipped Heaven with sacrifices of oxen and sheep every three years at a gathering of their clans (三年一聚會殺牛羊以祭天。 *Sui shu*, Ch. 83, p. 3). Any Buddhist or Indian influence is here excluded in view of the period in question (589—618).

this word was of Sanskrit origin, is highly improbable. The Chinese account shows us that the Tibetans, in the same manner as the Turkish, Mongol, and other tribes of Asia, in times prior to Indian influence, had a well established worship of Heaven and Earth (as well as of the astral bodies), and this implies the fact that an indigenous word for "heaven" was theirs. This word was *gnam*, *nam*, or *nam-ka*, and there is no reason, from its phonetic make-up, why it should not plainly be a Tibetan word. The Tibetan lexicographers are very familiar with Sanskrit loan-words, and never fail to point them out in every case; this is not done, however, in the case of the word for "heaven." The archaic form *nam-ka* bears out the fact that *nam* is a good native word, for the suffix *ka* is never attached to a Sanskrit loan-word. [1]) In the same manner as the prefix *g* is noteworthy in *gnam*, so the prefix *m* must not be overlooked in the word *mkᶜa*; the spelling *nam-kᶜa* (but frequently enough also *nam-mkᶜa*) is a purely graphic expediency, and the outward resemblance to Skr. *kha* is accidental. SCHIEFNER [2]) compared Tib. *mkᶜa* with Chin. *kᶜi* 氣; this equation is untenable chiefly for the reason that Tib. *a* cannot correspond to Chin. *i*, but it shows that Schiefner had sense enough to regard *mkᶜa* as a truly Tibetan word. It is widely diffused in the allied languages. [3]) Lolo *mukᶜiai* (*ai* = *ä*) [4]) presents a counterpart to Tib. *namkᶜa*.

The word *žaṅ lon* occurs three times in the Table. In one passage (IV, 3), M. BACOT takes it in the sense of "minister" and accordingly accepts it as an equivalent of *žaṅ blon*. In III, 5, he translates it "news;" and in XI, 5, we read "indique que l'oncle viendra

---

1) On the suffix *ka* (*kᶜa*, *ga*) see SCHIEFNER, *Mélanges asiatiques*, Vol I, p. 380.

2) *L. c.*, p. 340.

3) Compare the list of words for "heaven" in Mission D'OLLONE, *Langues des peuples non chinois de la Chine*, p. 24, Paris, 1912, particularly such forms as *hé ka*, *mu ko*, *m'keuk*, *nakamu*, *mongkele*.

4) P. VIAL, *Dictionnaire français-lolo*, p 83 (Hongkong, 1909).

aux nouvelles," where the text offers *bdag žaṅ lon-du oṅ-bar ston*. In the latter case, M. BACOT separates the compound, and assumes *žaṅ-(po)* = maternal uncle, and *lon* = tidings, message; but this is no very possible. Further, the word *bdag* [1]) must not be overlooked in this sentence, and *oṅ-ba* in connection with the terminative means "to become;" so that I think the sense of the sentence is, "It indicates that I shall become a *žaṅ lon*." It goes without saying that in the three passages this word is one and the same, and can but have the same significance. The word *lon*, accordingly, is written without the prefix *b*. This way of writing cannot be considered an anomaly, but exactly corresponds to the pronunciation of the word at that period, as we established on the basis of the transcription *lun* 論 (= Tib. *blon*) furnished by the Annals of the Tᶜang Dynasty (*Kiu Tᶜang shu*, Ch. 196 上, p. 1; *Tᶜang shu*, Ch. 216 上, p. 1) and the inscription of 822. The word *btsan*, the title of the kings of Tibet, was likewise sounded *tsan*, as evidenced by the Chinese transcription *tsan* 贊. [2]) The prefixed media

---

1) The word *bdag*, the personal pronoun of the first person, occurs several times in the answers of the Table (VII, 1; VIII, 7, 8; XI, 7). In this connection it should be remembered that *bdag sgrog*, "crying *bdag*," is one of the synonyms of the raven (given in the Dictionary of the French Missionaries, p. 86); it is evidently an imitation of Skr. *ātmaghosha*, a synonym of the crow, which is rendered in the Tibetan version of *Amarakosha* (ed. *Bibl. ind.*, p. 134) *sgrogs-pai bdag-ñid-can*.

2) It has been asserted that the Chinese term *tsan-pᶜʋ* 贊普 corresponds to Tibetan *btsan-po* (BUSHELL, *The Early History of Tibet*, p. 104, note *a*; CHAVANNES, *Documents*, pp. 150, 186). But this identification is not exact; the Chinese words very accurately reproduce the Tibetan form *(b)tsan-pᶜo*, as is evidenced first by the presence of the labial aspirate in the Chinese word *pᶜu*, and secondly by the gloss expressly given in *Tᶜang shu* (Ch. 216 上, p. 1): 丈夫曰普 "a man is called in Tibetan *pᶜu*." This explanation leaves no doubt that the Tibetan noun *pᶜo* "man," and not the mere suffix *po*, is intended, which, by the way, is transcribed in Chinese *pu* 逋, as shown by many examples in *Tᶜang shu*; for instance, in the titles of the ministers, as *nang lun chᶜö pu* 囊論掣逋 = Tib. *naṅ blon čᶜe-po*, "great minister of the interior." This reading *(b)tsan-pᶜo* is confirmed by a Lhasa inscription of the ninth century published by Mr. WADDELL (*J. R. A. S.*, 1909, pp. 1269, 1280), where the word is written twice *btsan-pᶜo*; it

b, accordingly, is not an integral part of these two stems, but an additional prefix which must have a grammatical function; and this, in my opinion, is that it forms *nomina actionis*, in a similar manner as it designates a past action in connection with verbal roots. The stem *tsan* means "powerful, warlike, heroic;" *b-tsan*, "one having the title or dignity of *tsan*"; *b-lon*, "one who has the function of, or acts as, minister." What is a *žaṅ lon*? [1]) Mr. WADDELL (*J. R. A. S.*, 1909, p. 1274) explains that this term means "uncle-minister," and designates "a sort of privy councillor, a title previously borne apparently only by the highest ministers, some or

---

certainly does not mean, as alleged by Mr. WADDELL, "the mighty father" (father is *p̄a*; *p̄o* never means "father," but only "male, man"), but "the martial man," "the male hero" (*tsan*, as *Tᶜang shu* says, means *kiang hiung* 疆 雄). The stress laid on the word "male" is very natural, as there always were, and still are, also queens ruling Tibetan tribes (compare the account of the Tibetan Women's Kingdom in *Sui shu*, Ch. 83). The contrast is clearly enough expressed in the *Tᶜang shu*, which adds, "The wife of the Tsan-p̄u is called *mo mung* 末 蒙." Whatever the latter element may represent, it is evident that the first is the Tibetan word *mo*, "woman." (A royal consort is called in Tibetan *lcam-mo*, *btsun-mo*, or *cᶜuṅ-ma*; probably the Chinese *mo-mung* represents an ancient Tibetan word still unknown to us, which would be *mo·moṅ*; Chinese *mung* phonetically corresponds to Tib. *moṅ*, as proved by *Hua i yi yü* [Ch. 13, p. 65], where Tib. *rṅa-moṅ* ["camel"] is transliterated in Chinese 兒 阿 蒙; in *rGyal rabs* [fol. 79] one of the wives of King Sroṅ-btsan is styled *Moṅ bza kᶜri lcam*, which indeed goes to prove that a word *moṅ* in the sense of "royal consort" must have existed in ancient Tibetan.) The king is therefore styled the "male warrior" in opposition to the attribute "female" appearing in the title of his queen. The inscription of 822 (see the facsimile in BUSH-ELL's paper, pl. II, line 2) writes the word *btsan-po*; WADDELL sets the date of his inscription on inward evidence in 842—4; so that it must be granted that both ways of writing co-existed at that period. The writing *btsan-p̄o* doubtless is the older one, and appears as the index of the ancient matriarchal conditions of Tibet at a stage when masculine power gradually emerged from the institution of female preponderance. When the sway of the Central Tibetan kings was ultimately established in the male line of succession, the plain *btsan-po*, without emphasis of sex, was allowed to take its permanent place. Note that according to *Tᶜang shu* (BUSHELL, *l. c.*, p. 98) the inhabitants of the Women's Kingdom elected a man as their ruler from 742.

1) JÄSCHKE (*Dictionary*, p. 471) quotes the word from *rGyal rabs*, saying that it seems to be a kind of title given to a minister (or magistrate); wisely enough, he makes it a separate heading, and does not link it with the word *žaṅ-po*, "uncle." So do also the French Missionaries (p. 845).

most of whom were of the blood-royal." This is a surmise which is not founded on any evidence.

The Tibetan administrative system is entirely based on Chinese institutions; and the official style of the Tibetan chancery, as clearly demonstrated by the Tibetan inscriptions of the T'ang period, is modelled on that of China.[1]) For the explanation of Tibetan terms relating to officialdom, we have in the majority of cases to look to China. What a žan lon is, is plainly stated in T'ang shu (l. c.), where we meet it in the garb shang lun 尚論. The nine Tibetan

---

1) A feature to which Mr. WADDELL in his *Lhasa Edicts*, and Mr. A. H. FRANCKE in his rendering of the inscription of 822, did not pay attention, wherefore they missed the meaning of several phrases which cannot be derived from a literal translation of the Tibetan words in their ordinary sense, but which must be viewed through Chinese spectacles, and taken as imitations of Chinese documentary and epigraphical style. But this subject calls for a special investigation. To this Chinese official terminology belongs, for example, the Tibetan designation of the people as "black-headed" (*mgo nag*), which is purely and simply copied from Chinese phraseology, as it is likewise when it occurs in the Orkhon inscriptions and among the Mongols. Mr. WADDELL (*J. R. A. S.*, 1909, p. 1255) remarks on this term that it "probably may denote that in those days the Tibetans did not wear caps; indeed, the caps at the present day are all of Chinese pattern and manufactured in China." In this case, Mr. WADDELL must unfortunately forego the claim to originality, for the present writer was the first to advance this explanation, but with reference to ancient China (*T'oung Pao*, 1908, p. 40), and supported it also with good reasons based on the peculiar ceremonial character of Chinese head-gear. With regard to Tibet, however, this interpretation is out of place. There, it is plainly a loan-word, an artificial imitation of Chinese official speech. Further, Mr. WADDELL's observation that all Tibetan caps are of Chinese pattern and manufacture is erroneous, as a glance at ROCKHILL's *Notes on the Ethnology of Tibet* (pp. 688—689, *Report U. S. Nat. Mus.*, 1893) and his plates 3—4 will convince one. The Tibetan nomads living on the high and cold plateaus naturally always wore fur caps and manufactured them themselves, and there is a large variety of types of indigenous head-gear, without Chinese affinities, everywhere in eastern Tibet and in the Kukunōr region (so also F. GRENARD, *Mission scient. dans la haute Asie*, Vol. II, p. 340, Paris, 1898); even the round felt caps made in Peking for the Mongol and Tibetan market do not at all represent a Chinese but a Mongol-Tibetan style of cap. As in so many other cases, the Chinese have taken into their hands an industry of their subjected neighbors, and cater to their taste. Tibetan officials certainly wear the caps of the Chinese official costume made in, and imported from, China, but that is all. And the manifold styles of priestly head-gear, partially like the *pan žva* traced to Indian traditions, certainly do not come from China.

Boards of Ministry are there enumerated, which it is said are designated with the general name 尙論掣逋突瞿 *shang lun ch*°ö *pu t*°u *kiū* (*du gu*), which, as stated, may be taken as transcription of Tibetan *žaṅ* (*b*)*lon c*°*e-po dgu*, "the Nine Great Boards." The word *žaṅ* cannot be explained through Tibetan, and indeed is nothing but the Tibetan transcription of Chinese *shang* 尙; and *žaṅ* (*b*)*lon*, "chief minister," corresponds in meaning to *shang shu* 尙書, "President of a Board," a term rendered in the inscription (above, No. 20) by Tib. *c*°*en-po*. Tib. *žaṅ* is a strictly phonetic transcription of 尙, as both agree in tone, *shang*[4] having the sinking lower tone, and *žaṅ* being low-toned; the Tibetans cannot write Chinese *shang*[4] with the voiceless palatal sibilant *š*, as all words with this initial sound have the high tone, but for this reason must resort to the deep-toned *ž*.[1]) The tone, as pointed out before, is a matter

---

1) In the Tibetan vocabulary contained in Ch. 11 of *Hua i yi yü* (Hirth's copy in the Royal Library of Berlin), the Tibetan words are all transliterated in Chinese characters according to their Tibetan spelling (the transliterations do not reproduce the Tibetan pronunciation), and the rule is usually observed to transcribe a Tibetan word with initial *ž* by means of a Chinese syllable in the lower tone; for example, Tib. *žiṅ* to be read *shéng* 繩, Tib. *žag* to be read *hia* 廈, Tib. *žu* to be read *jo* 熱 or *shu* 熟. If Tib. *šu* renders Chin. *shui* 水 in the inscription of 822 (see above, p. 79), this exception is only seeming, and confirms the rule; for *shui* has the rising upper tone, consequently the Tibetans rendered it with *šu* in the high tone, being their tone nearest to the Chinese, while Tib. *žu* has the deep tone. *Vice versa*, Chinese *š* is transcribed by the Tibetans *ž*, for example, *shéng* 省 "province" being transcribed Tib. *žiṅ* in *Shambhalai lam yig* (regarding this work compare *T*°*oung Pao*, 1907, p. 403), and Tib. *š* is transcribed by Chin. *ž*, for instance, Tib. *žo* = 若 *žo* at the end of royal names, occurring in three names of King *Sroṅ btsan's* ancestors (*T*°*ang shu*, Ch. 216 上, p. 2 a): 揭利失若 *Kie* (*kat*, Korean *kal*) *li ži*(*t*) *žo* = Tib. *Gal*(?) *ri* ("mountain") *žid*(?) *žo*; 勃弄若 *P*°*o lung žo* = Tib. *P*°*o sroṅ žo*, and 詎素若 *Ku* (*gio*, *gu*) *so*(*k*) *žo* = Tib. *Go*(?) *zug*(?) *žo*. There is no doubt of the identification of Chin. *žo* with Tib. *šo*, as this Tibetan word is indeed found with four of the so-called "six terrestrial *Legs*" (*sa-i legs drug*): *,O šo, De šo, T*°*i šo, ,I šo* (*dPag bsam ljon bzaṅ*, p. 150, l. 12). Then we have allied words in both languages: as Tib. *ša*, "flesh, meat" = Chin. *žou* (*jou*) 肉; Tib. *šes*, "knowledge, to know" = Chin *či* 知 and 智; Tib (*b*)*cu*, "ten" = Chin. *ši*

of importance in the study of Tibeto-Chinese and Chinese-Tibetan transcriptions. The fact that Tibetan *žaṅ* really corresponds to Chinese 尙 is evidenced by the inscription of 822, where the word *žaṅ* in the titles of the Tibetan ministers repeatedly occurs, being rendered in each case by Chin. *shang* (above, Nos. 9, 10, 13, 15, 19). It is therefore beyond any doubt that the equation Tib. *žaṅ* = Chin. *shang* 尙 belonged to the permanent equipment of the Tibeto-Chinese chancery in the first part of the ninth century. [1])

The most interesting phonetic phenomenon of our text is the writing *dmyig* for *mig*, "eye." There cannot be any doubt of this identification, as the word is required by the context, as it is determined by the adjective *rno*, "sharp," and the phrase *dmyig rno* is a parallelism to the following *sñan gsan*, "to have a sharp ear."

十; Tib. *lce*, "tongue" = Chin. *šé* 舌. The words *žo*, "milk" = Chin. *žu* 乳, and *šig*, "louse" = Chin. *ši(t)* 虱, seem to belong to an earlier stage of relationship between the two languages.

1) The word *shang* appears as the first element in the names of three Tibetan generals who attacked China in 765 (*Kiu T'ang shu*, Ch. 196 上, p. 10 a; BUSHELL, *The Early History of Tibet*, p. 45): *Shang kie si tsan mo* 尙結息贊磨, who died in 797; *Shang si tung tsan* 尙息東贊 (Tib. *Žaṅ ston btsan*); and *Shang ye si* 尙野息 (a fourth is called *Ma chung ying* 馬重英). Under the year 768 (*ibid.*, Ch. 196 下, p. 1) a general *Shang si mo* 尙悉摩 (BUSHELL, p. 48: *Shang tsan mo*) is mentioned. *T'ang shu* (Ch. 216 下, p. 6 b) has a Tibetan commander-in-chief *Shang t'a tsang* 尙塔藏 (Tib. *Žaṅ t'aʾ bzaṅ*). In these cases Chin. *shang* corresponds to Tib. *žaṅ*, which is a well-known clan name based on the district of this name in the province of *gTsaṅ* (CHANDRA DAS, *Dictionary*, p. 1065). One of the ministers of King *K'ri-sroṅ* was *Žaṅ ñams bzaṅ* (*dPag bsam ljon bzaṅ*, p. 170); in *rGyal rabs* we meet a minister *Žaṅ dhu riṅ* and the well-known translator *Bandhe Ye-šes sde* with the clan name *Žaṅ sna-nam*, that is, from *Sna-nam* in *Žaṅ* (CHANDRA DAS, p. 765, is wrong to refer in this case to Samarkand; as a clan name *Sna-nam* relates to a place in the district *Žaṅ* in the province *gTsaṅ*). — In the iconographical work "The Three Hundred Gods of Narthang" (section *Rin abyuṅ*, fols. 112, 113) a deity is represented in three forms under the name *Žaṅ blon rdo-rje bdud adul*. *rDo-rje bdud adul* (with the title and office of *žaṅ blon*), "the Subduer of Māra by means of the Vajra," appears as a sorcerer at the time of *K'ri-sroṅ lde-btsan* (*Roman*, p. 122). — Also the T'u-yü-hun had the office of *shang shu* (*Sui shu*, Ch. 83, p. 1 b).

Also the Lama bsTan-pa du-ldan (p. 448, line 3) has perfectly understood the word in the sense of *mig*. The spelling *dmyig* is neither erroneous nor arbitrary, but proves that at the time when, and in the locality where, our text was written, the word was actually articulated *dmyig*, as here spelled; for in the dialect of the Jyarung, [1]) inhabiting the northwestern part of the present Chinese province of Sze-ch'uan, I actually heard the word articulated *dmye*. The form *dmig* is still found in modern popular texts; for instance, twice in the small work *Sa bdag klu gñan-gyi byad grol*, along with the orthography *mig* four times (*Ein Sühngedicht der Bonpo, l. c., p. 21*). It is therefore patent how important it is to observe carefully such

---

1) The Jyarung styled Kin-ch'uan 金 川 by the Chinese (see M. Jametel, *L'épigraphie chinoise au Tibet*, p. 31, Paris, 1880) are a group of Tibetan tribes inhabiting the high mountain-valleys of Sze-ch'uan Province. The name is written in Tibetan *rgya-roñ* which is explained as "Chinese ravines." Of their language we possess only scant vocabularies. B. Hodgson (*Essays on the Languages*, etc., *of Nepal and Tibet*, pp. 65—82, London, 1874) offers a vocabulary of 176 words. T. de Lacouperie (*Les langues de la Chine avant les Chinois*, pp. 78—80, Paris, 1888) has some remarks on the language. A. v. Rosthorn has published a vocabulary in a volume of *Z. D. M. G.* (owing to a misplacement of my notes referring to it, I regret being unable for the present to give an exact reference). Jyarung is one of the most archaic Tibetan dialects in which not only the ancient prefixes are still articulated (*rgyal* "king," *stoñ* "thousand," *šta* "horse"), but also single and even double prefixes appear where literary Tibetan has none at all; they are supermen in prefixes, or, if it is permissible to coin the word, super-prefixists. They say, for example, *drmi* for common Tib. *mi*, "man"; the prefix *d* largely enters the names for the organs of the body; as *dmye* "eye," *dešnu* (*sna*) "nose," *desi'e* (*so*) "tooth," *drnu'* (*rna*) "ear," *dešmi'* (*lce*) "tongue," *demjä'* "chin," *demki'* (*ske*) "neck." This corroborates my opinion that the prefixes are survivals of ancient numeratives; for this reason they are not stable, but variable, in the various dialects. The Jyarung language not only had numeratives different from standard Tibetan, but also arranged its words under different categories, so that they appear with prefixes entirely at variance with other dialects: thus, *tayák*, "hand" (*p'yag*), *poñi'*, "silver" (*dñul*). The stems, accordingly, are *ñi*, *ñul* (Hakka *ñ'in*, Burmese *ñwe*), *po* and *d* being prefixes. The Jyarung numerals are 1 *ktig* or *kti'*, 2 *knis*, 3 *ksam*, 4 *kbli*, 5 *kmu*, 6 *kčo*, 7 *kšnis*, 8 *vryad*, 9 *knu*, 10 *šči*. The numerals 4—7 and 9, at variance with standard Tibetan, have been raised into the *k*- category in analogy with 1—3, which agree with standard Tibetan. It is of especial interest that in the numeral 3, *ksam*, Jyarung agrees in the *a* vowel with Chinese *sam* where standard Tibetan has *u* (*gsum*), and that in the numeral 5, *knu*, Jyarung agrees in the *u* vowel with Chinese *ñu* where standard Tibetan has *a* (*lña*).

variations of spelling, even in recent manuscripts and prints, and it is obvious also that they cannot always be laid down as clerical errors. This has likewise a bearing upon ancient manuscripts; the mere occurrence of abnormal, obsolete, or dialectic forms is not sufficient evidence for pronouncing the verdict that the said manuscript or work is old, while certainly the total evidence presented by archaisms will always influence our judgment in favor of a greater antiquity. It would be, for example, perfectly conceivable to me that a Jyarung Lama who, owing to the far-reaching divergence of his tongue from the written language, is forced to study the latter thoroughly, as we, for instance, would study Latin, will be inclined to write the word *mig* in the form *dmyig* or *dmig*. Analogous to the latter is the form *dmag-pa* (Table XI, 1) for the more common *mag-pa*; and as the prefix *d* before *m*, in cases where the written language is without a prefix, is a characteristic of the Jyarung dialect, the conclusion may be hazarded that the *document Pelliot* was composed either in a locality where a dialect identical with, or allied to, Jyarung was spoken, or that, regardless of the locality where the composition took place, the author of the document was conversant with a language related in phonology to Jyarung.

What is the meaning of the prefixed dental *d*? In the written language we find such formations as *ma*, "below," — *dma*, "to be low;" *mañ*, "many," — *dmañs*, "multitude," and *dmag*, "army;" *mig*, "eye," — *dmig*, "hole." The formations with the prefix *d* apparently are secondary derivatives from the stem beginning with *m*. Comparison with the allied languages tends to confirm this opinion; *mig* is the Tibetan stem-word, as shown by Lepcha *a-mik*, Burmese *myak* (*myet*), Kuki-Chin *mit*, *mi(k)*, [1]) Chinese *muk* 目 . In all Indo-Chinese languages we observe that nouns are clas-

---

1) STEN KONOW, *Z. D. M. G.*, Vol. LVI, 1902, p. 506.

sified into certain categories, and that each of these categories is associated with a particular numerative. The numerative is the index or outward symbol of the mental association underlying these categories of ideas. These numeratives, with a few exceptions, have disappeared from modern Tibetan, but they are preserved in many of the so-called prefixes which represent their survivals, and this is the usual function of prefixes in nouns (though they certainly have also other origins and functions). The original significance of the majority of them can no longer be made out, and will probably remain obscure; the numerous variations of prefixes in the dialects indicate that there has been a large number of differing numeratives from remote times. A few examples may serve as illustration. The prefix *m* appears in connection with words denoting organs of the body, and it is curious that there are groups with the same initial sounds. There is a *mc^c* group, — *mc^ced* "body," *mc^ce-ba* "tusk," *mc^cer-pa* "spleen," *mc^cin-pa* "liver," *mc^can* "side of the breast," *mc^cu* "lip," *mc^ci-ma* "tear," *mc^cil-ma* "spittle;" there is a *mg* group, — *mgo* "head," *mgur, mgul* "throat," *mgrin-pa* "neck," a *mk^c* group, — *mk^cal-ma* "kidneys," *mk^cris-pa* "bile," *mk^crig-ma* "wrist," *mk^cur-ba* "cheek." The occurrence of the prefix *m* in these fifteen words belonging to the same category of idea cannot be accidental, and the supposition of a former numerative *m* joined to names of bodily parts seems a plausible explanation for its presence. The following groups are also suggestive: *ldad-pa* "to chew," *ldan-pa* "cheek," and *ldag-pa*, "to lick;" *lte-ba* "navel," *lto-ba* "stomach," and *ltogs-pa* "hunger;" *rkań-pa* "foot," *rke* "waist," *rkan* "palate," *rkub* "anus."

The laws of *sandhi*, as established by the Tibetan grammarians,[1])

---

[1]) The generally adopted metrical versions are given in text and translation in *Studien zur Sprachwissenschaft der Tibeter (Sitzungsberichte der bayerischen Akademie,* 1898, pp. 579—587).

are not strictly observed. The indefinite article *žig* is correctly employed after nouns ending in a vowel, *n* and *m*: *dgra žig, gñen žig, mye ñan žig, gtam žig*; *rton cig* in V. 23 is correct owing to the existing *da drug*;[1]) *cig* correctly in *myi rgod cig*; *ri-dags žig* instead of *šig*. Of designations of the genitive, we find -*i*, *kyi*, and *gi*, but not *gyi*: *lhai, pᶜyogs-kyi, bud-myed-kyi, dguñ-gi*; but *yul-gi, žañ-lon-gi*, instead of *gyi*; likewise in the instrumental case, *gñen-gis, gcan-zan-gis*. The termination of the terminative is *du*: *žañ-lon-du, ᵠbriñ-du, pᶜyag du* (instead of *tu*), *mtᶜo du* (V. 19) instead of *mtᶜo-ru* or *mtᶜor*, but *dgu-r* (V. 11), *bzañ-por* (V. 25), *riñs-par* (V. 27); also *ltas-su* (V. 12) is a regular formation. The suffix *tu* after vowels occurs in modern manuscripts likewise.[2]) The particle *te* of the gerund, with its variants, is utilized according to rule: *ᶜyer-te, kᶜrid-de, rñed-de, ši-ste*.

---

1) Compare the rule as formulated in *Za-ma-tog, l. c.*, p. 584; and above, p. 61, note 2.
2) *Ein Sühngedicht der Bonpo, l. c.*, p. 22.

ADDITIONAL NOTES. — Regarding the crow of orientation employed by the navigators (p. 11, note), see now also R. OTTO FRANKE (*Dīghanikāya*, p. 166, Göttingen, 1913). FRANKE claims for himself the priority in having established the fact of this practice of mariners; but MINAYEV, at any rate, was the first to explain correctly the term *disākāka*.

On p. 29, after line 21, the following was omitted through an oversight of the printer: In *K.* we meet the sentence *tᶜag riñs-su agro-bar agyur-ro*, "you will set out on a distant journey;" the same is expressed in the Table in genuinely popular style by *lam riñ-por dgos-pa*.

Note on p. 95. In regard to Tᶜu-po see also HIRTH, *Sinologische Beiträge zur Geschichte der Türk-Völker* (*Bull. Ac. St.-Pét.*, 1900, p. 242). The sole object of the note above referred to was to discuss the relation of the Chinese to the Tibetan and alleged Tibetan names.

104

和德理到过西藏吗？

T'OUNG PAO

通 報

ou

ARCHIVES

CONCERNANT L'HISTOIRE, LES LANGUES,
LA GÉOGRAPHIE ET L'ETHNOGRAPHIE
DE
L'ASIE ORIENTALE

———

Revue dirigée par

Henri CORDIER
Membre de l'Institut
Professeur à l'Ecole spéciale des Langues orientales vivantes
ET
Edouard CHAVANNES
Membre de l'Institut, Professeur au Collège de France.

———

VOL. XV.

———

LIBRAIRIE ET IMPRIMERIE
CI-DEVANT
E. J. BRILL
LEIDE — 1914.

WAS ODORIC OF PORDENONE EVER IN TIBET?

BY

BERTHOLD LAUFER.

————◆——◆◀▆▶◆——◆————

In January of this year the Hakluyt Society began to re-issue Colonel Sir Henry Yule's memorable work *Cathay and the Way Thither*, published in 1866. We cannot be grateful enough to Professor H. Cordier for having taken upon himself the difficult task of thoroughly revising and re-editing this learned work, the republication of which meets a long-felt want, the first edition having been exhausted for at least fifteen years. The additional up-to-date information supplied by the wide erudition of Professor Cordier has increased the bulk of the work to such an extent that in its new garb it will comprise four, instead of the former two, volumes. Volume II, containing the description of the journey of Friar Odoric of Pordenone, is the one that has now been issued. In Chapter 45, "Concerning the Realm of Tibet, where dwelleth the Pope of the Idolaters," Professor Cordier has added a note, taken from L. A. Waddell's *Lhasa and its Mysteries* (p. 425), to this effect: "As to Friar Odoric's alleged visit, as the first European to enter Lhasa, it seems to me very doubtful whether the city he visited in the fourteenth century A. D. could have been this one at all, as his description of the place is so different from Lhasa as we now find it... Now none of the streets of Lhasa are paved, although plenty of stones are locally available for the purpose, and it seems unlikely that a city which was formerly 'very well paved' should have so entirely given up this practice and left no trace of it." The weight of this argument is not very cogent, and hardly presents a sufficient basis in favor of the desired proof. Still more categorically P. LANDON [1] has given vent to his feelings in the words, "It seems clear that he never reached Lhasa." This conclusion, however, is not backed up by any evidence; and for the rest, LANDON adheres to the general view that "Odoric appears to have visited Tibet about 1328." Whatever the foundation of these contentions may be, on reading Odoric's account of Tibet carefully, it seems to me a fitting time now to raise the broader question, Was Odoric of Pordenone ever in Tibet?

It is with a considerable amount of reluctance and *gêne* that this challenge is advanced. No lesser geographer than Colonel Sir Henry Yule has indorsed the tradition that Odoric, starting from Peking, turned westward through Tenduc, the Ordos country, and Shen-si, to Tibet, and its capital Lhasa, where "we lose all indication of his further route, and can only conjecture on very slight hints, added to general probabilities, that his homeward journey led him by Kabul, Khorasan, and the south of the Caspian, to Tabriz, and thence to Venice." [2]   F. v. RICHTHOFEN [3] and C. R. MARKHAM [4] have shared this opinion, and Friar Odoric has thus acquired the fame of having been the first European traveller to visit Tibet and Lhasa. Mr. ROCKHILL [5] opens a history of the exploration of Tibet in the words: "As far as my knowledge goes, the first European

---

[1] *Lhasa*, Vol. I, p. 4 (London, 1905).

[2] *Cathay*, Vol. II, p. 10; again, on p. 23, "We are ignorant of his route from Tibet westward."

[3] *China*, Vol. I, p. 617.

[4] *Narratives of the Mission of George Bogle*, p. XLVI

[5] *Diary of a Journey through Mongolia and Tibet*, p. IX.

traveller who entered Tibet was Friar Odoric, who, coming from northwestern China, traversed central Tibet on his way to India in or about 1325, and sojourned some time in its capital, Lhasa." It would seem almost cruel to destroy this nimbus, and to depose the good Friar from the throne which he has so long occupied in the history of geographical discoveries.

His claim to the honor of being the first Tibetan traveller, however, is deserving of serious scrutiny. It means a great deal to strip him of this glory, and such a denial should certainly be placed on more weighty arguments than the mere matter of street pavement. Lhasa may have undergone manifold changes from the fourteenth century down to 1904, and the lack of stone pavement in recent times does not yet exclude the possibility of better street conditions during the middle ages. It seems unfair, at any rate, to throw this stone at the poor Friar; and while I am not inclined to believe that he ever was in Lhasa, I feel perfectly convinced that he was given the information regarding the well-paved streets. There is no doubt that the Tibetans understand the art of cutting stones and making pavement. The excellent Jesuit Father Ippolito DESIDERI, who lived in Tibet from 1715 to 1721, relates that the halls, main rooms, galleries, and terraces, in most houses of Lhasa, were covered with a very fine pavement made from small pebbles of various colors, and well arranged; between these they put resin of pine-trees and various other ingredients, and then for several days they continually beat stones and ingredients together, till the pavement becomes like a veritable porphyry, very smooth and lustrous, so that when cleansed with water it is like a mirror. [1]

---

[1] Le sale, le camere principali, le logge e i terrazzi della maggior parte delle case, hanno un bellissimo pavimento, fatto di pietruzze minutissime, di diversi colori e ben disposte, fra le quali pongono della ragia di pino, e varii altri ingredienti, e di poi battono per varii giorni continuamente e pietruzze e ingredienti, fin che il pavimento viene come un intero porfido, molto liscio e molto lustro; di modo che lavato con acqua, diventa come

If the people of Lhasa were capable of work of this sort within their habitations, they could have accomplished the same on their public high-roads. [1] The technical term for this kind of mosaic work is *rtsig ṅos bstar-ba*. [2]

All authors seem to be agreed on the one point that Odoric's account of Tibet is a rather thin fabric woven of slender threads, and that it is certainly not what we ought to expect from a man who is reported to have traversed Tibet from one corner to the other, and to have even sojourned at Lhasa. The first question to be brought on the tapis, then, is this, — Is his information that of an eye-witness, or of one who drew it second-hand from the interviews of Chinese or Mongols regarding Tibet? If it contains such striking features as could only come to the notice of a personal observer of things and events, we are compelled to admit that Odoric did dwell within the boundaries of Tibet proper. Odoric, however, imparts nothing that would immediately force upon us such a conclusion: his scant notes could have been gathered at that time in China or as well in Mongolia. Tibet then was subject to the sway of the Great Khan; and Tibetans, those of the clergy and the laity, swarmed at his Court. Plano Carpini, who was not in Tibet, nevertheless had occasion to see Tibetan people, and to observe their custom of plucking out the hairs of their beards with iron tweezers. [3] Marco Polo's notice of Tibet is succinct, yet more graphic

---

uno specchio. — C. Puini, *Il Tibet*, p. 59 (Roma, 1904). To Yule's note on p. 249, lamenting the loss of the records of Desideri and Samuel van de Putte, it should now be added that Desideri's manuscript has been rediscovered and edited by Puini under the title quoted, and that the remains of Putte's diary have been published by P. J. Veth (*De Nederlandsche reiziger Samuel van de Putte*, *Tijdschrift van het Aardrijkskundig Genootschap*, 1876, deel II, pp. 5—19).

[1] Desideri (l. c., p. 58) speaks of a "cammino largo, e ben fabbricato."

[2] Schiefner, *Mélanges asiatiques*, Vol. VII, p. 524, note 5.

[3] Hi pilos in barba non habent, imo ferrum quoddam in manibus, sicut vidimus, portant, cum quo semper barbam, si forte crinis aliquis in ea crescit, depilant (*Libellus historicus*, Cap. X).

and lively than Odoric's, and presents the result of border information, presumably picked up at Ya-chou fu or thereabout.[1] Half of Odoric's chapter on Tibet is devoted to a description of the burial-practice; and he tells with manifest interest the story of how the corpses are cut to pieces by the priests and devoured by the eagles and vultures, how the son cooks and eats his father's head and makes his skull into a goblet, from which he and all of the family always drink devoutly to the memory of the deceased father; and they say that by acting in this way they show their great respect for their father.[2] The same is reported in substance by his predecessors, Plano Carpini (1246) and William of Rubruk (1253), the latter honestly adding that an eye-witness had told it to him.[3] Certainly these two writers were not copied by Odoric, but each of the three independently reported a tradition which he had heard from the Mongols. Here we are allowed to apply the same verdict as pronounced by YULE[4] in regard to Odoric and Marco Polo having in common the story of the Old Man of the Mountain, — "Both related the story in the popular form in which it spread over the East." Their peculiar burial-practice was that characteristic trait of the Tibetans by which their neighbors were most deeply struck, and which also was

---

[1] YULE and CORDIER, *The Book of Ser Marco Polo*, Vol. II, p. 45.

[2] This is in striking agreement with what HERODOTUS (IV, 26) relates concerning the Issedonians, who have been identified by W. Tomaschek with the forefathers of the Tibetans (compare also HERODOTUS, IV, 65). Regarding skulls as drinking-cups in the country of Chao and among the Hiung-nu, see CHAVANNES (*Les Mémoires historiques de Se-ma Ts'ien*, Vol. V, pp. 50, 485). Compare R. ANDREE, *Z. d. Vereins für Volkskunde*, 1912, p 1—33.

[3] Plano Carpini (Cap. XI) says, "Hi consuetudinem habent mirabilem, imo potius miserabilem. Cum enim alicuius pater humanae naturae solvit debitum, congregant omnem parentelam, et comedunt eum." And William of Rubruk relates, "Post illos sunt Tebet homines solentes comedere parentes suos defunctos, ut causa pietatis non facerent aliud sepulchrum eis nisi viscera sua. Modo tamen hoc dimiserunt, quia abominabiles erant omni nationi. Tamen adhuc faciunt pulchros ciphos de capitibus parentum, ut illis bibentes habeant memoriam eorum in iocunditate sua. Hoc dixit mihi qui viderat."

[4] *Cathay*, Vol. II, p. 257, note 3.

doubtless exaggerated by them. Carpini's and Rubruk's versions show us that this report was a current story circulating among the Mongols, and Odoric must have derived it from exactly the same source. He simply relates it as "another fashion they have in this country;" but he does not say that this custom came under his own observation, or that it was communicated to him directly by Tibetans. The fact that Odoric shares this part of his information concerning Tibet with Carpini and Rubruk, who had never been in Tibet, constitutes evidence that this account cannot be utilized for a plea in favor of his personal experience with Tibetan affairs.

In analyzing the remaining portion of his chapter, we have to discriminate between statements which are correct, and data which are inexact or out and out wrong. Odoric is perfectly correct on three points: he is acquainted with the geographical location of Tibet on the confines of India proper; he is familiar with the law of Lhasa, prohibiting bloodshed within the precincts of the holy city;[1] and he knows that "in that city dwelleth the *Abassi*, that is, in their tongue, the Pope, who is the head of all the idolaters, and who has the disposal of all their benefices such as they are after their manner."[2] All this is true, but rather general; at any rate,

---

[1] This law, of course, is merely theoretical, as intimated also by M. Cordier by the insertion of a note culled from a letter of Desgodins. King gLañ-dar-ma was assassinated in Lhasa by a Lama, and civil war shook the city under the rule of the Dalai Lamas. "Executions are conducted in the open street before the people, and apparently culprits suffer not far from the temple, and not outside the city, Buddhist injunctions notwithstanding. When Nain Singh visited Lhasa, he saw a Chinaman beheaded in public" (G. Sandberg, *Tibet and the Tibetans*, p. 191, London, 1906). The holy city has also its meat-market (śa k'rom). According to E. Kawaguchi (*Three Years in Tibet*, p. 286), there is a special place near the monastery ḥBras-spuñs, some miles west from Lhasa, where yaks, goats, and sheep, seven in number, are daily butchered for the table of the Dalai Lama.

[2] It is a debatable point whether the Sa-skya hierarchs really took up their permanent residence in Lhasa. The famed ḥP'ags-pa Blo-gros rgyal-mts'an, the spiritual adviser of Kubilai, on his return from China, wended his way back in 1265 to "the great residence dpal-ldan Sa-skya" in southern Tibet, and there he returned again in 1276 after his second

it is not of such a specific or intimate character that it could be explained only through an actual visit to Tibet. All this, and more, could have been learned at that time from the Chinese and the Mongols. It is somewhat a matter of regret that Colonel Sir Henry Yule's note on Odoric's *Abassi* has been allowed to remain. This word has no connection whatever with *lobaes, ubashi,* or *bakshi,* [1] nor is it necessary to resort to such extravagances. Odoric plainly states that the word is of the Tibetan language; and it has to be sought, therefore, in Tibetan only. KOEPPEN's [2] explanation, overlooked by YULE, remains the only one that is admissible. The Sa-skya hierarchs, who practically ruled Tibet in the age of the Mongols, bore the Tibetan title *ɑPᶜags-pa* ("eminent, excellent"), [3]

---

journey (HUTH, *Geschichte des Buddhismus in der Mongolei*, Vol. II, pp. 154, 157). His biography makes no allusion to his residing in Lhasa. Our knowledge of Tibetan historical sources is still so limited that we cannot be positive on this point. The greater probability seems to be that the abode of the Sa-skya was their ancestral seat, the monastery of Sa-skya. Lhasa, nevertheless, may have continued as the capital of political administration.

[1] The word *bakshi* is not, as stated by YULE (also *Marco Polo*, Vol. I, p. 314), connected with Skr. *bhikshu*. The Tibetans are acquainted with both words, translating the latter by the term *dge-sloṅ*, and writing the former *paŋ-ɣi* (JÄSCHKE's spelling *pa-ɣi* is inexact) The Tibetan dictionary *Li-ɣii gur kʻaṅ*, fol. 23a (see *Tʻoung-Pao*, 1914, p. 65), explains this word by *btsun-pa* ("respectable, reverend"), and states that it is derived from the language of the Turks (*Hor*). The word seems to be, indeed, of Turkish origin (VÁMBÉRY, *Primitive Cultur*, p. 248; RADLOFF, *Wörterbuch der Türk-dialecte*, Vol. IV, col. 1445).

[2] *Die lamaische Hierarchie*, p. 105. It is notable that BOLLAND's text in the *Acta Sanctorum*, as quoted by Koeppen, "Abbassi, quod sonat Papa in illa lingua" (M. CORDIER quotes the same reading from the manuscript of Berlin), differs from the texts of Yule (Latin version, "Lo Abassi, id est Papa in linguâ suâ;" Italian version, "il Atassi, che viene a dire in nostro modo il Papa"). It seems to me much more probable that Bolland has preserved the true, original reading. Odoric means to say that the Tibetan word which is written *apʻags-pa* (varying in its pronunciation) was heard by him *ba-se, ba-si*, and sounds in their language also like *papa* (*pʻaʻ-pa*). The comparison with the Pope would almost savor of a heresy in the mouth of the pious Friar, and "the Pope" was no doubt dragged in by the later copyists. — N. KÜNER (*Description of Tibet*, in Russian, Vol. I, 1, p. 30) attempts to explain *Abassi* as an inexact transcription of *bLo-bzaṅ śes-rab*, "a common title of the highest Buddhist clergy." I see no possibility of such an interpretation; this term, moreover, is neither a common title nor a title at all, but merely a personal name.

[3] ɑJigs-med nam-mkʻa tells a little anecdote in explanation of this title (HUTH,

and were spoken of as the *qP^{c}ags-pa bLa-ma*. This word, variously articulated *p^{c}ags-pa*, *p^{c}aɡ-pa*, *p^{c}as-pa*, [1] *p^{c}a^ɔ-pa*, is the source of Odoric's Abassi. [2]

A striking assertion made by the Friar is that "they have in it great plenty of bread and wine as anywhere in the world." Such a statement cannot possibly be advanced by any one who has had but the slightest contact with the Tibetan borderlands and the most superficial acquaintance with Tibetan people. First of all, there is nothing like bread in Tibet, where even the preparation of dough is unknown. Parched barley-flour mixed with tea or milk into a porridge forms the staple food; and the alcoholic beverage called *č^{c}añ*, obtained from fermented barley, is neither wine nor beer, but a liquor *sui generis*. [3] Even granted that Odoric simply committed a mistake in the choice of his words, and merely intended to say that food and drink abound in Tibet, his statement nevertheless remains very strange. The majority of Tibetans eke out a wretched living as poor shepherds or farmers, and earn enough to be kept from starvation; but emphasis on the food-supplies being as abundant as anywhere in the world is thoroughly out of place for a poor country like Tibet.

The assertion that the women have a couple of tusks as long as those of wild boars has been attributed by Yule to an error of the scribe. I am rather under the impression that it is a bit of information misunderstood on the part of Odoric. Boar's tusks are

---

Geschichte des Buddhismus in der Mongolei, Vol. II, p. 141). The same is narrated in *Yuan shi* (Ch. 202, p 1), where the word is written 帕克斯巴.

[1] The spelling *Passepa* appears in the *Lettres édifiantes*, nouv. éd., Vol. XXIV, p. 9 (Paris, 1781). The Mongols pronounce the word *papa* (PALLAS, *Sammlungen*, Vol. II, p. 87).

[2] Koeppen's theoretical *ap'ags-çri*, which does not exist, must be discarded.

[3] The grape-wine mentioned by Mr ROCKHILL (*J. R. A. S*, 1891, p. 227, note 1) as being made in small quantities, and high-priced, is almost restricted to religious offerings, and plays no part in the life of the people. No foreign traveller has ever seen or tasted it.

generally employed by Tibetan women for making the parting of their hair. [1] Odoric's remark that the women have their hair plaited in more than a hundred tresses applies only to the pastoral tribes of northern and north-eastern Tibet; [2] and if he had really crossed Tibet to Lhasa and beyond, he could not have failed to notice that quite different styles of hair-dressing prevail in other parts of the country. This matter is not very serious, but an error of grave account is the observation that "the folk of that country dwell in tents made of black felt." Certainly the Tibetans understand the art of making felt; [3] but the tents inhabited by the pastoral tribes of Tibet, throughout the country, are covered with a black cloth woven from yak-hair. [4] In this respect, and in its quadrangular structure, the Tibetan tent represents a dwelling-type of its own, which is plainly distinguished from the Mongol circular felt tent. It is impossible to assume that in the days of Odoric there may have been Tibetan nomads living in felt tents, and thus come to the Friar's rescue. The mode of habitation is one of the most permanent and enduring factors in the life of all peoples, which is but very seldom sacrificed to outward influences. The conclusion

---

[1] I doubt very much the correctness of Yule's statement that the women in Tibet commonly use boar's tusks as ornaments, both attached to the head and hung round the neck. I paid particular attention to ornaments in Tibet, and never saw a woman wearing boar's teeth on her head or neck. Among the nomads of Derge I observed now and then a man wearing a perforated boar's tooth as a protective amulet; sometimes two such teeth are joined together at their bases and held by a brass hoop.

[2] See, for instance, the plate opposite p. 18 in GRUM-GRŽIMAILO's *Description of a Journey in Western China* (in Russian, Vol. II, St. Petersburg, 1907)

[3] The process is described by ROCKHILL (*Notes on the Ethnology of Tibet*, p. 700). F GRENARD (*Mission scient. dans la haute Asie*, Vol. II, p. 372) is certainly right in saying that Tibetan felt is rather mediocre, and very inferior to the Chinese and Kirgiz specimens.

[4] See ROCKHILL, *l. c.*, p. 701, and *The Land of the Lamas*, pp. 75—77; GRENARD, *l. c.*, p. 337. I do not concur with Grenard in the view that the Tibetan tent is in every respect much inferior to the Mongol one; for myself, I prefer the Tibetan tent as more practical and durable, and a more efficient means of protection against heat and cold.

prompted by the ethnological point of view, that the Tibetan tents of yak-hair stuffs go back to a venerable age, is fully corroborated by the records of the Chinese. Both *Sui shu* and *T'ang shu* tell in regard to the Tang-hiang 党項, a Tibetan tribe living in south-western Kan-su and in the vicinity of the Kuku-nōr, that their habitations are made from weavings of the hair of yak-tails and sheep. [1] The Annals of the T'ang Dynasty relate, in regard to the

---

[1] 織犛牛尾及粘羺毛以爲屋 (*Sui shu*, Ch. 83, p 2b) 織犛尾羊毛覆屋歲一易 (*T'ang shu*, Ch. 221 上, p 1b). — When *Kiu T'ang shu* (Ch. 196 上, p. 1 b) asserts that the Tibetan nobles dwell in large felt tents called *fu-lu* (貴人處於大氈帳名爲拂廬 [see *T'oung Pao*, 1914, p. 92]), it is not contradictory to the fact, as stated above. In the sentence preceding this one the question is of the houses in which the Tibetan people ordinarily live, covered with flat roofs and reaching a height up to ten feet. In this case, accordingly, it is the sedentary agricultural portion of the populace which is spoken of, but not the pastoral tribes These Tibetan nobles were not nomads, but warriors, with a stationary residence among the sedentary farmers, and they undoubtedly imitated the custom of the Turkish chieftains (at a later date adopted by the Mongols) of residing in felt tents (*sbra*) as a mode of living better suiting their warlike occupation (compare Tib. *p'yiṅ gur* ("felt tent, a Tartar hut") in JASCHKE's *Dictionary*, p 350). The probability that Odoric might have struck such war-tents is so slight that it merits no discussion. His statement, moreover, is generalized to the effect that the folk of that country dwell in tents made of black felt. — The above word *ku-li* (*ku* is written also 毀) is recorded both in *Erh ya* and *Shuo wén*. Li Shi-chên (*Pén ts'ao kang mu*, Ch. 50 上, p. 11 b) defines it as a sheep with plenty of hair. K'ou Tsung-shi, in his *Pén ts'ao yen i* of 1116, says that its habitat is in Shen-si and Ho-tung 河東 (Shan-si), and that its hair is very strong, long, and thick. Chinese authors, in their descriptions of Sikkim (*Pai mu jung* 白木戎, transcription of Tib. *aBras-mo ljoṅ*, pronounced *bu* or *drà mo juṅ*, "Land of Rice;" the identity of the names has been recognized neither by KLAPROTH, *Description du Tubet*, p. 275, nor by ROCKHILL, *J. R. A. S*, 1891, p. 131; the latter's identification with Pari-djong is untenable), speak of a species of sheep styled "big *kü-lü* sheep" 大羭羖 羊 (these two characters are not recorded in K'ang-hi) (*Wei Tsang t'ung chi* 衞藏通志, Ch. 15, p. 13b, 1896, anonymous, not mentioned by Rockhill). The *Wei Tsang t'u chi* (Ch. 下, p 32) calls this animal *kü chao* 羭羒 (ROCKHILL writes 羭羚, and transcribes *chü-shao*; KLAPROTH, "des moutons ou chèvres appelés *kiu tchao*;" both without explanation). This is doubtless the burrel sheep (*Ovis nahura*), found in considerable flocks at high altitudes in Sikkim (RISLEY, *Gazetteer of Sikkim*, p. 239) and throughout Tibet, and called by the Tibetans *gna-ba*, colloquially *na-po, nao, nau*, Nepalese *nāhur*: hence the zoölogical *nahura* (Jäschke's explanation "antelope" is wrong, Chandra Das is correct), Lepcha *na-wo*. The

Tibetans, that, although they have towns formed by huts, they are loath to live there, but prefer to dwell in tents made from pieces of soft animal hair joined together, and that those styled "big tents" (ta fu-lu) are capable of holding several hundred men. [1]

From whatever point of China Odoric may have transgressed the Tibetan boundary, he could not have failed to observe the peculiar tents which have struck the eyes of all subsequent travellers, and at none of these points are felt tents to be seen. [2] It is obvious beyond any doubt that Odoric's observation refers, not to Tibetan, but to Mongol tents, which he may have encountered in the Ordos country [3] or while crossing Mongolia on his way back to Europe. It seems to me infinitely more probable that Odoric, coming out of the Ordos and Kan-su, returned by way of Mongolia, on a similar route as Carpini and Rubruk, than that he should

---

species was first described by B. H. Hodgson (*On the Two Wild Species of Sheep inhabiting the Himalayan Region*, *J. A. S. B.*, Vol. X, 1841, p. 231), then by W. T. Blanford (*Fauna of British India*, p. 499, with illustration). This *kü-lü*, as a word, is perhaps related to *ku-li*, though the two certainly refer to different animals. The *ku-li* mentioned above in *Sui shu* must be a domestic sheep, its wool being utilized, while *kü-lü* is a wild sheep. Mr. Rockhill remarks that "these characters are used phonetically, they have no meaning in Chinese;" but I can trace no Tibetan or Lepcha word which they could be intended to transcribe.

[1] 有城廓廬舍不肯處聯氊帳以居號大弗廬容數百人 (*T‘ang shu*, Ch. 216上, p. 1).

[2] It is a gratuitous speculation of C Puini (*Il Tibet*, p. xxv) when he makes Odoric descend from Tenduc to Si-ngan fu, "e di là, per entrare nel Tibet, seguì probabilmente la via percorsa da Marco Polo, o se ne tenne forse più a settentrione; ma il nostro frate francescano si spinse assai più oltre, giungendo fino a Lhasa." If Odoric should have taken this beaten track, which is so familiar to me, I should be very positive in denying that he could have found any felt tents on this route From Ta-tsien-lu to Ba-t'ang and beyond, from Ta-tsien-lu to Derge and Chamdo, further, in north-western and northern Sze-ch'uan, in southern and western Kan-su, and in the region of the Kukunör, — nowhere is there to be met with a single felt tent. Also Küner (*l. c.*) has Odoric travel through Shan-si, Shen-si, Sze-ch'uan, and Tibet.

[3] There felt tents are now scarce, the Mongols usually living in houses of plaited wicker-work plastered with clay (see Potanin, *The Tanguto-Tibetan Borderland of China*, in Russian, Vol. I, p. 108, St. Petersburg, 1893).

have performed the long and fatiguing journey across Tibet. True
it is, he himself tells us that he came to a certain great kingdom
called Tibet, and there is no reason whatever to question his vera-
city. Odoric was earnestly and honestly convinced of having come
to Tibet, but coming to Tibet does not yet mean entering and
crossing Tibet. The geographical notion "Tibet" was always con-
ceded a liberal interpretation on the part of travellers; the days are
not so far behind us when men nearing the outskirts of Tibet,
touching Ladakh, Darjeeling, Ta-tsien-lu, Ba-t'ang, or Si-ning, had
all been "to Tibet;" and the books on Tibet whose authors were
around but never in the country are numerous. No doubt Odoric
came in contact with Tibetans somewhere in Kan-su [1] or on its
borders, but this is the utmost concession that can be made to him.
It is incredible that he should have traversed Tibet, nor does he
himself make any statement to this effect. He makes no pretence
whatever to having been in Lhasa. All these allegations are preposterous inferences of his overzealous admirers. The fact remains
that the diary of his travels abruptly closes and absolutely terminates with the first sentence of Chapter 45. What follows it, down
to the end of the book, consists, not of observations of the traveller,

---

[1] I believe that his province called *Kansan* is rather Kan-su than Shen-si, as explained
by Yule; though Yule also is inclined to regard it as Shen-si and Kan-su united, as the
two provinces were indeed under the Sung; the name Kan-su appears only from under the
Yüan. Odoric's reference to rhubarb as growing in this province, "and in such abundance
that you may load an ass with it for less than six groats," fits Kan-su far better than
Shen-si. True it is, that rhubarb grows also in Shen-si (PARENNIN, *Lettres édifiantes*, nouv.
éd., Vol. XIX, p. 307; BRETSCHNEIDER, *Bot. Sin*, pt. 3, p. 230; *List of Chinese Medicines*,
p. 480, Shanghai, 1889), but the output is not so large that it would strike the casual
traveller. Kan-su, the adjoining Amdo region, then Sze-ch'uan and Tibet, were always the
classical land of rhubarb; and it is in the mountains of Tangut that, according to MARCO
POLO (ed. of YULE and CORDIER, Vol. I, p. 217), rhubarb is found in great abundance, and
where merchants come to buy it and carry it all over the world. Hence we may take it
for granted that likewise Odoric did not hear about rhubarb before reaching the territory
of Kan-su.

but of stories reproduced from hearsay. The story of Tibet moves along the same line as the following stories of the rich man in Manzi, the Old Man of the Mountain, the devil exorcisms in Tartary, and the valley of terrors. [1] No principle of geographical order is observed in the arrangement of these concluding chapters, which is sure evidence of the fact that Odoric terminated the narrative of his journey at the moment when he turned his back to Cathay. In Chapter 46 he reverts to the province of Manzi as the theatre of action for the plot of the rich man; and in the next chapter we are told that he reached a certain country which is called Millestorte, the residence of the Old Man of the Mountain, but, very curiously, after he had left the lands of Prester John and was travelling toward the west. Where, then, is Tibet? If he had ever crossed Tibet, he would naturally have located Millestorte to the west of, or beyond, Tibet; but he has forgotten Tibet, and takes us back to Prester John. Tibet has left no profound or lasting impression upon his mind, because he rubbed elbows but superficially with its north-eastern borderland. If the case were further supported by negative circumstantial evidence, it would lead to no end of discussions: he lisps not a word as to the nature and physical conditions of Tibet, and whoever enters Tibet from China is soon aware of being transferred into another world. There is no need, however, of invoking this striking lack of personal experience and observation. Odoric of Pordenone has never traversed Tibet proper,

---

[1] It is certainly out of the question to utilize the alleged localities of these stories for reconstructing the stages of Odoric's return journey, as attempted, for instance, by PUINI (*l c.*, p. XXVI), who remarks that Odoric, coming out of Tibet, tells us that, leaving that country, he betook himself to Millestorte. Odoric, of course, does not even express himself in this manner; but he came to Millestorte by journeying towards the west, after leaving the lands of Prester John. — KÜNER (*l. c.*, notes, p. 25) reads much between the lines when he distils out of Odoric the inference that, according to him, Tibet is situated between the possessions of Prester John and the Old Man of the Mountain.

has never been at Lhasa, [1] — a feat with which he has been unduly credited for so long, and to which he himself lays no claim. The honor of being the first Europeans to have reached Lhasa is justly due to the two Jesuit Fathers Grueber and Dorville, who spent two months there in 1661.

---

[1] He does not even make mention of the very name Lhasa, but speaks only of "the chief and royal city," and "in this city." Only the French version adds, "Elle est appelée Gota;" and M. CORDIER justly annotates that there is no city called Gota. This name certainly is mere fancy. Is it credible that a man who has visited Lhasa should not even record the name of the city? And where does Odoric say that he visited it at all? How did modern writers ever get at the assuring statement that he sojourned there for some time? Surely this is a repetition of the miracles attributed to the good Friar after his death, and of which he himself was innocent.

105

方法敛讣告

NÉCROLOGIE.

FRANK H. CHALFANT †.

On the 14th January of this year, the Rev. Frank H. Chalfant, D. D., passed away at Pittsburgh. Pa., after a long and lingering illness of most distressing character which for two years he had borne with heroic patience and fortitude. Born on May 29, 1862, in Mechanicsburgh, Pennsylvania, the son of an eminent clergyman, he graduated from Lafayette College at Easton, Pa., in 1881 and studied for the ministry in Western Theological Seminary where he was graduated in 1886. He was ordained to the ministry in the same year by the Presbytery of Pittsburgh and appointed by the Board of Foreign Missions of the Presbyterian Church in the U. S. A. to the West Shantung Mission at Wei-hien, Shantung, China, March 21, 1887, sailing for his field October 20, 1888. His evangelistic career was one of devotion and efficiency, and for a quarter of a century, during indefatigable itinerations, brought him into close contact with the rural population of Shantung. He acquired the language both oral and written, and amid his mission labors found time for intelligent study and research. Mr. Chalfant kept aloof from the popularizing tendencies to which so many missionaries in China too easily succumb, and remained an earnest student of scientific problems. His interest was particularly aroused in the early development of Chinese writing. The first fruit of his studies was published in 1906 by the Carnegie Museum of Pittsburgh under the title "Early Chinese Writing" [1]. He became deeply interested in Chinese archaeology [2] and was a coin collector and connoisseur of distinction. A well illustrated treatise from his pen on "Ancient Chinese Coinage" is embodied in the work "Shantung, the Sacred Province of China" edited by R. C. FORSYTH

---

[1] Reviewed by E. S. MORSE, *Science*. 1906, pp. 758—60, and H. MASPERO, *B.E.F.E.O.*, Vol. VIII, 1908, pp. 264—7.

[2] He wrote *Standard Weights and Measures of the Ch'in Dynasty* (*J.Ch.Br.R.A.S.*. Vol. XXXV, 1903—4, pp. 21—24). He excavated several graves of the Sung period near Wei-hien and donated the ceramic finds yielded by them to the American Museum of New York; they are figured and described in the writer's *Chinese Pottery*, pp. 312—320.

(Shanghai, 1912) [1]). Owing to its intrinsic value his numismatic collection including 690 coins, among these many rare and unique specimens, has recently been acquired by the Field Museum of Chicago. Mr. Chalfant's name will forever be connected with the discovery and original decipherment of the inscriptions carved in bone and tortoise-shell first exhumed in Honan Province in 1899 [2]). The greater part of these finds was bought by Mr. Chalfant and S. Couling, who acted very wisely in the distribution of these little treasures. They are now deposited in the Museum of the Royal Asiatic Society of Shanghai, Royal Scottish Museum of Edinburgh, British Museum of London, private collection of Mr. L. C. Hopkins [3]), Carnegie Museum of Pittsburgh, and Field Museum of Chicago. These collections comprise not only inscribed fragmentary bone and tortoise-shell pieces but also wonderful carvings of inscribed bone. According to his statement, the 150 specimens in the Field Museum secured from Mr. Chalfant, include many of the finest specimens which passed through his hands, as, for exemple, the largest perforated disk *pi* 璧 ever found, and a masterly carving of a charm showing twin phenixes connected by two serpent-heads. of bone stained turquois-blue and inscribed on three faces. For the last seven years Mr. Chalfant was zealously engaged in the study of these inscriptions on which he leaves a voluminous work in manuscript, which he planned to have issued in two volumes. The first is to consist of over 400 plates containing facsimiles of all the bone documents which came under his notice, amounting to 4812, of which 929 are carved amulets, together with an introduction upon the methods of divination by the tortoise and copious notes upon decipherment; the second was to embrace a syllabary of all the characters found, — some 3000 in all including variants, but this is left incomplete. In April of last year I spent a day with Mr. Chalfant in Pittsburg, going with him over the pages of his manuscript. It is hoped that no effort will be spared toward its publication. It is very deplorable that he was not allowed to live to himself give his important work to the world. As an autodidact and self-made sinologue in America, Mr. Chalfant will always command respect and leave pleasant memories among his friends for the seriousness and unselfishness of his aspirations, for the tenacity of his purpose, for his modesty and the sterling qualities of his character. I wish there were more men of his type in this country.

B. LAUFER.

---

1) He contributed to the same work a gazetteer of the prefecture of I-chou (pp. 337—50).

2) Compare CHAVANNES, *Journal asiatique*, 1911, Janv.-Févr., pp 127—137.

3) Compare HOPKINS, *J. R. A. S.*, 1911, p. 1026.

106

再论六十干支

MÉLANGES.

The Sexagenary Cycle Once More.

I have to apologize for having given in my notice (*T'oung Pao*, 1913, p. 569) wrong references on p. 594 to the plates in FRANKE and LAUFER (*Lamaistische Klosterinschriften aus Peking, Jehol und Si-ngan*). My references were based on a set of proofs sent to me at that time from Berlin; meanwhile, however, technical reasons demanded a rearrangement and renumbering of the plates. I therefore beg to enter the following corrections on p. 594 of the preceding volume: "plates 2 and 3" should read "plates 4 and 7;" "plates 30, 31, 42, 43" should read "plates 27, 28, 52, 53;" "plates 45, 47" should read "plates 48, 50;" and "plates 22, 23" should read "plates 24, 25."

I avail myself of this opportunity to add a few remarks to my previous notes. The attitude of Schiefner toward the application of the Tibetan sexagenary cycle seems to me to merit a renewed examination. In his translation of Tāranātha SCHIEFNER has made three independent chronological calculations based on the *Reu mig*, and all three turn out to be erroneous. He states (on p. VI) that Dol-bu šer rgyan was born in 1290 and died in 1353. Chandra Das in his translation of *Reu mig* (p. 57) gives the same year 1290 as the date of his birth, which is sufficient to show that 1291 is intended. Indeed, the cyclical date indicated in *Reu mig* is *c'u qbrug* ("water dragon") and answers to the year 1291. The year 1353 is not only a formal, but a substantial error; the year intended is 1354, but in that year Dol-bu šer rgyan was fully alive: what the *Reu mig* states under this year is that the monastery Ñam-riñ was repaired with his approbation. According to *Reu mig*, he died only in 1361 (Chandra Das: 1360). On p. VII Schiefner allows Kun-dga grol-mc'og to live from 1493 to 1566; in fact, however, he lived from 1495 to 1565 (Chandra Das: from 1494 to 1564). Neither this nor the previous date is found in Vasilyev's introduction to the Russian translation of Tāranātha; accordingly, we here have examples of Schiefner's own computations. On p. 60, note 2, Schiefner makes gŽon-nu dpal from ạGos die in 1480, and again agrees in this date with Chandra Das (p. 68); in fact, he died in 1481, the year being *lcags glañ* ("iron ox"). It is not Vasilyev who made this wrong calculation, for

- 270 -

Vasilyev in his translation (p. 65), while giving the same date, remarks that he adopted it from Schiefner, although he should have known better. These three cases settle the question definitely and show Schiefner's inability to convert Tibetan dates correctly. They further demonstrate that he applied not one but two (or even three) wrong methods (the case of 1493 for 1495 being identical with his wrong date 1573 for 1575 of Tāranātha's birth), and it is difficult to say by what principles he was guided. Thus, also, my previous impression that the correct computation of the years of Tāranātha's birth and History of Buddhism is solely due to Vasilyev is fully confirmed. [1]

My statement in regard to Kanjur and Tanjur (p. 587) should have been made with the modification that cyclical dates do not appear in the colophons of the older translations, that is, those made prior to the year 1027. Many translations incorporated in the Tanjur having been made after this date, it is not only possible that such dates are employed in the colophons, but these, though rarely, do indeed occur. Thus, HUTH [2] has indicated a "female earth hog" year in the colophon of Tanjur, Sūtra, Vol. 123, No. 17; the cycle, however, not being determined, the date is beyond computation. In the seventy volumes of the Tanjur analyzed by P. Cordier no dates seem to be given. Different from the case mentioned is the reference to a "tiger" year in the colophon No. 11 of the same volume of the Tanjur. [3] This case I had intentionally left out of consideration, because the plain "tiger" year is characteristic of a duodenary cycle, and the subject of my article was the sexagenary cycle only. The time for discussing the former is not yet ripe, but the discussion is bound to come in the near future when the Tibetan documents discovered by A. Stein on his last journey will be laid before us. A. H. FRANCKE [4] has asserted that dates expressed in a duodenary cycle frequently appear in these, and quite recently repeats the same statement. [5] But not a single example of such a date has as yet been given us. It remains to be seen when reproductions of the documents in question will be published.

In regard to the Tibetan reckoning of days and months I should have mentioned that M. JAMETEL [6] had already ventilated this question by pointing to a comment of Wei Yüan in his *Shêng wu ki* and to a passage in a work styled by him *Chu êrh hui.*   B. LAUFER.

---

[1] Among the adherents of the old chronology not mentioned by M. Pelliot or me, there are also T. DE LACOUPERIE (*Beginnings of Writing*, p. 59, London, 1894, and *The Silver Coinage of Tibet* in *Numismatic Chronicle*, 1881, p. 346) and P. CORDIER who derived his dates from Chandra Das as established in his translation of *Reu mig* (*B. E. F. E. O.*, Vol III, 1903, pp. 617, 627).

[2] *Sitzungsberichte der preussischen Akademie*, 1895, p. 274.

[3] *L. c.*, p. 273.

[4] *Anthropos*, 1912, p. 264.

[5] *J. R. A. S.*, 1914, p. 47.

[6] In his article *Histoire de la pacification du Tibet* (*Revue de l'Extréme-Orient*, Vol. I, 1882, p. 588, note).

107

中国陶俑（一）：护甲史导论
附：书评三则

FIELD MUSEUM OF NATURAL HISTORY

PUBLICATION 177

ANTHROPOLOGICAL SERIES   VOL. XIII, No. 2

CHINESE CLAY FIGURES

PART I

PROLEGOMENA ON THE HISTORY OF DEFENSIVE ARMOR

BY

BERTHOLD LAUFER
Associate Curator of Asiatic Ethnology

64 Plates and 55 Text-figures

The Mrs. T. B. Blackstone Expedition

CHICAGO
1914

CONTENTS

CHINESE CLAY FIGURES

PART I

PROLEGOMENA ON THE HISTORY OF DEFENSIVE ARMOR

I. HISTORY OF THE RHINOCEROS.

An extensive collection of ancient clay figures gathered in the provinces of Shen-si and Ho-nan during the period from 1908 to 1910 is the basis of the present investigation. As the character of this material gives rise to research of manifold kinds, it has been thought advisable to publish it in two separate parts. Many of the clay statuettes which form the nucleus of our study are characterized by the wear of defensive armor, hence this first part is devoted to an inquiry into the history of defensive armor,—a task of great interest, and one which heretofore has not been attempted. It will be recognized that this subject sheds new light on the ancient culture of China and her relations to other culture zones of Asia. The second part of this publication will deal in detail with the history of clay figures, the practice of interring them, the religious significance underlying the various types, and the culture phase of the nation from which they have emanated.

Before embarking on our subject proper, a preliminary question must be decided. It is the tradition of the Chou period that the cuirasses[1] employed at that time were manufactured from the hides of two animals designated by the words *se* (No. 10,298) and *si* (No. 4218).[2] It is imperative to have a clear understanding of what these two animals were in the early antiquity of China. As this problem is still pending, and as a close and coherent investigation of the matter has never been made, I have decided to treat it from the very beginning by means of all accessible methods, with the possible hope of a final solution.

The present state of the problem is as follows: EDOUARD BIOT,[3]

---

[1] "Cuirass" or "cuirbouilly" is the right term for this kind of armor, as these words (like French *cuirasse*, Italian *corazza*) go back to Latin *coratium* ("a breastplate of leather"), derived from the word *corium* ("leather").

[2] These figures refer to the numbers of the Chinese characters in the Chinese-English Dictionary of H. A. GILES.

[3] Le Tcheou-li, ou Rites des Tcheou, Vol. II, p. 507 (Paris, 1851).

73

the ingenious translator of the *Chou li*, has expressed his opinion in these words: "I translate by *buffalo* the character *si*, and by *rhinoceros* the character *se*.   These two characters[1] denote in the *Shi king* a rhinoceros or a wild buffalo, without the possibility of distinguishing between them. The skin of the rhinoceros being very thick, it seems difficult to believe that it could have been sliced, and that the pieces were sewed together, in order to make cuirasses.   In this case the two characters of the text[2] would designate here two species of buffalo."[3]   PALLADIUS, in his Chinese-Russian Dictionary, treats the matter in the opposite way, and renders *se* by (1) "an animal resembling a wild ox," (2) "Malayan rhinoceros," and *si* by "rhinoceros."   COUVREUR credits the word *se* first with the latter meaning, secondly with that of *bœuf sauvage*.[4]

CHAVANNES[5] has clearly and sensibly expressed the opinion that

---

[1] It should properly read, "words."

[2] Referring to the passage of the *Chou li* where the hide cuirasses are mentioned.

[3] In his essay on the Manners of the Ancient Chinese (in LEGGE, Chinese Classics, Vol. IV, Prolegomena, p. 148), BIOT says that "they hunted also herds of deer, of boars, of wild oxen," on which LEGGE annotates, "These wild oxen would seem to be rhinoceroses."   But in his original article (*Journal asiatique*, 1843, p. 321), BIOT has added the following comment: "Le caractère *si* est traduit ordinairement par rhinocéros, et c'est, en effet, son sens actuel.   Lacharme a traduit, tantôt *bos sylvestris*, tantôt rhinocéros.   Il me semble que les grandes chasses devaient être dirigées surtout contre des troupeaux de bœufs sauvages ou buffles."   The objections raised by Biot in the above passage are not valid; it is certainly possible to slice rhinoceros-hide, and to sew the pieces together.   Cuirasses and shields have been made from it, as may be seen from many specimens in the collections of our museums.   A shield of rhinoceros-hide is illustrated in Plate XXVII.   In accordance with the above definition, BIOT, likewise in his translation of the Annals of the Bamboo Books (Extrait du *Journal asiatique* 1841 and 1842, pp. 41, 46), rendered *se* by "rhinoceros" and *si* by "bœuf-*si* (rhinocéros)," while LEGGE (Chinese Classics, Vol. III, Prolegomena, pp. 149, 153) in both cases has "rhinoceros."   It will be seen in the course of this investigation how Biot's error was caused, and that his opinion is untenable.   W. R. GINGELL (The Ceremonial Usages of the Chinese, p. 81, London, 1852) treated the two words in a way opposite to that of Biot, translating in the passage of *Chou li* the term *si kia* by "rhinoceros-hide armor" and *se kia* by "wild buffalo's-hide armor."   No one of those who from purely philological points of view proposed the rendering "wild buffalo" has ever taken the trouble to raise the question whether anything like wild buffalo exists in China, anciently or in modern times.   BUSHELL (The Stone Drums of the Chou Dynasty, *Journal China Branch R. As. Soc.*, Vol. VIII, 1874, p. 154) was of the opinion that the ancient Chinese hunted the rhinoceros in the low swamps.

[4] The passage in *Lun yü* (XVI, 7) is translated by COUVREUR (Les quatre livres, p. 250), "Si un tigre ou un bœuf sauvage s'échappe de sa cage."   Nevertheless in the glossary (p. 664) the word *se* is rendered by "rhinoceros."   LEGGE (Chinese Classics, Vol. I, p. 307) translates here "rhinoceros," despite Chu Hi's (undoubtedly wrong) interpretation of *se* being a *ye niu* ("wild bull").   In his first edition of *Lun yü* (which is not accessible to me, but this may be gleaned from PLATH, Die Beschäftigungen der alten Chinesen, p. 56), LEGGE translated *se* by "wild ox."   In the text of *Mêng-tse* (III, 2, IX, 6), LEGGE (Classics, Vol. II, p. 281) and COUVREUR (*l. c.*, p. 452) are in mutual accord in translating the word *si* by "rhinoceros," and this is likewise the case with reference to the word *se* in *Li ki*, II, I, III, 40 (LEGGE in *Sacred Books of the East*, Vol. XXVII, p. 158; COUVREUR, *Li ki*, Vol. I, p. 181).   In *Tso chuan*, VII, 2, LEGGE (Classics, Vol. V, p. 289) renders *si se* by "rhinoceroses and wild bulls."

[5] Les Mémoires historiques de Se-ma Ts'ien, Vol. III, p. 502.

se niu and *si* appear to be two different species of rhinoceros. Also G. Devéria[1] has translated *se* and *si* by "rhinoceros."

Bretschneider, both a naturalist and an eminent sinologue, upheld the opinion that the rhinoceros, and goblets made from rhinoceros-horn, are repeatedly mentioned in the Chinese classics, and that the latter has been reputed from time immemorial for its antipoisonous virtues. He refers the saying that rhinoceros-horn cures all poisons, to the *Shên-nung pên ts'ao king*, attributed by tradition to the mythical Emperor Shên-nung, at all events the most ancient Chinese materia medica in existence.[2]

In the first edition of his Chinese-English Dictionary, Professor Giles, the eminent sinologue at the University of Cambridge, England, attributed to both *se* and *si* the meaning of "rhinoceros," without establishing a distinction between the two. In the second edition, however, we read under *se* (No. 10,298), "A bovine animal, figured as a buffalo with one horn, known as the *se niu*. Another name for the *si* 4128; see 8346 for its confusion with the rhinoceros." Under the last-named heading it is said that the term *si niu* is "a bovine animal, figured as a buffalo with a single horn;" with the addition that the traditional "rhinoceros" of foreigners seems to be wholly wrong. Further, the reader is requested to correct No. 4128 *si*, where the meanings "tapir" and "rhinoceros" had been given. In his "Adversaria Sinica" (p. 394), Mr. Giles has expounded more in detail the reasons which induced him to make these alterations. The arguments advanced by him are briefly three: 1. The rhinoceros is known to the Chinese as *pi kio*, "nose-horn." 2. In two passages of Chao Ju-kua (translation of Hirth and Rockhill, pp. 118, 233), rhinoceroses are spoken of as being shot with arrows, while Giles finds it stated in the *T'u shu tsi ch'êng* that arrows cannot pierce the hide of the rhinoceros. 3.. The *si* and the *se* are figured in the latter work as slightly differing

---

[1] Histoire des relations de la Chine avec l'Annam, p. 88 (Paris, 1880).

[2] *Chinese Recorder*, Vol. VI, 1875, p. 19, and Mediæval Researches, Vol. I, p. 153. Regarding the materia medica current under the name of Shên-nung see Bretschneider (Botanicon Sinicum, pt. 1, pp. 27–32). Bretschneider, though believing that in India the people from time immemorial attribute the same antipoisonous virtues to the rhinoceros-horn as the Chinese do, says he cannot believe that the Chinese have borrowed this practice from the Hindu or *vice versa*. The Hindu conception is not attested by any passage in Sanskrit literature, but only by Ctesias and Aelian who state that drinking-vessels made from the horn of the unicorn safeguard from poison and various diseases. The belief is likewise absent among the Greeks and Romans, in whose records the number of references to rhinoceros-horn is exceedingly small (H. Blümner, Technologie und Terminologie der Gewerbe und Künste, Vol. II, p. 358). There is no evidence that the Chinese notions are due to any stimulus received from outside; they appear, on the contrary, as legitimate offshoots grown on Taoist soil. The Chinese likewise conceived the idea of carving rhinoceros-horn into cups, girdle-plaques, and fanciful ornaments. We shall come back to these various points in detail. Compare p. 154, note.

獨角獸圖

坤輿圖說　獨角獸

亞細亞州印度國產獨角獸形太如馬極輕快毛色黃頭有角長四五尺其色明作飲器能解毒角

銳能觸大獅獅與之鬥避身樹後若悞觸樹木獅反嚙之

FIG. 1.
Monoceros of European Armorial Style, introduced into China by the Jesuit Father Ferdinand
Verbiest (from *T'u shu tsi ch'ĕng*).

鼻角獸圖

古今圖書集成

博物彙編禽蟲典第一百二十五卷異獸部彙考三之六

坤輿圖說　鼻角獸

亞細亞州印度國剛霸亞地產獸名鼻角身長如象足稍短遍體皆紅黃斑點有鱗介矢不能透鼻

上一角堅如鋼鐵將與象鬪時則于山石磨其角觸象腹而斃之

FIG. 2.

Rhinoceros, Design of European Origin, introduced into China by the Jesuit Father Ferdinand Verbiest (from *T'u shu tsi ch'êng*).

bovine animals,[1] with a single horn on the head. Says Mr. GILES, "The *Erh ya* says: the latter is like an ox, and the former like a pig, while the *Shan hai king* speaks of both as occurring in many parts of China. There is thus hopeless confusion, of which perhaps the explanation is that a term which originally meant a bovine animal was later on wrongly applied to the rhinoceros."

The first argument advanced by Mr. Giles is not admissible as good evidence in the case. "The rhinoceros is known to the Chinese as *pi kio*, 'nose-horn,' and is approximately figured in the *T'u shu*." By referring to the Chinese cyclopædia we find, however, that this name with the illustration is extracted from the *K'un yü t'u shuo*. The latter is not the production of a Chinese author, but of the Jesuit FERDINAND VERBIEST, born in 1623, and who arrived in China in 1659 and died in 1688.[2] This section of the *T'u shu tsi ch'êng* alluded to by Mr. Giles and devoted to "strange animals" contains quite a number of illustrations and texts derived from the work of Verbiest; and neither his zoölogical nomenclature nor his descriptions and illustrations, which are based on European lore, can be laid at the door of the Chinese. The evidence is here produced in Figs. 1 and 2. In Fig. 1, Verbiest pictures a "single-horned animal" (*tu kio shou*), saying, "India, situated on the continent of Asia, is the habitat of the single-horned animal which is as big as a horse, very light and swift, and yellow in color. On its head it has a horn, four to five feet long, of bright color. It is made into drinking-vessels which are capable of neutralizing poison. As the horn is pointed, the animal can charge a big lion. The lion, while struggling with it, takes refuge behind a tree; and when missing its aim, it butts the tree, while the lion bites it at this moment." In Fig. 2, the *pi kio shou* referred to by Mr. Giles is pictured. Verbiest comments, "The locality Kang-pa-ya[3] in India, situated on the continent of Asia, is the habitat of an animal called 'nose-horn' [rendering of 'rhinoceros']. Its body is as powerful as that of the elephant, but its feet are somewhat shorter. Its trunk is covered all over with red and yellow spots, and is overlaid with scales. Arrows cannot pierce it. On its nose there is a single horn as strong as steel. It prepares for its battles with the elephant by whetting its horn on the rocks; and hitting

---

[1] This is a debatable point. The two illustrations do not resemble bovine animals, but deer (see Figs. 9 and 10 on pp. 102 and 103). The "bovine animal with one horn" first appears in LIONEL GILES, An Alphabetical Index to the Chinese Encyclopaedia, p. 5 (London, 1911).

[2] WYLIE, Notes on Chinese Literature, p. 58; M. COURANT, Catalogue des livres chinois, p. 95; H. CORDIER, L'imprimerie sino-européenne en Chine, p. 59; P. PELLIOT *Bulletin de l'Ecole française d'Extrême-Orient*, Vol. III, 1903, pp. 109, 115.

[3] That is, Khambāyat or Cambay, in the western part of the province of Gujarāt.

the elephant's paunch, it kills it." The alleged combats of the rhinoceros with the lion and elephant are classical reminiscences (see p. 84) which are absent from Chinese folk-lore. Verbiest repeats the popular traditions current at his time in Europe, and like Cosmas Indicopleustes, still discriminates between the *monoceros* or *unicornis* (*tu kio*) and the *rhinoceros* (*pi kio*), illustrating the former by the unicorn of European heraldry. Consequently the terms employed by Verbiest are literal translations of European nomenclature into Chinese, made by Verbiest for his purpose; and the word *pi kio* cannot be claimed, as has been done by Mr. Giles, as a genuine term of the Chinese language. It is a foreign term not employed by the Chinese. Indeed, in a long series of Chinese texts dealing with the rhinoceros, and given below, not any use of this name is made. Only a single case is known to me: the Manchu-Chinese dictionary *Ts'ing wên pu hui* of 1786 (Ch. 4, p. 23) explains the Manchu word *sufen* by the said *pi kio*, adding the definition, "a strange animal bred in Cambaya in India, like an elephant, with short feet," etc., the same as given by Verbiest. This, accordingly, is a mere repetition of the latter's statement, and is not conclusive. Curiously enough, that expression which Mr. Giles credits as the only authentic word for "rhinoceros" is given a quite different meaning in the Polyglot Dictionary of K'ien-lung (Appendix, Ch. 4, p. 75), where we find the series Chin. *pi kio shou*, Manchu *sufen*, Tibetan *ba-men*, Mongol *bamin*. The Tibetan word *ba-men*, reflected in Marco Polo's *beyamini*,[1] denotes the gayal wild ox (*Bos gavaeus*). Whether this equation, as a matter of fact, is correct, is certainly a debatable question; but this point does not concern us here. The point to be brought out is that *pi kio* in the sense of "rhinoceros" is a term coined by Verbiest, and that it has not yet been pointed out in any Chinese text prior to his time.[2] Simultaneously Mr. Giles's argument directed against Hirth —"the *T'u shu* expressly

---

[1] See the writer's Chinese Pottery, p. 260, note 4.

[2] The general Chinese expression for rhinoceros-horn which is even now traded to Canton and there made into carvings is still *si kio;* hence it follows that at the present day the designation of the animal itself, as it has been for several millenniums, is the word *si*. The English and Chinese Standard Dictionary of the Commercial Press, issued by a commission of Chinese scholars, who must know their language, renders the word "rhinoceros" into *se niu* and *se* (Vol. II, p. 1919). COUVREUR (Dict. français-chinois, 2d ed.) has likewise *se niu*. DOOLITTLE (Hand-Book of the Chinese Language, Vol. I, p. 411) gives under "rhinoceros" *si, se niu,* and *si niu.* SCHLEGEL (Nederlandsch-chineesch Woordenboek, Vol. III, p. 622) renders the word by *se, si,* and *si niu.* True it is that in recent times the words *se* and *si* have been transferred to bovine animals, and the Chinese themselves are well aware of this fact. Thus Li Shi-chên, in his *Pên ts'ao kang mu*, remarks that the term "hairy rhinoceros" is *at present* referred to the yak (see p. 150). This, however, as will be established by abundant evidence, was not the case in former times. In fact, these recent adjustments prove nothing for conditions which obtained in earlier periods. The question as to how the word *se* became transferred to the buffalo is discussed on p. 161, note 5.

says that arrows cannot pierce the hide of the rhinoceros "— falls to the ground. This is a verdict of Verbiest, and not to be encountered in any Chinese report regarding the rhinoceros. It is, moreover, an argument of no meaning and no value; it is simply a popular notion of fabulous character.

The numerous stories formerly current anent the rhinoceros chiefly culminated in three points,— its ferocity, the use of its horn as a weapon of attack, and its invulnerability. These notions have been refuted by close observation. We quote an authority, R. LYDEKKER:[1] "Fortunately, in spite of stories to the contrary, the creature in its wild state appears to be of a mild and harmless disposition,[2] seeking rather to escape from

---

[1] The Game Animals of India, Burma, Malaya, and Tibet, p. 31 (London, 1907).

[2] Certainly; it is easily kept in confinement and tamed, and has often been transported over vast tracts of water and land. A good example of the overland transportation of a tamed rhinoceros or several animals is furnished by Se-ma Ts'ien, in the chapter on the Imperial Sacrifices to Heaven and Earth, when this animal together with an elephant was conducted as far as the foot of Mount T'ai in Shan-tung with a possible view to their being sacrificed; but the Emperor spared their lives, and the animals were allowed to return (see CHAVANNES, Les Mémoires historiques de Se-ma Ts'ien, Vol. III, p. 502). The following tributes of living rhinoceroses are on record. In the year 2 A.D. the country Huang-chi (south of Tonking, 30,000 li from the capital of China) sent a living rhinoceros as tribute to the Court of China, as mentioned three times in the Ts'ien Han shu (Ch. 27 B, p. 17 b). These texts have recently been studied by PAUL PELLIOT (T'oung Pao, 1912, pp. 457–460), who has revealed their fundamental importance for the history of Chinese relations with the countries of the Indian Ocean in the first century of our era. On the basis of Pelliot's translations, the country Huang-chi has recently been made the object of an interesting geographical study on the part of A. HERRMANN (Ein alter Seeverkehr zwischen Abessinien und Süd-China bis zum Beginn unserer Zeitrechnung, Zeitschrift der Gesellschaft für Erdkunde zu Berlin, 1913, pp. 553–561). This author identifies Huang-chi with Abyssinia mainly on the ground that the rhinoceros occurs there. This argument is not cogent, since the home of the animal is in all parts of both Indias, Borneo, Java, and Sumatra as well. Also for other reasons this identification is unfortunate. The transportation of a live rhinoceros from Abyssinia to China over a maritime route would have been a feat impossible in those days, in view of the imperfect state of navigation, while it could easily have been accomplished, if Huang-chi, as assumed by me, was located on the Malayan Peninsula; and as shown by the Chinese records, the live rhinoceroses all hailed from Indo-China or Java. The name Huang-chi, moreover, cannot be derived from Aghāzī, as HERRMANN thinks. His decisive argument in support of this theory is, of course, the statement in the Chinese text that Huang-chi is 30,000 li distant from Ch'ang-ngan, the then capital of China. Mr. Herrmann unreservedly accepts this as a fact, and is in this manner carried away to eastern Africa. We have known for a long time (in fact, the Jesuits of the eighteenth century knew it) that the Chinese definitions of distances over maritime routes must not be taken at their surface value. Nor have we any reason to be more Chinese in this respect than the Chinese themselves. The following is expressly stated in the Sung shu, the History of the Liu Sung Dynasty (420–478 A.D.; Ch. 91): "The southern and south-western barbarians, generally speaking, live to the south and south-west of Kiao-chi (northern Annam), and also inhabit the islands in the great ocean; the distance is about three to five thousand li for those that are nearer, and twenty to thirty thousand li for those that are farther away. When sailing in a vessel it is difficult to compute the length of the road, and therefore we must recollect that the number of li, given with respect to the barbarians of the outer countries, must not be taken as exact" (see GROENEVELDT, in Miscellaneous Papers relating to Indo-China, Vol. I, p. 127). It is plainly indicated in this passage that the distances

its enemies by flight than to rout them by attack. When badly wounded, or so hustled about by elephants and beaters as to become be-
wildered, a rhinoceros will, however, occasionally charge home. In
such onslaughts it is the common belief that the animal, like its African
cousins, uses its horn as its weapon of offence; but this is an error, the
real weapons being the triangular, sharp-pointed low tusks." The
same author states in another work[1] on the skin of the animal, "From
the immense thickness and apparent toughness of its enormous folds,
it was long considered that the hide of the Indian rhinoceros was bullet-
proof, and that the only places where the animal was vulnerable were
the joints of the armor. . . . As a matter of fact, the skin of the
living animal is quite soft, and can readily be penetrated in any place
by a bullet, or easily pierced by a hunting knife. When dried it becomes,
however, exceedingly hard; and it was formerly employed by the
Indian princes in the manufacture of shields for their soldiery."

---

given for the routes in the southern ocean are not exact, and that a description of
twenty to thirty thousand *li* is nothing but a convention to denote the very remote
barbarians of the south. Compare, on Chinese calculations of sea-routes, particularly
G. SCHLEGEL (*T'oung Pao*, Vol. III, 1892, pp. 161–5). In *Hou Han shu* (Ch. 116,
p. 3 a) the location of Huang-chi is positively indicated as being south of Ji-nan (Tonking), which means that it was situated on the Malayan Peninsula. In 84 A.D. the
Man I beyond the boundary of Ji-nan offered to the Court a living rhinoceros and
a white pheasant (*Hou Han shu*, Ch. 116, p. 3 b). In 94 A.D. the tribes in the south-
west of Sze-ch'uan sent an envoy and interpreter presenting a rhinoceros and a big
elephant (*ibid.*, Ch. 116, p. 8 b). At the time of the Emperor Ling (168–188 A.D.)
of the Later Han dynasty, Kiu-chên in Tonking despatched a living rhinoceros to the
Chinese Court (*Huan yü ki*, and *Ta Ming i t'ung chi*, ed. of 1461, Ch. 90, fol. 5, where
it is said also that at the time of the Yüan dynasty [1260–1367] Annam presented a
rhinoceros). In 539 Fu-nan sent a live rhinoceros (*Liang shu*, Ch. 54, p. 4). A similar
report in regard to the country of Ho-ling (Java) occurs in 819 A.D. at the time of the
T'ang dynasty (*Kiu T'ang shu*, Ch. 197, p. 2 b). Finally the poets Yüan Chên
(779–831; GILES, Biographical Dictionary, p. 964) and Po Kü-i have celebrated in
verse a tame rhinoceros which had been sent as tribute in the year 796; it was housed
in the Shang-lin palace, and an official was appointed to care for it; but in the winter
of the following year when great cold set in, the poor creature died. In 1009 Kiao-chi
(Annam) presented a tame rhinoceros to the Court (*Sung shi*, Ch. 489), and there
are other similar reports by the essayists of the Sung period.—TAVERNIER (Travels
in India, ed. V. BALL, Vol. I, p. 114) saw a rhinoceros eating stalks of millet presented
to it by a small boy; encouraged by this sight, the traveller seized some stalks, and
the rhinoceros at once approached him, opening its mouth four or five times; he
placed some stalks in it, and when the animal had eaten them, it continued to open
its mouth to receive some more. Tame rhinoceroses, to which a good deal of freedom
was allowed, were formerly not uncommonly kept by the Rajas of India. Surely, not
only men, but also animals, are usually better than their reputation among men. One
of the most notable facts about the behavior of the rhinoceros in captivity, as already observed by DARWIN (The Variation of Animals and Plants under Domestication, Vol. II, p. 165, Murray's edition, 1905), is that under this condition it breeds in
India far more readily than the elephant. The captive elephants, in contrast to the
rhinoceros, as pointed out by Darwin and confirmed by others (E. HAHN, Kulturgeschichte der Haustiere, p. 37), but very rarely breed; as a rule, they do not even
copulate. There is no doubt that the rhinoceros possesses the qualities fitting it for
domestication, and that only the lack of promising advantages has prevented man
from embarking on such a plan.

[1] The New Natural History, Vol. II, pp. 1055–1056.

Naturally the skin of the animal is as soft and sensitive as that of any other living creature, and arrows are certainly painful to it. Only when properly prepared and dried does the skin assume that iron-like hardness which has achieved its reputation and probably caused the fable of its being impenetrable in the live beast. The account of the Arab envoy given in 993 to the Chinese Emperor, that "to capture a rhinoceros, a man with a bow and arrow climbs a big tree, where he watches for the animal until he can shoot and kill it," as narrated by Chao Ju-kua, is entirely trustworthy.[1] The fable lies entirely in the "arrows cannot pierce the hide," to which Mr. Giles gives credence. When it is said, "he rips up a man with his horn," Chao Ju-kua simply accepts the belief of all his contemporaries, eastern and western; and the remark certainly proves that he speaks of the rhinoceros, while it is no argument in favor of Mr. Giles's opinion that the animal in question is not the rhinoceros.

While the general result at which Mr. Giles has arrived is not novel, being partly anticipated, as we have seen, by Biot, Palladius, and Couvreur, his arguments, as summed up above under No. 3, are original, and deserve serious consideration and discussion. What appears to Mr. Giles as the most weighty evidence in favor of his view are the queer Chinese illustrations of the two animals. Queer they are, but we must make an attempt at understanding and explaining them. For this reason, we shall first enter on a somewhat lengthy digression into the iconography of the rhinoceros; and it will be seen that this, as every-

---

[1] The effect of arrows on the rhinoceros is well illustrated in the following story of GASPAR CORREA, who went to India in 1512, and wrote a detailed chronicle of the Portuguese possessions there. He describes a battle of King Cacandar, who availed himself of elephants fighting with swords upon their tusks, and in front of them were arrayed eighty rhinoceroses (*gaṇḍas*) "carrying on their horns three-pronged iron weapons with which they fought very stoutly . . . and the Mogors with their arrows made a great discharge, wounding many of the elephants and the *gaṇḍas*, which as they felt the arrows, turned and fled, breaking up the battles" . . . (quoted by YULE and BURNELL, Hobson-Jobson, p. 363). In India rhinoceroses were hunted with sabre, lance, and arrows. Timur killed on the frontier of Kashmir several rhinoceroses with sabre and lances, although this animal has such a hard skin that it can be pierced only by extraordinary efforts (PETIS DE LA CROIX, Histoire de Timur Bec, Vol. III, p. 159, quoted by YULE and BURNELL, Hobson-Jobson, p. 762). In Baber's Memoirs (quoted *ibid.*) a rhinoceros-hunt is described in these words: "A she rhinoceros, that had whelps, came out, and fled along the plain; many arrows were shot at her, but . . . she gained cover." The hunters of Java hide sickle-shaped knives under the moss on steep mountain-paths; the animal, dragging its paunch almost close to the ground, rips up itself, and is then easily mastered (P. J. VETH, Java, Vol. III, p. 289, Haarlem, 1903). HOSE and McDOUGALL (The Pagan Tribes of Borneo, Vol. I, p. 145, London, 1912) have this observation to report: "Punans, who hunt without dogs (which in fact they do not possess), will lie in wait for the rhinoceros beside the track, along which he comes to his daily mud-bath, and drive a spear into his flank or shoulder; then, after hastily retiring, they track him through the jungle, until they come upon him again, and find an opportunity of driving in another spear or a poisoned dart through some weak spot of his armor."

thing else connected with the animal, is an attractive subject of great culture-historical interest. It should be stated at the outset that the Chinese sketches pointed out by Mr. Giles, and other Chinese illustrations as well, can never have been intended for any bovines, whatever the alleged bovine character in the animal may be; for there is in this world no bovine animal with a single horn and three toes which, as will be shown, appear in the early Chinese definition, and are plainly outlined in the sketch of the rhinoceros said in the *Erh ya* to be of hog-like appearance (Fig. 6).[1] The single horn and the three toes, however, are thoroughly characteristic of the rhinoceros, and of this animal exclusively. But we are first going to study the psychology of the case.

On the first day of May of the year 1515 the first live rhinoceros was brought to modern Europe from India by Portuguese, and presented to King Emanuel of Portugal.[2] In commemoration of this event, Albrecht Dürer, who took a deep interest in exotic animals and people, sketched in the same year a likeness of this rhinoceros, published as a wood-engraving, with a somewhat lengthy description in German. Dürer's original drawing is still preserved in the British Museum (Plate IX).[3] It is so weak that, as already pointed out by Dr. Parsons,[4] the first serious

---

[1] See likewise Fig. 9, p. 102.

[2] The history of this event is narrated in the Decadas de Asia of J. DE BARROS (quoted by YULE and BURNELL, Hobson-Jobson, p. 363): "And in return for many rich presents which this Diogo Fernandez carried to the King, and besides others which the King sent to Affonso Alboquerque, there was an animal, the biggest which Nature has created after the elephant, and the great enemy of the latter . . . which the natives of the land of Cambaya, whence this one came, call Ganda, and the Greeks and Latins Rhinoceros. And Affonso d'Alboquerque sent this to the King Don Manuel, and it came to this Kingdom, and it was afterwards lost on its way to Rome, when the King sent it as a present to the Pope."

[3] I am indebted to Mr. Laurence Binyon of the British Museum for his courtesy in favoring me with a copy of this wood-engraving, from which our reproduction is made. The particulars of the history of this engraving are discussed by C. DODGSON (Catalogue of Early German and Flemish Woodcuts in the British Museum, Vol. I, p. 307, London, 1903).

[4] Die natürliche Historie des Nashorns, welche von Doctor PARSONS in einem Schreiben an MARTIN FOLKES, Rittern und Präsidenten der Königlich-Englischen Societät abgefasset, mit zuverlässigen Abbildungen versehen, und aus dem Englischen in das Deutsche übersezt worden von Doctor GEORG LEONHART HUTH, Nürnberg, bey Stein und Raspe, 1747. The English original of this interesting pamphlet of 16 pages in quarto is not known to me. It is accompanied by three plates engraved on copper representing the first fairly exact figures of the rhinoceros in various views, its horn and other organs of its body. An anonymous copper-engraving was published in 1748 under the title, "Vera effigies Rhinocerotis qui in Asia, et quidem in terris Mogolis Magni in regione Assam captus et anno 1741 tertio aetatis anno a capitano Douvemont van der Meer ex Bengala in Belgium translatus est." This rhinoceros, a three years old animal, was exhibited in Holland in 1741, and styled on the placards the *behemoth* of the Bible (JOB, 40) and the unicorn of mediæval times. It proved an overwhelming sensation. In 1747 it made its appearance at Leipzig where GELLERT set it a literary monument in the poem with the beginning, "In order to behold the rhinoceros, I was told by my friend, I resolved to stroll out." In

student of the anatomy of the rhinoceros, it is impossible to assume that he had ever seen the animal. This fact is quite certain, for it is known that the King of Portugal despatched the animal to the Pope, and that it was drowned off Genova when the vessel on board which it was being carried was foundered. The only supposition that remains, therefore, is that some one of Lisbon near King Emanuel must have sent on to Dürer a rough outline-sketch of the novel and curious creature, which was improved and somewhat adorned by the great artist. But to what sources did he turn for information on the subject? Naturally to that fountain-head from which all knowledge was drawn during that period, the authors of classical antiquity. The fact that Dürer really followed this procedure is evidenced by the very description of the animal, which he added to his sketch, and in which he reiterates the story of the ancients regarding the eternal enmity and struggle of rhinoceros and elephant.[1] The most curious feature about Dürer's rhinoceros is that, besides the horn on

FIG. 3.

Marble Relief of Two-Horned Rhinoceros in Pompeii
(from O. Keller, Antike Tierwelt).

1748 it reached Augsburg where Johann Ridinger made a drawing and etching of it with the title as stated (L. REINHARDT, Kulturgeschichte der Nutztiere, p. 751, München, 1912). The rhinoceros is a subject which for obvious reasons has seldom tempted an artist. It should be emphasized that no artist has ever made even a tolerably good sketch of it, and that only photography has done it full justice.

[1] According to the tales of the ancients, the feuds between the two animals were fought for the sake of watering-places and pastures; and the rhinoceros prepared itself for the combat by sharpening its horn on the rocks in order to better rip the arch-enemy's paunch which it knows to be its softest part (compare DIODOR, I, 36; AELIAN, Nat. animalium, XVII, 44; PAUSANIAS, IX, 21; and PLINY, Nat. hist., VIII, 20: alter hic genitus hostis elephanto cornu ad saxa limato praeparat se pugnae, in dimicatione alvum maxime petens, quam scit esse molliorem). The same story is still repeated by JOHAN NEUHOF (Die Gesantschaft der Ost-Indischen Gesellschaft [1655–57], p. 349, Amsterdam, 1669) in his description of the Chinese rhinoceros, which is based on classical, not Chinese reports: "It makes permanent war on the elephant, and when ready to fight, it whets its horn on stones. In the struggle with the elephant it always hits toward its paunch where it is softest, and when it has opened a hole there, it desists, and allows it to bleed to death. It grunts like a hog; its flesh eaten by the Moors is so tough that only teeth of steel could bite it." The Brahmans allowed the flesh of the rhinoceros to be eaten as a medicine (M. CHAKRAVARTI, Animals in the Inscriptions of Piyadasi, Memoirs As. Soc. of Bengal, Vol. I, p. 371, Calcutta, 1906); according to al-Bērūnī (SACHAU, Alberuni's India, Vol. I, p. 204), they had the privilege of eating its flesh. CTESIAS stated wrongly that the flesh is so bitter that it is not eaten.

its nose, it is provided with another smaller horn on its neck.   This proves that he must have read about a two-horned rhinoceros, for the specimen shipped to Portugal was the single-horned species of India. MARTIAL, in one of his epigrams (*Spect. Ep.* XXII), has the verse, "*namque gravem gemino cornu sic extulit ursum.*"   As long as the fact of a two-horned rhinoceros was not yet scientifically established,— and Dr. Parsons was one of the first to point it out,— the critics of Martial felt greatly embarrassed over the statement that a rhinoceros with double horn[1] should have lifted a bear, and arbitrarily changed the verse in various ways to get around the double horn.   Dürer no doubt had this passage in mind, and accepted it as a fact.   Nobody at that time, however, knew the location of the second horn: thus it found its place on the neck.[2]   This case is very instructive, for the Chinese

---

[1] The two-horned African rhinoceros is figured on the bronze coins of Emperor Domitian and on Alexandrian coins of the same emperor (IMHOOF-BLUMER and KELLER, Tier- und Pflanzenbilder auf Münzen und Gemmen, Plate IV, 8), and unmistakably referred to by PAUSANIAS (*l. c.*), who describes it as having the one horn on the extremity of its nose, the other, not very large, above the latter.   The struggle between bear and rhinoceros is represented on a pottery lamp from Labicum, which is reproduced in Fig. 7 after O. KELLER (Tiere des classischen Altertums, p. 118, Innsbruck, 1887), in order to illustrate the affinity of this creature with the "hog-like" rhinoceros of the Chinese (Fig. 6).   Dürer's picture formerly led astray many a student of classical antiquity by giving the impression that a horn was really growing up from the animal's back.   Thus S. BOCHART, in his Hierozoicon (p. 931, Lugduni Batavorum, 1692), a learned treatise on the animals mentioned in the Bible, makes the following observation with reference to the verse of Martial above quoted: "Frustra etiam id observatur, Rhinocerotem geminum habere cornu.   Alterum enim est in dorso, quo ursum extulisse dici non potest.   Itaque ad illud cornu non pertinent haec poetae: gemino cornu sic extulit ursum."   It was Bochart who proposed several conjectures tending to ameliorate Martial's text.   JOHANNES BECKMANN (De historia naturali veterum libellus primus, p. 129, Petropoli et Goettingae, 1766) was the first to point out emphatically the actual truth in the matter, in these words: "Sed non soli philologi, verum etiam physici duo cornua neglectis illis veterum locis [i.e., the passages of Martial and Pausanias] negarunt Rhinoceroti; uti Scheuchzerus, Peyerus. Consultius fuisset nec affirmare nec negare.   Hodie enim auctoritatibus gravissimorum virorum satis probatum est, esse Rhinocerotes etiam bicornes, qui cornu alterum non in fronte, non in dorso, sed etiam in nare habent."   In view of our subject, it is of especial interest to us to note that this truth was generally recognized in Europe as late as the latter part of the eighteenth century, while Chinese authors were well informed on the subject from the beginning of our era.

[2] It has recently been asserted (compare the notice of S. REINACH, *Revue archéologique*, 1913, p. 105) that the rhinoceros on a marble relief of Pompeii (Fig. 3; reproduced also by REINACH, Répertoire de reliefs, Vol. III, p. 93; and O. KELLER, Die antike Tierwelt, Vol. I, p. 388) is an exact copy of the wood-engraving by Dürer and accordingly the work of a forger.   This point of view seems to me inadmissible, and I concur with Reinach in the view that a common antique model may have been handed down by the illustrators of the bestiaries.   The most striking coincidence between the rhinoceros of Pompeii and that of Dürer is the location of the second horn on the neck.   This argument, however, is not cogent in establishing a close interdependence of the two; for also in China, on a picture of Yen Li-pên of the T'ang period (Fig. 11), the rhinoceros appears with a horn on its neck, and with scales on its body.   As the artists all over the world were so much puzzled as to where to place the horn or horns, it is perfectly conceivable that Dürer, solely guided by his reading of ancient writers, even without having recourse to an antique pictorial representation, worked out his

draughtsmen who had set before them the task of portraying a rhinoceros saw themselves in the same predicament as Dürer, in that they were lacking all personal experience of the animal, and for this reason were actuated by the same psychological factors. They, on their part, resorted to the classical definitions of the animal, as laid down in the ancient dictionaries *Erh ya* and *Shuo wên;* they did not intend to picture a rhinoceros true to nature and directly from nature, simply because they were deprived of this opportunity, but they composed and pieced together the creature from certain notions which they formed from bits of information gathered from their literary records. Whatever caricatures their achievements may be, however, there cannot be the slightest doubt that they intended to represent a rhinoceros, not some other animal. Dürer's work, from a scientific viewpoint, is in details highly inaccurate and untrue; the modern naturalist may even pronounce the verdict that what he represented is far from resembling a rhinoceros at all; but the bare fact remains — and this is the essential point — that the artist, as expressly stated in the legend by his own hand, had the intention of representing in this work a rhinoceros. As in most cases, the artist does not reproduce an object as it appears in the world of reality, but conveys to us his own notions of things as they are projected in his mind. Exactly as it happened in China, so Dürer's model found many adherents and followers, even among the naturalists who copied him again and again, and who surpassed him in fanciful additions of scales, wrinkles, and other decorations. Even BONTIUS,[1] who pretends that he saw the animal in exotic forests and stables, and boasts of furnishing a figure of it free from Dürer's defects, represents it, instead of with hoofs, with a paw very similar to that of a dog, only that it is somewhat larger.

---

own theory in regard to the second horn. But it is desirable that, as suggested by Reinach, the iconographic question should be studied in detail. Neither should the differences between the two be overlooked. Dürer's posterior horn is directly behind the ears; in the Pompeiian picture it is far behind the ears, above the front legs; in the same spot Dürer has a small triangular point, the significance of which is not clear. It is certainly astonishing that the artists of Pompeii could commit this error, as the two-horned African rhinoceros was perfectly known in the Roman circus, and is correctly represented on the coins of Domitian mentioned above.—ULYSSES ALDROVANDUS (Quadrupedum omnium bisulcorum historia, p. 354, Francofurti, 1647) has the figure of a rhinoceros, with an additional horn in the shape of a corkscrew placed on the shoulders.

[1] JACOBI BONTII, Historiae naturalis et medicae Indiae Orientalis libri sex, p. 51 (Amsterdam, 1658). The horn is correctly drawn. Bòntius avails himself of the word *abada*, which was used by old Spanish and Portuguese writers for a rhinoceros, and adopted by some of the older English narrators. The word is probably connected with Malayan *badak*, "rhinoceros" (see YULE and BURNELL, Hobson-Jobson, p. 1). In G. DE MENDOZA (Dell' historia del gran regno della China, 1586, p. 437) the word *abada* is identified with the rhinoceros.

Archæologists are agreed that the rhinoceros (Fig. 4)[1] is represented on the black obelisk of Salmanassar (B.C. 860—824) in company with an elephant, human-looking apes, and long-tailed monkeys.   This tribute-picture suggests to I. KENNEDY[2] the first certain evidence of Baby-lonian intercourse with India.   The animals formed part of the tribute of the Muzri, an Armenian tribe living in the mountains to the north-east of Nineveh.[3]   The rhinoceros is called in the inscription an "ox of the river Sakeya," and Kennedy criticises its repre-sentation as "very ugly and ill-drawn."   Indeed, it is no more and no less than a bull, and, as far as natural truth is concerned, much in-ferior to the Chinese sketches.   It even has cloven bull-feet, while one of the Chinese drawings has correctly three toes,[4] and the single clumsy horn rises on its forehead

FIG. 4.

Rhinoceros from Obelisk of Salmanassar II
(from O. Keller, Antike Tierwelt).

---

[1] After O. KELLER, Die antike Tierwelt, Vol. I, p. 386 (Leipzig, 1909).

[2] The Early Commerce of Babylon with India (*Journal R. As. Soc.*, 1898, p. 259).

[3] According to J. MARQUART (Untersuchungen zur Geschichte von Eran, II, p. 101, Leipzig, 1905), who discusses the same passage in the inscription of Salmanas-sar II, Muzri is the name of a country and mountain-range (Muzûr Mountains) west of the Euphrates, and comprising also a part of the mountainous region south of the river.   MARQUART translates "cattle of the river Irkea."   Others, like Schrader, Hommel, and W. Max Müller (see B. MEISSNER, Assyrische Jagden, p. 20, Leipzig, 1911) identify Muzri with Egypt.   KENNEDY does not explain how the rhinoceros could have gotten into that region from India; and it may have been, after all, an African species, although the single horn would rather point to India; the elephant, however, in his opinion, came over the passes of the Hindu Kush.   There is, of course, the possibility that the lower Euphrates region may have harbored the rhinoceros, if we can depend upon the report of the *Hou Han shu* regarding the country of T'iao-chi (HIRTH, China and the Roman Orient, p. 38); and I am in full accord with what HIRTH remarks on this point in the passage (pp. X–XII).   However this may be, I agree with KENNEDY, F. HOMMEL (Die Namen der Säugetiere bei den südsemiti-schen Völkern, p. 324), MEISSNER, and KELLER that the animal figured on the black obelisk of Salmanassar is intended for a rhinoceros, and not merely for an ox, for there is no ox with single horn as here represented.   The Assyrian name for the rhinoceros is *kur-ki-za-an-nu* = *kurkizannu* (F. DELITZSCH, Assyrische Tiernamen, p. 56, Leipzig, 1874), which, according to HOMMEL (*l. c.*, p. 328), is a loan-word received from Ethiopic *karkand* (compare Arabic *karkadan*, Persian *kerk*).   The trade-relations of India with Babylon are well established (see particularly G. BÜHLER, Indian Studies III, p. 84).

[4] The ancients did not notice this fact, nor did the Hindu, who classified the rhi-noceros, owing to a confusion with the elephant, among the five-toed animals (M. CHAKRAVARTI, Animals in the Inscriptions of Piyadasi, *Memoirs As. Soc. Bengal*,

between the eyes, as it occurs in the armorial unicorns. It is very instructive to compare this Babylonian representation with those of the Chinese; and whoever will view them together will certainly grant attenuating circumstances to the latter. The Babylonian production is the more surprising, as the supposition is granted that the live animal was sent as tribute; and the "artist," we should think, had occasion to actually see it. The outcome is such a caricature, however, that this point of view seems impossible; the "artist" simply acted on hearsay, or had been instructed to represent a queer foreign animal of the appearance of an ox, but with only a single horn on its forehead. And here we are again landing right at the threshold of the psychology of the Chinese draughtsman who, most assuredly, had never throughout his life viewed any living specimen of a rhinoceros, but merely reconstructed it in a vision of his mind from what he had heard or read. Nevertheless his product is not what it may seem to us on the surface, but it is and remains what it is intended for, — the rhinoceros.

Another instructive example for the iconography of the rhinoceros is furnished by Cosmas Indicopleustes, the Egyptian monk and traveller of the sixth century A.D. COSMAS[1] discriminates between the unicorn (monokeros) and the "nose-horn" (rhinokeros), and has handed down to us sketches of both. In regard to the former, he remarks that he has not seen it, but that he had had occasion to notice four brazen figures of it set up in the four-towered palace of the King of Ethiopia, from which he was able to draw it. His figure[2] looks somewhat like a missing link between a horse and a giraffe, carrying on its head a straight, long horn. "In Ethiopia," Cosmas assures us, "I once saw a living rhinoceros from a great distance and saw also the skin of a dead one stuffed with chaff, standing in the royal palace, and thus I was able to draw it accurately." The result of this "accurate" drawing is the figure of a maned horse with bushy tail, with two horns planted upright on its nose.[3] Nobody, as far as I know, has as yet inferred from this figure that the Greek word rhinokeros relates to an equine animal and should be translated by "horse."

An interesting example of a Persian conception of the rhinoceros is depicted in the Burlington Magazine.[4] This is derived from an

Vol. I, p. 371, Calcutta, 1906). In the commentary of Kuo P'o to the dictionary Erh ya (see below, p. 94) and in the Kiao chou ki of the fifth century A.D. it is clearly stated that the rhinoceros has three toes. Compare p. 95, note 6.

[1] Ed. MIGNE (Patrologia, Vol. 88), p. 442.

[2] Christian Topography, translated by MacCRINDLE, Plate IV, No. 28 (Hakluyt Society, 1897).

[3] Ibid., No. 23.

[4] Vol. XXIII, July, 1913, Plate III.

illustrated "Description of Animals," the *Manafi-i-heiwan*, translated from Arabic into Persian and completed between 1295 and 1300. Here we have the interesting case that the author of this article, C. ANET, who evidently does not read Persian, mistakes the rhinoceros for "a horned gnu." But the picture is entitled in Persian *kerkeden* (or *kargadan*), "the rhinoceros," and it is therefore superfluous to discuss the point that it cannot represent a gnu.[1] Although the creature has the shape of an ox, exactly as on the Assyrian obelisk and in the Chinese woodcut (Fig. 5), with the additional hump of a zebu[2] and black antelope-like stripes on its body, it is unmistakably characterized by a single horn in the form of a crescent.[3]

In order to understand how the early Chinese illustrations of the rhinoceros alluded to by Mr. Giles were made, it is imperative to study the ancient definitions of the two words *se* and *si*. These definitions are sufficiently clear to place us on the right track in nicely discriminating between the two words, which plainly refer to two distinct species of rhinoceros. The weak point in Mr. Giles's definition of "bovine animal"[4] is that it is somewhat generalized, and leaves us entirely in the dark as to the difference between the two words *se* and *si*. They are physically differentiated words, and are expressed by different symbols in writing.

Se-ma Ts'ien[5] mentions the two species of rhinoceros and elephant as inhabitants of the country of Shu (Sze-ch'uan).[6] The commentator

---

[1] A species of antelope restricted to Africa, which could hardly be expected in Persian art.

[2] This hints at the square-mouthed or white rhinoceros of Africa. One of the peculiarities of this species is the prominent, rounded, fleshy hump on the nape of the neck, just forward of the withers (E. HELLER, The White Rhinoceros, p. 20, Washington, 1913).

[3] A representation of the rhinoceros in sculpture is spoken of in a Persian description of the province of Fars from the beginning of the twelfth century; in Iṣṭakhr the portrait-statue of King Jamshīd was erected in stone, with his left hand grasping the neck of a lion, or else seizing a wild ass by the head, or again he is taking a unicorn (or rhinoceros) by the horn, while in his right hand he holds a hunting-knife, which he has plunged into the belly of the lion or unicorn (G. LE STRANGE, *Journal R. As. Soc.*, 1912, p. 27). In the Annals of the T'ang Dynasty it is on record that in 746 A.D. Persia offered a rhinoceros and an elephant (CHAVANNES, *T'oung Pao*, 1904, p. 76).

[4] What wild bovine animal should be understood has never been indicated.

[5] *Shi ki*, Ch. 117, pp. 3 b, 7 b.

[6] Our historians of Japan have been greatly puzzled by the fact that the Japanese Buddhist monk Tiao-jan (Japanese Chōnen), who came to China in 984, stated in his report embodied in *Sung shi* (Ch. 494, p. 4 b) that there were in his native country water-buffalo, donkeys, sheep, and plenty of — thus it has been translated — rhinoceroses and elephants (for example, by P. A. TSCHEPE, Japans Beziehungen zu China, p. 89, Yen-chou fu, 1907). O. NACHOD (Geschichte von Japan, Vol. I, p. 22) went so far as to appeal to a misunderstanding on the part of the Japanese informant, which he believes cannot be surprising, as Tiao-jan, though well versed in the written characters of the Chinese, did not understand their spoken language. This argu-

states, "The animal *se* is built like the water-buffalo.   The elephant is a large animal with long trunk and tusks ten feet long; it is popularly styled 'river ape' (*kiang yüan*, No. 13,741).   The animal *si* has a head resembling that of the ape *yüan* and a single horn on its forehead."[1]

mentation is entirely inadmissible.   It is certain that neither rhinoceros nor elephant exists in Japan: consequently Tiao-jan, in using the expression *si siang* (Japanese *sai-zō*) cannot be understood to convey to it its literal meaning, but he is sure to employ it in a different sense.   Chinese expressions (and Japanese are largely based on them) do not always mean what they seem to imply on the surface, but are often literary allusions or reminiscences of a metaphorical significance.   The Japanese monk indeed avails himself of a Chinese phrase of classical origin traceable to *Mêng-tse* (LEGGE, Classics, Vol. II, p. 281), and in my opinion, simply means to say that Japan produces "extraordinary wild animals."   Yen Shi-ku, defining the word *shou* ("wild animals") in the Annals of the Han (*Ts'ien Han shu*, Ch. 28 A, p. 4 b), explains it as embracing such kinds as rhinoceros and elephants, whence it follows that this compound *si siang* is capable of rendering the general notion of wild animals.   *Si siang* has thus become a stereotyped term occurring in many authors, although the literal meaning usually remains, as, for example, in *Ts'ien Han shu* (Ch. 28 B, p. 17), *Erh ya* (see p. 94, note 3), *Nan shi* (Ch. 78, p. 7), *T'ang shu* (Chs. 43 A, p. 1, and 221 A, p. 10 b), and in the History of Shu (*Shu kien*) written by Kuo Yün-t'ao in 1236 (Ch. 10, p. 1, ed. of *Shou shan ko ts'ung shu*, Vol. 23).   HIRTH and ROCKHILL (Chau Ju-kua, p. 174) have taken a different view of the matter and suppose that the document utilized in the Sung Annals, and partially copied by Chao Ju-kua (inclusive of the statement that Japan produces *si siang*), contained a number of clerical errors; they are convinced that Tiao-jan's statement really was to the effect that there are neither rhinoceroses nor elephants in Japan.   There is certainly no direct objection to be raised to such a point of view, but I am inclined to believe that with the indication as given there is no necessity of resorting to such a conjecture.

[1] This universal notion could have emanated only from the two-horned species with reference to the rear horn, which anatomically is indeed placed over the frontal bone, while the front horn is situated over the conjoined nasal bones (FLOWER and LYDEKKER, Introduction to the Study of Mammals, p. 403).   The posterior horn immediately follows the anterior one, and is somewhat beneath the eyes.   Curiously enough, this idea of the position of the horn on the forehead was transferred also to the single-horned species, and became a well-established tradition, which one author copied from another.   It is found in the classical world as well as among the Arabic authors.   CTESIAS (ed. BAEHR, p. 254) seems to be the most ancient writer in whom this tradition has crystallized: he describes the wild white asses of India as "having on the forehead a horn a cubit and a half in length."   The fact that he speaks of the rhinoceros, above all, is evidenced by his reference to the horn being made into drinking-cups which were a preventive of poisoning (compare also LASSEN, Indische Altertumskunde, Vol. II, p. 646).   The *monoceros* of India, in the description of PLINY (*Nat. hist.*, VIII, 21), had a single black horn projecting from its forehead, two cubits in length (*uno cornu nigro media fronte cubitorum duum eminente*).   The horn of the rhinoceros sculptured in Assyria, as we have seen, is planted on its forehead.   Of course, when describing a rhinoceros which he saw at the games in the circus, PLINY (VIII, 20) states correctly that it has a single horn on its nose (*unius in nare cornus*); so does AELIAN (XVII, 44), and so does likewise Kuo P'o.   The Arabic merchant Soleiman, writing in 851 (M. REINAUD, Relation des voyages faits par les Arabes, Vol. I, p. 28), attributes to the rhinoceros of India a single horn in the middle of its forehead, and is duly seconded by his copyist Mas'ūdi (RUSKA, *Der Islam*, Vol. IV, p. 164).   Ibn al-Faqīh, describing the two-horned species of Africa, states that it has on its forehead a horn, by means of which it inflicts mortal wounds; and another minor one is beneath the former and placed between its eyes (E. WIEDE- MANN, Zur Mineralogie im Islam, p. 250).   Even al-Bērūni (E. SACHAU, Alberuni's India, Vol. I, p. 204), who imparts a sensible account of the Indian rhinoceros, asserts from hearsay that the African species has a conical horn on the skull, and a second and longer horn on the front.   Early European observers also believed that the horn of the rhinoceros was growing on its forehead.   BARKER, as quoted by YULE

In the other passage, the definition of Kuo P'o (276–324), the editor of the dictionary *Erh ya*, is quoted.

The following definitions of the words *se* and *si* are given in the ancient dictionary *Shuo wên* (about 100 A.D.), and are here reproduced from an edition of this work printed in 1598, which is an exact facsimile reproduction of the Sung edition of the year 986. In all probability, this one faithfully mirrors the text of the original issue. The definition of *se* consists of only five words: "It is like a wild ox and dark-colored."[1] The character is then explained as a pictorial symbol (compare the reproduction of the Chinese text on p. 92).

It is doubtless on this enigmatic and incomplete definition that the explanations of PALLADIUS and COUVREUR (above, p. 74) are based. In order to reach a satisfactory result, however, it is always necessary to consult all records relating to a case; and it will always be unsafe to rely upon a single statement, which, after all, may have been curtailed, or incorrectly handed down. Let us note at the outset that the *Shuo wên* by no means says that the animal in question is a wild ox, but only that it is *like* one; a comparison with a wild ox is not yet proof of identity with it. Hing Ping (932–1010), the commentator of *Shuo wên*, annotates on the above passage as follows,—"Its skin is so strong and thick that armor can be made from it,"—and quotes the *Kiao chou ki*[2] to the effect that "the horn is over three feet long and shaped like the handle of a horse-whip."[3] The fact that this author means to speak of a single horn becomes evident from the statement of Kuo P'o to be cited presently.[4] The

---

and BURNELL (Hobson-Jobson, p. 1),wrote in 1592, "Now this Abath [*abada, bada* =rhinoceros] is a beast that hath one horne only in her forehead, and is thought to be the female Unicorne, and is highly esteemed of all the Moores in those parts as a most soveraigne remedie against poyson."

[1] K'ang-hi's Dictionary quotes the *Shuo wên* as saying that "the animal *se* has the shape or body of a wild ox and is dark-colored."

[2] Records of Annam, of the fourth or fifth century, by Liu Hin-k'i (BRETSCHNEI-DER, Bot. Sin., pt. I, p. 159).

[3] In a somewhat different way, the *Shuo wên* is cited in *Yen kien lei han* (Ch. 430, p. 16 b), where original text and commentary are blended together: "The animal *se* resembles a wild ox and has a dark-colored skin which is so strong and thick that it can be worked up into armor. Among the animals on the mountain Po-chung, there is a large number of *se*." The latter name, according to PALLADIUS, is an ancient designation for a mountain in the west of Shan-si. The fact that the rhinoceros should have occurred there in ancient times is not at all surprising (see the notes below on the distribution of the animal in ancient times). It is noteworthy that we meet here the reading, "it resembles a wild ox," in agreement with the wording of the *Erh ya*, whence it follows that the *se* was not straightway looked upon as a wild ox, but as something else; it was merely likened to it—a phraseology which is echoed in Babylonia and in the classical authors. This simile seems to account for the erroneous attempt of later commentators, like Chu Hi, to interpret *se* as identical with a wild ox.

[4] The *Kiao chou ki* is credited in the *Yen kien lei han* with the words, "The *se* has a single horn which is over *two* feet long and shaped like the handle of a horse-whip."

十六

如野牛而青象形與禽离頭同凡㕥爲之屬皆从㕥徐姊切

犀　南徼外牛一角在鼻一角在頂似豕从牛尾聲先一切

animal *si* is defined in the *Shuo wên* as "an ox occurring beyond the southern frontier. It has a horn on its nose and another one on the crown of its head; it resembles a pig."[1] This definition fits no other animal than the two-horned species of rhinoceros, and has great historical value as a piece of evidence in determining the former geographical distribution of the species. The passage shows us that in the first century A.D. it no longer existed in northern China, where its habitat had been prior to that time, and that it was then driven back beyond the southern border, speaking roughly, south of the Yangtse. It was then naturalized in Yün-nan, in the country of the Ai-lao,[2] and in Tonking.[3]

To the author of *Kiao chou ki* we owe the following interesting description of the Annamese rhinoceros:[4] "The rhinoceros (*si*) has its habitat in the district of Kiu-tê (in Tonking). It has hair like swine, three toes, and a head like a horse. It is provided with *two* horns,— the horn on the nose being long, the horn on the forehead short." It is clearly manifest that this description comes from an eye-witness, or one well informed by the native hunters, and that it perfectly fits the two-horned so-called Sumatran rhinoceros (*Rhinoceros sumatrensis*), the only living Asiatic species with two horns, and also the most hairy one.[5] Its essential characteristics are well observed and briefly set forth in this definition.

The dictionary *Erh ya*, edited by Kuo P'o (276-324), defines the animal *se* as resembling the ox, and the animal *si* as resembling swine. The commentary by Kuo P'o explains that the *se* has a single horn, is dark in color, and weighs a thousand catties;[6] and "the *si* resembles in form

---

[1] MARCO POLO (edition of YULE and CORDIER, Vol. II, p. 285) says regarding the rhinoceros of Java that its head resembles that of a boar.

[2] *Hou Han shu*, Ch. 116, p. 8 b.

[3] The question of the former geographical distribution of the rhinoceros in China is studied in detail below, pp. 159-166.

[4] *Yen kien lei han*, Ch. 340, p. 1. In Annamese the rhinoceros is called *hui* (written with the Chinese character for *se*) and *tây* or *tê* (written with the character for *si*).

[5] Hair grows sparsely all over the head and body, but attains its maximum development on the ears and the tail, its color varying from brown to black. The longest known specimen of the front horn is in the British Museum, and has a length of 32½ inches, with a basal girth of 17⅜ inches; a second specimen in the same collection measures 27⅛ inches in length, and 17⅛ in circumference (R. LYDEKKER, The Game Animals of India, p. 38). The statement of the *Kiao chou ki* that the horn is two or three feet long is therefore no exaggeration. Concerning the two horns in the *si*, there is consensus of opinion between that work and the *Shuo wên*.

[6] This may not be an exaggeration, though merely based on a rough estimate. The average weight of the rhinoceros, for reasons easy to comprehend, has never been ascertained. But if the weight of the skin alone may come to three hundred pounds (E. HELLER, The White Rhinoceros, p. 10), the complete animal may easily total a thousand and more. *K'ang-hi* and the modern editions of the *Erh ya* write "thousand

the water-buffalo,[1] but has the head of a pig, a big paunch, short legs, and three toes on its feet; it is black in color and has three horns, one on the head, another on the forehead, and the third on the nose.   The horn on the nose is the one by means of which it feeds [that is, uproots shrubs and trees];[2] it is small and not long; it likes to eat thorny brambles; there is also a kind with but a single horn."   Kuo P'o, accordingly, is fully acquainted with the single-horned rhinoceros (his three-horned species is discussed farther on), and renders it plain enough that in his opinion neither the *se* nor the *si* is a bovine animal, as he treats them in a different section; while in his section on bovines, with twelve illustrations of such, no hint is made at *se* or *si*.[3]   The last doubt which might still exist as to the acquaintance with the single-horned rhinoceros on the part of Kuo P'o and Hü Shên, the author of *Shuo wên*, will be banished by another word, *tuan*[4] (or *kio tuan*), of which *Shuo wên* (Ch. 11, p. 2) says that it is an animal of the shape of swine, with a horn which is good for making bows, and which is produced in the country Hu-siu.[5]

---

catties." *Yen kien lei han* (*l. c.*) has the erroneous reading "ten," which is impossible. Also Chang Yü-si, the author of the *Pu chu pên ts'ao* of the year 1057, as may be seen from the *Chêng lei pên ts'ao*, quotes the *Erh ya* as saying that "the *se* resembles an ox and has a single horn."   Kuo P'o, accordingly, concurs with Liu Hin-k'i in the view that *se* is the single-horned rhinoceros.

[1] *Yen kien lei han* (Ch. 430, p. 1) offers the variant, "The *si* resembles swine, but is in shape like an ox;" then the same text as above is given, but the clause in regard to the three horns is wanting.

[2] While feeding, the point of the horn of the animal may come in contact with the ground, so that the point is sometimes worn flat on its outer face (E. HELLER, The White Rhinoceros, p. 31).   According to Ibn al-Faqîh, the African rhinoceros tears herbage out with the anterior horn, and kills the lion with the posterior one (E. WIEDEMANN, Zur Mineralogie im Islam, p. 250).

[3] The rhinoceros is incidentally mentioned in another passage of *Erh ya* (Ch. B, fol. 29), where nine mountains with their famed productions are enumerated: "The finest productions of the southern region are the rhinoceros (*si*) and elephant of Mount Liang" (*Liang shan*, in Chung chou, Sze-ch'uan; PLAYFAIR, 2d ed., No. 3790, 2; BRETSCHNEIDER, Bot. Sin., pt. 3, p. 575, No. 187).   Kuo P'o adds, "The rhinoceros furnishes hide and horn, the elephant ivory and bones."   It follows therefrom, as is also confirmed by other sources, that in the third century A.D., the lifetime of Kuo P'o, the rhinoceros still existed in Sze-ch'uan, as seen above; its existence was attested there by Se-ma Ts'ien several centuries earlier.

[4] Composed of the classifier *kio* ('horn') and the phonetic element *tuan* (No. 12,136).   Not in GILES; see PALLADIUS, Vol. I, p. 189.   A unicorn is represented on the Han bas-reliefs (CHAVANNES, Mission archéologique, Vol. I, p. 60, Paris, 1913).

[5] Nos. 4930 and 4651.   Other editions write Hu-lin.   A horn bow is not a bow exclusively made from horn, which is technically impossible; but horn is only one of the substances entering into its manufacture.   Technically the Chinese bow belongs to the class of composite bows, the production of which is a complicated process and requires a large amount of toil and dexterity.   The foundation of the bow is formed of flexible wood connected with a bamboo staff.   Along the back a thick layer of carefully soaked and prepared animal sinew is pressed, which, after drying, stiffens into a hard elastic substance.   The inner side of the bow is then covered with two long horn sticks joining each other in the centre.   The opposite of the horn bow is the wooden (or simple) bow (*mu kung*), as it is mentioned, for instance, as being used by

Kuo P'o states in regard to the same animal, "The horn is on the nose and capable of being made into bows. Li Ling presented ten such bows to Su Wu.[1] The animal mentioned in the Life of Se-ma Siang-ju in the *Shi ki* (Ch. 117) is the *k'i-lin*[2] *kio tuan*."

The animal with a horn on its nose is the single-horned rhinoceros; and the term *tuan* or *kio tuan* is a counterpart of the word *monoceros* of the ancients, as alluded to by Ctesias, Aristotle, Pliny, Aelian, and others, and which, according to the general consensus of opinion, relates to the one-horned rhinoceros of India. Bows manufactured from the horn are mentioned also in the Annals of the Kin Dynasty.[3] The allusion to armor by Hing Ping is additional proof for *se* being a rhinoceros, for, as we shall see, armor was not made in ancient China from the hides of bovine animals.[4]

It is beyond any doubt that in those various definitions there is plainly the question of a rhinoceros. We cannot get over the single horn, whether placed on the nose, the head, or the forehead;[5] we cannot get over the fact, either, that a conspicuous distinction between the single-horned (*se*) and two-horned (*si*) species is made, — a fact which will be discussed in full farther on when we have learned everything that Chinese authors have to report anent the two animals; nor can we get over the three toes which form a prominent characteristic of the rhinoceros,[6] but assuredly not of any bovine species. In fact, the Chinese definitions, without pretension to scientific accuracy, which could not be

---

the populace of Tonking (*Ts'ien Han shu*, Ch. 28 B, p. 17), which in connection with it availed itself of flint, bamboo, and sometimes bone arrowheads.

[1] See GILES, Biographical Dictionary, pp. 450, 684.

[2] Regarding the *k'i-lin* see below, p. 113.

[3] *Kin shi*, Ch. 120, p. 3. Fossil rhinoceros-horn (from *Rhinoceros tichorrhinus*) is still employed by the Yakut in the manufacture of bows (B. ADLER, *Int. Archiv für Ethnographie*, Vol. XIV, 1901, p. 11).

[4] Regarding other Chinese notions of monoceroses see p. 114. Of later descriptions of the rhinoceros, the one contained in *Ying yai shêng lan* of 1416 by Ma Kuan is the most interesting. It is the most concise and correct definition ever given of the animal outside of our modern zoölogy. "The products of Champa are rhinoceros-horn and ivory of which there is a large quantity. The rhinoceros is like the water-buffalo. Animals of full growth weigh eight hundred catties. The body is hairless, black in color, and covered by a thick skin in the manner of a scale armor. The hoofs are provided with three toes. A single horn is placed on the extremity of the nose, the longest reaching almost fifteen inches. It subsists only on brambles, tree leaves and branches, and dried wood."

[5] As already remarked by CUVIER, the only real animal with a single horn is the rhinoceros.

[6] This statement reflects much credit on the observational power of the Chinese, especially as it is not pointed out by any classical author in describing the rhinoceros or unicorn. Al-Bērūnī (SACHAU, Alberuni's India, Vol. I, p. 203) is the only early author outside of China to make the same observation. Al-Bērūnī gives two different and contradictory descriptions of the rhinoceros, apparently emanating from two different sources. First, the animal is sensibly described from personal observation

expected, are perfectly sound and to the point in stating what a primitive observer could testify in regard to an animal so difficult of access and so difficult to describe. Surely, the Chinese definitions are not worse, and in several points perhaps better, than anything said about the animal in classical antiquity, among the Arabs, or in Europe up to the eighteenth century. And we shall soon recognize that until the very recent dawn of our scientific era the Chinese were the nation of the world which was best informed on the subject.[1] The Chinese likened the rhinoceros to the ox, the water-buffalo, the pig,[2] and its head to that of an ape.

---

as follows: "The *gaṇḍa* exists in large numbers in India, more particularly about the Ganges. It is of the build of the buffalo [analogous to the Chinese definition], has a black scaly skin, and dewlaps hanging down under the chin. It has three yellow hoofs on each foot, the biggest one forward, the others on both sides. The tail is not long; the eyes lie low, farther down the cheek than is the case with all other animals. On the top of the nose there is a single horn which is bent upwards. The Brahmins have the privilege of eating the flesh of the *gaṇḍa*. I have myself witnessed how an elephant coming across a young *gaṇḍa* was attacked by it. The *gaṇḍa* wounded with its horn a forefoot of the elephant, and threw it down on its face." The other account of al-Bērūnī, which refers to the double-horned African species, is composed of the narrative of a man who had visited Sufāla in Africa, and of classical reminiscences freely intermingled with it; to the latter belong the beliefs in the mobility of the horn and in the sharpening of the horn against rocks, and here appears also the wrong notion that it has hoofs. — PLINY (*Nat. hist.*, VIII, 21, § 76) asserts that the single-horned oxen of India have solid hoofs (in India et boves solidis ungulis unicornes), a tradition which savors of the description of a unicorn after a sculpture (on the Assyrian obelisk the animal has bovine hoofs). Even ARISTOTLE (*Hist. an.*, II, 18; ed. of AUBERT and WIMMER, Vol. I, pp. 74, 254), who evidently speaks after Ctesias, characterizes the single-horned "Indian ass" as solid-hoofed (μώνυχα). This lacune in the descriptions of the ancients was aptly pointed out by BELIN DE BALLU (La chasse, poëme d'Oppien, p. 174, Strasbourg, 1787), who, in speaking of the familiarity of the ancients with the animal, concludes by saying, "Mais ce qui doit nous étonner c'est qu' aucun n'ait parlé d'un caractère particulier de cet animal, dont les pieds sont partagés en trois parties, revêtue chacune d'une sole semblable à celle du bœuf."

[1] The only reproach that can be made to the Chinese authors is that they never point to the peculiar skin-folds of the animal (with the only exception, perhaps, of Fan Chên of the Sung period, who describes the rhinoceros of Annam as "clad with a fleshy armor;" see p. 113), and that, despite the live specimens procured for the Imperial Court (p. 80), no attempt has ever been made at a more precise description based on actual observation. But we may address the same charge of omission to the authors of India, the Greek writers on India, and to Pliny and Aelian. PLINY is content with stating that he saw the animal in the Roman circus, but does not describe what he saw, while he is eager to reproduce all the fables regarding the monoceros, emanating from India or from former sources relative to India. AELIAN (*Nat. an.*, XVII, 44) thinks it superfluous to describe the form of the rhinoceros, since a great many Greeks and Romans have seen and clearly know it. In matters of description the animal presents as difficult a subject as in matters of art. Exact descriptions of it are due only to competent zoölogists of recent times.

[2] How very natural this comparison is, may be gleaned from the account contained in *Nan Yüe chi* (quoted in *T'u shu tsi ch'êng*, chapter on rhinoceros), that at the time of the Han a rhinoceros once stampeded from Kiao chi (Annam) into Kao-liang (the ancient name for Kao-chou fu in Kuang-tung Province), and that it was mistaken by the people for a black ox, while those acquainted with the animal asserted that it was a black rhinoceros. The resemblance of the rhinoceros to an ox or buffalo has indeed obtruded itself on the observers of all times; and this notion is so far from being restricted to the Chinese, that it may almost be called universal. As seen above (p. 87), the Assyrians called the animal "ox of the river Sakeya." PLINY (*Nat. hist.*,

This is all exceedingly good: it is simply the result of that mental process which classifies a novel experience under a well-known category,

---

VIII, 21, § 72, 76) speaks of the unicorn oxen of India. FESTUS calls the African rhinoceros the Egyptian ox, and PAUSANIAS tells of "Ethiopic bulls styled rhinoceroses" which he saw himself in Rome (O. KELLER, Die antike Tierwelt, Vol. I, p. 385). The Indian physician Caraka, who lived at the Court of King Kanishka in Kashmir, placed the rhinoceros in the class of buffalo (anūpa, Mem. As. Soc. Bengal, Vol. I, 1906, p. 371). The Arabic merchant Soleiman, who wrote in 851, compared the Indian rhinoceros with the buffalo (M. REINAUD, Relation des voyages, Vol. I, p. 29); and so did, as seen above, al-Bērūnī. Ibn al-Faqīh says regarding the African rhinoceros that it resembles a calf (E. WIEDEMANN, Zur Mineralogie im Islam, p. 250). The Talmud, in three passages, mentions the one-horned ox as an animal sacrificed by Adam (L. LEWYSOHN, Die Zoologie des Talmuds, p. 151, Frankfurt, 1858). The "sea-ox" mentioned by Leo Africanus (HIRTH and ROCKHILL, Chau Ju-kua, p. 145) certainly is the rhinoceros. The Malays designate the two-horned species badak-karbau, "the buffalo-rhinoceros," and the single-horned species badak-gājah, "the elephant-rhinoceros." It is difficult to understand, however, why some of the classical authors allude to the rhinoceros under the designation "the Indian ass" (ARISTOTLE, Hist. an., II, 18, ed. of AUBERT and WIMMER, Vol. I, pp. 74, 254). Aristotle's definition is traceable to CTESIAS (ed. BAEHR, p. 254), who states that there were in India wild white asses celebrated for their swiftness of foot, having on the forehead a horn a cubit and a half in length, and that they are colored white, red, and black; from the horn were made drinking-cups which were a preventive of poisoning (compare also LASSEN, Indische Altertumskunde, Vol. II, p. 646). The mention of these antipoisonous cups is good evidence for the fact that Ctesias hints at the Indian rhinoceros (HERODOTUS, IV, 191, speaks of horned asses of Libya, but they are not one-horned). Ctesias is an author difficult to judge. His account of India, said to have been written in B.C. 389, it should be borne in mind, was derived second-hand, while he resided in Persia as court-physician of King Artaxerxes Mnemon, so that his data may partially be based on Persian accounts of India, and misunderstandings of his informants may have crept in; moreover, his report is handed down in a bad and fragmentary condition, and may have been disfigured by Photias of Byzance of the ninth century, to whom the preservation of his work is due. The definition of Ctesias in the present case cannot be regarded as correct, as we do not find in India, or anywhere else in the East, a comparison of the rhinoceros with an ass, nor any tradition to this effect,— a tradition which is not likely ever to have existed. If the ass really was contained in his original text, it must go back, in my estimation, to a misunderstanding on his part of the word imparted to him by the authorities whom he questioned. With the exception of the horn, Ctesias does not seem to have entertained any clear notion of the animal; and his description of the skin as white, red, and black, is baffling. V. BALL (Proceedings Royal Irish Academy, Vol. II, 1885, and in his edition of Tavernier's Travels in India, Vol. I, p. 114) tried to show that the colors seen by Ctesias were artificial pigments applied to the hide, as they are on elephants at the present day; rhinoceroses kept by the Rajas for fighting-purposes were, according to him, commonly painted with diverse bright colors. This forced explanation, shifting quite recent affairs to the days of early antiquity, is hardly plausible. It seems to me that we are bound to assume that the text of this passage is not correctly handed down. The colors white, red, and black would seem rather to have originally adhered to the horn. The Eastern lore of the rhinoceros, as shown by the reports of the Chinese and Arabs, essentially clusters around the horn.— MARCO POLO (ed. of YULE and CORDIER, Vol. II, p. 285) says in regard to the Javanese rhinoceros that its head resembles that of a wild boar; and this characterization is quite to the point, as is that of Kuo P'o when he compares the two-horned si to swine. A glance at Fig. 8, representing the specimen of a Sumatran two-horned rhinoceros in the Field Museum, will convince every one of the appropriateness of this simile. The pig shape of the rhinoceros is apparent also in a Roman representation on a clay lamp from Labicum illustrating the struggle between that animal and a bear (Fig. 7), so that even the most skeptic critic of Chinese animal sketches will be compelled to grant a certain foundation of fact to the hog-like rhinoceros of the Erh ya (Fig. 6).

FIG. 6.

The Animal "*si* resembling Swine" (from the Illustrated Edition of *Erh ya*).

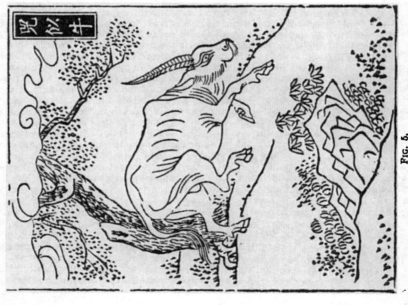

FIG. 5.

The Animal "*se* resembling the Ox" (from the Illustrated Edition of *Erh ya*).

and the comparisons could not be any better.   We should halt a moment to reflect by what class of people these observations had been made. Most certainly by the hardy hunters who chased the wild beasts.   We must distinguish between the original observer and story-teller, and the scholar closeted in his study who draughted the definitions for the consumption of the learned.   It was not the Chinese philologist who went out into the jungle to study the rhinoceros; he, indeed, never had occasion

FIG. 7.

Struggle of Bear and Rhinoceros, represented on a Clay Lamp from Labicum (after O. Keller, Tiere des classischen Altertums).

to see it, but he derived his knowledge from reports made to him by the sportsman.   The latter probably was plain and matter-of-fact; the

FIG. 8.

Sumatran Rhinoceros, Sketch from Museum Specimen (compare Elliot, Catalogue of the Collection of Mammals, Zoöl. Series, Vol. VIII, p. 105).

former added a bit of romance and exaggeration.   Have we any right to ridicule the Chinese over their embarrassment as to where to locate the horn or the horns, when we observe that this was still a matter of wild speculation amidst Europe in the seventeenth and eighteenth centuries?[1]

---

[1] Dr. PARSONS, in the pamphlet quoted, justly remarks, "Nothing could serve as a better proof of how easily men may fall into uncertainty through preconceived conclusions than this very topic of the horn of the rhinoceros."

Have we any right to look down upon their artists in their naïve attempts to sketch the rhinoceros in the shape of an ox with a horn on the forehead (Fig. 5), when we observe that the so-called "civilization" of Assyria and the painting of Persia committed the same error, or when we glance at the puerile drawings of Cosmas and recall Dürer's work with the horn on the animal's neck?

In the above definitions we recognize the elements and tools with which the subsequent Chinese illustrators worked. They set out to illustrate, not the rhinoceros, but the descriptions given of it in the ancient dictionaries. They studied, not the animal, but the ready-made definitions of it encountered in book-knowledge. They read, and their reading guided the strokes of their brush. "The *se* resembles in body a water-buffalo, the *si* a pig:" consequently such bodies were outlined by the illustrator of *Erh ya;* and long, curved, and pointed single horns were placed on the heads (Figs. 5 and 6).[1] He apparently shunned the three horns, as the matter was difficult to draw; and nobody knew how to arrange them. He carefully outlined the three toes

---

[1] Our illustrations are derived from a folio edition of the *Erh ya* printed in 1801 (3 vols.), which is designated as "a reproduction of the illustrated *Erh ya* of the Sung period" (*Ying Sung ch'ao hui t'u Erh ya*). The ancient illustrations of the *Erh ya* by Kuo P'o and Kiang Kuan are lost (see BRETSCHNEIDER, Bot. Sin., pt. 1, p. 34), and were renewed in the age of the Sung, presumably without any tradition connecting the latter with the former. This fact may account for the purely reconstructive work of some illustrations, and we may well assume that the earlier sketches were far better. Many other illustrations of the *Erh ya* have been brought about in the same manner as those of the rhinoceros. Compare, for instance, the picture of the fabulous horse *po* (No. 9393) surrounded by flamed fluttering bands and about to lacerate a tiger seized by its carnivora-like, sharp claws; while a panther is swiftly making for safety to escape a similar fate. Of course, the craftsman has never observed this scene, but faithfully depicts the definition of the book, "The animal *po* is like a horse with powerful teeth, devouring tigers and panthers." This notion, as indicated by Kuo P'o, goes back to the *Shan hai king*, which says, "There is a wild animal styled *po*, like a white horse with black tail and powerful teeth, emitting sounds like a drum and devouring tigers and panthers." (Here we have a parallel to, and presumably an echo of, the flesh-eating horses of Diomed and the man-devouring Bucephalus of the Alexander legend; see J. v. NEGELEIN, Das Pferd im arischen Altertum, pp. 43, 75, Königsberg, 1903.) Otherwise the horses pictured in the *Erh ya*, aside from their technical drawbacks, are quite realistic; and so are the oxen and other animals which came under the every-day observation of the Chinese. It is still a mystery, and a problem worth while investigating, why the Chinese were rather good at drawing some animals and completely failed in others. It may be pointed out that the tapir of the *Erh ya*, aside from the exaggerated trunk and wrong tail, is rather correctly outlined with its white saddle, and corresponds to a well-known species (*Tapirus indicus*). In view of the retrospective and reconstructive sketches of this work, we have the same state of affairs as in the illustrations accompanying the *Shan hai king*, and as formerly shown by me in *Jade*, in the *San li t'u*, and to a certain extent in the *Ku yü t'u p'u*. The illustrators of the ancient Rituals did not directly picture the actual, ancient ceremonial objects, most of which were lost past hope in their time, but reconstructed them from the descriptions supplied by the commentators of the ancient texts, and for better or worse, based their illustrations on these artificial reconstructions, which to a large extent are erroneous or imaginary.

in the animal *si;* and this feature, combined with the single horns, is sufficient flavor of the rhinoceros to guard from any rash conclusion even one who has not considered the psychological foundation of these sketches.

From the fact that the animal *se* is drawn in the shape of an ox, Mr. Giles infers that the word *se* does not denote the rhinoceros, but "a bovine animal." Then, how about the word *si?* The animal *si* (Fig. 6) is undeniably represented in the *Erh ya t'u* with the body of a hog,— why not, to be consistent, also translate the word *si* by "swine"? If a child who was invited to make a sketch of a whale should delineate it in the shape of a fish, should we conclude for this reason that the whale is a fish? To make use of an illustration for a far-reaching philological and zoölogical conclusion, it is indispensable to ascertain the real value of such an illustration, and to make a somewhat critical study of its origin and basis. Mr. Giles is right in stating that there are illustrations of the animal *se* that are purely those of an ox. The ill-reputed *San li t'u,* for instance, stooped to this wisdom when the difficult task arose of illustrating in the shape of a rhinoceros the target used by the lords and ministers in the practice of archery, and spoken of in the *Chou li* and *I li.* But what wonder! Those illustrators who employed the pure-ox design simply stood on the platform of the sober and incomplete definition of the *Shuo wên,* "The animal *se* is like a wild ox." Nothing could be more convenient to the unthinking and mechanical craftsman; this plain recipe freed him from the responsibility for the horn. Anybody could outline an ox with two regular horns; and by inscribing it *se,* the satisfaction at this achievement was naturally the greater.

It is incorrect, however, to say that the animal *se,* as outlined in *T'u shu tsi ch'êng* (Fig. 9), is the picture of an ox. In its general features it resembles a kind of deer, as does likewise the animal *si* (Fig. 10). A lengthy discussion of the "deer-like" rhinoceros follows below (p. 109). Again, in Fig. 9, the draughtsman has taken particular pains to set off distinctly three toes in the left front foot; and where is the bovine animal with three toes? And where is the bovine animal with a single horn, and with this peculiar shape of horn? As to Fig. 10, it presents itself as an illustration of the legend that, while the rhinoceros is gazing at the moon, the peculiar designs within its horn are formed (p. 147). This notion exclusively refers to rhinoceros-horn, so that the animal here intended can be no other than the rhinoceros.[1]

---

[1] The two illustrations of *T'u shu tsi ch'êng* are derived, with a few slight alterations, from *San ts'ai t'u hui* (section on Animals, Ch. 3, p. 7; Ch. 4, p. 12), where, curi-

古今圖書集成

博物彙編禽蟲典第六十八卷犀兕部彙考之二

兕圖

FIG. 9.
The Animal *se* (from *T'u shu tsi ch'êng*).

FIG. 10.
The Animal *si* gazing at the Moon (from *T'u shu tsi ch'ĕng*).

The three-horned rhinoceros described by Kuo P'o is perhaps not so fabulous as it may appear at first sight; for it is known to naturalists that the animal has also the tendency of developing three horns. E. HELLER[1] states in regard to the black rhinoceros covering the whole of Africa with the exception of the Congo Basin that, although the species is almost invariably two-horned, occasional variations of one and *three-horned* specimens are met with. In the light of this observation, PLINY's (*Nat. hist.*, VIII, 21) notice of oxen of India, some with one horn, and others with three (Indicos boves unicornes tricornesque), is apt to lose much of the legendary character with which it was formerly charged. As far as I know, a three-horned specimen has not yet been pointed out among the species of the Indo-Malayan region; notwithstanding, the possibility remains that such may have occurred in times of antiquity. However this may be, whether we assume that the notion of a three-horned species was founded on a natural observation or not, the fact of the coincidence between Kuo P'o and Pliny remains, and hints at the existence of a tradition anent a three-horned variety in the beginning of our era.[2] At any rate, whether real or imaginary, the latter is but a variation of the two-horned species; and by omitting Kuo P'o's illusory "horn on the head," we arrive at a fairly accurate description of it, and then Kuo P'o exactly agrees with Hü Shên's definition of the word *si*. And there can be no doubt of the point that

---

ously enough, they are separated and dispersed in two different chapters. In the latter work, the horn of the *se* is decorated with different designs, which are white on black, while they are black on white in *T'u shu*. The *si* of *San ts'ai* is adorned with flamed and fluttering bands, and the crescent of the moon is absent.

[1] The White Rhinoceros, p. 35 (Washington, 1913). Again on p. 17: "The number of dermal horns on the snout is of less importance. These have been found to show some individual variation in the African species varying from one to three in number in the same species. The front horn, however, is nearly always the better developed and is never wanting."

[2] The case could certainly be argued also from a purely philological point of view. Kuo P'o's creation might be explained as an ill-advised combination of the single-horned and two-horned species, or even regarded as a subsequent interpolation in his text, due to a scribe who meant to be sure of his definition being as complete as possible. Pliny's *tricornis* might be rationally interpreted as the result of an arithmetical process, providing the rhinoceros as a species of ox with two bovine horns, and adding the nose-horn as the third. In this manner Damīrī's three-horned rhinoceros must have arisen (RUSKA, *Der Islam*, Vol. IV, 1913, p. 164), for it has one horn between the eyes and two above the ears. The natural explanation based on zoölogical observation appeals to me to a much higher degree, for we must not be forgetful of the fact that it is impossible for the human mind to invent spontaneously such an observation; a feature of this kind, in order to be observed by man, must have somehow pre-existed in nature. It means nothing, of course, to say that the three horns are a fable; if fable it is, then how did the fable come into existence? It is not the question of a mythological conception, or of a mythical monster, but plainly of a really existing animal described in sober words. I feel confident that the three-horned variation in a living or extinct species will be found some day also in Eastern Asia.

what Kuo P'o intends to describe is the two-horned species of rhinoceros, not any other animal: his statement in regard to "the horn on the nose" excludes any other idea, and the bovine animal with such a horn remains as yet to be discovered. Li Shi-chên of the sixteenth century, as will be seen below (p. 150), rejects the definition of Kuo P'o as erroneous; that is to say, he did not know of any three-horned variety, and recognized in it the two-horned species. An illustration of this three-horned creature may be viewed in the *Wa-Kan San-sai-zu-e*, the Japanese edition of the Chinese cyclopædia *San ts'ai t'u hui*.[1] The definition runs thus: "The rhinoceros has the hair of swine and three toes on each foot; it has the head of a horse and three horns, on the nose, the forehead, and on the skull, respectively." The three toes and three horns are exactly drawn in accordance with this prescription; curiously enough, however, the head is not that of a horse, but of a bull. The old tradition of the draughtsmen is retained in spite of the definition.

Kuo P'o, in all probability, is not the first or the only author to speak of a three-horned variety. A work *Kiao Kuang chi*,[2] Account of Kiao chou (northern part of what is now Annam) and Kuang-tung, reports, "In the territory of the Barbarians of the South-west occurs a strange rhinoceros with three horns emitting light at night like big torches at a distance of a thousand paces. When it sheds its horns, it hides them in a remote and dense jungle to prevent men from seeing them. The sovereigns hold this strange product in high esteem, and make it into hair-pins. These are capable of checking evil and rebellion." Here we have the testimony of an eye-witness or one reproducing a hearsay account; and, quite correctly, he points out this variety as a freak of nature. The exact date of the work in question is unfortunately not known to me; but as the quotation is placed between one from *Kuang-chi* by Ku Yi-kung, who according to BRETSCHNEIDER[3] belonged to the Liang dynasty (502–556), and one from *Kuang chou ki*, a work of the Tsin period (265–419), the inference may be justifiable that *Kiao Kuang chi* likewise is a production of the Leu-ch'ao period. However remote from truth all these Chinese illustrations may be, most of them are fairly correct as to the outlines of the horn, naturally because

---

[1] The illustration is easily accessible in L. SERRURIER, Encyclopédie japonaise, le chapitre des quadrupèdes, Plate VIII (Leiden, 1875). This cut is not contained in a recent edition of this Japanese work (Tōkyō, 1906), but is replaced by a rhinoceros with two horns, the one on the forehead, the other on top of the skull. These attempts clearly prove that Japanese as well as Chinese illustrators did not draw the animal from life, but from the definitions of the books. In the Chinese *San ts'ai t'u hui* (Ch. 4, p. 32) only a three-horned animal (*san kio shou*) is depicted.

[2] Quoted in the chapter on Rhinoceros in *T'u shu tsi ch'êng*.

[3] Bot. Sin., pt. I, p. 164.

the horn as an article of trade was always known, but not the animal itself.[1]

The rôle played by the rhinoceros in Chinese art is limited. As shown by the symbol illustrated in the *Po ku t'u lu* (Fig. 18), it was pictured in early antiquity; and other representations of that period mentioned in Chinese records are discussed on p. 160. The animal lacks those æsthetic qualities of form which tempt the brush of the painter; and this may be the reason why despite the living rhinoceroses sent up as tribute to the capital (see p. 80) it has never been immortalized on any Chinese scroll known to us.[2] There is, however, one case on record. Chang Shi-nan, who wrote the book *Yu huan ki wên* early in the thirteenth century,[3] narrates that he once saw in Sze-ch'uan (Shu) the painting of an unknown artist showing the outlines of a rhinoceros with a horn on its nose.[4] The inhabitants of Sze-ch'uan, accordingly, were familiar with the animal, and for this reason represented it correctly. On some Buddhist pictures it may owe its existence to a mere lucky chance; that is, to the fact that it was so copied from an Indian-Buddhist model. On Yen Li-pên's picture showing Samantabhadra's elephant,[5] the rhinoceros is unmistakably contrasted with the elephant as the smaller animal with scaly body, and head surmounted by a single horn. Another illustration of the same subject is reproduced in Fig. 11 from *Ch'êng shi mo yüan* (Ch. 6 B, p. 16) published in the Wanli period, after 1605. Possibly it occurs also on the later typical paintings of Buddha's Nirvāṇa in the group of wailing animals.[6] On the sculptures of Angkor-Vat the rhinoceros is represented as the vehicle of the god Kārttikeya.[7]

The Mongol emperors made practical use of the typical, conventional designs of the rhinoceros on the standards of the army: there was a standard with the picture of the animal *se*, "resembling an ox, with a single horn, and of dark color," and another with a picture of the

---

[1] A modern Chinese school-book published at Shanghai in 1901, and illustrated by Wu Tse-ch'êng of Su-chou, illustrates the word *si* with the cut of a rhinoceros of European origin, and the word *se* with a jovial ox of his own invention; while the text accompanying it, imbued with the spirit of the *Shuo wên* and *Erh ya*, speaks of one horn on the nose and three toes.

[2] It is likewise absent from classical Greek art. The marble relief of Pompeii, the lamp from Labicum, and the coins of Domitian referred to, are the only known examples of its representation in late Roman art.

[3] WYLIE, Notes, p. 165.

[4] The text is reprinted in *T'u shu tsi ch'êng*, chapter on rhinoceros, *hui k'ao*, p. 5.

[5] Reproduced in the writer's *Jade*, p. 342.

[6] See for example A. GRÜNWEDEL, Buddhistische Kunst in Indien, p. 114, or Buddhist Art in India, p. 124 (in the right lower corner).

[7] According to M. G. COEDÈS, Les bas-reliefs d'Angkor-Vat, p. 12 (Paris, 1911).

FIG. 11.

"Brushing the Elephant." Rhinoceros with Scaly Armor in Front. Wood-engraving from *Ch'êng-shi mo yüan.*

rhinoceros *si niu*, which is not described. They had also standards with designs of a three-horned animal (*san kio shou*) and the unicorn (*kio tuan*), which was outlined "like a sheep, with a small tail and a single horn on its crest."[1]

In plastic art,[2] the rhinoceros has been carved from jade either as the handle of a paper-weight or as the knob of a seal.[3] An example of either kind is illustrated in *Ku yü t'u p'u* (Ch. 74, p. 1, reproduced in

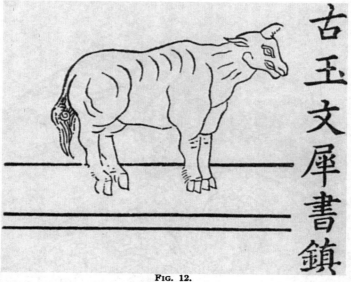

FIG. 12.
Ancient Paper-Weight of Jade surmounted by Figure of Rhinoceros (from *Ku yü t'u p'u*).

Fig. 12; and Ch. 37, p. 11). The traditional reconstructions of the animal are here faithfully preserved; the three toes (the third, of course, is not visible) and the shape of the horn, though it is wrongly placed, come somewhat near the truth. The manufacturers of ink-cakes availed themselves of the same design for printing on the surface of their products. The *Ch'êng shi mo yüan* (Ch. 13, p. 30) illustrates "a spiritual rhinoceros" (*ling si*) with body of an ox, hump of a zebu, cloven feet, snout of a pig, and horn on the front.

---

[1] *Yüan shi*, Ch. 79, p. 10 (K'ien-lung edition).

[2] BUSHELL (Chinese Art, Vol. I, p. 91) figures a bronze vessel of the type styled *hi ts'un*, and describes it as being "shaped in the form of a rhinoceros standing with ears erect and a collar round the neck." But this explanation conflicts with Chinese tradition, according to which the animal *hi* is a sacrificial ox; and an ox is apparently represented in this bronze. Neither is there a single or double horn, which would be necessary to establish such a case.

[3] Seals surmounted by the full figure of a rhinoceros seem to make their first appearance in the Han period (see *Hou Han shu*, Ch. 40, p. 5).

The most curious item in the history of the iconography of the rhinoceros is the illustration of the animal in the *Chêng lei pên ts'ao* published in 1208 by the physician T'ang Shên-wei[1] (reproduced in Fig. 13). Here we see the animal represented as a hairy and spotted deer, its head being surmounted by a single curved horn, peacefully chewing a bunch of leaves with a most innocent expression on its face. The legend is *si kio* ("rhinoceros-horn"), all illustrations of animals

FIG. 13.
Deer with Single Horn, labelled Rhinoceros-Horn, being an Echo of the Indian Legend of Ekaçriṅga
(from *Chêng lei pên ts'ao*, edition of 1523).

in this work being named for the product yielded by them; and the illustration is immediately followed by the description of the two animals *se* and *si*, so that there can be no doubt that this figure, in the mind of the author, is intended for the rhinoceros. It will certainly not induce us to propose for the word *si* the new translation "cervine animal;" but a rhinoceros of cervine character has really existed in the imagination of the ancient world. The idea started from India, has taken a footing in the classical authors, and long survived even down to our middle ages. It is a fascinating story, deserving full discussion, the more so as it has never been clearly and correctly set forth. Two classical texts may first be quoted which fit well as an explanation to our Chinese woodcut. PLINY (*Nat. hist.*, VIII, 21) tells regarding the Orsaean Indians that "they hunt the indomitable, fierce *monoceros* (unicorn) which has the head of a *stag*, the feet of the elephant, the

---

[1] Regarding this work and its history see *T'oung Pao*, 1913, p. 351. In the edition of 1523 from which our illustration is taken it is in Ch. 17, fol. 20 b.

tail of a boar, while the rest of the body is like that of the horse; it emits a deep roar, and has on the middle of its forehead a single black horn two cubits in length.   This beast, it is asserted, cannot be captured alive."[1]   In the *Cyranides*, a curious Greek work written between 227 and 400 A.D.,[2] it is said, "The rhinoceros is a quadruped resembling the *stag*, having a very large horn on its nose.   It can be captured only by means of the perfume and the beauty of well dressed women; it is indeed much inclined toward love."[3]   The importance of this passage, first of all, rests on the fact that the single-horned cervine animal is here clearly identified with the rhinoceros, an identification not yet made by Pliny, who speaks of rhinoceros and monoceros as two distinct species; and we remember that Cosmas Indicopleustes makes the same distinction in regard to India.   In his introduction, F. DE MÉLY[4] observes that the *Cyranides* is the first work to reveal to us the starting-point of the legend of the chase of the unicorn which is nothing but the rhinoceros.   This, however, is very inexact.   The first Occidental source relating this legend is the *Physiologus* which is older than the *Cyranides*. The *Physiologus*[5] tells of the monoceros that it is a small animal resembling a buck, but very cunning; the hunter cannot approach it, as it possesses great strength; the horn grows in the centre of its head; it can be captured only by a pure virgin who suckles it; then she seizes it, and carries it into the palace of the king; or according to another version, the unicorn falls asleep while in the lap of the virgin, whereupon the hunters gradually approach and fetter it.   The *monoceros* is located by Pliny in India; and the western legend of the unicorn ensnared by a virgin was first traced by S. BEAL[6] to the ancient Indian legend of Ekaçriṅga, the hermit Single Horn.   H. LÜDERS,[7] who has traced with great ingenuity the development of the legend in the sources of Indian

---

[1] Orsaei Indi . . . venantur asperrimam autem feram monocerotem, reliquo corpore equo similem, capite cervo, pedibus elephanto, cauda apro, mugitu gravi, uno cornu nigro media fronte cubitorum duum eminente. hanc feram vivam negant capi. (Ed. of C. MAYHOFF, Vol. II, p. 104.)

[2] F. DE MÉLY, Les lapidaires grecs, p. LXXI; DE MÉLY is the first editor and translator of this work.

[3] *L. c.*, p. 90.

[4] *L. c.*, p. LXV.

[5] F. LAUCHERT, Geschichte des Physiologus, pp. 22, 254 (Strassburg, 1889); F. HOMMEL, Die aethiopische Übersetzung des Physiologus, p. 68 (Leipzig, 1877); E. PETERS, Der griechische Physiologus und seine orientalischen Übersetzungen, p. 34 (Berlin, 1898); K. AHRENS, Das "Buch der Naturgegenstände," p. 43 (Kiel, 1892).

[6] The Romantic Legend of Çakyamuni Buddha, p. 125; see also his Buddhist Records of the Western World, Vol. I, p. 113.

[7] Die Sage von Ṛṣyaśṛṅga (*Nachrichten d. k. Ges. d. Wiss. zu Göttingen*, 1897, pp. 1–49), p. 29; an additional study from his pen on the same subject *ibid.*, 1901, pp. 1–29.

literature, justly points out that all our mediæval versions of the story,[1] as a last resort, go back to the Greek *Physiologus*, and that the last clause of the Greek text contains a visible trace of the old Indian legend of the king's daughter who carries away the penitent into the palace of her father. Lüders rises also against the view of Lauchert, who interprets the story in *Physiologus* from a misunderstood passage of AELIAN (XVI, 20); and I am in full accord with the criticism of Lüders, to which the argument should be added that this alleged influence of Aelian on the *Physiologus* is out of the question, as Aelian is in time posterior to the latter.[2] F. W. K. MÜLLER studied the same question in connection with a Japanese Nō play, the plot of which is the legend of Ekaçriṅga.[3] Müller likewise thinks Lauchert's explanation to be hardly plausible, and admits, with excellent arguments, the dependence of the *Physiologus* story on the tradition of India. There is but one point in which my opinion differs from the one expressed by Müller. Müller, at the close of his highly interesting study, advances the theory that the real unicorn, as already recognized by Marco Polo, may always have been the

---

[1] Of the mediæval versions, that of JOHN TZETZES, the Byzantine poet and grammarian, who flourished during the twelfth century, in his *Chiliades* (v, 398), deserves special mention: "The monoceros carries a horn on the middle of its forehead. This animal is passionately fond of perfumes. It is hunted in this manner. A young man disguised as a woman exhaling the odor of the most exquisite perfumes takes his position in the places frequented by this quadruped. The hunters lie in ambush at a short distance. The odor of the perfumes soon attracts the monoceros toward the young man; it caresses him, and he covers its eyes with perfumed woman's gloves. The hunters hasten to the spot, seize the animal which does not offer resistance, cut off its horn, which is an excellent antidote to poison, and send it back, without inflicting on it further harm."

[2] Claudius Aelianus flourished under Septimius Severus, and probably outlived Elagabalus (218–222 A.D.). His writings come down from the beginning of the third century (BAUMGARTEN, POLAND, and WAGNER, Die hellenistisch-römische Kultur, p. 615, Leipzig, 1913), while the *Physiologus* was written in Alexandria as early as the second century (*ibid.*, p. 622). Little is known about Aelian's life; only Philostratus and Suidas have some brief notes regarding him. He availed himself of the writings of Athenaeus, who wrote at the time of Elagabalus, or in the first years of Alexander Severus](222–235); Philostratus mentions his death in his Lives of Sophists composed between 222 and 244. As regards the *Physiologus*, it is necessary to discriminate between the final Greek recension clothed in a Christian-theological garb, as we have it now, and the primeval source or sources of animal stories without the allegories, from which the former was extracted. LAUCHERT (*l. c.*, p. 42) certainly is quite right in rejecting the hypothesis of an "*Urphysiologus*" in the sense that it was a literary production serving as model to our *Physiologus;* but a primeval *Physiologus* must be presupposed for about the beginning of the first century, in the sense that it simply was an assemblage of verbal stories current in Alexandria, and some of which were imported from India (compare *T'oung Pao*, 1913, pp. 361–4).

[3] Ikkaku sennin, eine mittelalterliche japanische Oper (*Bastian Festschrift*, pp. 513–538, Berlin, 1896). Lüders, whose work appeared in 1897, did not take note of Müller's investigation; it seems that the treatises of both scholars originated about the same time, and independently of each other. Compare also J. TAKAKUSU, The Story of the Ṛṣi Ekaśṛṅga (*Hansei Zasshi*, Vol. XIII, 1898, pp. 10–18); and K. WADAGAKI, Monoceros, The Rishi (*ibid.*, pp. 19–24).

rhinoceros. Also O. KELLER[1] has arrived at the same result, and
reduced all ancient traditions and representations of the unicorn to the
Indian rhinoceros. This opinion seems to me fundamentally wrong.
Not one of the numerous variants of the ancient Indian tradition regarding the Hermit Single-Horn alludes in this connection to the
rhinoceros; he is miraculously born from a gazelle, and has received his
horn from the latter.[2] Single-Horn is not even his original name, but
this one was Antelope-Horn (Ṛishya-çrĩnga); and according to LÜDERS,[3]
the name Single-Horn has arisen from the latter, owing to popular
etymological re-interpretation caused by the tradition, already appearing
in the Mahābhārata that the penitent had a single horn on his head. In
other texts, the Padmapurāṇa, Skandapurāṇa, and Kanjur, he is even
equipped with *two* horns, while the versions of the Rāmāyaṇa and the
Pāli Jātaka make no statement with regard to the horn. The Greek
Physiologus, in the story alluded to, avails itself of the word *monokeros*
("unicorn"), which literally corresponds in meaning to Sanskrit Ekaçrĩnga, and describes the creature as a small animal resembling a buck,
without any qualities inherent in the rhinoceros; and this is plainly
corroborated by the illustration accompanying the *Physiologus*, in

---

[1] Die antike Tierwelt, Vol. I, pp. 415–420; this is presumably the weakest chapter
of an otherwise intelligent and excellent book. I do not understand how Keller arrives
at the opinion that the ancients in general treat *monoceros*, *unicornis*, and *rhinoceros*
as identical notions, and in most cases conceive them as the African rhinoceros. The
historical connection of the unicorn legend with Ekaçrĩnga has escaped Keller entirely.

[2] The iconography of Ekaçrĩnga in Indian art has been traced by LÜDERS and
MÜLLER. It is notable that any suggestion of a rhinoceros is absent. As proved by
the masks of the hermit used in the dramatic plays of Japan and Tibet (Plate X),
he was conceived as a human being with a single, short, forked horn, or with
a very long, curved horn. The illustration of the Japanese mask is derived from
the work *Nōgaku dai-jiten* (Dictionary of Nō Plays) by Masada Shōjirō and Amaya
Kangichi (Tōkyō, 1908; compare *Bulletin de l'Ecole française d'Extrême-Orient*,
Vol. IX, 1909, p. 607). The Tibetan mask, much worn off by long use, was obtained
by me from a monastery of Bagme, in the western part of the province of Sze-ch'uan.
It is very striking that the rhinoceros hardly plays any rôle in the culture-life, folk-
lore, or mythology of India. The allusions to it in literary records are exceedingly
sparse. The word *khaḍga* appears but a few times in Vedic literature, a rhinoceros-
hide being mentioned in one passage as the covering of a chariot (MACDONELL and
KEITH, Vedic Index, Vol. I, p. 213, London, 1912). The animal is mentioned in the
inscriptions of King Açoka (third century B.C.); and the consumption of its flesh,
blood, and urine plays a certain rôle in Indian pharmacology (see CHAKRAVARTI,
Mem. As. Soc. Beng., Vol. I, p. 370, Calcutta, 1906; and HOOPER, *J. As. Soc. Beng.*,
Vol. VI, 1910, p. 518). It is very curious that no Indian record regarding rhinoceros-
horn cups and their antipoisonous virtues has as yet been pointed out; our information
on this point rests on Ctesias, Aelian (see below, p. 115), some Arabic authors, and
more recent observers like Linschoten and GARCIA AB HORTO (Aromatum et simplicium aliquot medicamentorum apud Indos nascentium historia, p. 66, Antverpiae,
1567), who says, "Illud tamen scio Bengala incolas eius cornu adversus venena usur-
pare, unicornu esse existimantes, tametsi non sit, ut ii referunt qui se probe scire autu-
mant." It remains to be pointed out also that the literatures of India contain no
accounts of unicorns.

[3] *L. c.*, p. 28.

which the animal is outlined as a long-tailed antelope with a large single horn curved like that of a gazelle.[1] PLINY, as we saw, credits the *monoceros* of India with the head of a stag and a single horn on its forehead (that is, the gazelle-horned Ekaçriṅga), but does not identify it with the rhinoceros, which was well known to him from the circus. For the first time, as far as the West is concerned, the identification of the single-horned cervine animal with the rhinoceros is made in the *Cyranides*.[2] In the East, the first intimation of it leaks out in our Chinese illustration from *Chêng lei pên ts'ao*, which depicts the rhinoceros in the form of a deer with one horn on its forehead, and which, without any doubt, is an offshoot of the Indian conception of Ekaçriṅga. Now, we encounter the curious fact that at a much older date also the Chinese mention a single-horned deer under the name *p'ao* (No. 9104), described in the *Erh ya* as an animal "with the tail of an ox and one horn." PALLADIUS[3] straightway translated the word by "rhinoceros," but this venture is not justified by Chinese tradition; the Chinese, in this case, make no reference whatever to the rhinoceros. On the contrary, Kuo P'o, the editor and interpreter of *Erh ya*, states that the animal *p'ao* is identical with the deer called *chang* (No. 407); and Yen Shi-ku (579–645), as quoted in K'ang-hi's Dictionary, maintains that it resembles in shape the deer *chang*. The very definition shows that the animal *p'ao* is a near cousin of the *k'i-lin*[4] which has likewise "the tail

---

[1] Figured by STRZYGOWSKI, Der Bilderkreis des griechischen Physiologus, Plate XII (*Byzantinische Zeitschrift*, Ergänzungsheft 1, 1899), and KELLER (*l. c.*, p. 419). Regarding the illuminated editions of the *Physiologus* see also O. M. DALTON, Byzantine Art, p. 482 (Oxford, 1911).

[2] Neither LÜDERS nor MÜLLER has consulted these two important passages of Pliny and the Cyranides.

[3] Chinese-Russian Dictionary, Vol. I, p. 58.

[4] At times a temptation was felt to identify the animal *lin* with the rhinoceros. Shên Kua, the versatile author of the *Mêng k'i pi t'an* of the twelfth century, narrates that in the period Chi-ho (1054–56) the country Kiao-chi (Annam) offered a *lin* like an ox, having the entire body covered with large scales and a single horn on its head. There is no question that this animal was a rhinoceros; this follows also from the further observation of the author that it did not resemble the *lin*, as described in ancient records, and that there were people designating it as a mountain-rhinoceros (*shan si*, a variety recognized also by Li Shi-chên). But as Shên Kua could not trace any report in which scales are attributed to the rhinoceros (for explanation see p. 149), he formed the erroneous theory that the animal in question was identical with the T'ien-lu cast in bronze by the Emperor Ling in 186 A.D., a specimen of which he had beheld at Nan-yang in Têng chou in Ho-nan. In a similar manner, Fan Chên of the Sung period, in his work *Tung chai ki shi* (Ch. 1, p. 8; in *Shou shan ko ts'ung shu*, Vol. 84), tells the story of two *K'i-lin* sent as tribute from Kiao-chi in the period Kia-yu (1056–63), which he had occasion to see in the imperial palace. He describes them as having the shape of water-buffalo clad with a fleshy armor, and equipped with a single horn on the extremity of the nose; they subsisted on grass, fruit, and melon, and every time before feeding had to be beaten on their horns with a stick. This writer likewise concludes with a discussion, in which serious doubts of the identification of these animals with the *lin* are expressed.

of an ox and a single horn."[1] Indeed in the *Erh yu t'u*, both creatures are figured almost alike, and agree in their essential characteristics. It is obvious that, as iconographic types, these creatures are not derived from any rhinoceros, but point in the direction of the fabulous one-horned monsters (known in archæology as "Oriental animals") developed in the art of Mesopotamia.[2]   In regard to the type of *k'i-lin*, this has been aptly pointed out by A. GRÜNWEDEL;[3] and as the same West-Asiatic forms found their way into the art of India, we here have the basis for the origin of the single-horned gazelle (deer or antelope) transferred to, or personified in, the person of Ekaçriṅga.   In Babylonia, these types of unicorn are very ancient, going back to the third millennium B.C.,[4] and could not have been developed there from a rhinoceros.   The conclusion therefore presents itself that the notion of a unicorn cervine animal which was developed in Western Asia from remote times spread together with artistic motives into India and China,[5] while the identification of this fabulous creature with the

---

[1] Regarding the *k'i-lin* see Yen Shi-ku (in *Ts'ien Han shu*, Ch. 6, p. 5 b); MAYERS (Chinese Reader's Manual, p. 127); F. W. K. MÜLLER (in *Feestbundel aan P. J. Veth*, p. 222, Leiden, 1894); DE GROOT (The Religious System of China, Vol. II, pp. 822–4); and H. DORÉ (Recherches sur les superstitions en Chine, pt. 1, Vol. II, pp. 446–8). I do not subscribe to everything that the last two authors say about the subject.   The Chinese illustrations are reproduced in C. GOULD (Mythical Monsters, pp. 350, 353, 354, London, 1886).

[2] A distinction must be made between iconographic or archæological type or artistic representation, and traditions or speculations regarding such a type.   The *lin*, as early mentioned in *Shi king* and *Li ki*, may very well be an indigenous Chinese thought.   Nevertheless its subsequent portrayal in art rests on a borrowed type, which has again fertilized native ideas as to form and behavior of the creature.   An interesting example of the fact that iconography and literary tradition may move along lines widely different and emanating from diverse sources is afforded by the unicorn of Europe.   The unicorn tradition of the *Physiologus* is traceable to India; the iconography of the creature, however, has no connection with Indian art, but leans in the beginning toward the ancient West-Asiatic types.   Throughout the middle ages, there is not a trace of the rhinoceros in the representations of the unicorn (compare Marco Polo's astonishment when he saw the ugly beast on Java, "not in the least like that which our stories tell of as being caught in the lap of a virgin, in fact, altogether different from what we fancied"); now it is an antelope, now an ox, now a narwhal, now a hybrid formation composed of various creatures.   My opinion in this respect deviates from the one expressed by STRZYGOWSKI (*l. c.*) that there may be interaction between the animal types of the earliest Buddhist art in India and those of the *Physiologus*.   It is not there the question of interaction, but of affinity, solely caused by West-Asiatic productions which both have in common as their source.

[3] Bemerkungen über das Kilin (*Feestbundel aan P. J. Veth*, pp. 223–5, Leiden, 1894), and Buddhist Art in India, p. 19.

[4] E. SCHRADER, Die Vorstellung vom monokeros und ihr Ursprung (*Abhandlungen der preussischen Akademie*, 1892, pp. 573–581).

[5] In order to dispel the doubts of those who may not feel inclined in this case to link China with the West, another striking analogy may be indicated, which will show that Chinese ideas regarding unicorns coincide with those entertained in the West, and which crop up in the classical authors.   In the *Erh ya* is defined an animal called *chui* (written with the classifier 'horse' and the phonetic complement *sui*, No. 10,388), 'like a horse with a single horn; those without horn are spotted." Kuo P'o comments,

rhinoceros — owing to the single horn — is the product of a much later period; this is not the starting-point, but the final result of the matter. It is, of course, necessary to assume that this result was brought about in India itself; [1] otherwise it would be unintelligible why it appears on the surface in the *Cyranides* and in China.[2] In my opinion, we are even

---

"In the eighth year of the period Yūan-k'ang (298 A.D.) it was in the territory of Kiu-chên (in Tonking) that hunters captured a wild animal of the size of a horse with one horn, the horn being soft as the core of the young antlers of the deer (*lu jung*). This is identical with the animal *chui*. At present men sometimes meet it in the dense mountainous jungles, and there are among them also those without horn." Kiu-chên is situated in Tonking; and on p. 81 mention has been made of the tribute of a live rhinoceros sent from there to the Emperor Ling (168–188 A.D.); indeed, that region was always famed for this animal, which is apparently intended in the text of Kuo P'o. The same conception of the rhinoceros as a horse or horse-like animal with a single horn is met likewise in the West. The ancients enumerate altogether five animals as having single horns, the Indian ass first traceable to Ctesias, the single-horned ox, the monoceros, the single-horned horse, and the oryx of Africa. STRABO (XV, 56) quotes from Megasthenes' remarks upon Indian animals that there are horses in India with one horn. AELIAN (*Nat. anim.*, III, 41) says, "India, it is reported, produces horses with a single horn, likewise single-horned asses. Cups are made from these horns; and if a mortal poison is poured into them, it will do no harm to him who drinks it, for the horn of both animals seems to be an antidote against poison." In another chapter (XVI, 20) AELIAN describes the unicorn of the Indians, "called by them *kartazonos* [a word apparently connected with Assyrian *kurkizannu*, mentioned above, p. 87], said to equal in size a full-grown horse." HORACE (*Serm.*, I, 5, 58–60) speaks of a wild horse having a single horn in the midst of its forehead. As a matter of fact, the rhinoceros has no similarity to a horse; and it is difficult to see how the simile could ever arise. The bare fact remains, however, that it did; but it is inconceivable that this notion, not founded on a natural observation, could spontaneously spring up in the West and East alike. There is no other way out of this puzzle than to presume that India, to which the account of Megasthenes reproduced by Strabo and Aelian refers, is responsible for this idea, and disseminated it to the West and to China.

[1] It may be pointed out in this connection, though it is not wholly conclusive for the present case, that the Sanskrit word *vārdhrāṇasa* means a rhinoceros and an old white goat-buck.

[2] We meet also in ancient China a unicorn conceived of as a wild goat. This is the animal termed *chai* (No. 245) and *hiai* (No. 4423) *chai*. The fundamental passage relating to it is in the Annals of the Later Han Dynasty (*Hou Han shu*, Ch. 40, p. 3), where a judicial cap in the shape of this animal, and worn by the censors, is mentioned. The definition given of the animal in the text of the Annals is, "A divine goat (*shên yang*) which is able to discriminate between right and wrong, and which the king of Ch'u used to capture." Huai-nan-tse is quoted in K'ang-hi (under *hiai*) as saying that King Wên of Ch'u was fond of wearing *hiai* caps; the un-Chinese word *hiai chai*, therefore, will probably be a word of the language of Ch'u (T. DE LACOUPERIE, Les langues de la Chine avant les Chinois, p. 17, Paris, 1888), as above all proved by the vacillating modes of writing (FORKE, Lun-hêng, pt. II, p. 321). The comment added to the text of *Hou Han shu* is extracted from *I wu chi*, which may be read in SCHLEGEL's Uranographie chinoise, p. 587 (it is, of course, impossible, as proposed by Schlegel, to identify the animal with the Tibetan *chiru;* see below, p. 120). It is not stated in *Hou Han shu* nor in *I wu chi* (nor in K'ang-hi) that "it eats fire in its ravenous fury, even to its own destruction" (GILES). This is a subsequent addition which arose under the influence of Buddhist art. F. W. K. MÜLLER (*Feestbundel aan P. J. Veth*, p. 222, Leiden, 1894) has recognized correctly that this explanation is derived from the iconography of the animal, which is represented as being surrounded by flames. Müller, however, omits to state that this is a secondary development, which has nothing to do with the previous pre-Buddhistic conception of the creature on Chinese soil, when it was not equipped with flames, nor set in relation with a lion. The

forced to admit that the counterpart to the illustration of the *Chêng lei pên ts'ao* has already pre-existed in India, and was transmitted from there to China; for neither the author of that work, nor any other Chinese source, as far as I know, furnishes any explanation for this picture.   An unexpected confirmation of this opinion comes to us from another quarter, — Tibet.

In the Tibetan language we meet the word *bse-ru* which at present denotes two animals, — first, the rhinoceros, and second, a kind of antelope.   The former is the original and older significance, the latter is secondary.   The second element of the compound, *ru*, means "horn," and may be dropped; the proper word is *bse* (pronounced *se*).   The stem is *se*, the prefixed labial *b-* not being part of the word-stem, and like most prefixes in Tibetan nouns, representing the survival of an ancient numerative.   This is corroborated by the corresponding Lepcha word *sa* and the Chinese word *se*, all three referring to the rhinoceros. This linguistic coincidence leads to the conclusion that the Chinese and Tibetans as stocks of the large Indo-Chinese family of peoples were acquainted with the rhinoceros in prehistoric times, for otherwise they could not have the word for it in common; and this conclusion will be fully upheld by our historical inquiry into the subject.   This fact of comparative philology is also apt to refute the supposition of Mr. Giles that "a term which originally meant a bovine animal was later on wrongly applied to the rhinoceros."   As proved by comparison with the Tibetan and Lepcha words, the Chinese term originally must have designated the rhinoceros.[1]   Above all it is incumbent upon me to demonstrate that the Tibetan word *bse* really designates the rhinoceros, and that the Tibetans were familiar with this animal.   The ancient

---

translation "lion-unicorn" adopted by Müller is not to the point, as far as the time of Chinese antiquity is concerned.   The *hiai chai* is not explained as a lion (nor could this be expected, as the lion was unknown in ancient China), but as a divine wild goat (*shên yang*).   The fact that the conception of the animal existed among the Chinese in times prior to the contact with India is clearly proved by the occurrence of the word in Huai-nan-tse, in *Tso chuan* (Süan Wang 17th year: Legge, Chinese Classics, Vol. V, p. 332), Se-ma Ts'ien's *Shi ki* (Ch. 117), *Lun hêng, Hou Han shu, Erh ya*, and *Shuo wên*.   Only in such late compilations as the Japanese version of the *San ts'ai t'u hui* do we meet the statement that the animal resembles a lion, merely because it is sketched like a lion crowned with a single horn (see L. Serrurier, Encycl. japonaise, le chapitre des quadrupèdes, Plate III; or E. Kaempfer, The History of Japan, Vol. I, p. 195, Glasgow, 1906).   The connection of this creature with the rhinoceros, and its transformation into a goat, will be discussed below (p. 171).

[1] The hypothesis of such "confusions," which are usually assumed to suit one's own convenience, is untenable also for other reasons obvious to every ethnologist: people in the primitive stages of culture, being nearer to nature than we, are surely the keenest observers of animal life and habits, and will most assuredly never confound a bovine animal with a rhinoceros; they may, by way of explanation, compare the one with the other, but from comparison to confusion is a wide step.

Sanskrit-Tibetan dictionary Mahāvyutpatti[1] renders the Tibetan word *bse* by the Sanskrit word *gaṇḍa* which refers to the rhinoceros.[2] Wherever this word appears in the works of Sanskrit Buddhist literature, it is faithfully reproduced in the Tibetan translations by the word *bse*. An interesting example of its application appears in a Tibetan work from the first part of the ninth century.[3] It is well known that in India the Pratyeka-Buddha was styled Single-Horn Hermit and compared with the solitary rhinoceros;[4] and this simile is explained in that Tibetan book in the words that the Pratyeka-Buddha, who in the course of a hundred eons (*kalpa*), through the accumulation of merit, is no longer like ordinary beings, resembles the rhinoceros in his habit of living in the same solitary abode. It is interesting to note that in this early Tibetan text the word *bse-ru* is used for the designation of the rhinoceros. This comparison has passed into Tibetan poetry, and is frequently employed by the mystic and poet Milaraspa, who speaks of himself as being "lonely like a rhinoceros."[5] This meaning of *bse* is confirmed by two Chinese lexicographical sources, — the *Hua i yi yü*, which in its Tibetan-Chinese vocabulary[6] renders *bse-ru* by Chinese *si niu;* and the Polyglot Dictionary of the Emperor K'ien-lung (Ch. 31, p. 4 a), where *bse* is explained by Chinese *si* ("rhinoceros"). The national Tibetan word *bse*, akin to Lepcha *sa* and Chinese *se*, naturally bears out the fact that the ancient Tibetans were familiar with the

---

[1] Tanjur (Palace edition), Sūtra, Vol. 123, fol. 265 a. This work was written in the first part of the ninth century.

[2] Al-Bērūnī (SACHAU, Alberuni's India, Vol. I, p. 203) knew this word, and correctly described under it the rhinoceros of India (p. 95). It is likewise mentioned by GARCIA AB HORTO (*l. c.*) and other early European travellers enumerated by YULE and BURNELL (Hobson-Jobson, p. 363). The rhinoceros brought to Portugal in 1515 (mentioned above, p. 83) was labelled "rhinocero, called in Indian *gomda.*"

[3] Entitled *Sgra sbyor bam-po gñis-pa* (Tanjur, Sūtra, Vol. 124, fol. 14 a, 4), correctly dated by G. HUTH (*Sitzungsberichte der preussischen Akademie*, 1895, p. 277) in the first part of the ninth century. Compare also the application of the word in Tāranātha (SCHIEFNER's translation, p. 245): the sorcerer Ri-ri-pa summoned the fierce beasts of the forest, the rhinoceros and others, and mounted on their backs.

[4] EITEL, Hand-book of Chinese Buddhism (pp. 76, 123, 197); F. W. K. MÜLLER, Ikkaku sennin (*l. c.*, p. 530); and H. KERN, Manual of Indian Buddhism (pp. 61 and 62, note 1).

[5] G. SANDBERG (Tibet and the Tibetans, p. 297), who is ignorant of the fact that *bse* or *bse-ru* means "rhinoceros," and who merely carries the modern popular meaning of the word, "antelope," into the sphere of literature, makes Milaraspa say that he is "lonely as a seru" (antelope). The antelope, however, is not a lonely, but a highly social animal living in herds. Nowhere in Buddhist literature has *bse-ru* the significance of "antelope," but only that of "rhinoceros." The Tibetan poet, who in every line is imbued with the language and spirit of India, most obviously intends with this simile a literary allusion to the Buddhist comparison of the Pratyeka-Buddha with the rhinoceros.

[6] Copied by me from the manuscript deposited by HIRTH in the Royal Library of Berlin. Regarding the work see HIRTH (*J. China Branch R. As. Soc.*, Vol. XXII, 1888, pp. 207 *et seq.*), and *Bull. Ecole française*, 1912, p. 199.

animal. We know that the primeval habitat of the Tibetan stock was located along the upper course of the Huang-ho (where Ptolemy knows them as *Bautai*, derived from the native name *Bod*, "Tibetans;" the Yellow River is styled by him *Bautisos*), as well as along the upper Yang-tse. There they lived in close proximity to the ancient Chinese; and in that locality, as will be established from Chinese records, the rhinoceros was their contemporary. Large parts of the present Chinese provinces of Kan-su and Sze-ch'uan are still settled by Tibetan tribes; and we shall see that the rhinoceros occurred there in the times of antiquity, and long survived, even down to the middle ages. The Pai-lan — a tribe belonging to the Tibetan group of the K'iang, and border-ing in the north-east on the Tu-yü-hun — in 561 A.D. sent an embassy to China to present a cuirass of rhinoceros-hide (*si kia*) and iron armor.[1] Whether they had made this cuirass themselves, or had received it from an outside source (this fact is not indicated), this tribute, at any rate, shows that they were acquainted with this material and its manu-factures.[2] The *Pên ts'ao yen i* of 1116 extols the horns of the Tibetan breed of rhinoceros for the fine quality of the natural designs displayed in them (see p. 148). Li Shi-chên, in his *Pên ts'ao kang mu* (see p. 149), expressly names as habitats of the rhinoceros the regions of the Si Fan and Nan Fan; that is, the western and southern Tibetans, — the former scattered over Sze-ch'uan and Yün-nan with their borderlands, the latter peopling the valley of the Tsang-po (Brahmaputra) and the Himalayan tracts adjoining India. Indeed, down to the middle of the nineteenth century, or even later, the rhinoceros was to be met with along the foot of the Himalaya as far west as Rohilkund and Nepal; and it survived longer still in the Terai of Sikkim.[4] J. CH. WHITE[4] notes the

---

[1] *Chou shu*, Ch. 49, p. 5 b.

[2] In the year 824 the Tibetans offered to the Chinese Court silver-cast figures of a rhinoceros and a stag (*T'ang shu*, Ch. 216 B, p. 6 b). BUSHELL (The Early History of Tibet, p. 88) translates the word *si* in this passage by "yak," but this point of view is not admissible. True it is that some modern Chinese writers on Tibet call the yak *si niu*, but this usage of the word is not earlier than the eighteenth century. The T'ang Annals, however, persistently designate the Tibetan yak by the word *li niu* (No. 6938); and in the very passage alluded to, the gift of the rhinoceros and stag silver figures is immediately followed by the words, "and they brought as tribute a yak" (*kung li niu*), which BUSHELL correctly interprets likewise as yak. The words *si* and *li niu* in the same sentence cannot possibly refer to the same animal; and it becomes evident from a consideration of all Chinese sources concerned that down to the end of the Ming dynasty the Chinese word *si* with reference to Tibet and Tibetan tribes invariably denotes the rhinoceros, and nothing else. Rhinoceros-horn was formerly included among the tribute gifts which the Dalai Lamas of Tibet were obliged to send to China; it took its place between coral, genuine pearls, precious stones, amber, etc. (*Wei Tsang t'u chi*, 1792, Ch. A, p. 17).

[3] R. LYDEKKER, The Game Animals of India, p. 30.

[4] Sikhim and Bhutan, p. 322 (London, 1909).

rhinoceros in a few of the lower valleys of Bhūtan, though not common. In Tibet proper, the animal does not occur at present, but fossil remains of it were discovered at high elevations by Sir R. Strachey near the source of the Tsang-po.[1] The early Tibetan translators, when they correctly rendered the Sanskrit word *gaṇḍa* by *bse*, must have entertained an exact notion or reminiscence of the rhinoceros; but the animal, as everywhere, became rapidly exterminated in those territories where Tibetans had occasion to behold and to hunt it, while the inhabitants of Central Tibet seldom or never had this opportunity. For this reason, also in Tibet, the rhinoceros underwent the process of fabulous "unicornization." Reports of a Tibetan unicorn greatly stirred the imagination of European explorers, and gave rise to wild speculations. Captain S. Turner,[2] I believe, was the first to circulate such a report, being informed by the Rāja of Bhūtan that he was in possession of a unicorn, a sort of horse, with a horn growing from the middle of its forehead; it was kept at some distance from Tassisudon, the capital, and the people paid it religious respect, but Turner had no occasion to see it. The Lazarist fathers Huc and Gabet, who reached Lhasa in 1846, are said to have even claimed the discovery in Tibet of the unicorn of Scripture. Major Latter, in the first part of the nineteenth century, was very sanguine of being able to find a veritable unicorn in the interior of Tibet: he was advised by a native that he had often seen these animals, which "were fierce and exceedingly wild and seldom taken alive, but frequently shot;" and that they are commonly met with on the borders of the great desert, about a mile from Lhasa. From a drawing which accompanied Major Latter's communication, the presumed unicorn was something like a horse, but with cloven hoofs, a long, curved horn growing out of the forehead, and a boar-shaped tail. Under the heading "Unicorns in Asia,"[3] a writer revived the opinion of the existence of veritable unicorns, such as were reported to Major Latter: the animal in question was of the deer kind, having a single horn at the top of the head; it was known by the name of the *Seru*.[4] Then

---

[1] A. R. Wallace (The Geographical Distribution of Animals, Vol. II, p. 214; also Vol. I, p. 122) refers to this in the words that more than twenty species of extinct rhinoceroses are known, and that one has even been found at an altitude of 16,000 feet in Tibet. Mr. L. A. Waddell (Lhasa and its Mysteries, p. 315) has this suggestive remark: "The dense rank growth of wildflowers and weeds along the borders of the fields was such as to make this part of the Tsang-po oasis a quite suitable habitat for the rhinoceros, and to bring the discovery of the fossil remains of that animal by Sir R. Strachey near the source of this river into harmony with present-day facts."

[2] An Account of an Embassy to the Court of the Teshoo Lama, p. 157 (London, 1800).

[3] *Asiatic Journal*, Vol. II, 1830.

[4] Compare W. Haughton, On the Unicorn of the Ancients (*Annals and Magazine of Nat. Hist.*, Vol. IX, 1862, pp. 368, 369).

the famous J. D. Hooker[1] took the matter in hand, and published a sketch of the Chiru Antelope with the addition "unicorn of Tibet," a name which he thought was suggested by the animal when viewed in profile. It is identified as *Antilope* or *Pantholops Hodgsoni*, having been described by Hodgson.[2] It remains a mysterious creature, and little is known about it.[3] P. Landon[4] denies that this antelope, as pointed out by Hooker, occurs near the Cholamu Lake at the present day. L. A. Waddell[5] reports under *Chiru*, "None were seen and the people did not appear to know of any."

In Anglo-Indian nomenclature we now find two words in use, *chiru* and *seru*, the latter also Anglicized as *serow*, on which Yule, in his "Hobson-Jobson," unfortunately has not commented. *Serow* has become a household stock-word of the Anglo-Indian sportsman to denote a large variety of different Indian, Burmese, and Tibetan antelopes.[6] G. Sandberg[7] recognizes in it the Tibetan word *bse-ru*, and identifies the latter with the species *Nemorhaedus bubalinus*. Jäschke[8] says under *bse* or *bse-ru*, "Unicorn, 'tchiru,' an antelope, probably the same as *gtsod*," with reference to Hooker. Chandra Das,[9] who has fully

---

[1] Himalayan Journals, 2d ed., p. 401 (London, 1893).

[2] *Journal As. Soc. Bengal*, 1846, p. 338.

[3] N. Kuehner, Description of Tibet, in Russian (Vol. I, pt. 2, p. 157; and notes p. 77).

[4] Lhasa, Vol. I, p. 393.

[5] Lhasa and its Mysteries, p. 483.

[6] R. Lydekker, The Game Animals of India, pp. 139 *et seq.* M. Dauvergne (*Bull. Musée d'hist. nat. de Paris*, Vol. IV, 1898, p. 219) describes the animal as follows: "Serow; Ramu de Kashmir, ou chèvre-antilope, *Nemorhaedus bubalinus* Hodgs. Habite les rochers escarpés et broussailleux des montagnes, à une hauteur de 3,000 mètres, dans l'Himalaya et Kashmir. Très difficile à chasser, il tient tête aux chiens, qu'il fait rouler dans les précipices. C'est généralement l'hiver qu'on le chasse, car alors il se détache sur la neige, grâce à la teinte noire de sa robe, et comme il est très lourd, il s'effondre et se fait prendre par les chiens."

[7] Tibet and the Tibetans, p. 297. On p. 298 he points out that the word *chiru* should be written *gcig ru* ("one horn"). This derivation is impossible, as "one horn" can be in Tibetan only *ru* (or *rva*) *gcig*, or *ru žig*. The name Ekaçriṅga is rendered into Tibetan *Rva gcig-pa*. (Compare also *Hor c'os byuṅ*, ed. Huth, p. 16, l. 14.) *Chiru* is simply a local or dialectic variation of *se-ru*. Strange words exert a singular fascination upon the human mind. The Anglo-Indian *chiru* has had several good fortunes. Thanks to the imaginative powers of G. Schlegel (Uranographie chinoise, p. 587), it has found cheerful hospitality in Chinese astronomy, the Chinese animal *hiai* being wrongly identified with it. A few years ago the *chiru* was deemed worthy of the honor of being admitted into the sanctum of classical philology. O. Keller (Die antike Tierwelt, Vol. I, p. 293) identifies the Indian Oryx mentioned by Aelian, and the Oryx on the Hydaspes mentioned by Timotheus, with the Tibetan *chiru*,— a venture which has no foundation; in fact, the oryx of Aelian is located in India, and corresponds to the Indian black-buck.

[8] Tibetan-English Dictionary, p. 593. Skr. *khaḍga* rendered by Jäschke "a certain animal" is the rhinoceros.

[9] Tibetan-English Dictionary, p. 1319.

recognized the original meaning of *bse-ru* as "rhinoceros," proceeds to state that in Tibet the word is applied to the clumsy-looking deer known to sportsmen as the "serow." Both lexicographers, in this respect, rely on the statements of the European sportsmen, but leave us in the dark as to the opinion of the Tibetans on the point. The question arises, — Do those European speculations on a Tibetan unicorn identified with an antelope styled *se-ru* have any foundation in a Tibetan tradition? The French Missionaries, in their Tibetan Dictionary (p. 1056), give a slight intimation of the existence of such a tradition by remarking that the animal *bse-ru* is believed in Tibet to belong to the genus of goats (*ex genere caprarum*), but that nobody has ever seen it; the latter clause doubtless means that nobody has encountered this wild goat in the shape of a unicorn which it is fabled to be. I. J. SCHMIDT[1] had a certain presentiment of the matter when he annotated a passage in his translation of the Geser Saga, that the Tibetan and Mongol name of the unicorn is *seru*, that the existence of this animal in the wild mountains of Tibet is asserted in Tibetan books, but that the description given of it does not at all fit the rhinoceros. The unicorn which stopped Chinggis Khan on his expedition to Tibet and induced him to return,[2] judging from the description given by the Tibetan historian,[3] is identical with the Chinese *k'i-lin*, as already recognized by G. SCHLEGEL.[4] Another association of the unicorn with Tibet appears on the tribute painting ascribed to Li Kung-lin (Li Lung-mien), where BONIN[5] has pointed it out among the envoys from the Kingdom of Women. In the Polyglot Dictionary of the Emperor K'ien-lung[6] we find the Tibetan

---

[1] Die Thaten Bogda Gesser Chan's, p. 56 (St. Petersburg, 1839). Compare also p. 125.

[2] G. HUTH, Geschichte des Buddhismus in der Mongolei, Vol. II, p. 25.

[3] "An animal of green color with the body of a stag, the tail of a horse, and a single horn on its head."

[4] T'oung Pao, Vol. VI, 1896, p. 433. According to Chinese tradition, however (see the texts of *Kui sin tsa chi* and *Ch'o keng lu*, in *T'u shu tsi ch'êng*, Chapter *kio tuan, ki shi*, p. 1 b), the marvellous animal opposing the conqueror belonged to the class of unicorns (*kio tuan*), and is described as a hundred feet high, with a single horn like that of the rhinoceros, and able to speak a human language.

[5] Le royaume des neiges, pp. 40, 299 (Paris, 1911). M. Bonin's description of this painting is based on a copy of it in the Musée Guimet, which is certainly not the original from the hand of Li Kung-lin; it is a much later and somewhat weak copy, as stated also by TCHANG YI-TCHOU and HACKIN (La peinture chinoise au Musée Guimet, p. 59). On Plate V of the latter publication, the portion of the picture illustrating the envoys of the Kingdom of Women is reproduced; the unicorn is a wretched production. Mr. Freer of Detroit owns two copies of the same painting, both far superior to the one in the Musée Guimet. One of these offers such high qualities as come very near to an original. The other is a copy of the Yüan period, executed in 1364.

[6] Appendix, Ch. 4, p. 53.

FIG. 14.
Se-ru as Emblem of Long Life (from Tibetan Wood-engraving).

word *bse-ru* rendered by Chinese *shên yang* ("divine goat");[1] and this is thus far the only literary indication which I am able to trace in regard to a Tibetan unicorn of goat-like character.[2]

Such a *bse-ru* is represented on a Tibetan woodcut as an emblem of long life (*bse-ru ts'e riṅ;* Fig. 14). The picture, of which it forms a

---

[1] The Manchu has the artificial formation *šengkitu,* and three other words besides,— *šacintu, tontu,* and *tubitu* (see SACHAROV, Manchu-Russian Dictionary, p. 734),—for the designation of this unicorn. It will be remembered that the term *shên yang* occurs in *Hou Han shu* in defining the unicorn *hiai chai* (p. 115, note 2).

[2] The Mongols have adopted *seru* as a loan-word from Tibetan in the sense of "rhinoceros," as stated by KOVALEVSKI and GOLSTUNSKI in their Mongol dictionaries; but they take the word also in the sense of a "deer," as shown by the Mongol translation of the Tibetan medical work translated into Russian by A. POZDNÄYEV (Vol. I, p. 288). The Mongol equivalent of Tibetan *bse-ru* and Chinese *si kio* is here *bodi gürügäsün* ("the animal of the bodhi," Sanskrit *bodhimṛiga*); that is, the gazelle. Besides, the Mongols have a seemingly indigenous word for "rhinoceros," — *kiris, keris,* or *kers-un ābär.*

part, is known as "the six subjects of long life" (*ts'e riṅ drug skor*). These are,— the Buddha Amitāyus (the Buddha of Endless Life), the long-lived wishing-tree (*dpag bsam śiṅ ts'e riṅ*) figured as a peach-tree in Chinese style, the long-lived rocks (*brag ts'e riṅ*), the Chinese God of Longevity Shou-sing (in Tibetan *Mi ts'e riṅ*) seated on a mat and holding a rosary, a pair of cranes (*kruṅ kruṅ ts'e riṅ*) pecking at some peaches (*k'am-bu*) that are planted in a jar, and a pair of *bse-ru*. Though apparently inspired by the deer, which is the emblem of the Chinese God of Longevity, their outlines considerably differ from the latter, and approach the Tibetan notion of the appearance of a *bse-ru;* [1] but, curiously enough, they are without any horns. There can be little doubt, accordingly, that in recent times, when the rhinoceros had almost vanished from the memory of the Tibetan people, the word *bse-ru* was transferred to a species of deer or antelope; and, as the ancient tradition of the *bse-ru* being a single-horned animal had persisted through the centuries, the single horn, in popular imagination, was fixed on the antelope. When we inquire why it was just the antelope, and not any other animal on which the idea of a unicorn was projected, the story of Ekaçriṅga presents itself again as the happiest solution. We know that this legend, in a Tibetan translation, has been incorporated in the Kanjur; and A. SCHIEFNER [2] has translated it from this version. It is likewise extant in Kshemendra's Avadānakalpalatā, of which a literal versified rendering, and an abridged prose edition made for children by order of the Fifth Dalai Lama, exist in the Tibetan language. This plain version has rendered the story immensely popular among Tibetans; and, as pointed out, it is current also in a dramatized form. The Tibetan mask of Ekaçriṅga (Plate X) is equipped with an unmistakable antelope-horn. [3] The psychological process is therefore quite clear. The rhinoceros was grad-

---

[1] My explanation is based on the interpretation of this woodcut given me by an intelligent Lama. A. GRÜNWEDEL, in his Russian Description of the Lamaist Collection of Prince Uchtomski (*Bibl. Buddhica*, No. 6, p. 26), has figured a similar woodcut, but without explanation. The God of Longevity bears the Mongol legend *Tsaghan Ābughän* ("The White Old Man"), who is certainly, as stated on p. 117, a national Mongol deity; but from an iconographic point of view, as he appears in Grünwedel's drawing, he is nothing but a copy of the well-known Chinese God of Longevity.

[2] In RALSTON, Tibetan Tales, p. 253.

[3] On the lid of a Tibetan censer in the Field Museum (Cat. No. 122,522) are represented the full figures of two gazelles opposite and turned away from each other (the wheel of the law being placed between them), the well-known Buddhist motive symbolizing Buddha's first sermon in the Deer-Park (GRÜNWEDEL, Buddhist Art in India, p. 143). One of these is provided with a single horn on its forehead; the other, apparently conceived as the doe, is hornless. The former seems suggested again by a reminiscence of Ekaçriṅga, but it is not known to me whether the Tibetans would name it *bse-ru*. Other Tibetan censers are surmounted by a monster of Chinese style, showing a horn on its nose and another on its forehead,— manifestly derived from the two-horned rhinoceros.

ually forgotten by the people, the word *bse* or *bse-ru* of this meaning continued in literature; the people retained the recollection of its being a single-horned animal, and in their attempts at finding this creature, the legend of Hermit Single-Horn, the son of an antelope or gazelle, flashed into their minds; so that the unicorn *bse-ru* was finally identified with a species of antelope named for this reason *bse-ru*. This unicorn *bse-ru* we now recognize also in the Chinese drawing of *Chêng lei pên ts'ao* (Fig. 13). Since the proof is now established that the interaction and intermingling of deer and rhinoceros have taken place in China, in Tibet, and in the West with the first conspicuous allusion in the *Cyranides*,[1] and that this process of adjustment and affiliation has radiated from the Indian legend of Single-Horn born from a gazelle, we are justified in concluding that the foundation, or at least the commencement, of this transformation, must have arisen in India. The development of the matter in Tibet shows sufficiently that Ekaçriṅga is disguised also under our Chinese illustration. So much about the latter.

A most interesting psychological parallel to the representations of the rhinoceros in China is formed by the ostrich. We now know from the reproductions of CHAVANNES[2] that in the T'ang period the ostrich was chiselled in stone in a very naturalistic manner on the imperial burial-places (Fig. 15).[3]

---

[1] A counterpart of the rhinoceros of cervine character occurs also among the Arabs. In Ethiopic, the word *charīsh* corresponds to the *monokeros* of the Septuaginta (JOB, XXXIX, 9), and in all probability signifies the "rhinoceros." According to Qazwīnī, *charīsh* is an animal of the size of a ram, of great strength and swiftness, with a single horn on its forehead like the horn of the rhinoceros (*karkadan*). Some Arabic lexicographers even take it for a marine animal, others identify it directly with the rhinoceros. HOMMEL (Die Namen der Säugetiere bei den südsemitischen Völkern, p. 333, Leipzig, 1879), to whom this information is due, regards the Arabic word as a loan from Ethiopic. Damīrī, in his Lexicon of Animals, avails himself of this word in translating the text of the *Physiologus* regarding the unicorn (K. AHRENS, Das Buch der Naturgegenstände, p. 43). What escaped Hommel is the fact that Cosmas Indicopleustes (McCRINDLE, Ancient India as described in Class. Lit., p. 157) states that the Ethiopians, in their language, call the rhinoceros *arou* or *harisi*. G. JACOB (Studien in arabischen Geographen IV, p. 166, Berlin, 1892) holds that Qazwīnī is the only Arabic author to discriminate between *charīsh* and the rhinoceros, and identifies the former with the Saiga-antelope of southern Russia. The rendering "unicorn" by the Seventy and the English Bible is erroneous. The Hebrew word, thus translated, is *reem*, corresponding to Assyrian *rīmu*. It is now generally interpreted as a wild buffalo, and on the basis of Assyrian monuments is ingeniously identified with *Bos primigenius* by J. U. DÜRST (Die Rinder von Babylonien, pp. 8–11, Berlin, 1899). The animal, called in Hebrew *behemoth* (JOB, XL, 15–24), and formerly taken for the rhinoceros (p. 83), is the hippopotamus of the Nile. The Bible does not mention the rhinoceros or the unicorn.

[2] *Mission archéologique*, Nos. 458, 459, 472, 481.

[3] These ostriches belong to the very best ever executed in the history of art. They are much superior to any representations of the bird by the Egyptians (O. KELLER, Die antike Tierwelt, Vol. II, p. 170), the Assyrians (P. S. P. HANDCOCK, Mesopotamian Archaeology, p. 307), and the classical nations (IMHOOF-BLUMER and O. KELLER, Tier- und Pflanzenbilder auf Münzen und Gemmen, Plates V, 52; XXII, 33–36).

It was the great general and explorer Chang K'ien, the first modern Chinese, who during his peregrinations to the west, among many other novel things, discovered also the ostrich for his compatriots. After he had negotiated his treaties with the countries of the west, the King of Parthia (An-si) sent an embassy to the Chinese Court and presented large bird's eggs,[1] which most probably were ostrich eggs. A live

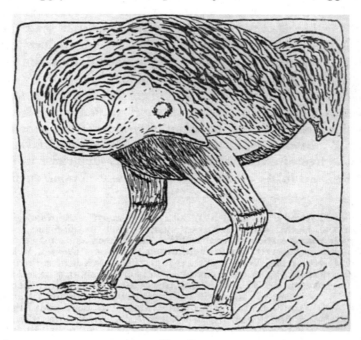

FIG. 15.
Ostrich sculptured in Stone, T'ang Period (Sketch after Chavannes, Mission, No. 472).

specimen (or specimens) of the "large bird of T'iao-chi" was despatched as tribute from the same country in 101 A.D., and termed in China "Parthian bird."[2]

They are not made after any western artistic models, but constitute invincible proof for the fact that the Chinese artists in the T'ang era observed and studied nature, and worked after natural models. This case may be recommended for due consideration to the adherents of the preconceived dogma that all Chinese art is copied from that of the west, and that no art is possible outside of the sanctum of classical art.

[1] *Shi ki*, Ch. 123, p. 6; HIRTH, China and the Roman Orient, p. 169. FORKE (*Mitteilungen des Seminars*, Vol. VII, 1904, p. 139) wrongly says that the *Shi ki* mentions "large birds (ostriches) with eggs as large as earthen pots as a peculiar feature of T'iao-chi;" this is not in the text of the *Shi ki*, which speaks only of large bird's eggs, but it is found in *Ts'ien Han shu* (Ch. 96 A, p. 6 a). The trade in ostrich eggs in the west is of very ancient date (O. KELLER, *l. c.*, p. 168).

[2] *Hou Han shu*, Ch. 118, p. 9; CHAVANNES, *T'oung Pao*, 1907, p. 178. M. CHAVANNES advances the theory that the Chinese erroneously applied to the ostrich the

It was styled also "great horse bird."[1]   Its resemblance to the camel was emphasized, and hence the name "camel-bird" was formed. Living ostriches were sent to China again in the T'ang period.   In 650 Tokhāra offered large birds seven feet high, of black color, with feet resembling those of the camel, marching with outspread wings, able to run three hundred *li* a day, and to swallow iron; they were styled camel-birds.[2]   The T'ang artists, accordingly, were in a position to witness and to study live specimens of the bird; and the fact that they really did so leaks out in the realistic high-relief carvings referred to above. But what do we find among the latter-day draughtsmen who endeavored to illustrate the creature for books?

Fig. 16 shows the woodcut with which the *Pên ts'ao kang mu* of Li Shi-chên is adorned.   BRETSCHNEIDER (*l. c.*), in a somewhat generous spirit, designated it as "a rude, but tolerably exact drawing of the camel-bird."   FORKE [3] holds that this ostrich is pictured like a big goose, but with the feet of a mammal; and he comes far nearer to the truth. Li Shi-chên, born in K'i chou in the province of Hu-pei, spent his life-

---

name "bird of Parthia" (An-si, Arsak), but that in fact these birds originated from T'iao-chi, that is, Desht Misan or Mesene, where ruled Arabic princes who had all facilities for obtaining ostriches from Arabia.   This theory does not seem necessary to me.   As already observed by BRETSCHNEIDER (*Notes and Queries*, Vol. IV, p. 53; and Mediæval Researches, Vol. I, pp. 144–145), the ostrich is described in *Wei shu* as a bird indigenous to Persia (compare also *Sui shu*, Ch. 83, p. 7 b; *Pei shi*, Ch. 97, p. 8), and is again mentioned in the T'ang Annals as a Persian bird; there is, on the other hand, the testimony of the Persian authors and of Xenophon (*Anabasis*, 1, 5), who saw the bird on the banks of the Euphrates; and up to the present time, ostriches are met with, though not frequently, in western Asia.   HANDCOCK (*l. c.*, p. 25) observes that the ostrich appears in Mesopotamian art at a late period, though in Elam rows of ostriches are found depicted on early pottery, closely and inexplicably resembling the familiar ostriches on the pre-dynastic pottery of ancient Egypt; it sometimes, however, assumes a conspicuous position in the embroidery of an Assyrian king's robe, and is found also on a chalcedony seal in Paris.   Further references to Assyrian representations are given by O. KELLER (*l. c.*, pp. 172, 594).   In ancient Syria, the ostrich is well attested by the interesting description in JOB (XXXIX, 13–18), — Moses prohibited the flesh of the bird as unclean food,— and by reliefs at Hierapolis of Roman times.   It further occurs in the Syrian version of the *Physiologus*. BREHM (Tierleben, Vol. III, p. 692) sums up, "In Asia, the area of the habitat of the ostrich may formerly have been much more extended than at present; but even now, as established by Hartlaub with as much diligence as erudition, it occurs in the deserts of the Euphrates region, especially the Bassida and Dekhena, in all suitable localities of Arabia, and finally in some parts of *southern Persia*.   Vámbéry even learned that it is still sometimes found on the lower course of the Oxus, in the region of Kungrad (?), and is named there camel or coffer bird."   Also in the Encyclopædia Britannica (Vol. XX, p. 362) it is said, "It is probable that it still lingers in the wastes of Kirwan in eastern Persia, whence examples may occasionally stray northward to those of Turkestan, even near the lower Oxus."

[1] *Ts'ien Han shu*, Ch. 96 A, p. 6 b.   In this passage the bird is noticed as a native of Parthia, and commented on by Yen Shi-ku.

[2] CHAVANNES, Documents, p. 156.   In the period K'ai-yüan (713–741) ostrich eggs were sent from Sogdiana (*ibid.*, p. 136).

[3] *L. c.*, p. 138.

time as magistrate of the district of P'êng-k'i in the prefecture of T'ung-ch'uan, province of Sze-ch'uan. The chances are that he had never seen the sculptures of ostriches in the mausolea of the T'ang emperors near Li-ts'üan, Shen-si Province; but, be this as it may, his woodcut proves that the T'ang tradition of the representation of the ostrich was wholly unknown to him, and moreover, that he himself had never beheld an ostrich. We have no records to the effect that ostriches were transported to China during the Ming period; and they were then probably known merely by name. Li Shi-chên's production is simply a reconstruction based on the definitions of the texts ("marching with outspread wings, feet of a camel," etc.); the only exact feature is the two toes, which are mentioned also in the older descriptions of the bird; everything else, notably the crane's head, is absurd, and a naturalist of the type of Bretschneider should have noticed this.

FIG. 16.
Ostrich (from *Pên ts'ao kang mu*).

In the great cyclopædia *T'u shu tsi ch'êng*, published in 1726, we find a singular illustration of the ostrich, which is reproduced in Fig. 17 as an object-lesson in Chinese psychology. This accomplishment must open every one's eyes: here we plainly see that the illustrator had not the slightest idea of the appearance of an ostrich, but merely endeavored, with appalling result, to outline a sketch of what he imagined the "camel-bird" should look like. He created a combination of a camel and a bird by illustrating the bare words, as they struck his ears, without any recourse to facts and logic; he committed the logical blunder (so common among the Chinese from the days of the Sung period) of confounding a descriptive point of similarity with a feature of reality. All Chinese texts are agreed on the point that the bird is just like a camel, or conveys that impression. This case is most instructive in disclosing the working of the minds of the recent Chinese illustrators, and in exhibiting the value due to their productions. It would not do in the present case to deny that this figure is intended for an ostrich, to define it as a new animal species, a "bird-shaped biped camel" (something like an *Avi-camelus bipes*), and to conclude that the Chinese term *t'o niao* does not denote the ostrich. On the contrary, we have to conclude that illustrations of this character are out and out valueless for our scientific purposes, that definitions of an animal cannot be deduced from them, but that all reasoning on the nature of the respective animal

FIG. 17.
Alleged Ostrich (from *T'u shu tsi ch'êng*).

can be based solely on the texts.[1] The illustrations are posterior in time and mere accessories, and, even if fairly sensible, of sheer secondary importance; in each and every case, however, if utilized as the basis for any far-reaching conclusion, their history, sources, and psychological foundation must be carefully examined. Another impressive lesson to be derived from the case of the ostrich is that China, which by virtue of a widely accepted school opinion appears to us as the classical soil of ultra-conservative perseverance of traditions, is very liable also to lose traditions, and even rather good ones. The excellent ostrich representations of the T'ang have not been perpetuated, but have remained as isolated instances. Indeed, they seem to have remained unknown to Chinese artists, archæologists, and naturalists, and hidden away in seclusion and oblivion until discovered by M. Chavannes. It is this very China unknown to the Chinese, which, as research advances, will become our most attractive subject of study.

We referred above (p. 100) to the fact that the ancient illustrations to the *Erh ya* are lost, and that Kuo-P'o's sketches of the rhinoceros may have been nearer to the truth. In now raising the question whether any representations of the animal are handed down in the ancient monuments of China, we naturally remember the primeval form of writing that mirrors the stage of her primitive culture. The celebrated Catalogue of Bronzes, the *Po ku t'u lu*, published by Wang Fu in the period Ta-kuan (1107–1111), has preserved to us (Ch. 9, p. 23) two ancient symbols which are veritable representations of the single-horned rhinoceros *se* (Fig. 18). They are placed on the ends of a handle of a bronze wine-kettle attributed to the Shang dynasty (B.C. 1766–1154). The explanatory text runs as follows: "The two lateral ears of the vessel are connected by a handle, on which are chased two characters in the shape of a rhinoceros (*se*). When it is said in the *Lun yü* that 'a tiger and rhinoceros escape from their cage,'[2] it follows that the rhinoceros is

---

[1] And these must certainly be handled with a critical mind, as, for instance, a glance at the chapter "Ostrich" in the *T'u shu tsi ch'êng* will convince one. The first extract there given from the *Ying yai shêng lan* of 1416 deals with the "fire-bird" of Sumatra, which is the cassowary (see GROENEVELDT, in *Miscell. Papers relating to Indo-China*, Vol. I, pp. 198, 262). *Mo k'o hui si*, a work written by P'êng Ch'êng in the first half of the eleventh century (BRETSCHNEIDER, Bot. Sin., pt. 1, p. 174), is quoted as making a contribution to the subject in question, because a bird able to eat iron and stone is mentioned there; this bird, however, called *ku-t'o*, occurs in Ho-chou, the present Lan-chou fu in Kan-su, is built like an eagle, and over three feet high! Accordingly we here have a wrong association of ideas, and the subject has nothing to do with the ostrich. The editors of the cyclopædia blindly follow the uncritical example of Li Shi-chên, who embodied the same in his notes on the ostrich. Finally, Verbiest's *K'un yü t'u shuo* is laid under contribution, as he describes the "camel-bird" of South America. This is the Rhea belonging to the Ratite family, but distinguished from the true ostrich by its possession of three toes.

[2] LEGGE, Chinese Classics, Vol. I, p. 306; and above, p. 74, note 4.

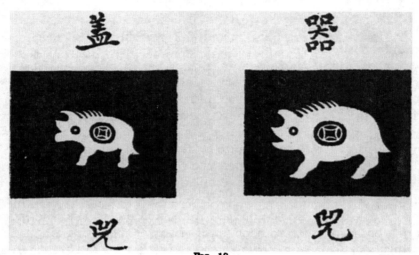

FIG. 18.
Single-Horned Rhinoceros on a Bronze Kettle attributed to Shang Period (from
Po ku t'u lu, edition of 1603).

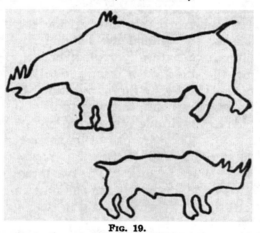

FIG. 19.
Bushman Sketches of Rhinoceros (from E. Cartailhac and H. Breuil,
La caverne d'Altamira, pp. 180, 189).

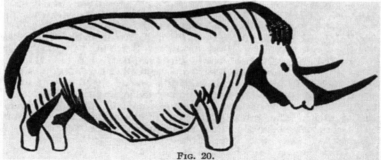

FIG. 20.
Red Drawing of a Two-Horned Rhinoceros, from Font-de-Gaume (after Capitan and Breuil).

not a tame animal. Indeed, it inflicts injury on man; and for this reason the ancients availed themselves of it to fine a person a cup of wine, which is expressed by the phrase 'to raise the goblet of rhinoceros-horn.'[1] This goblet receives its name from the rhinoceros, and so it is proper also that there should be wine-kettles with the emblem of the rhinoceros. On the two ends of the handle of this vessel is pictured a rhinoceros with head and body complete, the latter having the shape of a glutton (*t'ao t'ie*). This certainly indicates that it symbolizes a warning. In this manner all vessels were decorated during the Shang dynasty, and it is by such symbolic forms that they are distinguished from those of the Chou." Whatever the rough character of these two sketches transmitted by the *Po ku t'u lu* may be,[2] the single-horned rhinoceros is here clearly outlined with a naïve and refreshing realism, such as could be spontaneously produced only by the hand of primitive man, who with a few forceful outlines recorded his actual experience of the animal. Here we do not face the narrow-breasted academic and philological construction of the scholars of the Sung period, but the direct and vigorous impression of the strong-minded hunter of past ages, who was formed of the same stuff as the Bushman of southern Africa and palæolithic man living in the caves of Spain and France. No bridge spans the chasm yawning between the Shang and Sung productions. The Shang rhinoceros breathes the same spirit as its companions on the rock paintings of the Bushman (Fig. 19), and in the palæolithic cave of Font-de-Gaume in France (Fig. 20). The general form of the

FIG. 21.
Inscription on Bronze Kettle attributed to Shang Period, showing Pictorial Form of Sacrificial Bull (from *Po ku t'u lu*).

---

[1] Quotation from *Shi king* (see LEGGE, Chinese Classics, Vol. IV, p. 233). The rhinoceros-horn goblets are discussed below, p. 167.

[2] Another cruder and more conventionalized symbol of the rhinoceros *se*, in which, however, the single horn is duly accentuated, is figured in the same work (Ch. I, p. 25 b), as occurring in the inscription on a round tripod vessel (*ting*) attributed to the Shang period.

animal is well grasped in the Chinese sketch, and the shape of the horn is correctly outlined. For the sake of comparison, and in order to show that the primitive Chinese man knew very well how to discriminate between a rhinoceros and an ox, the contemporaneous symbol for the sacrificial bull (*hi niu*), and designs of recumbent oxen (explained as such in the *Po ku t'u lu*) on the lid of a bronze vessel, are here added (Figs. 21 and 22). We arrive at the result, which will be corroborated by other evidence, that in the earliest stage of Chinese culture the animal *se* was the single-horned rhinoceros.[1]

FIG. 22.
Lid of Bronze Kettle attributed to Shang Period, with Designs of Recumbent Oxen (from *Po ku t'u lu*).

Before plunging into the Chinese sources relative to the rhinoceros, it will be well to remember that all living species of rhinoceros are by most naturalists referred to a single genus, which is found living in Africa and south-eastern Asia, while formerly it was widely distributed over the entire Old World (with the exception of Australasia), ranging as far north as Siberia.[2] Three species exist in Asia, — *Rhinoceros unicornis*, the great one-horned rhinoceros, at the present day almost entirely restricted to the Assam plain, but formerly extensively distributed over India;[3] *Rhinoceros sundaicus*, called also the Javan rhinoceros, the smaller one-horned rhinoceros, found in parts of eastern Bengal (the Bengal Sunderbans near Calcutta), in Assam, throughout Burma, the Malay Peninsula, Sumatra, Java, and Borneo; and *Rhinoceros* (or *Dicerorhinus*) *sumatrensis*, the Asiatic two-horned rhinoceros, rare in Assam, ranging from there to Burma, Siam, the Malay Peninsula,

[1] The later developments of the early forms of the symbol *se* may be viewed by those who are debarred from Chinese sources in F. H. CHALFANT, Early Chinese Writing, Plate II, No. 17 (*Memoirs Carnegie Museum*, Vol. IV, No. 1, Pittsburgh, 1906). According to a communication of the late Mr. CHALFANT (Dec. 18, 1913), the ancient bone inscriptions twice reveal a character which may be identified with the word *se*, while the character for *si* has not yet been traced in them.

[2] Hornless species formerly occurred in North America, where the group has existed since the latter part of the Eocene period.

[3] Chiefly after W. T. BLANFORD, The Fauna of British India; Mammalia, pp. 471–477.

Sumatra, and Borneo.[1] Judging from this remarkable case of discontinuous distribution[2] and from historical records, there is every reason to believe that in ancient times this animal, like all the large mammals now facing extinction, was distributed over a much larger geographical area; and this fact is fully confirmed by palæontological research, as well as by the records of the Chinese.

For the purpose of our inquiry it should be particularly borne in mind that it is in the territory of Assam where we meet the three species together. "The Imperial Gazetteer of India"[3] states, in the chapter on Assam, "Rhinoceros are of three kinds: the large variety (*unicornis*), which lives in the swamps that fringe the Brahmaputra; the smaller variety (*sondaicus*), which is occasionally met with in the same locality; and the small two-horned rhinoceros (*sumatrensis*), which is now and again seen in the hills south of the Surmā Valley, though its ordinary habitat is Sumatra, Borneo, and the Malay Peninsula." Assam is inhabited by numerous tribes, a large portion of which ranges among the Indo-Chinese family. What now holds good for Assam, as will be recognized from a survey of Chinese sources, two millenniums and more ago was valid for the south-western and southern parts of China, the Tibeto-Chinese borderlands, and Indo-China in its total range; in short, the historical fact will be established that in the past the rhinoceros in its two main varieties, the single-horned and two-horned, had occupied the whole territory of south-eastern Asia.

The greater part of the knowledge possessed by the Chinese in regard to the rhinoceros has been digested by Li Shi-chên in his materia medica *Pên ts'ao kang mu* (Ch. 51 A, p. 5) completed in 1578 after twenty-six years' labor. He first quotes a number of authors beginning from the fifth century, and then sums up the argument in his own words. This discourse is also of value for zoögeography, in that it contributes materially to the possibility of reconstructing the early habitats of the rhinoceros in China. The text of this work is here translated *in extenso*, but rectified and supplemented from the materia medica of the Sung period, the *Chêng lei pên ts'ao*, first printed in 1108.[4]

---

[1] Al-Bērūnī (973–1048) states that the rhinoceros existed in large numbers in India, more particularly about the Ganges (SACHAU, *l. c.*, Vol. I, p. 203). In the sixteenth century it occurred in the western Himālaya and also in the forests near Peshāwar (YULE and BURNELL, Hobson-Jobson, p. 762). LINSCHOTEN found it in great numbers in Bengal (*ibid.*, p. 1); so also GARCIA AB HORTO (*l. c.*, p. 66): multos in Cambaya Bengala finitima, et Patane inveniri tradunt. ABUL FAZL ALLAMI (1551–1602), in his *Ain I Akbari* written in 1597 (translation of H. S. JARRETT, Vol. II, p. 281, Calcutta, 1891), mentions the occurrence of the rhinoceros among the game in the Sarkár of Sambal (near Delhi).

[2] Compare E. HELLER, The White Rhinoceros, p. 39.

[3] Vol. VI, p. 20 (Oxford, 1908).

[4] See *T'oung Pao*, 1913, p. 351.

禽蟲典第六十八卷

犀兜部彙考

釋名

犀 周禮	兒 爾雅	通天犀 抱朴子	駭雞犀 抱朴子
水犀 廣州記	却塵犀 述異記	辟寒犀 開元遺事	蠲忿犀 杜陽雜編
奴角 酉陽雜俎	辟塵犀 嶺表錄異記	胡帽犀 嶺表錄異記	喧羅犀 嶺表錄異記
光明犀 嶺表錄異記	骨咄犀 雲煙過眼錄	竭伽 湄水燕談錄	毗沙堅 犀角 湄水燕談錄

Other texts of importance apt to throw light on the matter have been added from the *T'u shu tsi ch'êng* and several other works, so that the result is a fairly complete digest of what Chinese authors of the post-Christian era have to say about the rhinoceros and its horn. After this survey, we shall turn to the times of early antiquity, and discuss the subject in the light of such information as has been handed down to us from those days.

Li Shi-chên opens his discourse on the rhinoceros with the explanation of the name. "The symbol for the word *si* still has in the seal character *chuan wên* the form of a pictograph,[1] and is the name for the female rhinoceros. The *se* is styled also 'sand rhinoceros' (*sha si*). The *Erh ya i*[2] says that the words *se* and *tse* (female) approach each other in sound like the two words *ku* ('ram,' No. 6226) and *ku* ('male'). In general, *si* and *se* are one and the same. The ancients were fond of saying *se*, the people of subsequent times inclined toward the word *si*. In the northern dialects the word *se* prevails, in the southern dialects the predilection is for *si*. This is the difference between the two. In Sanskrit literature the rhinoceros is called *khaḍga*." [3]

Li Shi-chên then proceeds to quote the ancient work *Pie lu*,[4] which makes the following important statement in regard to the former localities where the rhinoceros occurred: "The habitat of the rhinoceros

---

[1] This is indeed the case in the *Shuo wên* (see p. 92). The names of the rhinoceros and the various kinds of its horn are here reproduced from *T'u shu tsi ch'êng* (p. 134).

[2] An appendix to the *Erh ya* by Lo Yüan of the twelfth century (BRETSCHNEIDER, Bot. Sin., pt. 1, p. 37).

[3] Written with Nos. 1456 and 1558 (*k'et-ga*); compare EITEL, Hand-book of Chinese Buddhism, p. 76. (Other Sanskrit words for "rhinoceros" are *gaṇḍa, gaṇḍaka, gaṇḍānga*.) The work *Sheng shui yen t'an lu*, written by Wang P'i-chi about the end of the eleventh century (WYLIE, Notes, p. 195), seems to be the first to impart this Sanskrit name (see the Chinese text opposite); it further gives a Sanskrit word for the horn in the Chinese transcription *pi-sha-na* corresponding to Sanskrit *vishāṇa* ("horn"). The latter and the word *khaḍga* were among the first Sanskrit words in Chinese recognized by Abel Rémusat (see S. JULIEN, Méthode, p. 3).

[4] The *Pie lu* is not identical with the *Ming i pie lu*, as first stated by BRETSCHNEIDER (Bot. Sin., pt. 1, p. 42), but later rectified by him (in pt. 3, p. 2). It is an independent work, which must have existed before the time of T'ao Hung-king, and which was known to the latter and commented on by him. This is quite clear in the present case, as Li Shi-chên first introduces the *Pie lu*, and then proceeds, "T'ao Hung-king says." And since the latter starts with the phrase "at present," it is apparent that he had the words of the *Pie lu* before his eyes, and gave his definition in distinction from the older work. This is also proved by the text of the *Chêng lei pên ts'ao* published in 1108 by the physician T'ang Shên-wei (edition of 1523, Ch. 17, fol. 21), where the two quotations are separated and marked by type of different size. As in Bretschneider's opinion nearly all the geographical names occurring in the *Pie lu* refer to the Ts'in (third century B.C.) or Han periods, although some of them can be traced to the Chou dynasty (B.C. 1122-249), the above passage surely relates to a time antedating our era by several centuries; and it goes without saying, that as a matter of fact, in the age of the Chou and at a far earlier date, the two-horned rhinoceros must have been a live citizen in the south-western parts of China.

(*si*) is in the mountains and valleys of Yung-ch'ang and in Yi-chou;[1] Yung-ch'ang is the southern part of the present country of Tien (Yün-nan)."[2]

The next author invoked by Li Shi-chên is T'ao Hung-king (452–536), a celebrated adept of Taoism and a distinguished physician, author of the *Ming i pie lu*, a treatise on materia medica.[3] He states, "At present the rhinoceros (*si*) inhabits the distant mountains of Wu-ling,[4] Kiao-chou,[5] and Ning-chou.[6] It has two horns; the horn on the forehead is the one used in fighting.[7] There is a kind of rhinoceros styled 'communicating with the sky' (*t'ung t'ien*), whose horn is intersected by a white vein running clear through from the base to the tip; the night dew does not moisten it. It is employed as a remedy, whereby its wonderful properties are tested. In the opinion of some, this is the horn of the water-rhinoceros, which is produced in the water.[8] The Annals of the Han Dynasty speak of the horn of 'the rhinoceros frightening fowl' (*hiai ki si*): when it was placed in the rice that served as food for the chickens, they were all scared and did not dare to peck;

---

[1] PLAYFAIR, The Cities and Towns of China, No. 8596 (2d ed., No. 7527, 1). In the Han period, Yi-chou was the name of a province occupying the territory of the present province of Sze-ch'uan, a part of Kuei-chou and Yün-nan (BRETSCHNEIDER, Bot. Sin., pt. 3, p. 565), while the southern part of Yün-nan is understood by the designation Yung-ch'ang. The *Pie lu*, accordingly, locates in south-western China the rhinoceros *si*, which, as follows from the comment of T'ao Hung-king, is the two-horned species.

[2] This last clause is not contained in the text of the *Chêng lei pên ts'ao*, and is doubtless a later comment, presumably derived from T'ao Hung-king's edition of the *Pên ts'ao king*, which is listed in the Catalogue of the Sui Dynasty, and according to Bretschneider's supposition, embraced likewise the text of the *Pie lu*.

[3] His biography is in *Nan shi* (Ch. 76, p. 4 b) and *Liang shu* (Ch. 51, p. 12).

[4] PLAYFAIR, No. 8112 (2d ed., No. 7080): district forming the prefectural city of Ch'ang-tê, Hu-nan Province.

[5] Northern part of the present Tonking (see HIRTH and ROCKHILL, Chau Ju-kua, p. 46).

[6] PLAYFAIR, No. 5239, 2 (4672, 2): in Lin-an fu, Yün-nan Province. Under the Tsin it was a province comprising Yün-nan and part of Kuei-chou (compare *Hua yang kuo chi*, Ch. 4, p. 1, ed. of *Han Wei ts'ung shu*).

[7] Thus the two-horned (so-called Sumatran) rhinoceros is here clearly mentioned.

[8] The rhinoceros is fond of spending the hot hours of the day immersed in water, and thence the Chinese designation "water-rhinoceros" may take its origin. In this position particularly, the animal calls to mind the water-buffalo. In ancient times it was therefore dreaded as being able to overturn boats, which is quite believable; and soldiers crossing a river were encouraged to prompt action by their commander shouting the name of the animal (CHAVANNES, Les Mémoires historiques de Se-ma Ts'ien, Vol. I, p. 225, Vol. IV, p. 37; FORKE, Lun-Hêng, pt. II, p. 322; according to FORKE, the reading of the text is *ts'ang kuang*, but as quoted in *T'u shu tsi ch'êng* and *P'ei wên yün fu* it is *ts'ang se*, as in *Se-ma Ts'ien*). The water-rhinoceros (*shui si*) is mentioned in *Kuang chou ki* (see BRETSCHNEIDER, Bot. Sin., pt. 1, No. 377) as occurring in the open sea off the district of P'ing-ting, resembling an ox, emitting light when coming out of, or descending into, the water, and breaking a way through the water (quoted in *T'u shu tsi ch'êng*).

when it was placed on the roof of a house, the birds did not dare to assemble there.[1] There is also the horn of the female rhinoceros, which is very long, with patterns resembling those of the male, but it is not fit to enter the pharmacopœia." [2]

---

[1] The allusion to the *hiai ki si* occurs in Ch. 108 of *Hou Han shu* (compare CHAVANNES, Les pays d'Occident d'après le Heou Han Chou, *T'oung Pao*, 1907, p. 182; and HIRTH, China and the Roman Orient, p. 79), where this kind of horn is ascribed to the country of Ta Ts'in (the Roman Orient). The legend given in explanation as above is derived from the famous Taoist writer Ko Hung, who died about 330 A.D.; and it is not accidental that the Taoist T'ao Hung-king here copies his older colleague, for the legend is plainly Taoistic in character. It is quoted in the commentary to *Hou Han shu*, but not in the text of the Annals. The view of Hirth, that it has arisen in consequence of a false etymology based on the Chinese characters transcribing a foreign word, seems to me unfounded. First, as Chavannes remarks, the foreign word supposed to be hidden in *hiai-ki* has not yet been discovered, and in all probability does not exist. Second, as will be seen from *P'ei wên yün fu* (Ch. 8, p. 87 b), the term *hiai ki si* does not occur in *Hou Han shu* for the first time, but is noted as early as the *Chan kuo ts'e* at the time of Chang I, who died in B.C. 310, when the King of Ch'u despatched a hundred chariots to present to the King of Ts'in fowl-scaring rhinoceros-horns and jade disks resplendent at night (*ye kuang pi*). It is certainly somewhat striking to meet here these two names, which are identical with those in *Hou Han shu*, and occur there close together; and it cannot be denied that the passage of *Chan kuo ts'e* might be an interpolation. Huai-nan-tse, who died in B.C. 122, alludes to a rhinoceros-horn frightening foxes (*si kio hiai hu*, quoted in *P'ei wên yün fu, l. c.*, p. 89 a, "when placed in the lair of a fox, the fox does not dare return"), which is a case analogous in word and matter to the fowl-frightening horn. These notions must be taken in connection with the other legends regarding the rhinoceros, which all seem to spring from indigenous Taoist lore. The text of Ko Hung, as quoted in *P'ei wên yün fu* and translated by Hirth and Chavannes, is fuller than cited above in the *Pên ts'ao*, while the final clause in regard to placing the horn on the roof does not occur in Ko Hung. The latter links the *hiai ki si* with the *t'ung t'ien*, which Hirth and Chavannes translate "communicating with Heaven." This is certainly all right; but I prefer to avoid this term, because it may give rise to misunderstandings, as we are wont to think of Heaven as the great cosmic deity. A comparative study of all passages concerned renders it clear that the rhinoceros is not associated with spiritual, but with material heaven; that is, the sky. It is the stars of the sky which are supposed to be reflected in the veins of the horn. This means that the designs of the horn gave the impetus to the conception of connecting the rhinoceros with the phenomena of the sky,— again a thoroughly Taoistic idea, in which no trace of an outside influence can be discovered. Father ZOTTOLI (Cursus litteraturae sinicae, new ed., Vol. I, p. 301) renders the term *t'ung t'ien si tai* by "penetrantis coelum rhinocerotis cingulum."— Chao Ju-kua (HIRTH's and ROCKHILL's translation, p. 103) attributes *hiai ki si* or *t'ung t'ien si* also to Baghdad (but I see no reason why these words should denote there a precious stone, instead of rhinoceros-horn). On p. 108 (note 10) the two authors represent the matter as though this reference might occur in *Ling-wai tai ta*, but in fact it is not there (Ch. 3, p. 1 b); it must therefore be due to Chao Ju-kua, who seems to indulge in a literary reminiscence taken from *Hou Han shu*. The passage, accordingly, affords no evidence for a trade in rhinoceros-horns from Baghdad to China, which *per se* is not very likely.—In the illustrations to the *Fêng shên yen i* (ed. of *Tsi ch'êng t'u shu*, p. 9, Shanghai, 1908), T'ung t'ien kiao chu (see W. GRUBE, Die Metamorphosen der Götter, p. 652) is seated astride a rhinoceros (outlined as a bull with a single striped horn), apparently because his name T'ung t'ien has been identified with *t'ung t'ien si*.

[2] There are several additions to this text as edited in the *Chêng lei pên ts'ao*, the most interesting of which is that "only the living horns are excellent." This means the horn of a live animal slain in the chase, which was believed to be superior in quality to a horn cast off and accidentally found (compare HIRTH and ROCKHILL, Chau Ju-kua, p. 233). Similar beliefs prevailed in regard to ivory. That coming from the tusk of an elephant killed by means of a pike was considered the best; next in quality

Li Shi-chên does not refer to Ko Hung, the famous Taoist adept of the fourth century,[1] who is the first author to impart a fantastic account in regard to rhinoceros-horn. He is likewise the first to set forth its quality of detecting poison. His text is here translated, as given in *T'u shu tsi ch'êng*.[2]

"Mr. Chêng[3] once obtained a genuine rhinoceros-horn of the kind 'communicating with the sky,' three inches long, the upper portion being carved into the form of a fish. When a man carries such a piece in his mouth and descends into the water, the water will give way for him and leave a vacant space three feet square, so that he has a chance to breathe in the water.[4] The horn 'communicating with the sky' has a single red vein like a silk string running from the base to the tip. When a horn filled with rice is placed among a flock of chickens, the chickens want to peck the grains. Scarcely have they approached the horn to within an inch when they are taken aback and withdraw. Hence the people of the south designate the horn 'communicating with the sky' by the name 'fowl-frightening horn.' When such a horn is placed on a heap of grain, the birds do not dare assemble there. Enveloped by a thick fog or exposed to the night dew, when placed in a courtyard, the horn does not contract humidity. The rhinoceros (*si*) is a wild animal living in the deep mountain-forests. During dark nights its horn emits a brilliant light like torch-fire. The horn is a safe guide to tell the presence of poison: when poisonous medicines of liquid form are stirred with a horn, a white foam will bubble up, and no other test is necessary; when non-poisonous substances are stirred with it, no foam will rise. In this manner the presence of poison can be ascertained. When on a journey in foreign countries, or in places where contagion from *ku*

---

was the ivory of an animal which was found shortly after it had died a natural death; least esteemed was that discovered in mountains many years after the animal's death (PELLIOT, *Bulletin de l'Ecole française d'Extrême-Orient*, Vol. II, 1902, p. 166). In Siam, the rhinoceros is still killed with bamboo pikes hardened in the fire and thrust into its jaws and down the throat, as described by Bishop PALLEGOIX (Description du royaume Thai ou Siam, Vol. I, p. 75, Paris, 1854).

[1] He died in 330 A.D. at the age of eighty-one; see GILES (Biographical Dictionary, p. 372); MAYERS (Chinese Reader's Manual, p. 86); BRETSCHNEIDER (Bot. Sin., pt. 1, p. 42); and PELLIOT (*Journal asiatique*, 1912, Juillet-Août, p. 145).

[2] Chapter on Rhinoceros (*hui k'ao*, p. 3), introduced by the author's literary name Pao-p'u-tse, and the title of his work *Têng shê p'ien*, which is not included in the Taoist Canon.

[3] Presumably Chêng Se-yüan, a relative and spiritual predecessor of Ko Hung (L. WIEGER, Taoisme, Vol. I, Le canon, p. 16; PELLIOT, *l. c.*, p. 146).

[4] It is interesting to note that this belief is still upheld in the modern folk-lore of Annam: "Celui qui peut se procurer une corne de rhinocéros et la sculpte en forme de poisson, s'il la met entre ses dents, peut descendre sans danger, comme le rhinocéros ou le poisson, tout au fond de l'eau" (P. GIRAN, Magie et Religion Annamites, p. 104, Paris, 1912).

poison [1] threatens, a man takes his meals in other people's houses, he first ought to stir his food with a rhinoceros-horn. When a man hit by a poisonous arrow is on the verge of dying, and his wound is slightly touched with a rhinoceros-horn, foam will come forth from his wound, and he will feel relief.[2] This property of the horn 'communicating with the sky' of neutralizing poison is accounted for by the fact that the animal, while alive, particularly feeds on poisonous plants and trees provided with thorns and brambles,[3] while it shuns all soft and smooth vegetal matter. Annually one shedding of its horn takes place in the mountains, and people find horns scattered about among the rocks;[4] in this case, however, they must deposit there, in the place of the real one, another horn carved from wood, identical with that one in color, veins, and shape. Then the rhinoceros remains unaware of the theft. In the following year it moves to another place to shed its horn.[5] Other kinds of rhinoceros-horn also are capable of neutralizing poison, without having, however, the wonderful power of the *t'ung-t'ien* variety."

Su Kung, the editor of the *T'ang sin pên ts'ao* (the revised edition of the materia medica of the T'ang dynasty) states as follows: "The *tse* (No. 12,325) is the female rhinoceros. The patterns on its horn are smooth, spotted, white, and clearly differentiated. It is ordinarily called the 'spotted rhinoceros' (*pan si*). It is highly esteemed in pre-

---

[1] See *T'oung Pao*, 1913, p. 322.

[2] The belief that the horn will check the effects of poisoned arrows is repeated in the *Pei hu lu*, written by Tuan Kung-lu around 875 in the T'ang period (PELLIOT, *Bulletin de l'Ecole française*, Vol. IX, 1909, p. 223). The notes of this book regarding the horn are all based on the text of Ko Hung; instead of *t'ung t'ien si*, the term *t'ung si* is employed.

[3] The animal feeds, indeed, on herbage, shrubs, and leaves of trees.

[4] The supposition of the rhinoceros shedding its horn regularly has not been ascertained by our zoölogists; but it is not very probable that it does so, nor have the Chinese made the actual observation. It is clear that their conclusion is merely based on the circumstantial evidence of detached horns occasionally found and picked up in the wilderness, which suggested to them the notion of a natural process similar to the shedding of cervine antlers.

[5] A similar story is told in regard to the elephant by Chên Kūan, who wrote two treatises on the medical virtues of drugs, and who died in the first part of the seventh century (BRETSCHNEIDER, Bot. Sin., pt. I, p. 44): "The elephant, whenever it sheds its tusks, itself buries them. The people of K'un-lun make wooden tusks, stealthily exchange them, and take the real ones away." K'un-lun is the Chinese designation for the Malayan tribes of Malacca, and was extended to Negrito, Papua, and the negroes of Africa (see HIRTH and ROCKHILL, Chau Ju-kua, p. 32). In this connection we should remember also the words of PLINY (*Nat. hist.*, VIII, 3, §7), that the elephants, when their tusks have fallen out either accidentally or from old age, bury them in the ground (quam ob rem deciduos casu aliquo vel senecta defodiunt). It is not impossible that the great quantity of fossil ivory mentioned as early as by THEOPHRAST (*De lapidibus* 37, Opera ed. F. WIMMER, p. 345; compare the interesting notes of L. DE LAUNAY, Minéralogie des anciens, Vol. I, pp. 387–390, Bruxelles, 1803) may have given rise to this notion.

scriptions, but is not such an efficient remedy as the horn of the male rhinoceros."[1]

Ch'ên Ts'ang-k'i, who lived in the first half of the eighth century, states in his work *Pên ts'ao shi i* ("Omissions in Previous Works on Materia Medica") as follows: "There are not two kinds of the rhinoceros, called the land and water animal. This distinction merely refers to finer and coarser qualities of horns.[2] As to the rhinoceros 'communicating with the sky,' the horn on its skull elongates into a point after a thousand years. It is then adorned, from one end to the other, with white stars, and can exhale a vapor penetrating the sky; in this manner it can communicate with the spirits,[3] break the water, and frighten fowl. Hence the epithet 'communicating with the sky' is bestowed on it. Pao-p'u-tse[4] says, 'When such a rhinoceros-horn is carved into the shape of a fish, and one holding this in his mouth descends into water, a passage three feet wide will open in the water.'"[5]

Su Sung, author of the *T'u king pên ts'ao*, published by imperial order in the age of the Sung dynasty, has the following: "Of rhinoceros-horn, that coming from the regions of the Southern Sea (*Nan hai*) takes the first place; that from K'ien and Shu[6] ranks next. The rhinoceros resembles the water-buffalo, has the head of a pig, a big paunch, short legs, the feet being similar to those of the elephant and having three toes. It is black in color, and has prickles on its tongue. It is fond of eating thorny brambles.[7] Three hairs grow from each pore in its skin,

---

[1] Li Shi-chên's text exactly agrees with that given in the *Chêng lei pên ts'ao*. It is an interesting coincidence that the horn of the female rhinoceros (*tse si kio*) is mentioned in the Annals of the T'ang Dynasty (*T'ang shu*, Ch. 40, p. 6 b) as the tribute sent from the district of Si-p'ing in Shen chou, the present territory of Si-ning in Kan-su. The Annals therefore confirm the statement of the contemporaneous *Pên ts'ao*.

[2] It will be seen below that Li Shi-chên does not share this opinion.

[3] The same paragraph is found in Li Shi, the author of the *Sü po wu chi* (Ch. 10, p. 8 b; ed. of *Pai hai*), ascribed by tradition to the T'ang period, but in fact coming down from the Sung. He interprets the expression *t'ung t'ien* by the words, "It is capable of communicating with the spirits" (*nêng t'ung shên*). According to him, "the horn communicating with the sky" is a thousand years old, long and pointed, overstrewn with white stars, the tip emitting a vapor.

[4] Surname of Ko Hung, a famous Taoist writer, who died at the age of eighty-one about 330 A.D. (see p. 138).

[5] The text in the *Chêng lei pên ts'ao* is somewhat fuller. It opens by saying that the flesh of the rhinoceros cures all poisons, especially poisoning caused by the bites of snakes and mammals. On Java bits of the horn are considered as an infallible antidote against snake-bites (P. J. VETH, Java, Vol. III, p. 289). At the close of Ch'ên Ts'ang-k'i's text it is added that the horn is called also *nu kio* (literally, "slave horn") and *shi kio* ("the horn, with which the animal feeds"); the word *nu* seems to be the transcription of a word from a non-Chinese language.

[6] Ancient designations for the present territory of the provinces of Kuei-chou and Sze-ch'uan.

[7] The entire definition, except the "prickles on the tongue," is derived from Kuo P'o (see p. 93). MARCO POLO (ed. of YULE and CORDIER, Vol. II, p. 285), speaking of

as in swine. There are one-horned, two-horned, and three-horned ones." [1]

---

the rhinoceros on Java, says, "They do no mischief, however, with the horn, but with the tongue alone; for this is covered all over with long and strong prickles [and when savage with any one they crush him under their knees and then rasp him with their tongue]." YULE comments that the belief in the formidable nature of the tongue of the rhinoceros is very old and widespread, though he can find no foundation for it other than the rough appearance of the organ. Dr. PARSONS (p. 9 in the pamphlet quoted above, p. 83) observes, "As to the tongue of the rhinoceros, the scribes assure us that it is so rugged that it can lick off with it the flesh from the bones of a man, but the tongue of the live animal examined by me is as soft and mild as that of a calf; whether it will grow rougher with the advancing age of the animal, I am unable to say." It is easy to see how the fable of the prickly tongue arose. The animal mainly feeds on herbage, and the alleged or real observation of its inclination for brambles led to the conclusion that its tongue must be thorn-proof and prickly. A similar belief seems to obtain in Siam: "On dit que ce monstrueux quadrupède fait ses délices des épines de bambou" (Mgr. PALLEGOIX, Description du royaume Thai ou Siam, Vol. I, p. 156, Paris, 1854).

[1] Now follows in the Pên ts'ao the quotation from the Erh ya translated above (p. 93). The text then following in the Pên ts'ao is purported to be a quotation from Ling piao lu i; but it is in fact abridged, and intermingled with extracts from Yu yang tsa tsu. For this reason I have abandoned at this point the text of the Pên ts'ao, and given separately translations of the two documents, as they are published in T'u shu tsi ch'êng (Chapter on Rhinoceros, hui k'ao, p. 4). In evidence of my statement, the text of the Pên ts'ao here follows; the main share in the confusion will probably be due to Su Sung, not to Li Shi-chên. "The Ling piao lu i by Liu Sün (of the T'ang period) says, 'The rhinoceros has two horns: the one on the forehead is called se si, the other, on the nose, is called hu mao si. The male rhinoceros also has two horns both of which are comprised under the name mao si ('hairy rhinoceros'). At present people uphold the opinion that it has but a single horn. These two kinds of horn are provided with grain patterns, and their price largely depends upon the finer or coarser qualities of these designs. The most expensive is the horn with floral designs of the rhinoceros 'communicating with the sky.' The animals with such horns dislike their own shadow, and constantly drink muddy water in order to avoid beholding their reflection. High-grade horns bear likenesses of all things. Some attribute the qualities of the t'ung t'ien horn to a pathological cause, but the natural reason cannot be ascertained. The term tao ch'a means that one half of the lines pass through in the direction downward; the term chêng ch'a means that one half of the lines pass through in the direction upward; the term yao ku ch'a means that the lines are interrupted in the middle, and do not pass through. Such-like are a great many. The Po-se designate ivory as po-ngan, and rhinoceros-horn as hei-ngan,— words difficult to distinguish. The largest rhinoceros-horn is that of the to-lo-si, a single horn of which weighs from seven to eight catties. This is identified with the horn on the forehead of the male rhinoceros. It has numerous decorations conveying the impression of scattered beans. If the specks are deep in color, the horn is suitable to be made into plaques for girdle-ornaments; if the specks are scattered here and there, and light in color, the horn can be made only into bowls and dishes. In the opinion of some, the animal called se is the female of the si. [It resembles the water-buffalo, and is of dark color. Its hide is so hard and thick that it can be worked into armor.] I do not know whether this is the case or not." (There is here a confusion in Li Shi-chên's text. The passage enclosed in brackets does not occur in the text of the Chêng lei pên ts'ao, where it runs, "In the opinion of some, the animal called se is the female of the si; I do not know whether this is the case or not." The rest is evidently interpolated, and is derived from the Shuo wên and its commentaries; at all events, it cannot be ascribed to Su Sung.) "Wu Shi-kao, a physician of the T'ang period, tells the following story: 'The people near the sea, intent on capturing a rhinoceros, proceed by erecting on a mountain-path many structures of decayed timber, something like a stable for swine or sheep. As the front legs of the rhinoceros are straight, without joints, it is in the habit of sleeping by leaning against the trunk of a tree. The rotten

The *Ling piao lu i ki*[1] says, "The rhinoceros, in general, resembles an ox in form. Its hoofs and feet are like those of the elephant. It has a double armor and two horns. The one on the forehead is styled *se si*; the other, on the nose, which is comparatively smaller, is termed *hu mao si*.[2] The designs and spots in the anterior horn are small; many have extraordinary patterns. The male rhinoceros likewise has two horns, both of which are designated *mao si* ('hairy rhinoceros'), and are provided with grain patterns.[3] They are capable of being worked into plaques for girdles.[4] Among a large number of rhinoceros-horns there

---

timber will suddenly break down, and the animal will topple in front without being able for a long time to rise. Then they attack and kill it.'" The conclusion is translated above in the text.

[1] In the *Pên ts'ao*, and otherwise, usually styled *Ling piao lu i*. According to BRET-SCHNEIDER (Bot. Sin., pt. 1, p. 170), it is an account of the natural productions of China by Liu Sün of the T'ang dynasty.

[2] HIRTH and ROCKHILL (Chau Ju-kua, p. 233), briefly alluding to this text, under-stand the terms *se si* and *hu mao si* as two different varieties of the rhinoceros. This point of view seems to me inadmissible, as Liu Sün distinctly speaks of the two-horned variety only, and then goes on to specify the two horns in the same animal, which differing in size and shape are, from a commercial and industrial standpoint, of dif-ferent value. The term *Hu mao* ('cap of the Hu'; the Hu in general designate peoples of Central Asia, Turks and Iranians) is a very appropriate designation for the anterior horn of this species, which is a low, flat, roundish knob, and indeed resembles a small skull-cap. In the *Ming kung shi* (Ch. 4, p. 8; new edition in movable types, 1910, in 8 chs.), a most interesting description of the life at the Court of the Ming dynasty (compare HIRTH, *T'oung Pao*, Vol. VI, 1895, p. 440), this cap is explained as coming down from the T'ang dynasty, and as having been used by the heir-apparent of the Ming; it was made from sable and ermine skins, and worn in the winter on hunting-expeditions to keep the ears warm. It is mentioned in *T'ang shu*, Ch. 24, p. 8 (and presumably in other passages).

[3] Li Shi-chên (p. 150) expands this theme. Fang I-chi, who graduated in 1640, in his *Wu li siao shi* (Ch. 8, p. 20 b), states that only the rhinoceros-horn of Siam has grain patterns, while they are absent in the hairy (that is, the double-horned) rhi-noceros of Annam, which has flower-like and spotted designs.

[4] In the Treasure-House of Nara in Japan are preserved objects carved from rhinoceros-horn coming down from the T'ang period, as leather belts with horn plaques, drinking-cups, Ju-i, and back-scratchers. The girdles studded with plaques carved from the horn seem to make their appearance in China under the T'ang dynasty; the assertion of BUSHELL (Chinese Art, Vol. I, p. 119) that they were the "official" girdles of the dynasty does not seem to be justified: at least, they are not enumerated in the class of official girdles, but seem to have been restricted to the use of princesses (compare the account of *Tu yang tsa pien*, translated below, p. 152). Interesting texts bearing on rhinoceros-horn girdles are communicated in *T'u shu tsi ch'êng* (Chapter on Girdles, *tai p'ei, ki shi*, p. 9 b). Such girdles were made also in Champa: the Sung Annals (*Sung shi*, Ch. 489, p. 2) relate a tribute sent from there in the period Hien-tê (954–962) of the Hou Chou dynasty; it was local products including rhinoceros-horn girdles with plaques carved in the form of cloud-dragons. A rhinoceros-horn girdle sent from the Court of the Sung to that of the Khitan is men-tioned in *Liao shi* (Ch. 10, p. 1). Under the Kin dynasty (1115–1234) the materials employed for official costume were ranked in the order jade, gold, rhinoceros-horn, ivory (*Kin shi*, Ch. 34, § 3, p. 7). The emperor wore a hat-pin of rhinoceros-horn, and a girdle of black horn (*wu si tai*); the imperial saddle was decorated with gold, silver, rhinoceros-horn, and ivory. Officials of the second rank and higher were en-titled to a girdle of the *t'ung si* horn; those of the third rank, to a girdle of the *hua si* horn; the rest, to plain rhinoceros-horn girdles (*ibid.*, Ch. 43). They were in vogue also

are few in which the lines pass through from one end to the other. These are pointed, and their designs are large and numerous. Those with small designs are styled *tao ch'a t'ung*.[1] These two kinds are called also 'bottomless jade cups.'[2] If there is not sufficient space for the lines to pass through, and the white and black designs are equally distributed, then the price is considerably increased, and the horn will become the treasure of numberless generations. When I lived at P'an-yü,[3] I made a thorough examination of what is current there concerning rhinoceros-horn. There is, further, the *to-lo-si*, the largest among the rhinoceros-horns, which may reach seven catties in weight.[4] This is the horn on the forehead of the male rhinoceros, which has numerous designs in the interior conveying the impression of scattered beans. If the stripes are deep in color, the horn is capable of being made into girdle-plaques and implements; if the stripes are dispersed and light in color, the horn may be employed to advantage for the making of cups,

---

at the Court of the Ming emperors (*Ta Ming hui tien*, Ch. 5, p. 30), and were allowed to alternate with tortoise-shell girdles (*Ming kung shi* by Liu Jo-yü, Ch. 4, p. 3 b, new ed. of 1910). Under the Yüan dynasty a bureau for works in rhinoceros-horn and ivory was established. This was a sort of court-atelier, in which couches, tables, implements, and girdle-ornaments inlaid with these materials were turned out for the use of the imperial household. An official was placed in charge of it in 1263, and he received an assistant in 1268; the force consisted of a hundred and fifty working-men (*Yüan shi*, Ch. 90, p. 5, K'ien-lung edition). According to Qazwīnī (1203–83), the inhabitants of Sandābil (Kan-chou in Kan-su Province) were clad in silk and adorned with ivory and rhinoceros-horn (J. MARQUART, Osteuropäische und ostasiatische Streifzüge, p. 87, Leipzig, 1903). DE GOEJE is inclined to think in this connection of rhinoceros-horn set with gold and worn as amulet; but an instance of such a mode of use is not known in China, and it rather seems that it is in this case likewise the question of girdles decorated with plaques of ivory and rhinoceros-horn. The Mohammedan authors were well aware of the fondness of the Chinese for this material and its employment for girdles, and during the middle ages became the most active importers of the horn into China. The Arabic merchant Soleiman writing in 851 relates that the inhabitants of China make from the horn girdles reaching in price to two and three thousand dinars and more, according to the beauty of the figure found in the design of the horn (M. REINAUD, Relation des voyages faits par les Arabes, Vol. I, p. 29). Hafiz el Gharb, who wrote at the end of the eleventh century, observed, "The most highly esteemed ornaments among the Chinese are made from the horn of the rhinoceros, which, when cut, presents to the eye singular and varied figures" (CH. SCHEFER, Relations des Musulmans avec les Chinois, p. 10, in *Centenaire de l'Ecole des langues orientales*, Paris, 1895).

[1] *Tao*, "to reverse;" *ch'a*, "to insert;" *t'ung*, "to pass through."

[2] Thus this phrase is explained in GILES's Dictionary, p. 1326 b (tenth entry).

[3] PLAYFAIR (2d ed.), No. 4927: one of the two districts forming the city of Kuang-chou (Canton).

[4] HIRTH and ROCKHILL (Chau Ju-kua, p. 233), relying on Gerini, identify the country *To-lo* or *To-ho-lo*, as written in *T'ang shu*, with a country situated on the Gulf of Martaban. The journey from Kuang-chou to that country takes five months. An embassy with tribute came from there to China in the period Chêng-kuan (627–650), and emphasis is laid on the great number of fine rhinoceroses. See also SCHLEGEL (*T'oung Pao*, Vol. IX, 1898, p. 282) and PELLIOT (*Bull. de l'Ecole française*, Vol. IV, 1904, p. 360).

dishes, utensils, platters, and the like.[1] Then there is the horn 'frightening fowl' with a white, silk-like thread; placed in the rice, it scares the fowl away. The 'dust-dispelling horn' is utilized to make hairpins and combs for women; it keeps dust out of the hair. As to the 'water-dispelling horn,' when brought into the water of a river or the sea, it has the power of breaking a way across it. Exposed to a fog, and in the evening, it does not contract moisture. As to the 'resplendent horn,' this one, when put in a dark house, emits its own light.[2] Of all these various horns, I know only from hearsay, for I have not been able to procure and see them."

The *Yu yang tsa tsu* by Tuan Ch'êng-shi of the ninth century[3] makes the following comments on the rhinoceros: "The variety of rhinoceros styled 'communicating with the sky' dislikes its own shadow, and is in the habit of drinking muddy water.[4] When the animal is immersed in the water, men avail themselves of this opportunity to capture it, as it is impossible for it to pull its feet out of the mud. The natural structure of the horn is such that it is filled with figures resembling objects of nature. It is asserted by others that the designs penetrating the rhinoceros-horn are pathological.[5] There are three varieties of design, styled *tao ch'a* ('lines inverted and inserted'), *chêng ch'a* ('straight and inserted'), and *yao ku ch'a* ('inserted like a barrel-shaped drum').[6] They are styled 'inverted,' if one half of the lines pass

---

[1] The colors indicated by the Chinese writers altogether answer the facts. In its exterior, the color of rhinoceros-horn is usually black or dark brown. A cross-section reveals various colors. A specimen kindly presented to the Museum by Mr. F. W Kaldenberg of New York exhibits in the interior a large black zone running through the centre and extending from the base to the tip, and filling the entire space of the extremity. In the lower, broad portion it is surrounded on the one side by a gold-brown section, about 3.5 cm wide and 21 cm long, and on the other side by a mottled light-yellow and greenish zone almost soap-like in appearance. This horn was found in the woods, and is in places eaten through by insects. The surface of the base exhibits the tips of the bristles, and appears like a coarse brush. The fibres running longitudinally, owing to the effect of weathering, can be easily detached.

[2] As shown above (p. 138), optic properties are attributed to the horn as early as the time of Ko Hung. The subject is discussed in detail below (p. 151).

[3] As now established by P. PELLIOT (*T'oung Pao*, 1912, pp. 373–375), this work was published about 860.

[4] The *Pên ts'ao* adds, "In order to avoid beholding its reflection." This notion is doubtless derived from the animal's predilection for a mud-bath; its favorite haunts are generally in the neighborhood of swamps (LYDEKKER, *l. c.*, p. 31).

[5] The *Pên ts'ao* adds, "But the natural reason cannot be ascertained." This is a comment of Su Sung.

[6] The meaning of these technical terms is not quite easy to grasp. The word *tao* (No. 10,793) is "to invert," *ch'a* (No. 205) means "to insert:" *tao ch'a*, accordingly, may mean "lines inserted in the horn in an inverted position;" and *chêng ch'a*, "lines inserted straight." *Yao* ('loins') *ku* (No. 6421; in *Pên ts'ao* erroneously No. 6227) is the former name for a barrel-shaped drum (*hua ku*, see A. C. MOULE, Chinese Musical Instruments, p. 57, where an example from a verse of Su Tung-p'o is quoted). Yao K'uan, the author of the *Si k'i ts'ung yü*, written about the middle of the twelfth

through in the direction downward. They are styled 'straight,' if one half of the lines pass through in the direction upward. They are styled 'drum-shaped,' if the lines are interrupted in the middle, without passing through. The *Po-se* designate ivory as *po-ngan*, and rhinoceros-horn as *hei-ngan*.[1] Wu Shi-kao, a physician from Ch'êng shi mên,

---

century (WYLIE, Notes, p. 160), makes the following remark: "The fundamental color of rhinoceros-horn is black. Is the color simultaneously black and yellow, the horn is styled 'standard throughout' (*chêng t'ou*). Is the horn yellow with black borders, it is styled 'inverted throughout' (*tao t'ou*). The horns of standard color are highly esteemed by our contemporaries. If the shape of the horn is round, it is designated as 'horn communicating with the sky' (*t'ung t'ien si*). In the south, there are counterfeits which may be recognized from gradually getting warm when rubbed. In view of the fact that rhinoceros-horn by nature is cold, it does not become warm when rubbed."

[1] Su Sung makes the addition, "words difficult to distinguish." *Po-ngan* means literally "white *ngan*" (No. 57), and *hei-ngan* "black *ngan*,"— evidently transcriptions of *Po-se* words. PALLADIUS, in his Chinese-Russian Dictionary (Vol. I, p. 7), has indicated *po-ngan* ("ivory") and *hei-ngan* ("rhinoceros-horn") as Persian loanwords. Ivory, however, is called in Persian *shirmāhī;* and rhinoceros, as well as the horn of it, *kerkeden*. It is true that *Po-se* is the Chinese name for Persia, which first appears in the *Wei shu;* but Persia is not meant in the above passage. *P'ei wên yün fu* (Ch. 8, p. 89 b) gives three quotations under the heading *hei-ngan si*. One from a book *Sheng shui yen t'an* says that the *Po-se* call rhinoceros-horn *hei-ngan;* the reference to the name of ivory is omitted, so that the clause "it is difficult to discriminate" makes no sense. The second is derived from the *Leng chai ye hua* of the monk Hui-hung, written toward the close of the eleventh century (WYLIE, Notes on Chinese Literature, p. 164), and says that "the men of the south (*nan jên*) designate ivory as *po-ngan*, rhinoceros-horn as *hei-ngan*." The third reference is taken from a poem of Tu Fu (712–770), who remarks that *hei-ngan* is a general article of trade of the *Man*. These texts render it probable that the country of *Po-se* here referred to is not Persia, but identical with the Malayan region *Po-se* mentioned by Chou K'ü-fei in his *Ling-wai tai ta*, written in 1178 (Ch. 3, p. 6 b; edition of *Chi pu tsu chai ts'ung shu*), and then after him in the *Chu fan chi*, written in 1225 by Chao Ju-kua (translation of HIRTH and ROCKHILL, p. 125). The two authors seek it in or near the Malay Peninsula, though Negritos are not necessarily to be understood: the mere statement that the inhabitants have a dark complexion and curly hair is not sufficient to warrant this conclusion. GERINI identifies the name Po-se with Lambesi below Atjeh on the west coast of Sumatra, which seems somewhat hypothetical. Mr. C. O. BLAGDEN (*Journal Royal As. Soc.*, 1913, p. 168) is inclined to regard *Po-se* as identical with Pase (or Pasai) in north-eastern Sumatra, but adds that there is no evidence that the place existed as early as 1178. The above text shows that the *Po-se* of the Chinese mediæval writers were a Malayan tribe speaking a Malayan language, for the two transcriptions *po-ngan* and *hei-ngan* can be interpreted through Malayan. In the Hakka dialect, *hei-ngan* is *het-am;* and *hītam* is the Malayan word for "black" (Javanese Ngoko *hireng*). *Pei-ngan* is in the Hakka dialect *p'ak-am* (compare Dictionnaire chinois-français dialecte Hac-ka by CH. REY), in Cantonese *pak-om*, in Yang-chou *puk-yā*. In Javanese Krămâ "white" is *peṭak*, in Javanese Ngoko *putih*, likewise in Batak, in common Malayan *pūteh*. We should expect that the two Malayan words, judging from the Chinese transcriptions, would terminate in the same syllable, which caused misunderstandings on the part of Chinese dealers. There is (or was) perhaps a certain Malayan dialect, in which the word for "white" ended in *-am*, or in which the words for "white" and "black" terminated in *-i* or *-ih* (compare Madagassy *inṭim*, *inṭi*, "black;" and *puti*, "white;" G. FERRAND, Essai de phonétique comp. du malais et des dialectes malgaches, pp. 24, 54, Paris, 1909). It is evident that neither the Malayan words for "ivory" (*gāding*, Javanese *gaḍing*) and "rhinoceros-horn" (*chula bādak* or simply *chula*), nor the words for "elephant" (*gājah*, Javanese *gajah*) and "rhinoceros" (*badak*, Javanese *warak*), are intended here, but only the color names "white" and "black," with which the traders distinguished ivory and rhi-

while he served in the district of Nan-hai (in Kuang-tung), had occasion to meet there a captain who told him this story: 'The people of my country, intent on capturing a rhinoceros, proceed to erect on a mountain-path many wooden structures like watch-houses or posts for tethering animals.[1] As the front legs of the animal are straight, without joints, it is in the habit of sleeping by leaning against a tree. The rotten timber will suddenly break down, and the animal is unable to rise.[2] Another

---

noceros-horn. The Malayan word *badak* seems to cover the entire Malayan area where the rhinoceros is found; it occurs on Borneo in the language of the Dayak (A. HARDE-LAND, Dajacksch-deutsches Wörterbuch, p. 24, Amsterdam, 1859), and on Sumatra (M. JOUSTRA, Karo-Bataksch Woordenboek, p. 59, Leiden, 1907). Among the Malayans, the rhinoceros-horn (*chula*) is supposed to be a powerful aphrodisiac; and there is a belief in a species of "fiery" rhinoceros (*badak api*) which is excessively dangerous when attacked (W. W. SKEAT, Malay Magic, p. 150, London, 1900). The horn is carefully preserved, as it is believed to be possessed of medicinal properties, and is highly prized by the Malays, to whom the Semang generally barter it for to-bacco and similar commodities (SKEAT and BLAGDEN, Pagan Races of the Malay Peninsula, Vol. I, p. 203, London, 1906). There is nothing in these Malayan beliefs showing that complex series of ideas, met with in China. They may be a weak echo of Chinese notions conveyed by Chinese traders bartering among them for the horn.

[1] *Chü yi* (Nos. 2974 and 13,205). I do not know but this may have to be taken as a compound with a more specific technical meaning. The two *Pên ts'ao* have changed this unusual term into "stables for swine or sheep." There is no doubt of what is meant, — posts of rotten timber, which will easily break to pieces under the burden of the animal leaning toward it.

[2] This story has passed also into the Arabic account of the merchant-traveller Soleiman, written in 851 A.D. (M. REINAUD, Relation des voyages faits par les Arabes et les Persans dans l'Inde et à la Chine, Vol. I, p. 29, Paris, 1845): "The *kerkeden* (rhinoceros) has no articulation in the knee, nor in the hand; from the foot up to the armpit it is but one piece of flesh." In T'oung Pao (1913, pp. 361–4) the historical importance of this tradition is pointed out by me inasmuch as this originally Indian story has migrated also to the West, where it leaks out in the Greek *Physiologus* (only the rhinoceros is replaced by the elephant), and in CAESAR's and PLINY's stories of the elk. I wish to make two additions to these remarks. AELIAN (*Nat. an.*, XVI, 20), describing the rhinoceros of India, called by him καρτάζωνος, asserts that its feet have no joints and are grown together like the feet of the elephant (τοὺς μὲν πόδας ἀδιαρθρώτους τε καὶ ἐμφερεῖς ἐλέφαντι συμπεφυκέναι: ed. of F. JACOBS). This passage, therefore, confirms my former conclusion that it was the rhinoceros which was credited in India with jointless legs; but we see that the same notion was like-wise attached to the elephant. It may be the case, accordingly, that the elephant with jointless legs was borrowed by the *Physiologus* straight from India. Mr. W. W. ROCKHILL (Diplomatic Audiences at the Court of China, p. 32, London, 1905) quotes a statement made to him by T. WATTERS on the kotow question with reference to Lord Macartney's embassy, as follows: "It was an opinion universal, and was told among the Chinese, that the *Kuei-tse* or foreigner was not built up like the *jen* [that is, man] or Chinaman, and particularly that he had no joints in his legs. So that, if the *Kuei-tse* was knocked down or otherwise put on the ground, he could not rise again. It was because the Emperor did not want to have possibly a death or at any rate an unseemly spectacle that he waived the kotow." Compare also Rubruck's story of "the creatures who have in all respects human forms, except that their knees do not bend, so that they get along by some kind of jumping motion" (W. W. ROCKHILL, The Journey of William of Rubruck, p. 199, London, 1900). The fabulous notion of the jointless legs of the rhinoceros may have arisen from the observation that the animal is indeed in the habit of sleeping in a standing position. Says E. HELLER (The White Rhinoceros, p. 41), "The hot hours of the day are spent by the white rhinoceros sleeping in the shade of the scattered clumps of trees or bushes which dot the grassy veldt. They seem to rest indifferently, either lying down or standing

name for the rhinoceros is *nu kio*. There is also the *chên ch'u*, which is presumably a rhinoceros. The rhinoceros has three hairs growing out of each pore.[1] Liu Hiao-Piao asserts that the rhinoceros sheds its horn and buries it, and that people exchange it for a counterfeit horn."

The story alluded to in the latter clause is better worded in the *Pên ts'ao*, which says, "It is told also that the rhinoceros sheds its horn every year, and itself buries it in the mountains. The people near the sea, with all secrecy, make wooden horns, and exchange these for the real ones, and so they go ahead continually. If they would go to work openly, the animal would conceal its horns in another place and defy any search."[2]

Li Sün, who wrote an account of the drugs of southern countries (*Hai yao pên ts'ao*) in the second half of the eighth century, expresses himself in these words: "The rhinoceros 'communicating with the sky,' during the time of pregnancy, beholds the forms of things[3] passing across the sky, and these are reproduced in the horn of the embryo: hence the designation 'communicating with the sky.'[4] When the horn, placed in a water-basin during a moonlight night, reflects the brilliancy of the moon, it is manifest that it is a genuine horn 'communicating with the sky.' The *Wu k'i ki*[5] says, 'The mountain-rhinoceros lives on bamboo and trees. Its urinating is not completed in the course of a day. The I Liao[6] get hold of it by means of bow and arrow. This is

---

up with lowered head. When at rest they stand with their noses almost touching the ground, their heads being elevated to a horizontal position only when alarmed."

[1] The same is said in the *Pên ts'ao* in regard to the seal (compare G. SCHLEGEL, *T'oung Pao*, Vol. III, 1892, p. 508). Compare p. 140.

[2] In the text of the *Chêng lei pên ts'ao*, Su Sung terminates, "I do not know whether at present they take horns in this manner or not." Compare the account of Ko Hung, p. 139.

[3] The *Chêng lei pên ts'ao* reads "the destiny of things" (*wu ming*) instead of "forms of things" (*wu hing*).

[4] In the notes embodied in the *Pên ts'ao* regarding the elephant (Ch. 51 A, p. 4) it is said that the patterns in the horn are formed while the rhinoceros gazes at the moon, and that the designs spring forth in the tusks of the elephant while the animal hears the thunder. A work *Wu têng hui yüan*, as quoted in *P'ei wên yün fu* (Ch. 21, p. 113 b), similarly says that the rhinoceros, while enjoying the moonlight, produces the designs in its horn, and that the floral decorations enter the tusks of the elephant when it has been frightened by thunder. These passages prove that it is material heaven to whose influence the formation of the natural veins in horn and tusk is ascribed. The rhinoceros gazing at the moon is represented in *T'u shu tsi ch'êng* (Fig. 10).

[5] A work listed in the *T'ai p'ing yü lan* as being published in 983; but, as it is quoted here by Li Sün, it must have existed in or before the eighth century.

[6] An aboriginal tribe belonging to the stock of the *Man*, according to *T'ang shu* (Ch. 43 A, p. 6 b) settled in Ku chou (PLAYFAIR, No. 3256) in the province of Kuei-chou. Compare p. 82 in regard to the possibility of killing a rhinoceros with arrows.

the so-called rhinoceros of K'ien.'[1]   The *I wu chi*[2] says, 'In the sea-water of Shan-tung there is a bull that delights in the sounds of string and wind instruments.   When the people make music, this bull leaves the water to listen to it, and at that moment they capture it.'[3]   The rhinoceros has a horn on its nose, and another on the crown of its head.   The nose-horn is the one best esteemed.   The natural histories (*pên ts'ao*) are acquainted only with the mountain-rhinoceros.   I have not yet seen the water-rhinoceros."[4]

K'ou Tsung-shi, a celebrated physician of the Sung period, reports in his *Pên ts'ao yen i* (completed in 1116)[5] thus: "The designs in the horns of the river-rhinoceros and the southern rhinoceros are fine.   The black rhinoceros-horn has designs clearly displayed, while the yellow rhinoceros-horn has very sparse designs.   None equals the patterns in the horn of the Tibetan breed, which are high, and come out clearly at both ends.[6]   If the forms of objects pictured in the horn are yellow, while the rest is black, the horn is 'standard color throughout' (*chêng t'ou*). If the forms of objects are black, while the rest is yellow, the horn is 'inverted throughout' (*tao t'ou*).   If the black color is taken as standard, and the forms of the design are imitative of real objects, the horn is a treasure; this horn is styled *t'ung si* ('penetrating rhinoceros').   It is an indispensable condition that the patterns come out clearly, and that the yellow and black be sharply differentiated.   If both ends are moist and smooth, the horn is of the first quality."[7]

---

[1] The territory of the province of Kuei-chou, where the rhinoceros formerly occurred, as already attested by Su Sung (above, p. 140).

[2] Several works of this title were in existence (see BRETSCHNEIDER, Bot. Sin., pt. 1, p. 154).

[3] The animal in question is certainly not a rhinoceros, and has crept in here by way of wrong analogy.   In his notes on cattle, Li Shi-chên mentions a variety "marine ox" (*hai niu*, Ch. 51 A, p. 7 a).   This creature is described after the *Ts'i ti ki* by Fu Ch'ên of the fifth century or earlier (BRETSCHNEIDER, Bot. Sin., pt. 1, p. 201) as follows: "Its habitat is around the islands in the sea near Têng-chou fu (in Shan-tung); in shape it resembles an ox, it has the feet of an alligator (*t'o* No. 11,397, not *iguana*, as GILES still translates, despite the correction of E. v. ZACH, *China Review*, Vol. XXIV, 1900, p. 197), and the hair of a bull-head fish.   Its skin is soft, and can be turned to manifold purposes; its blubber is good to burn in lamps."   The marine ox, accordingly, must be an aquatic mammal of the suborder of Pinnipedia (seals). There may be a grain of truth in the above story: the intelligence of seals is remarkable, they are easily tamed and susceptible to music.   There is an interesting chapter on tamed seals in the classical treatise of K. E. v. BAER, Anatomische und zoologische Untersuchungen über das Wallross (*Mémoires de l'Acad. imp. des sciences de St. Pétersbourg*, 6th series, Vol. IV, 1838, pp. 150–159).

[4] The last clause is not in the text of *Chêng lei pên ts'ao*.

[5] PELLIOT (*Bulletin de l'Ecole française d'Extrême-Orient*, Vol. IX, 1909, p. 217).

[6] The rhinoceros of Tibet has been discussed above, p. 116.

[7] The Arabic authors assert that the interior of the Indian rhinoceros-horn frequently presents designs of a human figure, a peacock, or fish, and that the price paid in China is raised according to the beauty of these designs (M. REINAUD, Relation

Li Shi-chên himself, the author of the *Pên ts'ao kang mu*, sums up as follows: "The habitat of the rhinoceros is in the regions of the *Si Fan*,[1] the southern Tibetan tribes (*Nan Fan*), the southern portions of Yünnan, and in Kiao-chou, and occurs there everywhere. There are three species,— the mountain-rhinoceros, the water-rhinoceros, and the *se si*. There is, further, a hairy rhinoceros resembling the mountain-rhinoceros, and living in hilly forests; great numbers of it are captured by men. The water-rhinoceros makes its permanent abode in water, and is therefore very difficult to capture. It has, in all, two horns. The horn on its nose is long, that on its forehead is short. The skin of the water-rhinoceros has a pearl-like armor,[2] but not so the mountain-rhinoceros.

---

des voyages faits par les Arabes, Vol. I, p. 29). REINAUD (Vol. II, pp. 68, 69) comments on this point that the Chinese are satisfied to compare the designs with flowers and millet-seeds, and do not discover in them half of the things which the Arabs saw in them. It seems to me that the Arabs, in this case, merely reproduce the ideas of the Chinese. The philosophy of these designs was fully developed in the T'ang period. K'ou Tsung-shi speaks of real objects visible in the horn; and Wang P'i-chi, in his *Shêng shui yen t'an lu* (p. 135), offers an elaborate contribution to this question. According to him, "the designs in the horn from Kiao-chi are like hemp-seeds, the horn being dry, a bit warm, and glossy; the horn imported on ships and coming from the Arabs has patterns like *chu yü* flowers [this name applies to three different plants: BRETSCHNEIDER, Bot. Sin., pt. 2, No. 498], is glossy and brilliant with colors, some resembling dog-noses, as if they were glossed with fat; others with floral designs and strange objects, these horns being styled *t'ung t'ien si;* some like sun and stars, others like clouds and moon; some like the corolla of a flower, some like scenery; some have birds and mammals, others dragons and fishes; some have deities, others palaces; and there are even costume and cap, eyes and eyebrows, staff and footgear [conveying the illusion of the picture of a wanderer], beasts, birds, and fishes. When the horn is completed into a carving, as if it were a veritable picture, it is highly esteemed by the people. The prices are fluctuating, and it is unknown how they are conditioned." There is assuredly an inward relation between the statements of this account and the Arabic texts of Damīrī quoted by REINAUD (Vol. II, p. 69). It is hardly necessary to insist on the chronological point that Damīrī (1344–1405) wrote his zoölogical dictionary *Hayāt el-haiwān* (C. HUART, Littérature arabe, p. 365, Paris, 1902) several centuries after Wang P'i-chi (end of eleventh century). From a psychological point of view, the dependence of the Arabs in this matter on the philosophy of the Chinese is self-evident. Neither the classical world nor ancient India has developed any similar thoughts; and this subject is decidedly Chinese, with a strong Taoist flavor of nature sentiment. It must not be overlooked, either, that al-Bērūnī (SACHAU, Alberuni's India, Vol. I, p. 204) merely states that "the shaft of the horn is black inside, and white everywhere else," and that he is entirely reticent about figures in the horn. The Arabs interested in the trade of the horn to China imbibed this lesson, and propagated it themselves in catering to the taste of their customers. The question is whether, in the interest of the business, they did not help nature by art, and may have produced several of the more fanciful designs artificially. This, however, is no matter of great concern; and the fact remains that bristly fibres of various tinges compose the horn, and result in a natural play of design and color which is apt to arouse the imaginative power of a susceptible mind.

[1] Western Tibetan tribes; from our standpoint, eastern Tibetans.

[2] I take this to be identical with what our zoölogists say in regard to the skin of the Asiatic species, which "has the appearance of a rigid armor studded with tubercles." The whole skin of the Javan species, as already remarked by B. CUVIER (The Animal Kingdom, Vol. I, p. 157, London, 1834), is covered with small compact angular tubercles. JOANNES RAIUS (Synopsis methodica animalium quadrupedum, p. 122, Londini, 1693) describes the skin of the rhinoceros thus: "Auriculae porcinae,

The *se si* is the female of the rhinoceros which is termed also 'sand-rhinoceros.' It has but a single horn on the crown of the head. The natural designs of the horn are smooth, white, and clearly differentiated, but it is useless as medicine, for the patterns on the horn of the male are big, those on the horn of the female too fine. In the beginning of the period Hung-wu (1368–1398) Kiu-chên[1] sent one as tribute, which was called one-horned (*monoceros*) rhinoceros. The view of Ch'ên Ts'ang-k'i that there are not the two kinds of land and water animals, the view of Kuo P'o that the rhinoceros has three horns, and the view of Su Sung that the hairy rhinoceros is the male rhinoceros, are all erroneous. The term 'hairy rhinoceros' is at present applied to the yak.[2] The designs of the rhinoceros-horn are like fish-roe. On account of their shape they are styled 'grain patterns.'[3] Inside of the latter there are eyes, styled 'grain eyes.' If yellow decorations rise from a black background, the horn is 'standard throughout.' If black decorations rise from a yellow background, the horn is 'inverted throughout.' If within the decorations there are again other decorations, the horn is 'double throughout.' The general designation for these is *t'ung si*, and they are of the highest grade. If the decorations are spotted, as it were, with pepper and beans, the horns are middle grade. The horn of the black rhinoceros, which is of a uniform black color and devoid of decorations, is the lowest grade.[4] If the horn of the rhinoceros 'communicating with the sky' emits light, so that it can be seen at night, it is

---

molli et tenui cute vestitae; reliquum corpus dura admodum et crassa, velut squamis quibusdam crustaceis rotundis aspera." This is the reason why in some Chinese and early European sketches the animal is covered with scales (see Figs. 3 and 11, and Plate IX).

[1] PLAYFAIR, No. 1295 (1278): in Annam (compare above, p. 81).

[2] Li Shi-chên refers to the notes on this subject contained in the same chapter. This remark renders it plain that it was the notion of "rhinoceros" which was transferred in recent times to the yak, and that the development was not in the reverse order, as assumed by Professor Giles.

[3] This and the following sentences, commenting on the natural designs of the horn, have been translated by S. JULIEN (in M. REINAUD, Relation des voyages faits par les Arabes, Vol. II, p. 68).

[4] In the Memoirs on the Customs of Cambodja by Chou Ta-kuan of the Yüan period, translated by P. PELLIOT (*Bulletin de l'Ecole française d'Extrême-Orient*, Vol. II, 1902, p. 167), it is said that the white and veined rhinoceros-horn is the most esteemed kind, and that the inferior quality is black. The List of Medicines exported from Hankow, published by the Imperial Maritime Customs (p. 15, Shanghai, 1888), is therefore wrong in stating that the black and pointed horns are considered the best. A valuation for the horn is not given there. According to a report of Consul-General G. E. ANDERSON of Hongkong (*Daily Consular and Trade Reports*, 1913, p. 1356), rhinoceros-horns are imported into Hongkong to some extent, the price ranging from $360 to $460 per picul, or from about $1.30 to $1.65 gold per pound; they are largely of African production, and imported from Bombay. According to L. DE REINACH (Le Laos, Paris, no date, p. 271), rhinoceros-horns have in the territory of the Laos a market-value of 111–137 fr. the kilo, and rhinoceros-skins 60–70 fr. a hundred kilo.

called 'horn shining at night' (*ye ming si*):[1] hence it can communicate with the spirits, and open a way through the water. Birds and mammals are frightened at seeing it. The *Shan hai king* speaks of white rhinoceroses.[2]

---

[1] This idea may have been borrowed from the precious stones believed to shine at night (HIRTH, China and the Roman Orient, pp. 242–244; CHAVANNES, Les pays d'occident d'après le Heou Han Chou, *T'oung Pao*, 1907, p. 181). Jade disks shining at night (*ye kuang pi*) are mentioned in *Shi ki* (Ch. 87, p. 2 b). The note of Li Shi-chên is doubtless suggested by the following passage of the *Tu yang tsa pien*, written by Su Ngo in the latter part of the ninth century (WYLIE, Notes on Chin. Lit., p. 194; ed. of *Pai hai*, Ch. B, p. 9, or *P'ei wên yün fu*, Ch. 8, p. 87 b): "In the first year of the period Pao-li (825 A.D.) of the Emperor King-tsung of the T'ang dynasty, the country of Nan-ch'ang [in Kiang-si; PLAYFAIR, No. 4562] offered to the Court a rhinoceros-horn shining at night (*ye ming si*). In shape it was like the 'horn communicating with the sky.' At night it emitted light, so that a space of a hundred paces was illuminated. Manifold silk wrappers laid around it could not hide its luminous power. The Emperor ordered it to be cut into slices, and worked up into a girdle; and whenever he went out on a hunting-expedition, he saved candle-light at night." We even hear of a luminous pillow (*ye ming chên*) lighting an entire room at night (*Yün sien tsa shi*, Ch. 6, p. 3 b, in *T'ang Sung ts'ung shu*, which quotes from *K'ai-yüan T'ien-pao i shi*). The story of *Tu yang tsa pien* may be connected with the curious tradition regarding Wên K'iao (*Tsin shu*, Ch. 67, p. 5), who by the alleged light emitted from a rhinoceros-horn beheld the supernatural monsters in the water (see PÉTILLON, Allusions littéraires, p. 227; S. LOCKHART, A Manual of Chinese Quotations, p. 280; and GILES, Dictionary, p. 794 b, — who translate 'to light a rhinoceros-horn,' which is not possible, as in this case the horn would burn down; the horn was shining through its alleged own light). An illustration of this scene by Ting Yün-p'êng is published in *Ch'êng shi mo yüan* and *Fang shi mo p'u*. The notion that the rhinoceros-horn is luminous at night, and is therefore styled "shining or bright horn" (*ming si*, or *kuang ming si*), and also "shadow horn" (*ying si*), is found in *Tung ming ki* (Wu-ch'ang print, Ch. 2, p. 2), embodied in a fabulous report on a country Fei-lo, said to be nine thousand *li* from Ch'ang-ngan in Indo-China (Ji-nan). This work relating to the time of the Han Emperor Wu, though purported to have been written by Kuo Hien of the Han, is one of the many spurious productions of the Leu-ch'ao period (fourth or fifth century), and teeming with anachronisms and gross inventions; some accounts in it are interesting, but devoid of historical value (see WYLIE, Notes, p. 191). The assertion there made, that the inhabitants of Fei-lo drive in carriages drawn by rhinoceros and elephant, is very suspicious; but the report that the horns sent from there were plaited into a mat, the designs of which had the appearance of reticulated silk brocade, is probably not fictitious; for this is confirmed by a passage of the T'ang Annals (Chapter *wu hing chi*, quoted in *T'u shu tsi ch'êng*), according to which a certain Chang Yi-chi had a mat made for his mother from rhinoceros-horn. Since the latter (the designation "horn," from a scientific standpoint, is a misnomer) is composed of agglutinated hair or bristles, it is possible to dissolve a horn into thread-like fibres; and the possibility of a technique employing these for the plaiting of mats must be admitted.

[2] According to the more precise wording of the passage, as quoted in *P'ei wên yün fu* (Ch. 8, p. 88 a), the white rhinoceros occurs in the mountains of Kin-ku, inhabited by large numbers of other wild animals, also hogs and deer. The *Shan hai king* is an apocryphal work teeming with fables, and has little value for scientific purposes. The *P'ei wên yün fu*, further, quotes the *Tung kuan Han ki* (completed about 170 A.D.; BRETSCHNEIDER, Bot. Sin., pt. I, No. 990) to the effect that in the first year of the period Yüan-ho (84 A.D.) of the Emperor Chang of the Han dynasty the country Ji-nan (Tonking) offered to the Court a white pheasant and a white rhinoceros. But this text, unreservedly accepted by HIRTH (Das weisse Rhinoceros, *T'oung Pao*, Vol. V, 1894, p. 392), must be taken with some caution, as it is identical with, and apparently derived from, the passage in *Hou Han shu* (Ch. 116, p. 3 b), according to which, in the first year of the period Yüan-ho (84 A.D.), the *Man I* beyond the boundary of Ji-nan offered to the Court a live rhinoceros and a white pheasant. The

"The work *K'ai-yüan i shi*[1] mentions the 'cold-dispelling' rhinoceros-horn (*pi han si*), whose color is golden, and which was sent as tribute by Tonking (*Kiao-chi*).[2]   During the winter months it spreads warmth, which imparts a genial feeling to man.   The *Po k'ung leu t'ie*[3] speaks of the 'heat-dispelling' rhinoceros-horn (*pi shu si*) obtained by the Emperor Wen-tsung (827–840 A.D.) of the T'ang dynasty.[4]   During the summer months it can cool off the hot temperature.   The *Ling piao lu i*[5] records the horn of the 'dust-dispelling' rhinoceros (*pi ch'ên si*), from which hairpins, combs, and girdle-plaques are made, with the effect that dust keeps aloof from the body.   The *Tu yang tsa pien*[6]

---

text of the official Annals is decisive, and it is easy to see that the word "live" could have been altered into "white" by the suggestion of the white pheasant.   The *T'ang leu tien*, a description of the administrative organization of the period K'ai-yüan (713–741) of the T'ang dynasty, ascribed to the Emperor Yüan-tsung (compare PELLIOT, *Bulletin de l'École française d'Extrême-Orient*, Vol. III, 1903, p. 668), says that "the white rhinoceros (*pai se*) is an auspicious omen of the first order" (*shang jui;* quoted in *Yen kien lei han*, Ch. 410, p. 17 b).   But as most of the creatures appearing in the category of such "auspicious omens" are imaginary, it is more than probable that this white rhinoceros owes its existence to pure fancy.   The white rhinoceros, therefore, does not rest on good evidence; and I am not convinced that the Chinese were ever acquainted with such a variety.   Moreover, the so-called White or Square-nosed Rhinoceros (*Rhinoceros simus cottoni*) has not yet been traced in Asia, but is restricted to Africa.   It is described and illustrated by A. NEWTON (*Proceedings of the Zoölogical Soc. of London*, Vol. I, 1903, pp. 222–224; see *ibid.*, Vol. II, 1903, p. 194), R. LYDEKKER (The Game Animals of Africa, p. 38, London, 1908), and E. L. TROUESSART (Le Rhinocéros blanc du Soudan, *Proceedings* etc., 1909, pp. 198–200, 3 plates).   A fine monograph is devoted to it by E. HELLER, The White Rhinoceros (*Smithsonian Misc. Collections*, Vol. 61, No. 1, Washington, 1913, 31 plates), embodying the results of Colonel Roosevelt's African expedition.   As to the "white" color, Mr. Heller observes, "The skins cannot under the most lenient circumstances be classed as white.   They are, however, distinctly lighter than those of the black species, and may on this account be allowed to retain their popular designation of white.   Their true color is smoke gray of Ridgway, a color conspicuously lighter than the dark clove-brown of their geographical ally, *Diceros bicornis*."

[1] Matters omitted in the Annals of the Reign of *K'ai-yüan* (713–742) by Wang Jên-yü, written during the Wu-tai period (907–960); see BRETSCHNEIDER, Bot. Sin., pt. 1, p. 156.

[2] The text is quoted in *P'ei wên yün fu* (Ch. 8, p. 87 b) as follows: "The country of Tonking sent a rhinoceros-horn of golden color, which was placed in a golden pan in a hall of the palace; the warmth caused by it was felt by every one; the envoy said that it was the cold-dispelling rhinoceros-horn."

[3] The complete title runs *T'ang Sung Po k'ung leu t'ie;* it is a cyclopædia in 100 chapters arranged according to subject-matters dealing with affairs of the T'ang and Sung periods (Ming edition in John Crerar Library, No. 786, in 96 vols.).

[4] The exact text is given in *P'ei wên yün fu*.   A sceptre of auspicious augury (*Ju i*), made from a "heat-dispelling horn" in the possession of the same emperor, is mentioned in *Tu yang tsa pien* (Ch. B, p. 12; see note 6).   Another *Ju i* of ordinary rhinoceros-horn is spoken of in *Yün sien tsa shi* (Ch. 3, p. 5 b; ed. of *T'ang Sung ts'ung shu*).

[5] See p. 142.

[6] An account of rare and curious objects brought to China from foreign countries from 763 to 872, by Su Ngp in the latter part of the ninth century (BRETSCHNEIDER, *l. c.*, p. 204; WYLIE, Notes on Chin. Lit., p. 194).   According to the passage in the original text (ed. of *Pai hai*, Ch. C, p. 9 b), this girdle was in the possession of the

refers to the 'wrath-removing' rhinoceros-horn (*küan*, No. 3141, *fên si*), from which girdles are made, causing men to abandon their anger; these are scarce and veritable treasures."

These extracts, ranging from the fifth to the sixteenth century, leave no doubt that during this interval the two words *se* and *si* invariably referred to the rhinoceros, that the two species of the single-horned and two-horned animal were recognized, that their geographical distribution was perfectly and correctly known,[1] and that the main characteristics of the animal were seized upon. Among these, the horn naturally attracted widest attention, and in most cases was the only part of the animal that came within the experience of the writers. The wondrous lore surrounding the horn, the supernatural qualities attributed to it, led also to fabulous stories regarding the animal itself, which in the midst of impenetrable forests was seldom exposed to the eye of an observer. A lengthy dissertation on the healing properties of the horn, and on its utilization in prescriptions, is added in the *Pên ts'ao kang mu*; but this matter has no direct relation to our subject.[2]

Princess T'ung-ch'ang, and consisted of small balls turned from horn, as shown by the description that they were round like the clay pellets used in shooting with the bow *tan* (No. 10,603). These bows, a combination of a sling with a bow, are still turned out in Peking, and used in slaying birds, to prevent the plumage from being damaged. In India they are known as *goolail* (YULE and BURNELL, Hobson-Jobson, p. 386), and are chiefly employed for exterminating crows, being capable of inflicting severe injuries. Every ethnologist is familiar with these sling-bows or pellet-bows, as they are called, and with the difficult problem presented by their geographical distribution over India, south-eastern Asia, and in the valley of the Amazon in South America (compare G. ANTZE, in *Jahrbuch des Museums für Völkerkunde zu Leipzig*, Vol. III, 1908, pp. 79–95; and W. HOUGH, *Am. Anthr.*, 1912, p. 42). It is further added in the *Tu yang*, that this horn, when placed in the ground, does not rot,—a notion presumably originated by occasional finds of fossil horns or those accidentally shed by the animal.

[1] The case is certainly such that the zoölogist, as in so many other cases, is obliged to learn from the historian in regard to the distribution of animals in former periods of history. Our zoögeographers trace the area of the two-horned rhinoceros to Sumatra, Borneo, Siam, and the Malay Peninsula, and from there extending northward through Burma and Tenasserim to Chittagong and Assam. Our investigation has taught us that it covered in ancient times a much wider geographical zone, including Cambodja, Annam, and southern China, in particular Kuei-chou, Hu-nan, Yün-nan, and Sze-ch'uan.

[2] The theory of Ko Hung or Pao-p'u-tse of the fourth century, as shown above (p. 139), is that the horn can neutralize poison, because the animal devours all sorts of vegetable poisons with its food. Li Shi-chên states that the horn is non-poisonous, and is forestalled in this opinion by T'ang Shên-wei. Shavings of the horn, the decoction of which is taken in fever, small-pox, ophthalmia, etc., are still to be had in all Chinese drug-stores. A specimen obtained by me at Hankow was said to come from Tibet. According to S. W. WILLIAMS (The Chinese Commercial Guide, p. 95, Hongkong, 1863), a decoction of the horn shavings is given to women just before parturition and also to frightened children. As stated by the same author, the skin of the animal is likewise employed in medicine. It is made into a jelly which is highly esteemed, and the same is done with the feet (SOUBEIRAN and THIERSANT, La matière médicale chez les Chinois, p. 47, Paris, 1874). This practice presumably originated in Siam. Monseigneur PALLEGOIX (Description du royaume Thai ou

The word *se* is presumably the older of the two, as the ancient Chinese seem to have been first acquainted with this species, while it was still alive in their country; at a somewhat later time, which, however, still ranged in a prehistoric period, they became familiar with the two-horned *si*. This theory would account for the statement of Li Shi-chên that the ancients were fond of saying *se*, while later on people inclined toward the word *si;* and that in the north (the ancient habitat of the *se*) the word *se* prevailed, in the south the word *si*. This came about

---

Siam, Vol. I, p. 156) reports the following: "On attribue beaucoup de vertus à sa corne, et (chose singulière!) sa peau, quelque épaisse et coriace qu' elle soit, est regardée comme un mets délicat et fortifiant pour les personnes faibles. On grille d'abord la peau, on la ratisse, on la coupe en morceaux et on la fait bouillir avec des épices assez longtemps pour la convertir en matière gélatineuse et transparente. J'en ai mangé plusieurs fois avec plaisir, et je pense qu'on pourrait appliquer avec succès le même procédé aux peaux de quelques autres animaux." The skin, as well as the horn, the blood, and the teeth, were medicinally employed in Cambodja, notably against heart-diseases (A. Cabaton, Brève et véridique relation des événements du Cambodge par Gabriel Quiroga de San Antonio, p. 94, Paris, 1914). In Japan rhinoceros-horn is powdered and used as a specific in fever cases of all kinds (E. W. Clement, Japanese Medical Folk-lore, *Transactions As. Soc. of Japan*, Vol. XXXV, 1907, p. 20). Ko Hung of the fourth century, as we observed, is the very first Chinese author to develop the theory of the horn as to its ability to detect poison, and as an efficient antidote against poison. He also reasons his theory out, and supports it with arguments of natural philosophy breathing a decidedly Taoist spirit. Nothing appears in his account that would necessitate a cogent conclusion as to his dependence on Indian thought. Indian-Buddhist influence on the Taoism of that period certainly is within the reach of possibility, but like everything else, remains to be proved; and for the time being I can only side with Pelliot (*Journal asiatique*, 1912, Juillet-Août, p. 149) when he remarks to L. Wieger, "Ici non plus, je ne nie pas la possibilité de semblable influence, mais j'estime qu'il faut être très prudent." If a Buddhist text translated from Sanskrit into Chinese in or before the age of Ko Hung, and containing a distinct reference to this matter, can be pointed out, I am willing to concede that Ko Hung is indebted to an Indian source; if such evidence should fail to be forthcoming, it will be perfectly sound to adhere to the opinion that Ko Hung's idea is spontaneous, and the expression of general popular lore obtaining at his time; and there is no valid reason why it should not be. No ancient Sanskrit text containing similar or any other notions concerning this subject has as yet come to the fore; and the evidence in favor of Indian priority is restricted to the slender thread of Ctesias' account (p. 97), which is insufficient and inconclusive. The light-minded manner with which Bushell (Chinese Art, Vol. I, p. 119) dealt in the matter (as if the lore of the horn and the horn itself had only been a foreign import in China!) must be positively rejected. Bretschneider (above, p. 75) no doubt was a saner judge. Neither in ancient India nor in the classical world do we find any trace of such beliefs as those expounded by Ko Hung and his successors, nor a particle of all that Chinese natural philosophy of the horn. Aelian merely reiterates Ctesias; Juvenal (VII, 130) mentions an oil-bottle carved from the horn; the *Periplus Maris Erythraei* (ed. Fabricius, pp. 40, 44, 56) refers to the export of the horn from African ports only, not from India. The Cyranides (F. de Mély, Les lapidaires grecs, p. 90) are ignorant of the poison-revealing character of the horn. But for Ctesias, we should be compelled to admit that this belief originated in China and spread thence to India. At any rate, the report of Ctesias stands isolated in the ancient world; the untrustworthy character of this author is too well known to be insisted upon, and it would be preposterous to build a far-reaching conclusion on any of his statements which cannot be checked by other sources. His text is handed down in poor condition, and as late as by Photius, patriarch of Byzance (820–891), so that I am rather inclined to regard the incriminated passage as an interpolation of uncertain date. The belief in rhinoceros-horn being an efficient antidote against poison prevailed in Europe until recent times.

naturally, as the south bordered on Indo-China, where the two-horned species abounded, and a lively trade in its horn was carried on at all times. Hence in the primeval period represented by the songs of the *Shi king* the rhinoceros is styled *se*.

The philological students of China will certainly feel somewhat uneasy at the thought that an animal like the rhinoceros should have been within the vision of the early Chinese. We are all wont to look at

---

It seems to have received a fresh impetus from India in the sixteenth century. The Portuguese physician GARCIA AB HORTO (Aromatum et Simplicium aliquot, p. 66, Antverpiae, 1567; or Due libri dell' historia dei semplici, aromati, et altre cose che vengono portate dall' Indie Orientali pertinenti all' uso della medicina, p. 58, Venetia, 1582) first reports from personal experience that rhinoceros-horn is employed in Bengal as an antipoisonous remedy, and goes on to tell that this is a fact established by experiments; his story is that of two poisoned dogs—the one who had swallowed double the dose was cured after taking in water a powder prepared from the horn, while the other dog, who had been given but a small quantity of poison and did not receive the remedy of the horn, was doomed to death. Doctor NICOLÒ MONARDES, physician in Sevilla (Delle cose che vengono portate dall' Indie occidentali pertinenti all' uso della medicina, p. 72, Venetia, 1582), has the following account: "L' Unicorno vero è cosa di maggiore effetto, che habbiamo veduto, e nella quale si trova maggiore esperienza; del quale poco si scrive. Solo Philostrato nella vita di Apollonio dice, essere contra il veneno; il que ampliarono molto i Moderni. Bisogna, che sia del vero; perche ne sono molti di falsi, e finti. Io vidi in questa città un Vinitiano, che ne portò un pezzo molto grande, e ne dimandava cinquecento scudi; delquale fece in mia presenza la esperienza. Prese un filo, e lo unse molto bene con Elleboro, e lo passò per le creste di due polli; all' uno de'quali diede un poco di Unicorno raso in un poco di acqua comune; e all' altro non diede cosa alcuna. Questo morì tra un quarto di hora; l'altro che prese l'Unicorno durò due giorni, senza voler mangiare, e alla fine di due giorni morì, secco come un legno. Credo io, che se si desse ad huomo, che non morrebbe; perche tiene le vie più aperte da potere scacciare da se il veneno; e gli si può ancho fare de gli altri rimedij, col mezzo de' quali, e coll' Unicorno potrebbe liberarsi. Di tutte queste Medicine compongo io una polvere, che cosi per qualità manifeste, come per proprietadi occulte ha gran virtù, e è di grande efficacia contra tutti i veneni, e contra le febbri Pestilentiali, ò che habbiano mala qualità; ò cagione venenosa." Then he describes the composition of this remedy. This European doctor was a contemporary of Li Shi-chên. Who, after reading the confession of his firm belief in the virtues of rhinoceros-horn, will blame the Chinese physicist? In the court ceremonial of France as late as 1789, instruments of unicorn-horn are said to have been employed for testing the royal food for poison.— Chinese lore of the rhinoceros is based on actual observation and speculation built thereon. Not only, as previously pointed out, are the observations of the Chinese in this line more complete, but even more accurate, than those of the classical peoples. In fact, the Chinese adopted nothing from the latter as to their notions of the animal. It is of especial interest that the fantastic belief of the ancients in the mobility of the horn is entirely absent in China. PLINY (*Nat. hist.*, VIII, 21, §73; ed. MAYHOFF, Vol. II, p. 103) observes in regard to the animal *eale*, which has been regarded by some authors as the two-horned rhinoceros, "It has movable horns several cubits long, which it can alternately raise in a combat and turn straightforward or obliquely, according to opportunity" (maiora cubitalibus cornua habens mobilia, quae alterna in pugna sistit variatque infesta aut obliqua, utcumque ratio monstravit). The mobility of the horn is insisted on by COSMAS: "When it is wandering about, the horns are mobile; but when it sees anything which excites its rage, it stiffens them, and they become so rigid that they are strong enough to tear up even trees with the roots — those especially which come in the way of the front horn" (MCCRINDLE, Ancient India, p. 156). In a similar manner al-Bērūnī (SACHAU, Alberuni's India, Vol. I, p. 204) says about the African rhinoceros that its second and longer horn becomes erect as soon as the animal wants to ram with it.

things in the dim candle-light of school traditions, and to think of the rhinoceros as an exclusively southern, tropical animal; but the fact remains that it is not, any more than the tiger, whose original home doubtless was on the Amur, and who is a comparatively recent intruder into Bengal. Climatic conditions and natural surroundings were different in ancient China from what they are at present; and the hills were still crowned by dense forests which were haunted by colossal pachyderms, like the elephant, the tapir, and the rhinoceros.[1]

The historical fact that the rhinoceros was a living contemporary of the ancient Chinese is fully confirmed by the investigations and results of palæontology. As early as 1871, F. PORTER SMITH[2] stated, "The teeth of the extinct rhinoceros of China, met with in the caves of Szech'uan, are sold as dragon's teeth." Specimens of teeth in the possession of the naturalist D. Hanbury, obtained in Shen-si or Shan-si, were examined by Waterhouse of the British Museum, and referred to *Rhinoceros tichorhinus* Cuv., *Mastodon, Elephas, Equus,* and two *Hippotheria.*[3]

Armand David discovered at Süan-hua fu, north-west of Peking, Chili Province, bones from the extremities of a mammal and a nasal bone fragment, which were sent to Paris and determined by GAUDRY[4] as belonging to *Rhinoceros antiquitatis;* and in 1903 M. SCHLOSSER[5] was able to show that this species had once been distributed as far south as the Yang-tse.

The famous naturalist A. R. WALLACE[6] wrote in 1876 that in northern

---

[1] The alligator is now extinct in the Yang-tse, but has risen to life again in the ancient bone carvings of Ho-nan, and is represented in several excellent specimens in the Field Museum obtained with many others from the late F. H. Chalfant.

[2] Contributions towards the Mat. Med. of China, p. 185. Not all "dragon-teeth" (*lung ch'i*), however, originate from the rhinoceros. A number of these gathered by me in a drug-store of Hankow and now in the American Museum of New York (Cat. No. 13,847) were examined by the palæontologist Mr. B. Brown, and contained five teeth of Rhinoceros, one tooth of Mastodon, two teeth of Hipparion (1 m²), and one tooth (P₃) of an undescribed Hipparion. The palæontologist M. Schlosser of Munich (see below) has devoted a careful study to these teeth with remarkable results. Rhinoceros-teeth were employed for medicinal purposes as early as the middle ages. In the Annals of the Sung Dynasty (*Sung shi*), Biography of Ts'ien Shu (929–988; GILES, Biographical Dictionary, p. 144), there is a record that in the year 963 this prince, ruler of Wu and Yüe, sent as tribute ten thousand ounces of silver, one thousand single rhinoceros-teeth (*si ya*), fifteen thousand catties of perfume and drugs, and a hundred wrought objects of gold, silver, genuine pearls, and tortoise-shell (*P'ei wên yün fu*, Ch. 21, p. 114 b). For the year 983, a tribute of rhinoceros-teeth is recorded in the same Annals as having been sent from San-fo-ts'i (Palembang on the north-east coast of Sumatra).

[3] *China Review*, Vol. V, 1876, p. 69.

[4] *Bulletin de la société géologique de France*, Vol. XXIX, 1871-72, p. 178.

[5] Die fossilen Säugetiere Chinas (see below), p. 56.

[6] The Geographical Distribution of Animals, Vol. I, p. 123.

China remains of *Hyæna, Tapir, Rhinoceros, Chalicotherium,* and *Elephas,* had recently been found, closely resembling those from the Miocene or Pliocene deposits of Europe and India, and showing that the Palæarctic region had then the same great extent from west to east that it has now. Of two species,—complete carcasses with the skin,—the two horns, hair, and well-preserved interior organs, were discovered in frozen soil between the Yenisei and Lena Rivers in Siberia.[1] They lived during the ice age, and were covered with a coarse hairy and finely curled coat, the skin being smooth and without the characteristic folds of the now living species. K. A. ZITTEL[2] defines the zone of these two species (*Rhinoceros mercki* and *antiquitatis*) as extending over the whole of northern and central Asia, inclusive of China, and over northern and middle Europe.[3] The best study of this subject, thus far, has been made by M. SCHLOSSER.[4] He records a new species from China (*Rhinoceros habereri*)[5] in two different types, and two others belonging to the forest fauna, one of which is referred to the two-horned Sumatran type,

---

[1] This first find was made in 1771 on the bank of the river Wilui near 64° N. lat. It was first described by the prominent naturalist P. S. PALLAS, in his treatise De reliquiis animalium exoticorum per Asiam borealem repertis complementum (in *Novi Commentarii Acad. Scient. Petropolitanae*, Vol. XVII, 1772, p. 576), and in his Reise durch verschiedene Provinzen des russischen Reichs (Vol. III, p. 97, St. Petersburg, 1776). Head and feet of this animal are still preserved in St. Petersburg. A fundamental investigation still remains that of J. F. BRANDT, De rhinocerotis antiquitatis seu tichorhini seu pallasii structura externa etc. (*Mémoires de l'Acad. de St. Pétersbourg*, series 6, Vol. V, 1849, pp. 161–416). A rich collection of rhinoceros-bones made in the western part of Transbaikalia is in the Museum at Troitskosavsk (compare MOLLESON, in *Papers of the Troitskosavsk-Kiachta Section of the Russian Geogr. Soc.*, in Russian, Vol. I, 1898, p. 71; and the detailed descriptions of Mme. M. PAVLOV, *ibid.*, Vol. XIII, 1910, pp. 37–44).

[2] Palæozoologie, Vol. IV, p. 296. For a restoration of the woolly rhinoceros found in Siberia see N. N. HUTCHINSON, Extinct Monsters, Plate XXI.

[3] We know that fossil rhinoceros-horn had attracted the attention of Siberian natives long before it came to the notice of European scientists. It was employed to strengthen their bows, and the belief was entertained that it exerted a beneficial influence on the arrow hitting its mark. (Compare A. E. v. NORDENSKIÖLD, Die Umsegelung Asiens und Europas auf der Vega, Vol. I, p. 367, Leipzig, 1882.) Now we read in the Annals of the Kin Dynasty (*Kin shi*, Ch. 120, p. 3 a) that the Niüchi, a Tungusic tribe, availed themselves of rhinoceros-horn for the same purpose; and it may therefore be presumed that they obtained it through the medium of trade from inner Siberia (compare above, p. 95). Fossil rhinoceros-horns have also been found in the valley of the Kolyma River. K. v. DITMAR (Reisen und Aufenthalt in Kamtschatka, Vol. I, p. 37, St. Petersburg, 1890) saw one from that region nearly three feet long, and emphasizes the co-existence there of numerous remains of rhinoceros, mammoth, and narwhal.

[4] Die fossilen Säugetiere Chinas (*Abhandlungen der bayer. Akademie*, Cl. II, Vol. XXII, 1903, pp. 1–221, 14 plates). This work is conveniently summed up by H. F. OSBORN (The Age of Mammals, pp. 332–335), where an interesting map (p. 505) is added, showing the former and recent distribution of the rhinoceros. The material described by Schlosser is derived from Chinese drug-stores, and was collected by K. Haberer. The author gives also a valuable summary of the localities in China where fossil remains of mammals have been found (pp. 9–19).

[5] *L. c.*, pp. 58–63.

and the other (*Rhinoceros brancoi*) possibly to the single-horned Indian species.   This fact is in striking agreement with the result of our historical investigation, according to which these two species were known to the ancient Chinese and distinguished by the two names *si* and *se*. In view of the acquaintance of the Chinese with these two species, the question as to the age of the fossil remains is, of course, important. According to the researches of Schlosser, the number of species of fossil rhinoceroses traceable in China amounts to at least seven, three of which originate from the Pleistocene, four from the Pliocene; and Schlosser was able to prove that *Rhinoceros sinensis Owen* does not represent a species from the Tertiary, as presumed heretofore, but should be rather one from the Pleistocene.[1]   There is, accordingly, from a geological viewpoint, good reason to believe that several species of rhinoceros could have survived on Chinese soil down to the historic period when man made his first appearance there;[2] and it is in the records of the Chinese that this fact has been preserved to us.   It even seems to me (but this is the mere personal impression of a layman, which may not be acceptable to a specialist in this field) that the Chinese records, in a highly logical manner, fill a gap between the palæontological facts of Siberia and the present-day existence of the hairy two-horned rhinoceros in south-eastern Asia.   If it is admissible to identify the Siberian *tichorhinus* with the latter species, or to consider the former as the primeval ancestor of the latter, it is conceivable that the Siberian animal, pressed by the advance of the ice, started on a migration southward, and first halted in northern China, where it became the *si* of the Chinese, and whence it finally proceeded south-east.   Whatever this fancy may be worth, there can be no doubt of two points,— first, that the ancient Chinese, from the very beginning of their history, were acquainted with two species of rhinoceros, the single-horned and the two-horned ones, distinguished as *se* and *si;* and, second, that the

---

[1] *L. c.*, p. 52.

[2] We owe to M. SCHLOSSER an interesting discovery in regard to the age of man on Chinese soil.   He describes (pp. 20–21) and figures a tooth, a molar (m_3) of the left upper jaw, which originates either from man or from a new anthropoid.   This tooth is perfectly fossilized, wholly untransparent, and shows between the roots a reddish clay, such as is found only in teeth really coming from the Tertiary, and not from the loess; so that the author is inclined to ascribe to it a tertiary origin, or at all events, a very great age, going back at least to old Pleistocene.   A definite solution of the problem cannot be reached at present.   "The purpose of this notice is," concludes SCHLOSSER, "to call the attention of subsequent investigators, who may have an opportunity of undertaking excavations in China, to the possibility that either a new fossil anthropoid or tertiary man, or yet an old-Pleistocene man, might be found." I agree with Schlosser on this point, and regard his discovery, which certainly so far remains entirely hypothetical, as highly suggestive, and pointing in the direction of a future possibility of a new Pithecanthropus being discovered in Chinese soil.

former is identical with the present *Rhinoceros indicus unicornis* (as proved above all by the linguistic relationship of the word *se* with Tibetan *bse* and Lepcha *sa*), and the latter with the present *Rhinoceros sumatrensis*.[1]

We may now attempt something like a reconstructive history of the rhinoceros in the historical era. At the time of the *Shi king*, the rhinoceros was known to the Chinese as a game-animal. In a song celebrating a hunting-expedition by King Süan, it is said, "We have bent our bows: we have our arrows on the string. Here is a small boar transfixed; there is a large rhinoceros (*se*) killed." [2] As a metaphor, the name of the animal is employed in another song, in which soldiers constantly occupied on the war-path complain of cruel treatment, and say, "We are not rhinoceroses, we are not tigers, to be kept in these desolate wilds." [3] Also cups carved from rhinoceros-horn (*se kung*) [4] make their début in the *Shi king;* and from the passages where it is mentioned, an apparent symbolism is connected with it. In the region of Pin it was customary for the people in the tenth month to visit the palace of their prince with offerings of wine, and "to raise the cup of rhinoceros-horn with wishes for numberless years without end." [5] In another song, a woman yearning for her absent husband takes a cup of wine poured out of a rhinoceros-horn, in the hope that her grief will not last forever. [6] The idea of the healing property of the horn is possibly here involved.

In the *Shu king*, embodying the most ancient historical records of the nation, the rhinoceros is not directly mentioned, but one of the two principal products yielded by it is alluded to. At least, this is the opinion of the Chinese commentators. In the chapter entitled Tribute of Yü (*Yü kung*), "teeth" and "hide" are stated to have been the produce of the two provinces Yang-chou and King-chou, — the former covering the littoral territories south and north of the Yang-tse delta; the latter, the present area of Hu-nan and Hu-pei. The term "teeth" is interpreted

---

[1] It would now be appropriate to introduce for the two extinct Chinese species the names *Rhinoceros unicornis* var. *sinensis* (Chinese *se*), and *Rhinoceros bicornis* var. *sinensis* (Chinese *si*).

[2] *Shi king*, ed. LEGGE, p. 292.

[3] *Ibid.*, p. 424.

[4] Nos. 6393 and 6398. The two characters are read *kung* (according to *T'ang yün*) and *kuang* (according to *Shuo wên*).

[5] *Ibid.*, p. 233. The rhinoceros belongs to the long-lived animals. "Individuals have lived for over twenty years in the London Zoölogical Gardens, and it is stated that others have been kept in confinement for fully fifty years. Consequently there is no doubt that the animal is long-lived, and it has been suggested that its term of life may reach as much as a century" (R. LYDEKKER, The Game Animals of India, p. 31).

[6] *Ibid.*, p. 9.

as ivory; the term "hide," as rhinoceros-hide.[1]   This inference is very reasonable, for the tributes or taxes of those territories cannot have been any ordinary animal teeth or hides of any kind, but they certainly were those teeth and hides most highly prized in the Chou period,— and these were ivory, and rhinoceros-hide desirable for body armor.[2]   The sovereigns of the Chou dynasty hunted the rhinoceros.   In B.C. 965, as recorded in the Annals of the Bamboo Books, Chao Wang invaded the country of Ch'u, and crossing the Han River, met with a large single-horned rhinoceros (or rhinoceroses).   Yi Wang, in B.C. 855, captured, when hunting in the forest of Shê, a two-horned rhinoceros, and had it carried home.[3]

The rhinoceros was also pictured at an early date.   When the emperor mounted his chariot, they posted on both sides of it the lords, whose chariots had red wheels, two crouching rhinoceroses being represented on each wheel; and they posted in front the lords, whose chariots had red wheels with a single tiger represented on each wheel.[4]   This

---

[1] LEGGE, Chinese Classics, Vol. III, pp. 111, 115; COUVREUR, Chou King, pp. 71, 73 (see also HIRTH, The Ancient History of China, p. 121). LEGGE remarks, "This view is generally acquiesced in. Are we to suppose then that the rhinoceros and elephant were found in Yang-chou in Yü's time? They may very well have been so. Hu Wei observes that from the mention or supposed mention of these animals some argue for the extension of the limits of the province beyond the southern mountain-range to Kuang-tung, Kuang-si, and Annam, and replies that the princes might be required to send articles of value and use purchased from their neighbors, as well as what they could procure in their own territories." This conclusion of Hu Wei is quite unnecessary. It is merely elicited by the school opinion that the geographical distribution of animals must have been the same anciently as at present. There can certainly be no more erroneous view. Nothing in nature remains unchangeable. All the large mammals formerly had a far wider range, gradually narrowed by natural events and human depredations. We are simply forced to admit that the rhinoceros, as well as the elephant, existed in Yang-chou and King-chou in the times of antiquity. This logically results from the Chinese records, and is a logical inference from a zoö-geographic point of view. No jugglery or sophistry, like extension of geographic provinces, misunderstanding of words, or introduction of bovines, is necessary to explain and to understand a fact of such simplicity as this one.

[2] The skin of the rhinoceros was utilized in the Chou period also for the manufacture of a yellow glue employed for the purpose of combining the wooden and horn parts of a bow (Chou li, XLIV, BIOT's translation, Vol. II, p. 586). The commentator Wang Chao-yü of the twelfth century justly adds that either skin or horn can be made into glue, but that, as far as the rhinoceros is concerned, only the skin is laid under contribution to this end. Naturally, since the horn is too valuable. Chêng K'ang-ch'êng assures us that in his time (second century A.D.) the stag-glue was exclusively made from the antlers. It is hardly conceivable that Yang-chou and King-chou should have sent as tribute bovine hides which could be obtained everywhere: the specification of these territories implies a specific material peculiar to them; of wild bovines there, nothing is known.

[3] LEGGE, Chinese Classics, Vol. III, Prolegomena, pp. 149, 153; BIOT's translation of Chu shu ki nien, pp. 41, 46 (Paris, 1842). Note that the idea of the monoceros hiai-chai originated in the country of Ch'u (above, p. 115, note 2). In the Ch'un-ts'iu period, as it appears from a passage of Tso chuan (LEGGE, Chinese Classics, Vol. V, p. 289), both se and si were still plenty.

[4] CHAVANNES, Les Mémoires historiques de Se-ma Ts'ien, Vol. III, p. 214.

juxtaposition of rhinoceros and tiger is noteworthy, for it turns up again in Chuang-tse: "To travel by water and not avoid sea-serpents and dragons, — this is the courage of a fisherman. To travel by land and not avoid the rhinoceros and the tiger, — this is the courage of hunters." [1] And in Lao-tse's *Tao tê king* (Ch. 50): "He who knows how to take care of his life, when travelling by road, never meets rhinoceros or tiger; when entering the army, he does not require defensive or offensive armor. The rhinoceros, therefore, finds no place where to insert its horn, the tiger where to lay its claws, the soldier where to pierce him with his sword." [2] Finally in the passage of *Lun yü* [3] already referred to.

The extermination of wild animals made rapid progress; the gradually advancing Chinese agriculturist cleared the hills and deforested the plains in order to till the ground and to yield the means of subsistence for the steadily increasing populace. The famous passage in *Mêng-tse* [4] is of primary importance: Chou-kung, the organizer of the government of the Chou dynasty, broke the rebellions and established peace throughout the empire; "he drove far away also the tigers, leopards, rhinoceroses, and elephants, — and all the people was greatly delighted." Toward the end of the Chou period (middle of the third century B.C.) the one-horned rhinoceros was, in all likelihood, extinct in northern China; and the two-horned species had gradually withdrawn, and taken refuge in the high mountain-fastnesses of the south-west. The strong desire prevailing in the epoch of the Chou for the horn of the animal, which was carved into ornamental cups, and for its valuable skin, which was worked up into armor, had no doubt contributed to its final destruction in the north. So there is no reason to wonder that to the later authors the extinct animal *se* was a blank, and offered a convenient field for fanciful speculations. [5]

---

[1] GILES, Chuang Tzŭ, p. 214.

[2] Compare S. JULIEN, Le livre de la voie et de la vertu, p. 183. It is noticeable that the word *kia*, which in Lao-tse's time designated a cuirass of rhinoceros-hide, appears here in close connection with the rhinoceros.

[3] LEGGE, Chinese Classics, Vol. I, p. 307.

[4] LEGGE, The Chinese Classics, Vol. II, p. 281.

[5] It is a well-known phenomenon in all languages that newly-discovered animals are named for those already known, for example, that sea-mammals are named for land-mammals to which they bear some outward resemblance, or insects for larger animals. Thus we know a rhinoceros-beetle (*Oryctes rhinoceros*) with horns or processes on its head (see *Science*, 1913, p. 883), and a rhinoceros-bird or hornbill (*Buceros rhinoceros*) noted for the extraordinary horny protuberance on the crest of its bill. These examples certainly do not mean that our word "rhinoceros" originally referred to an insect or a bird; but in our effort to coin a name for this beetle and bird, we happened to hit upon the rhinoceros, because certain characteristics of it were, by way of comparison, seen in the former. It is exactly the same when the Chinese, in literary

Se-ma Ts'ien, the father of Chinese history, who was born in B.C. 145, and died between B.C. 86 and 74, and who in his Historical Memoirs repeatedly mentions the two species, doubtless was personally familiar with them; for he locates them in Sze-ch'uan,[1] and we know that he, a great traveller and observer, accompanied the military expedition of the Emperor Wu sent in B.C. 111 into Sze-ch'uan and Yün-nan.[2] Again and again, Chinese authors in the beginning of our era point to that territory as the stronghold of the rhinoceros. We noticed that Kuo P'o of the third century alludes to Mount Liang in Sze-ch'uan as its habitat (p. 94); and we may add to this the weighty testimony of Ch'ang K'ü

---

style, sometimes designate the buffalo "the water-rhinoceros" (*shui se*). In the pre-Christian era the word *se* invariably applied to the single-horned rhinoceros, — a fact confirmed by the concordance of the word with Tibetan (*b*)*se* (p. 116). In times following the ultimate extermination of this species on Chinese soil, this word naturally fell into disuse and became open to other functions; while *si* is still retained as the general word for rhinoceros, whether single or two horned. The word *se* was transferred to the buffalo, because to a naïve and primitive mind the two animals, as has been demonstrated by the world-wide propagation of this notion, bear a striking similarity to each other. The attribute "water" fits both with their fondness for lying embedded for hours in mud and water. A sequel of this transfer in meaning, then, was the impression of recent Chinese authors that the word *se* had denoted also the wild buffalo or ox in the times of antiquity. This, of course, is a phantom. The most instructive passage where the words *si* and *shui se* are used together in close succession occurs in *Sung shi* (Ch. 489, p. 1), where it is said, in the chapter on Champa (Chan-ch'êng), that "the country abounds in peacocks and rhinoceros (*si niu*), that the people keep yellow oxen and buffalo (*shui niu*), and that those engaged in the capture of rhinoceros and elephant (*si siang*) pay a tax on them to the king; they eat the flesh of wild goats and buffalo (*shui se*)." In Siam, permission to capture wild elephants must still be obtained from the Government, and for each animal caught a royalty of $150 is paid (C. C. HANSEN, *Daily Consular and Trade Reports*, 1911, p. 751). In mediæval times when the rhinoceros became gradually scarcer on Chinese soil, and the supply of its skin no longer satisfied the demand for it, buffalo-hide was substituted for it. Chinese authors, with fair accuracy, indicate the time when this change went into effect. A book *Ts'e lin hai ts'o*, quoted in the cyclopædia *Yen kien lei han* (Ch. 228, p. 4), states in substance that what is designated rhinoceros-hide armor in the T'ang History is at present made from buffalo hide, but continues under the general name "rhinoceros" (*si*). The Chinese, accordingly, were perfectly aware of the fact that the ancient cuirasses were wrought from rhinoceros-hide, and that buffalo-hide was a later substitute. Ch'êng Ta-ch'ang, who wrote in the latter part of the twelfth century, says in a discourse on defensive armor (inserted in *Wu pei chi*, published in 1621 by Mao Yüan-i, Ch. 105, p. 4) that the skin of a domesticated animal like the ox is always handy, while the two rhinoceroses *si* and *se* cannot be reared, and their skins are not always obtainable; and that in his time armor was produced from buffalo-hide. In *T'ang shu* (Ch. 41, p. 1) the tribute sent by the district of Kuang-ling in Yang-chou (circuit of Huai-nan) is stated to have consisted of armor made from buffalo-hide (*shui se kia*). The rhinoceros is here out of the question, as it did not occur in that region; and the geographical chapters of the T'ang Annals give us the best clew to the tracing of the geographical distribution of the rhinoceros in the China of that period. It is worthy of note that the term *shui si* ("water rhinoceros") is still employed with reference to the rhinoceros only, not the buffalo. Chung Kia-fu writing in 1845 (*Ch'un ts'ao t'ang chi*, Ch. 30, p. 13) makes the remark that "the cups and dishes carved from rhinoceros-horn (*si kio*) in his time are not from the genuine rhinoceros (*shui si*), but from the horn of a wild ox (*ye niu*) in the countries of the foreign barbarians."

[1] *Shi ki*, Ch. 117, p. 3 b.

[2] CHAVANNES, Les Mémoires historiques de Se-ma Ts'ien, Vol. I, p. XXXI.

of the period of the Tsin dynasty (265–419), who in his interesting work *Hua yang kuo chi* ascribes colossal rhinoceroses to the country of *Pa*, the ancient designation for the eastern part of Sze-ch'uan, and further places the animal in the district of Hui-wu, the present Hui-li in the prefecture of Ning-yüan, province of Sze-ch'uan.[1] However doubtful the exact date of the work *Pie lu* may be, the fact remains that it plainly indicates south-western China in its whole range as the geographical area of the rhinoceros (p. 135).

With their victorious advance toward the south-east in the third and second centuries B.C., the horizon of the Chinese people widened; and they encountered the two-horned rhinoceros also in Tonking.[2] The tributes of live rhinoceroses sent to the Chinese Court from that region have been mentioned (p. 80). Liu Hin-k'i, author of the Records of Kiao-chou, of the fourth or fifth century, gives a perfectly correct description of the two-horned Annamese rhinoceros (p. 93). T'ao Hung-king, the universal genius of the fifth and sixth centuries, logically combines the ancient information relative to the south-west with the additional experience coming from the conquered south-east: Hu-nan, Yün-nan, and Kiao-chou in Tonking, according to him, represent the home of the rhinoceros (p. 136). This alliance of the two geographical zones is a fact of the greatest interest, for this observation of T'ao Hung-king incontrovertibly proves that the word *si* can but signify the rhinoceros, and particularly the two-horned species. When the Chinese first struck the rhinoceros of Annam, the matter is not reported as a novel experience; but they merely renewed an old experience which they had long before made in their own country, and applied the same familiar word to it. If the *si* of Tonking is the rhinoceros (and there is not an atom of doubt about it),[3] the *si* formerly recorded in Sze-ch'uan, Yün-nan,

---

[1] PLAYFAIR, No. 2480 (2d ed., No. 2341). The passages referred to are in *Hua yang kuo chi*, Ch. 1, p. 2 b; Ch. 3, p. 23 (ed. of *Han Wei ts'ung shu*).

[2] *Ts'ien Han shu*, Ch. 28 B, p. 17. Thus the pseudo-embassy of the Emperor Marc Aurel, presenting in 166 A.D. the Annamese products ivory, rhinoceros-horn and tortoise-shell, and mentioned in the Annals of the Later Han Dynasty (HIRTH, China and the Roman Orient, pp. 42, 176), was not the first to make the rhinoceros-horn of Annam known to the Chinese, who were acquainted with it at least two centuries earlier.

[3] The fact is still evidenced by present-day conditions and the continuous trade carried on at all times in rhinoceros-horn from Annam to China. Compare G. DEVÉRIA, Histoire des relations de la Chine avec l'Annam, pp. 41, 88 (Paris, 1880); S. W. WILLIAMS (The Chinese Commercial Guide, p. 94) states that the best sort of rhinoceros-horn comes from Siam and Cochinchina, selling at times for $300 apiece, while that from India, Sumatra, and southern Africa, represents an inferior sort, and sells for $30 and upwards apiece. For the middle ages we have the testimony of Chao Ju-kua (HIRTH's and ROCKHILL's translation, p. 46). As has been pointed out, the word *se* gradually sank into oblivion in the post-Christian era, and was superseded by the exclusive use of the word *si*, which was then applied also to the

etc., must likewise be the rhinoceros; and T'ao Hung-king is our witness in establishing the identity of the animal as occurring in the Chinese and Indo-Chinese zones.   This fact is borne out also by the coincidence of the definitions contributed by Kuo P'o and Liu Hin-k'i.

In the T'ang period (618–906) the animal must have been plentiful in many parts of China.   The geographical section in the Annals of that dynasty carefully enumerates the various articles sent up to the capital as taxes from every district; and it is the local products which come into question.   Besides, rhinoceros-horn, as far as I know, was not imported at that time from beyond the sea.   The present territory of the province of Hu-nan in central China seems to have then abounded in the animal,[1] for no less than eight localities within its boundaries are on record which furnished rhinoceros-horn to the Court: viz., Li-yang in Li chou, circuit of Shan-nan; Wu-ling in Lang-chou; K'ien-chung in K'ien-chou; Lu-k'i in Ch'ên-chou; Lu-yang in Kin-chou; Ling-k'i in K'i chou (modern Yung-shun fu); Kiang-hua in Tao-chou, circuit of Kiang-nan; and Shao-yang in Shao-chou.   Rhinoceros-horn was further supplied from Lung-k'i in Tsiang-chou, from T'an-yang in Sü-chou, Sze-ch'uan; from Ts'ing-hua in Shi-chou (now Shi-nan fu) in Hu-pei Province; from Yi-ts'üan[2] in Yi-chou, province of Kuei-chou; from Annam; and elephants and rhinoceroses were sent from Ling-nan (Kuang-tung), forming the southern part of Yang-chou.[3] Is it conceivable that the tribute of those regions should have consisted of bovine horns which have hardly any commercial value? From mediæval times onward, as the geographical knowledge of the Chinese more and more advanced, and their intercourse and trade with the nations of the southern ocean increased, they became cognizant of the existence of the rhinoceros in India,[4] Java,[5] and Sumatra, and even

---

single-horned rhinoceros.   The rhinoceros of India is indeed designated *si* (*Hou Han shu*, Ch. 118, p. 5 b; *Nan shi*, Ch. 78, p. 7; *T'ang shu*, Ch. 221 A, p. 10 b).   This proves again that the word *si* refers to the rhinoceros, and to this animal only.

[1] Hu-nan, as said before, is mentioned also by T'ao Hung-king.   In this province formerly occurred both the rhinoceros and the elephant, furnishing hide and ivory, respectively, at the time of the Chou dynasty (HIRTH, The Ancient History of China, p. 121, and above, p. 159).   In *Hu-nan fang wu chi*, "Records of the Local Products of Hu-nan" (Ch. 3, p. 14; edition of 1846), it is stated that there was rhinoceros-horn among the local products sent as tribute from Heng-chou; the text is quoted from *Kiu yü chi*, a geographical description of China, which, according to BRETSCHNEIDER (Bot. Sin., pt. 1, p. 162), was published in 1080 A.D.

[2] PLAYFAIR, Nos. 6381, 6713 (2d ed., No. 5701).

[3] PLAYFAIR, No. 8350 (2d ed. No. 3939).   Compare *T'ang shu*, Chs. 40, pp. 1 b, 6 b; 41, pp. 9 a, 9 b, 10 a; 43, p. 1 a.

[4] See note 3 on p. 163.

[5] As regards Java, rhinoceros-horn is listed among its products in *T'ang shu* (Ch. 222 C, p. 3; and GROENEVELDT, *Miscell. Papers relating to Indo-China*, Vol. I, p. 139).   The *Sung shi* (Ch. 489; GROENEVELDT, *ibid.*, p. 144) reports a tribute from Java

Africa. The interesting notes of Chao Ju-kua written in 1225,[1] eminently translated and interpreted by HIRTH and ROCKHILL, afford an
excellent view of all the localities from which rhinoceros-horn was
traded to China, during the middle ages;[2] he refers to the Berbera coast
as producing big horns (p. 128), and records them also for the island of
Pemba (p. 149).[3]

Returning to China, we find trustworthy accounts, according to
which the rhinoceros has persisted there in some localities at least
down to the thirteenth century. Kuo Yün-tao, who composed an elaborate history of Sze-ch'uan in the thirteenth century,[4] states that the
region of the aboriginal tribes of the south-west (Si-nan I) harbors a
great number of rhinoceroses and elephants; and this agrees with the
above statement of Su Sung (p. 140) that rhinoceros-horns came from
Sze-ch'uan at the same period. As the author includes also the province of Kuei-chou, we are allowed to presume that the two-horned
rhinoceros still inhabited the forests of Sze-ch'uan and Kuei-chou during
the age of the Sung dynasty (960–1278).[5] In the year 987, as narrated
in the Annals of the Sung Dynasty,[6] a rhinoceros penetrated from the
southern part of K'ien into Wan-chou[7] where people seized and slew it,

---

of short swords with hilts of rhinoceros-horn or gold, and records the word *ti-mi* as
the native name of the rhinoceros. This word is not Javanese, in which the animal
is called *warak*, but is presumably traceable to the Kawi language (compare the
discussions of this word by G. SCHLEGEL, *T'oung Pao*, Vol. X, 1899, p. 272; and P.
PELLIOT, *Bull. de l'Ecole française*, Vol. IV, 1904, p. 310).

[1] PELLIOT, *T'oung Pao*, 1912, p. 449.

[2] At least as early as the fifth century, carved objects of rhinoceros-horn were
traded to China from the Roman Orient and India (HIRTH, China and the Roman
Orient, p. 46). In the year 730 a tribute of rhinoceros-horn from Persia is mentioned
(CHAVANNES, *T'oung Pao*, 1904, p. 51).

[3] The Geography of the Ming Dynasty (*Ta Ming i t'ung chi*, ed. of 1461, Ch. 91,
fol. 20) lists rhinoceros-horn also among the products of Arabia (*T'ien-fang*). Under the Ming, rhinoceros-horn was imported to China from Champa, Cambodja,
Malacca, Borneo, Siam, Bengal, and rhinoceros-flesh from Java. These data are
derived from the *Si yang ch'ao kung tien lu* by Huang Shêng-tsêng, published in 1520
(reprinted in *Pie hia chai ts'ung shu*); this is the most convenient work on the countries of the Indian Ocean and on Chinese knowledge of them during the Ming, and
gives more information than the Ming Annals.

[4] *Shu kien* (Ch. 10, p. 1), reprinted in *Shou shan ko ts'ung shu*, Vol. 23. The preface of Li Wên-tse is dated 1236.

[5] It might seem that the rhinoceros was extinct in China proper at the time of
the Yüan period (1271–1367), judging from a remark made by Chou Ta-kuan, in
his Memoirs on the Customs of Cambodja, to the effect that the latter country harbors the rhinoceros, elephant, the wild buffalo, and the mountain-horse, which do not
occur in China (PELLIOT, *Bulletin de l'Ecole française*, Vol. II, 1902, p. 169); but the
passage is by no means conclusive, and may simply be interpreted in the sense that
the author had never seen or heard of a rhinoceros in China.

[6] *Sung shi*, Chapter *Wu hing chi*, quoted in *T'u shu tsi ch'êng* (Chapter on Rhinoceros).

[7] Now the district of Wan in K'uei-chou fu, Sze-ch'uan Province.

keeping its skin and horn.   It should be remembered that Li Shi-chên,
who lived in the sixteenth century, still assigned to the rhinoceros the
southern portion of Yün-nan and the adjoining Tibetan regions.   Even
at the present time the rhinoceros may still exist in isolated spots on
Chinese territory.

JOHAN NEUHOF[1] locates it in the province of Sze-ch'uan, particularly
near the small town of Po (P'a is presumably meant).

O. DAPPER[2] appropriates to the rhinoceros Sze-ch'uan and Chucheu-
fu (?) in Kuang-si.   DU HALDE[3] ascribes the rhinoceros to the prefecture
of Wu-chou in Kuang-si.   L. RICHARD[4] states, "On account of the
devastation prevailing in Kuang-si, a great number of wild animals are
found there: the tiger, rhinoceros, panther, tapir, wolf, bear, and fox."
The zoölogist W. MARSHALL,[5] in a general summary of the Chinese
fauna, observes that the south, and particularly the south-west, of China,
harbor decidedly Indian types of mammals, among these the Indian
tapir and the single-horned rhinoceros.

The products yielded by an animal, and the manner of their utilization, allow also conclusive evidence in regard to the nature of the animal
itself.   That rhinoceros-horn was worked in ancient times and well
differentiated from other ordinary horn, is evidenced by the curious
fact that three distinct verbs pertaining to the treatment of ivory,
ordinary horn, and rhinoceros-horn, are listed in the dictionary *Erh ya.*
The carving of ivory is designated by the word *ku* (No. 6248); the treating of ordinary horn (*kio*), by the word *hio*;[6] the carving of rhinoceros-
horn (*si*), by the word *ts'o* or *ts'uo* (No. 11,766).   In the latter case
Mr. GILES, in the second edition of his Dictionary, has justly retained
the meaning "to make rhinoceros-horn into cups; to carve."   The
word is apparently identical with *ts'o* (No. 11,778), meaning "to file,
trim, cut, plane, polish," etc., including all the various manipulations of
the carver.

At this point it may not be amiss to call to mind the fact that a

---

[1] Die Gesantschaft der ostindischen Geselschaft, p. 348 (Amsterdam, 1669).
[2] Beschryving des Keizerryks van Taising of Sina, p. 230 (Amsterdam, 1670).
[3] A Description of the Empire of China, Vol. I, p. 121 (London, 1738).
[4] Comprehensive Geography of the Chinese Empire, p. 198 (Shanghai, 1908).
[5] Die Tierwelt Chinas (*Zeitschrift für Naturwissenschaften*, Vol. 73, 1900, p. 73).
[6] Composed of the classifier *kio* ('horn') at the foot, and the phonetic complement *hio* ('to learn').   The character is not contained in our current Chinese dictionaries (not even in PALLADIUS); students of Chinese will easily find it in K'ang-hi's
Dictionary under classifier 148 (13 strokes, first character).   The definition of the
word *hio* given by the *Shuo wên* — *chi kio* ("to treat horn") — calls for attention,
any word like cutting or carving being avoided.   The ancient Chinese were familiar
with all processes of horn-work (soaking, slicing, welding, etc.), which are described
in the *Chou li.*

rhinoceros-horn is capable of being carved, but that the horn of a bovine animal cannot be carved. These horns, biologically, are entirely different in origin and structure. The Chinese were quite right in regarding the rhinoceros-horn as a marvel of nature, for it is a unique phenomenon of creation. It is composed of a solid mass of agglutinated hairs or bristles, and has no firm attachment to the bones of the skull, which are merely roughened and somewhat elevated so as to fit into the concave base of the solid horn. Ox, sheep, or antelope, however, have hollow horns; deer and giraffe, bony antlers. None of these is fit to be worked into a cup; and a cup carved from a horn can mean nothing but one carved from rhinoceros-horn. Horns of bovine animals, as we all know, may be utilized as drinking-vessels, or, as among primitive tribes, as powder-flasks, or, as among the Tibetans, even as snuff-bottles, or, as in India, to pour out holy water; but they are by nature made ready for use, and do not require any carving. The *se kung* of antiquity are certainly cups carved from rhinoceros-horn,[1] not cups of buffalo-horn, as Mr. GILES (No. 10,298) has it in the second edition of his Dictionary.

Naturally, none of those ancient drinking-horns has survived, but at a later time they were imitated in bronze. There are, at least, some bronze drinking-cups preserved, which are connected by Chinese archæologists with the drinking-horns of antiquity. In the *Po ku t'u lu* (Ch. 16, p. 16) an illustration (Fig. 23) is given under the title *Han hi shou pei* ("cup with the head of a sacrificial bull, of the Han period"). A similar bronze (Fig. 24) is figured in the *Kin shi so*, with the legend *Chou se kung* ("rhinoceros-horn cup of the Chou period").[2] The text of the *Po ku t'u lu* quotes the passage of the *Shi king* in which the *se kung* are spoken of (above, p. 159), and says that this bronze cup comes very near to them. The bull-head is certainly a feature which originated only subsequently in bronze-casting, when the accepted forms of the horn cups were imitated in bronze. It is noticeable that the cup, as figured in the Sung Catalogue of Bronzes, corresponds in a measure to the form of a rhinoceros-horn inverted and hollowed out from the base.

---

[1] Likewise PALLADIUS (Vol. I, p. 136) and COUVREUR (p. 451).

[2] The authenticity of the specimen of the *Kin shi so* seems somewhat contestable. The head is that of a stag, but is equipped with ox-horns. The dating in the Chou period is arbitrary and unsupported by evidence. It is remarked in the explanatory text that it is not known whether the piece is a rhinoceros-horn cup (*se kung*). The similarity of the two specimens (Figs. 23, 24) with the rhyton of the Greeks is apparent, but there is no necessity of assuming an historical interrelation of the two types. Both were independently developed from natural horns used as drinking-cups, which were subsequently imitated in more durable materials, like clay and metal. Moreover, the Greek rhyton has a feature lacking in the Chinese specimens,— a single oblong loop-handle.

As stated by a great number of commentaries,[1] the *se kung* were carved from wood if rhinoceros-horn were lacking. Certainly, there could have never been any want of bovine horns; and it is inconceivable that an ox-horn should have been ever reproduced in wood. Fan Ch'êng-ta, in his *Kui hai yü hêng chi*,[2] has a note to the effect that "the people on the seacoast make cups from ox-horn (*niu kio pei*) by splitting the horn

FIG. 23.
Bronze Rhyton attributed to Han Period (from *Po ku t'u lu*).

in two and smoothing the edges to enable them to drink wine from them, which appears as a survival of the ancient rhinoceros-horn goblets." They did not carve their cups from ox-horn, however: they merely split the latter, as the author advisedly says.[3]

---

[1] See *T'u shu tsi ch'êng, K'ao kung tien*, sect. 197, *kung pu.*

[2] Edition of *Chi pu tsu chai ts'ung shu*, p. 14 b.

[3] It may be stated positively that a confusion of rhinoceros and ox horns (or any other horns) is absolutely impossible, the two being entirely distinct organic substances of different origin and structure; and we are quite willing to believe Chang Shi-nan, the author of *Yu huan ki wên* early in the thirteenth century, that an artisan of Shuang-liu hien in Ch'êng-tu fu, who chanced upon the idea of making ox-horn into rhinoceros-horn, was not very successful in passing off his ware, because it did not exhibit any of the properties of rhinoceros-horn. The latter is indeed a unique product

The *Chou li* has a report on the office of the horn-collectors (*kio jên*) whose task it was to collect teeth, horns, and bones in mountains and marshy places.[1] Chêng K'ang-ch'êng of the second century A.D. comments that the big ones among these objects came from the elephant and rhinoceros, those of small dimensions came from Cervidae. They did not pick up ox-horns. The word *kio* ("horn") is

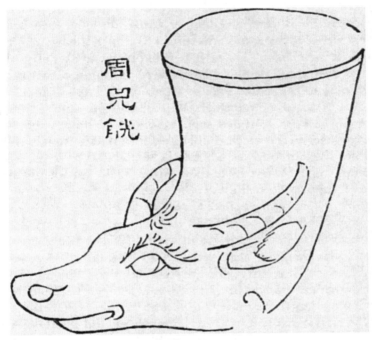

FIG. 24.
Bronze Rhyton attributed to Chou Period (from *Kin shi so*).

used also in the sense of a vessel carved from horn; and there are several types of ancient bronze vessels, the names of which are written with characters combined with the classifier *kio* ("horn"). This would hardly be the case if these various bronze forms did not go back to older vessels carved from horn. He who will study the illustrations of these cups in the *Po ku t'u lu*, or in the *T'u shu tsi ch'êng*, where they are reproduced after the former work, will be struck by the fact that they do not exhibit the slightest resemblance to ox-

---

of nature and has no substitute. A very interesting piece of ancient Japanese pottery in the Imperial Museum of Tōkyō (figured by N. G. MUNRO, Prehistoric Japan, p. 483) is made in imitation of an animal's horn, bearing a striking resemblance to a rhinoceros-horn.

[1] BIOT, Chou li, Vol. I, p. 378. The *Chou li* describes the rhinoceros-horn as yellow (Vol. II, p. 586).

horns, but display most elegant shapes of soft, rounded outlines, such as could have been carved only from rhinoceros-horn. Moreover, these horn vessels were differentiated according to their capacities: the vessel *kio* (No. 2218) containing one pint (*shêng*); the vessel *ku* (No. 6221), two pints; the vessel *chi* (No. 1925), three pints;[1] the vessel *kio* ("horn"), four pints;[2] the vessel *kung* or *kuang* (No. 6393), seven pints. All of these served the same purpose,—they were filled with wine; and the ancient tradition is that the bad or tardy disciple, or whoever had violated a rule or lost a game, was forced to empty the horn at a draught by way of punishment.[3]   Now, there could be no greater absurdity than to suppose that these drinking-horns were veritable ox-horns, whether from a wild or domesticated ox, and were emptied at a draught by those wretched fellows. Every former German student knows from experience that an ox-horn contains such a volume of liquor, that even the strongest drinker in the world could not empty it at a draught; and every one who has lived among the Chinese is acquainted with those tiny bits of porcelain cups from which they enjoy their hot rice-wine during meals, and knows how limited their abilities *in Baccho* are. The punishment of forcing a negligent student to do away with a quantity of wine contained in a buffalo-horn would certainly have been most efficient in killing him instantly and saving further trouble about him; that, however, was not the intention of the law-giver. Naturally, these drinking-cups of early antiquity were nothing but miniature cups carved from rhinoceros-horn. Indeed, it is the very horn of the rhinoceros, which renders this cup eligible as a fit means of correction, for "the horn of the rhinoceros is terrible to its enemies; and for this reason the holy emperors of old, in condemning a man to empty a cup by way of punishment, wanted it to be made from rhinoceros-horn."[4] The terror which the animal was able to inspire in man should be brought home to the mind of the culprit, and this was the essential point of his punishment. Similar was the idea when the rhinoceros-horn cup was emptied on the occasion of a vow; as in the case of the three lords who pledged fidelity to the King of Tsin, with imprecations of calamities to

---

[1] According to *Shuo wên* (Ch. 11, p. 4), four pints; while the vessel *shang* (No. 9744) held three pints.

[2] Compare the dictionary *Kuang ya* by Chang I, written in the first part of the third century (Ch. 8, p. 5 b; edition of *Han Wei ts'ung shu*).

[3] Compare BIOT, Chou li, Vol. I, p. 259; Vol. II, p. 17. In one passage of the *Li ki* (ed. COUVREUR, Vol. II, p. 618), horns (together with *kia*) appear as sacrificial cups, from which to pour out libations to the ancestors.

[4] According to *Yün hui*, as quoted by A. TSCHEPE (Histoire du royaume de Tsin, p. 308, Shanghai, 1910).

themselves should they break their word.[1]   As Wang Fu says in the
Po ku t'u lu (quoted above, p. 131), the rhinoceros represented on the
bronze wine-kettles of the Shang period was a fit emblem to serve as a
warning to the drinker, and to inculcate in him moderation: as the
rhinoceros is capable of doing injury to man, so excessive indulgence
in spirits might harm him.[2]

We now recognize that the rhinoceros, looked upon as a moral and
educational factor, moves on the same line as the monoceros *hiai-chai*
discussed above (p. 115), which is able to decide judicial proceedings.[3]
This inward affinity proves that this monoceros is a legitimate offshoot
of the rhinoceros.   We have seen that the single-horned rhinoceros *se*
existed in the country of Ch'u in the beginning of the Chou dynasty
(p. 160), and it was among the people of Ch'u that the notion and word
hiai chai originated (p. 115).   The transformation into a goat of what
originally was the rhinoceros was developed by the notion of "butting"
under the influence of a legend emanating from Ch'u, which unfortunately is lost.

In past times the rhinoceros was so plentiful in the home of the
Chinese, that carvings from its horn belonged to the common household
objects, especially at the period before the utilization of metals, when
wood, bone, horn, antler, and stone furnished the material for the making
of implements.

There are other objects stated to have been made of rhinoceros-
horn, where the supposition that ox-horn might be involved is again
out of the question.   In the biography of Li Se, who died in B.C. 208,[4]
objects carved from rhinoceros-horn and ivory (*si siang k'i*) are mentioned, and classed among *objets de vertu*.[5]   Implements of ox-horn
would certainly not rank in this category.   According to *Hou Han shu*,[6]
seals were cut out of rhinoceros-horn and ivory.   Everybody knows the

---

[1] TSCHEPE, *l. c.* The warlike character of the rhinoceros is still indicated by the literary designation *Si pu* for the Board of War (*Ping pu*) and the rhinoceros forming
the badge of the ninth grade of the military officials.

[2] The rhinoceros as a means of punishment appears also in the case of Wan of
Sung, who paid the penalty of his crimes by being bound up in a rhinoceros-hide (*Tso
chuan, Chuang kung*, twelfth year: LEGGE, Chinese Classics, Vol. V, p. 89).

[3] In the time of the philosopher Wang Ch'ung, who wrote his work *Lun hêng* in
82 or 83 A.D., Kao Yao and this creature were painted in the courtyards of public
buildings; the latter, in agreement with the ancient definitions, apparently as a goat
with a single horn, for it instinctively knew the guilty.   When Kao Yao administered
justice and entertained doubts of a man's guilt, he ordered this goat to disentangle
the case: it butted the guilty party, but spared the innocent (FORKE, Lun-hêng,
pt. II, p. 321).

[4] GILES, Biographical Dictionary, p. 464.

[5] *Shi ki*, Ch. 87, p. 2 b.

[6] Ch. 40, p. 5 a.

square and rectangular cubes in which Chinese seals are shaped, and to cut such a seal out of ox-horn is impossible.

Finally, the memorable passage in the *Chou li* from which we started, and that is discussed in the following chapter, regarding the manufacture of hide armor, is sufficient evidence in itself that the hide in question is only that of the rhinoceros. Mr. Giles renders the words *se* and *si* indiscriminately by "bovine animal;" it is manifest, however, from the text in question, that *se* and *si* are two distinct animals, but can by no means be two distinct bovine animals. It will be seen that the *Chou li* speaks of three kinds of cuirasses, — those made from the hide of the two-horned rhinoceros (*si*), which consist of seven layers, and will last a hundred years; those made from the hide of the single-horned rhinoceros (*se*), which consist of six layers, and will last two hundred years; and those made from a combination of both hides, which consist of five layers, and will last three hundred years. The skin of the rhinoceros was utilized for the manufacture of hide armor, because it was the thickest and strongest known in the animal kingdom,[1] and because the rhinoceros was justly considered a strong, warlike, and long-lived creature (see p. 159); and the qualities of the animal were believed to be transfused into the body of the wearer of the cuirass. The single-horned rhinoceros was the bigger and stronger of the two species known; and for this reason armor from its hide was believed to last twice as long as that of the two-horned kind. We notice that there is a close interrelation between the number of layers of the hide and the number of years that the armor is supposed to endure. All this becomes intelligible only if we interpret the two words *se* and *si* in the manner that has been proposed.[2] But what would the interpretation be if the armor of the Chou had been made from the hide of wild bovine animals? The passage, in this case, could receive no intelligent and convincing interpretation. That bovine hide can be utilized in the making of armor, nobody denies. It is utterly inconceivable, however, that the ancient Chinese should have taken the trouble to hunt wild bovine animals, in order to secure their skins for cuirasses, since they were in possession of plenty of domestic cattle from which leather was obtainable; and this one certainly could

---

[1] The toughness and durability of rhinoceros-hide are indicated also by its utilization in the coffin of the Son of Heaven, which was fourfold. The innermost coffin was formed by hide of water-buffalo and rhinoceros, each three inches thick. This leather case was enclosed in a coffin of white poplar timber; and this one, in two others of catalpa-wood (Couvreur, Li ki, Vol. I, p. 184; Legge's translation in *Sacred Books of the East*, Vol. XXVII, p. 158).

[2] The fact that the general notion of leather and hide (*p'i ko*) was closely associated with rhinoceros-skin is evidenced by Yen Shi-ku defining that term by the words *si se* (*Ts'ien Han shu*, Ch. 28 B, p. 16 b).

have been employed with greater facility and the same result for the purpose of defence. And if they had really employed cowhide to this end, why should the *Chou li* not simply state that cuirasses were made of this material (*niu p'i*)? Why should it introduce the story of two wonderful animals *se* and *si*, interwoven with religious beliefs of longevity, if nothing but a mere every-day cowhide was at issue? On the other hand, there is every reason to believe that the skin of ox or cow was never, for religious reasons, employed in ancient China in the making of armor. The ox was a sacred, and in a measure inviolable animal, looked upon as the helpmate in gaining man's daily bread. He was the animal sacrificed to the deities Heaven and Earth. There is no account to the effect that neat-leather was ever employed for cuirasses; while the tradition that rhinoceros-skin is a fit material for this purpose, as we saw, has been maintained even by later authors.

II. DEFENSIVE ARMOR OF THE ARCHAIC PERIOD

"Your subject has heard that the army of the Son of
Heaven is rather maintained for the assurance of peace
than for the purpose of aggressive war. The Empire and
all its inhabitants being your own, is it worth while wast-
ing a day's business on the land of the Barbarians, or driving a single horse to exhaustion on their behalf?"
Memorial of HUAI-NAN-TSE to the Emperor Wu.

Defensive armor, as employed in the epoch of antiquity, is characterized by the absence of any metal.[1] During the Chou period
(B.C. 1122–255) harness was exclusively made of hide (*lorica* of the
Romans). Ts'ai Ch'ên, in his commentary to the *Shu king* (published
in 1210), makes this correct general observation on the subject: "In
ancient canonical literature it is a question only of cuirasses (*kia*,
No. 1167) and leather helmets (*chou*, No. 2463). Prior to the time of the
Ts'in, metal armor (*k'ai*, No. 5798) and metal helmets (*tou mou*, Nos.
11,424, 8041) were not in existence. The ancients availed themselves of hide for the making of armor (*kia*). From the time of the

---

[1] It is not the object of the present investigation to give a detailed history of
Chinese defensive armor of all periods, or to describe each and every type of armor
mentioned in Chinese records. Such a task would require dwelling at great length on
the military organization and activities of every dynasty, and would swell into several
volumes of questionable practical value. It is merely my intention to outline the
principal and conspicuous features of the general development of the matter, and to
emphasize those types of armor which are of particular interest to the archæologist
and ethnologist. Only those Chinese records which have a real value for an historical
consideration of this subject are here exhibited. The theories of the philosophers
and the later legendary inventions are historically worthless, and only interesting
for what they are worth,— in their quality as philosophy, poetry, or folk-lore. A
pure fable it is, for example, when the philosopher Kuan-tse makes Ch'i Yu (alleged
B.C. 2698) the first inventor of metal armor (*k'ai*), and when as late a work as the
T'ai po yin king by Li Tsūan of the middle of the eighth century (WYLIE, Notes on
Chinese Literature, p. 90) is gracious enough to ascribe to the same also the honor of
having first cut hide into armor, and goes on to construct the evolutionary scheme
that Shên-nung made weapons of stone, Huang-ti of jade, and Ch'i Yu of bright metal. The famous Ts'ao Chi (192–232) is credited with the statement that the former
emperors bestowed on officials an armor (*k'ai*) called "brilliant like ink" (*mo kuang*)
and another called "brilliant like light" (*ming kuang*), one suit of armor with a
double seat in the trousers (*liang tang* [No. 10,727] *k'ai*), one suit of ring and chain
armor (*huan so k'ai*), and one suit of horse mail. This text is not well authenticated, and is hardly deserving of historical credence. The ring and chain armor is
an anachronism in view of Ts'ao Chi's time; and any armor of the designation *k'ai*
did not exist under the ancient emperors. The expression *huan so k'ai* occurring
in this passage is explained in the dictionary *Chêng tse t'ung* as identical with *so
kia* ("chain armor"). *T'u shu tsi ch'êng*, in reproducing this passage, writes *mo
kuang*, as above; *P'ei wên yün fu* has in its place *hei kuang* ("of black brilliancy");
and *Ko chi king yüan* has *li* (No. 6870) *kuang*, which seems to be a misprint. The
two latter works write the character *tang* in the phrase *liang tang k'ai* without the
classifier 145.

174

Ts'in and Han, iron armor and helmets (*t'ie k'ai mou*) gradually came into use.   These two characters (*k'ai mou*) are formed with the classifier 'metal' (*kin*), for these objects were made from iron." This chronological cal division of words and matters, indeed, corresponds to the facts as expressed in the documents of literature.   The comment of Ts'ai Ch'ên relates to the speech of the Prince of Lu, Po K'in, son of Chou Kung (*Shu king*, IV, 19), in which he admonished his soldiers to see that their cuirasses and helmets were well sewed together (that is, were in good order), and that the laces of their shields were well secured.   In this passage the three means for making the complete defensive armor of the primeval epoch are named; and these are followed by the three principal representatives of offensive armor,— the bow, the long and the short spears.

We meet in the early period essentially two varieties of hide armor, distinguished by two different words, *kia* (No. 1167) and *kiai* (No. 1518). The latter, as will be seen (p. 195), was scale armor, composed of imbricated leather pieces which were cut out in the shape of scales (compare Plate XIV).   The former was a cuirass made in imitation of a coat.   Our knowledge of this device is mainly founded on the State Handbook containing the ritual and institutes of the Chou dynasty, the *Chou li*.   A special office of armorers was instituted at the Court of the Chou dynasty; they were called *han jên*, "men who envelop (*han*, No. 3809) the body with a protective contrivance."   The manufacture of these military leathern costumes is minutely described in the *Chou li*.[1]

"The armorers make the cuirasses (*kia*).   Those made from the hide of the two-horned rhinoceros (*si*) consist of seven layers of hide; those made from the hide of the single-horned rhinoceros (*se*) consist of six layers; those made from a combination of both hides consist of five layers.   The first endure a hundred years; the second, two hundred years; the third, three hundred years.   In order to accomplish a cuirass, first, a form (dummy) is made,[2] and then the hide is cut in accordance with it.   The hide pieces are weighed; and two piles equal in weight are apportioned, the one for the upper, the other for the lower part of the cuirass.   The long strips, into which the hide has been cut up,

---

[1] BIOT, Vol. II, p. 506.   The work of Biot is here, as in other instances, quoted for easy reference, as by referring to Biot the Chinese text may readily be looked up; but my rendering is based on the original text, and on several points deviates from that of Biot, and fundamentally, in this passage descriptive of armor.

[2] The dummy was patterned according to the figure of the individual for whom the cuirass was intended, and the hide was tailored and adjusted in correspondence with the dummy.   It was left on the latter for some time, until it was thoroughly hardened and had assumed the required shape.   The process was the same as that still practised on a smaller scale by the Chinese hatters, who fashion their caps over wooden models.

are laid around horizontally. In general when the hide has not been properly cured, the cuirass is not strong;[1] when the hide is worn out, it will wrinkle. The method of inspecting cuirasses is as follows: the stitches, when examined, must be fine and close; the inner side of the hide must be smooth; the seams are required to be straight; the cuirass must perfectly fit into the case in which it is to be enclosed.[2] Then it is taken up,[3] and when examined, it must allow of ample space. When it is donned, it must not wrinkle. When the stitches are examined, and found to be fine and close, it is a sign that the hide is strong. When the inner side is examined, and found to be smooth, the material is well prepared and durable. When the seams are examined, and found to be straight, the cutting is perfect. When it is rolled up and placed in its case, it should fold closely. When, however, it is taken out, it should offer ample space to the wearer, and it is then beautiful.[4] When it is donned without wrinkling, it will gradually adjust itself to the form of the trunk."

We gather from this account that the ancient hide corselets were not downright primitive affairs, but testify to an advanced stage of culture. Armor, as early as that archaic period, was individual, and carefully adapted to the shape of the body. Its weight was equally balanced between the upper and lower portions, the former reaching from the shoulders to the loins, the latter from the loins to the knees. Apparently it was but one uniform coat, without sleeves, and without any separate parts for protection, as nape-guards, greaves, knee-covers, or

---

[1] BIOT translates, "En général, si la façon n'est point parfaite, la cuirasse n'est pas solide." And COUVREUR (Dictionnaire chinois-français, p. 799), "Toute cuirasse d'un travail imparfait n'est pas solide." My rendering is based on the comment of Chêng Ngo.

[2] The cuirass was rolled up and encased in a covering, presumably of hide. This case was styled *kao* (No. 5949), a word now used in the sense of "quiver." Hide bags in which to preserve armor are still used in Tibet, and there is one in the Museum's collection. The Chinese now avail themselves of trunks with a special compartment in the lid for the helmet (compare Plate XLIII).

[3] The first test that the cuirass is exposed to refers to its fitting into the case; the second, to its fitting on the wearer; for this purpose it is taken out of the case.

[4] As will be seen from Biot's comment, the K'ien-lung editors hold that the last two qualities are difficult to reconcile, as, on the one hand, the cuirass must fit like a coat without throwing folds, and, on the other hand, must have ample space and splendor. I do not believe that this objection is very serious. The conditions stipulated in the text could all, indeed, be fulfilled. The essential requisite was elasticity to grant full freedom of motion; the cuirass must be tight-fitting, but if the hide is sufficiently elastic, "ample space" is secured to the wearer. Owing to its flexible character it could be readily rolled up, and, when taken out of its case, immediately reverted to its original shape, so that it could be donned without loss of time. The word *ming* ("brilliant") translated by BIOT "alors elle a de l'éclat," I believe, means something like "it is then in evidence, it fulfils its purpose."

buskins.[1] The hide was well cured, and the inner side cleaned from all adhering impurities.

My conception of the technicalities in the construction of this armor is widely different from that of BIOT based on the opinions of the Chinese commentators. These interpret that the cuirass made from the hide of the two-horned rhinoceros consisted of seven pieces sewed together; that from the hide of the one-horned rhinoceros, of six; and that made from a combination of both, of five pieces. There is no sense in this point of view of the matter. The commentators of the Han and later ages were unable to form a clear idea of the cuirass peculiar to the Chou period, because it was lost in their time; and they merely applied to the latter the notions which they had gained from a consideration of contemporaneous armor. The armor terminology of the Han was read into Chou armor, and a purely philological reconstruction was reached, which hardly corresponds to a living reality. The armor, as interpreted by the Chinese scholars, in my opinion, is technically impossible, and beyond our experience: armor-suits of such requirements have been made nowhere in this world, and in all likelihood never could have been made.[2]

There is no *raison d'être* in assuming that the first should have been

---

[1] Red knee-covers and buskins are mentioned in the *Shi king*, but they were outfits belonging to the costume of ceremony, not of war (LEGGE, Chinese Classics, Vol. IV, *Prolegomena*, p. 157, and p. 402).

[2] For technical reasons it is highly improbable that the hide armor of the Chou was sewed together from different pieces, because such a process would considerably diminish its strength and capability of resistance, and a blow struck at the seams would have had dangerous consequences. On the contrary, wherever hide armor was made, the principle was quite naturally developed to make it, as far as possible, in one piece; and this is exactly the point where the chief purpose of defensive armor comes in. If the Chou cuirass had been patched together from odd pieces, as the later Chinese philologists would make us believe, it could not have been a defensive armor proper, but simply a skin garment. W. HOUGH (Primitive American Armor, *Report U. S. National Museum*, 1893, p. 641) informs us that "American skin armor was always made in one piece folded over, sewed above the shoulders, leaving an orifice for the head and with a hole cut out of the left side for the left arm, the right side of the garment remaining open; the skin was often doubled, but more frequently the coat was reinforced with pieces of thick hide." Indeed, our Chou armor, *cum grano salis*, can have been no other in type and appearance than the hide armor of the American Indians, as figured on our Plate XI and by HOUGH on Plates XVI–XIX, although it may have been somewhat more elegant in its fit to the individual wearer. HOUGH (pp. 645, 646) furnishes several examples of the fact that hide armor in America was worked in several layers; thus, two, three, or more folds of the strongest hides were employed by the Nass Indians of the Tsimshian stock; a great many folds of dressed antelope-skins by the Shoshoni; and the Navajo singer chants of suits of armor made of several layers of buckskin. Likewise A. P. NIBLACK (The Coast Indians of Southern Alaska, *Report U. S. National Museum*, 1888, p. 268) states that the leather jerkins formerly made in Alaska were of one, two, or three thicknesses of hide, and in itself offered considerable resistance to arrows, spears, or dagger thrusts. Armor of rhinoceros-hide, according to Nachtigall, is still made and employed by the Arabs of the Sudan (H. SCHURTZ, Grundzüge einer Philosophie der Tracht, p. 114).

made in seven, the second in six, and the third in five pieces; moreover, they double these figures, and conjecture that the upper portion (*shang lü*) and the lower portion (*hia lü*) each consisted of this number of pieces. But how can such an affair be realized? It is perfectly conceivable that a coat is composed of six pieces (two in front, two in the back, and two on the sides); any other even number — as four, eight, ten, or more — likewise is imaginable. It is not easily conceivable, however, as being incompatible with a normal state of affairs, that a cuirass should have consisted of seven or five pieces (or any larger odd number of pieces), as the Chinese commentators and Biot would have us believe. This supposition is not very reasonable. The symmetry of the human body inevitably results in principle in a strictly symmetrical style and technique of costume, and of armor especially: asymmetric armor nowhere exists.[1] Normal harness of the primitive stages of culture is usually composed of an even number of pieces; and for this reason, the Chinese interpretation is improbable. Even granted that another point of view is possible in theory, — that, for example, the harness of seven pieces may have had four in the back and three in front, or three in the back, two on the sides, and two in front, etc.,[2] — we still face the mystery of the threefold classification graduated according to age: what should be the reason that the cuirass of seven pieces is supposed to last a hundred years, that of six pieces two hundred years, and that of five pieces three hundred years? This is the salient point, to which no Chinese commentator has paid due attention; but it is obvious that this belief is associated with the two animals *si* and *se* furnishing the hide for the cuirasses, and that the supposed differentiation of the age of the two creatures is transferred to their products. Certain it is that the philological interpretation of the Chinese literati must be at fault. Their fundamental error lies in the misunderstanding of the word *shu;*[3] and in

---

[1] I am, of course, aware of the fact that in European armor, which is more or less artificial, a studied asymmetry is sometimes displayed (see, for instance, BASHFORD DEAN, Catalogue of European Arms and Armor, p. 64). The above remark refers only to the spontaneous productions of primitive cultures.

[2] Such an arrangement, moreover, I must confess, would appear to me as too sophisticated, and technically too complex for such a simple and primitive age as that of the Chou. In order to grasp the character of its culture-objects, we should collect experience from the life of primitive peoples as we actually observe it (compare Plate XI).

[3] The text unfortunately is very succinct, and merely contains the terms *ts'i shu*, *leu shu*, and *wu shu*. The Chinese commentators, accordingly, take the word *shu* (No. 10,061) in the sense of "hide pieces laid out side by side and then joined together," but this is a point which I venture to contest. In my opinion, the question can be satisfactorily decided, not only from a technological, but from a philological point of view as well, if we interpret the word *shu* in the sense of "strata, or layers of hide pressed together." The word *shu* is capable of assuming many significations; its original meaning is, "to adhere, to place one thing on another, to tie together,

the venture of dragging in the terms *cha* (No. 127) and *ye* ("leaf"), which are peculiar to the Han period, but which did not exist with this meaning and with reference to armor in the age of the Chou. These two terms refer to laminæ or plates of hide or metal reinforcing armor (see pp. 196, 210), and it will be seen that this type of armor springs up only from under the Han. It certainly had not come into existence under the Chou, as proved by the description of the armor given above after the *Chou li*, in which those terms are absent. Again, it is an absurdity to speak of an armor consisting of seven, six, or five laminæ or plates, as these are of small dimensions, and a very large number of them is required to make a suit of armor.[1] The verdict of the Chinese scholars must therefore be repealed. It is solely to the very text of the *Chou li*, which is sound and sane, that we must appeal for a correct understanding of the structure of this cuirass.

We can understand, in my estimation, only that the suits were composed of seven, six, and five superposed layers or thicknesses of hide, respectively, as in fact hide armor has been produced. Then the whole passage becomes intelligible. There is a sensible gradation of three coats, regulated according to the quality believed to inhere in the hide. That of the two-horned rhinoceros ranks lowest in strength, therefore requiring seven layers,[2] and lasts only a hundred years. That of the single-horned rhinoceros, which is the stronger animal, is superior, therefore requiring only six layers, and yet it will last two hundred years. That of both kinds combined is the best and strongest of all, therefore demanding only five layers, and will last three hundred years (see also p. 172). The hide, accordingly, was cut up in horizontal sec-

---

to unite, combine, to assemble," whence the significance "layer, stratum" is doubtless derived; whereas there is no evidence that it was ever understood in the sense of "piece." COUVREUR explains it as a numerative of the pieces of an armor, and cites from *Ts'ien Han shu*, "an armor composed of three pieces" (*san shu chi kia*). It is inconceivable that such a device ever existed. It certainly was a hide armor consisting of three layers of skin. A. CONRADY (Eine indochinesische Causativ-Denominativ-Bildung, p. 165) has succeeded well in tracing the etymology of the word *shu*. The ancient pronunciation, according to him, was *žuk* (Japanese *šuk*); the primeval form to be supposed is *grog*, identical with the Tibetan root *grog* in *s-grog-pa* ("to tie"), *s-grog* ("rope, strap"), and *grogs* ("fellow, friend"). This derivation also sheds light on the Chinese word *shu* assuming the significance "strip or layer of hide or leather."

[1] It is therefore an anachronism when the passage in the text of the *Chou li* (GILES, No. 4437) is translated, "In coats of mail, it is desirable for the plates to fit evenly." Anything like plates is then out of the question. What is meant in this passage is (and it is so understood by the Chinese commentators) that the hide used in the cuirasses should not wrinkle. BIOT very aptly translates, "On la revêt, et on demande qu'il n'y ait pas d'inégalités dans les coutures (qu'elles ne grimacent pas)."

[2] A cuirass of seven thicknesses is mentioned in the biography of I Shên (*T'ang shu*, Ch. 170, p. 2).

tions into large and thin sheets, such as could be weighed and divided into equal parts. It would be unreasonable to infer that a rhinoceros-skin in its natural state of thickness could be properly cured, and then utilized for the making of an armor: the skin was split into strata evenly thick, which were cured, probably boiled, and according to the number required were tightly pressed together. The fact that the harness was not composed of seven, six, and five pieces becomes sufficiently evident also from the rule that the long hide strips were laid around the trunk horizontally;[1] naturally, for this was the most rational and efficient use that could have been made of them. In all probability, the entire affair consisted of only two main parts,— the corselet enveloping the trunk, and the skirt protecting the thighs,—both being closely joined together. Either part could have been made from a single piece of hide. The sewing, of course, refers to the various layers of hide and the seams. How the garment was put on is not indicated in the text; but it seems plausible to infer that it was open in the middle of the front.

By a very similar process, cuirasses were still turned out in northern China and Mongolia in recent times. The American Consul BEDLOE[2] reported on this subject as follows: "The original armor of the north (Manchuria and Mongolia) seems to have been leather, and in shape was more like a blouse than a jerkin. In the course of years the skin was doubled, trebled, and quadrupled, and a Chinese lower garment that might be called leather greaves and cuirasses combined was added to the upper one. The Mongolian nomads learned at an early age that a coat or cuirass made of sheepskin in several thicknesses makes a very warm garment and would turn a spear, arrow, or sword. Apparel of this class is in use to-day and may be bought very cheaply in Shan-tung." In the same manner the cuirasses of the Mongols invading Europe were wrought. Thomas of Spalato, an historian of the thir-teenth century, describes their defensive armor as made of ox-hide, several layers of it being so tightly pressed together that the armor is quite impermeable, and affords considerable protection.[3] This is confirmed by MARCO POLO,[4] who relates that the Mongols wear on their backs armor of cuirbouly, prepared from buffalo and other hides, which

---

[1] BIOT translates with perfect correctness, "On prend leur longueur totale pour faire le contour de la cuirasse."

[2] *Consular Reports on Commerce, Manufactures*, etc., No. 147, p. 494 (Washington, 1892).

[3] G. STRAKOSCH-GROSSMANN, Der Einfall der Mongolen in Mitteleuropa, p. 28 (Innsbruck, 1893). The Tlingit cuirass on Plate XI consists of two superposed layers of elk-hide.

[4] Ed. of YULE and CORDIER, Vol. I, p. 260.

is very strong.[1] Japanese accounts of the Mongol attempt to invade Japan allude likewise to the cuirasses of the Mongols.[2]

The leather corselets *kia* seem to have been in general use, even at an early date, among the people of the state of Ts'in, who were prepared to don them in case of war, as mentioned in a song of the *Shi king*.[3] MÊNG-TSE[4] speaks of the strong armor and the sharp weapons of Ts'in and Ch'u. Siün K'ing, a philosopher of the third century B.C., ascribes armor of sharkskin and rhinoceros-hide to the people of Ch'u; both were hard like metal and stone.[5] This is the more remarkable, as the author goes on to say that the people of Ch'u possessed the iron and steel of Yüan, a place corresponding to the modern Nan-yang in Ho-nan Province, and that their lance and arrow heads, apparently of iron or steel, were sharp like the stings of wasps and scorpions. We may therefore infer that the people of Ch'u, despite their acquaintance with iron, had not yet advanced to the stage of iron armor. Their hide armor must have been light in weight; for they are reported to be "light and agile, fiery and swift, and rapid like a hurricane." In general, however, or in other states, these cuirasses seem to have been heavy and uncomfortable; for we hear that they were donned only during battle, but rolled up and carried by the soldiers during the march.[6] They did not allow the wearer to run; and when driven to flight, the soldiers threw them off, trailing their arms behind.[7]

From a text in *Tso chuan*[8] it appears that rhinoceros cuirasses were

---

[1] Buffalo-hide came up as a substitute for rhinoceros-hide in the making of armor during the T'ang period (p. 162).

[2] A. PFIZMAIER, Die Geschichte der Mongolen-Angriffe auf Japan (*Sitzungsberichte Wiener Akademie*, 1874, p. 151).

[3] LEGGE, Chinese Classics, Vol. IV, p. 202.

[4] *Ibid.*, Vol. II, p. 135.

[5] This passage is quoted also by Se-ma Ts'ien (CHAVANNES, Les Mémoires historiques de Se-ma Ts'ien, Vol. III, p. 217). The *Wu pei chi*, an extensive work on military science written under the Ming dynasty by Mao Yüan-i, and published in 1621, comments on this statement of Siün K'ing that sharkskin armor equals rhinoceros-hide armor in hardness, and is therefore styled *shui si* ("water-rhinoceros"), because the shark is produced in the water. Another instance of sharkskin armor occurs in the *T'ung kien kang mu* (quoted in *T'u shu tsi ch'êng*), where it is ascribed to the Mongols. Shagreen seems to have been utilized by the Chinese in olden times, especially in saddlery. The imperial "caparisons made of shagreen" (CHAVANNES, *l. c.*, p. 214), I believe, are identical with the modern saddles mounted with shagreen. It is used also for mounting the sheaths and handles of knives and swords, even for the decoration of snuff-bottles. A detailed investigation of the subject is contained in H. L. JOLY and I. HOGITARO (The Sword Book, pp. 3 *et seq.* of the appendix).

[6] As attested by Sun-tse (see L. GILES, Sun Tzǔ on the Art of War, p. 58, London, 1910). The case in which the rolled-up cuirass was enclosed was styled *kao* (No. 5949).

[7] As is evident from a passage of Mêng-tse (LEGGE, Chinese Classics, Vol. II, p. 130).

[8] LEGGE, Chinese Classics, Vol. V, p. 290.

also varnished with a red lacquer.   They are frequently alluded to in that work,[1] and were doubtless the usual means of body protection during the whole Ch'un-ts'iu period (B.C. 722–481).   The states drew up schedules of their weapons and defensive armor.   In one passage,[2] a distinction is made between soldiers wearing armor lashed with cords (*tsu kia*, No. 11,828) and those who had donned an armor of silken fabrics (*p'i lien*, Nos. 8769, 7151).   It is clear only that two kinds of armor are here discriminated, and that their diversity of technique and quality of material brought about a different effect: of the soldiers clad with the former armor, there were three hundred, of whom eighty escaped; of soldiers with the latter armor, there was a force of three thousand, of whom only three hundred escaped.   We do not exactly know, however, what these armors really were.   LEGGE interprets *tsu kia* as "buff-coats lacquered as if made of strings" (then again translating "the men whose buff-coats looked as if made of strings"), and *p'i lien* as "whose coats were covered with silk."   Neither is intelligible.   S. COUVREUR[3] has proposed to explain the term *tsu lien* as "cuirasse faite de cordons de soie, et tunique ouatée faite de grosse soie cuite,"[4] and the term *tsu kia* as "cuirasse faite de cordon de soie et enduite de vernis."   These definitions are helpful, yet they leave us in the dark as to the contrast between the armor *tsu* and the armor *lien*.   The latter, which proved so disastrous to their wearers, may have been made entirely from a coarse silken material; the former, however, as attested by the word *kia*, seem to have consisted essentially of hide, with the addition of silk cords (styled *tsu*), which I am inclined to think refer to the lashings of the hide armor.

A special protective contrivance employed by the archers was an arm-guard, called *han* (No. 3799), a leather cuff wrapped around the left arm, the bow being supported against it.[5]   From the Han period these objects were made of iron.

The utilization of rhinoceros-hide for armor persisted down to the T'ang period.   Li Wang of the Han makes mention of this material (*si se*) for that purpose.   A helmet of rhinoceros-hide is mentioned under the year 30 A.D. in the *Tung kuan Han ki*, completed about 170 A.D. In the biography of General Ma Lung,[6] who died in 300 A.D.,[7] we hear

---

[1] *Ibid.*, pp. 289, 397, 419, 517.

[2] Duke Siang, third year (LEGGE, p. 419).

[3] Dictionnaire chinois-français, pp. 494, 982.

[4] In *Li ki*, garments of coarse boiled silk worn after the first year of mourning are mentioned.

[5] COUVREUR, Li ki, Vol. I, p. 621.

[6] Inserted in the Annals of the Tsin Dynasty (*Tsin shu*, Ch. 57, p. 2 b).

[7] GILES, Biographical Dictionary, p. 568.

of a singular stratagem, in which iron mail (*t'ie k'ai*) versus rhinoceros-hide cuirasses (*si kia*) was at stake. Ma Lung defeated a hostile army by covering the sides of a narrow pass with loadstone,[1] so that the iron-clad enemies were unable to move, whereas his cuirassed men got the better of them. Whatever the basis of this anecdote may be, we recognize that hide armor still held its ground in the age of iron armor, and insured mobility of troops to such a degree that hide-clad soldiers could carry a victory over a heavy-mailed force struggling along under the burden of metal. In some other passages of *Tsin shu* and *Sung shu* we meet the term *si p'i k'ai* ("rhinoceros-hide metal armor"), which must have been a suit with a hide foundation reinforced by metal laminæ. We shall hear more of cuirasses in later periods, and likewise of metal armor.

The hide armor of the Chou is irretrievably lost, and there is little or no chance that any will ever come to light. To a certain degree, hide armor, as still manufactured not so long ago by native tribes of America, may serve as an object-lesson and substitute, and assist us in reconstructing in our minds the appearance of the ancient Chinese warriors. As the course of our investigation renders it necessary to touch also the subject of American defensive armor, these illustrations of American specimens not easily accessible will be welcome to many students. Plate XI illustrates an armor, in the form of a vest, made from extremely hard, heavy, tanned moose-skin of two thicknesses, the two layers being tightly pressed together. It is proof, against musket-balls fired at a reasonable distance. It opens in front, and is closed by means of three iron buckles of foreign make. The specimen comes from the Tlingit, Alaska. [2]

The armor figured in Plate XII is the work of Asiatic Eskimo from East Cape on the Chukotsk Peninsula. It is of particular interest in this connection as exhibiting the tendency toward making a cuirass of a single large piece of hide, as far as possible, thus avoiding the cutting of it. Extending in its total width to fully 1.55 m, two complete skins of seals are utilized in this specimen, the one forming the exterior, the other the interior, of the suit. They are sewed together along the edges

---

[1] Regarding the loadstone in China see J. KLAPROTH (Lettre sur l'invention de la boussole, pp. 66 *et seq.*, Paris, 1834), and F. DE MÉLY (Les lapidaires chinois, p. 106).

[2] Similar coats of hardened hide were turned out by the Haida, Chinuk, Hupa, Shoshoni, Navajo, Pawni, Mohawk, and others. There are in the Field Museum several other Tlingit cuirasses painted with the totemic emblems of the clans to which the chiefs wearing them belonged. The shields of the Plains Indians were made from buffalo-hide, with one or two covers of soft dressed buffalo, elk, or deer skin; the hide used for the purpose was taken from the neck of the buffalo bull, and was made exceptionally thick and tough by shrinking it, while wet, over a fire built in a hole in the ground (J. MOONEY, in *Handbook of American Indians*, Vol. II, p. 547).

with bands of seal-thongs, and enclose between them wooden slats. The central piece protecting the chest has incased in it a board of the same shape and size, while the gradually narrowing flaps have each four slats inserted to secure greater elasticity of movement.

On Plate XIII is illustrated an armor of hard tanned caribou-skin, of especial interest to students of China because it is covered all over with Chinese coins. It is of the same type of cuirass as the one in Plate XI and comes from the Tlingit, Tarku Tribe, on the Tarku River, Alaska. It was obtained by Lieutenant G. T. Emmons, who says that "the Chinese money was procured in trade from the early Russians, whose ships, exchanging the furs of the North Pacific with the Chinese for tea, plied constantly between the two countries, by which means many Chinese articles found their way to this coast." The coins (about a thousand in number) are arranged in regular vertical rows, and are fastened to the surface of the skin coat by means of leather strips, which pass through their square perforations. The coins are all carefully selected, and only well preserved specimens have been used. The obverse, containing the Chinese legend, is usually on the outside; only in a few cases does the reverse with the Manchu legend stand out. The bulk of these coins date from the beginning of the Manchu dynasty, and are those inscribed with the periods Shun-chi (1644–1661), K'ang-hi (1662-1722), and Yung-chêng (1723-1735). There are several coins of the period K'ien-lung (1736–1795) in this lot, but they form the minority, while the K'ang-hi coins outnumber all others. There is no coin later than the K'ien-lung period, so that it may well be supposed that this collection of coins was traded off in Alaska during or shortly after that period, say roughly at the end of the eighteenth century. We know, of course, that until a few years ago coins of the said description were still circulating in many parts of the interior of China, particularly in the country, though I understand that they have now been withdrawn from currency owing to the financial and monetary reform; it is not likely, however, that such a large number of those older coins would have arrived in Alaska in recent times without any additional modern coins. The conspicuous absence of any coins of the nineteenth century in a lot of a thousand speaks in favor of the assumption that they had been traded at the termination of the eighteenth century. A closer attempt at dating could be made, if it were possible to take off all the K'ien-lung coins, in order to read their reverses, which usually impart the place of the mint, and in some cases would allow of the establishment of a fixed year for the coinage. The last year thus determined would yield the *terminus a quo;* that is, the approximate date, after which this money may have left China *en route* to the north-east. It is not feasible to detach the coins from the armor, nor to lift them sufficiently to enable one to read the reverse, as they are fastened very tightly. Certainly, I do not mean to say that the armor itself originated at the end of the eighteenth century, though of course this might be possible; while it is conceivable also that the coins, on arrival in Alaska, were kept in a family; or bequeathed to some member of it, and were attached to the cuirass at a much later date. [1]

It is curious that in the *Chou li* no mention is made of helmets. A reference to them was presumably contained in the lost chapter *Se kia*, "the Superintendents of Armor," an office dealing with the business of defensive and offensive armor. In the *Shi king*, in one of the songs of the country of Lu, helmets adorned with shells (*pei chou*) are mentioned. The shells, as is explained by the commentaries, were connected, and attached to the helmets by means of strings of vermilion color. [2] The helmets were nothing but round leather caps, corresponding to the *galea* of the Romans.

Armor and helmet were designed to create the impression of strength and bravery, and to inspire such fear that the enemy did not dare to attack the wearer. [3] They were considered valuable objects and were presented as gifts. [4]

The regular force which a great state could at the utmost bring into the field consisted of a thousand chariots. [5] Each chariot contained

---

[1] F. RATZEL (Über die Stäbchenpanzer, *Sitzungsberichte der bayerischen Akademie*, 1886, p. 191), who mentions such coin armor among the Tlingit, derives it from the idea of armor-scales, and remarks that motives of protection and decoration here come into close contact with each other. The idea of a scale armor, however, is excluded in such specimens as the one figured by HOUGH (Primitive American Armor, Plate XXI, Fig. 1) where the coins are strung loosely and at some distance from one another, so that protection from them, if any at all, could only amount to a minimum. Further, the conspicuous absence of scale armor on the entire continent of America conflicts with the view that the comparatively recent coin armor might be the imitation of scale armor. The coins have a merely ornamental purpose, and possibly also the function of amulets or magic protection; as such, these two ideas being combined, we find Chinese coins sewed on to every-day garments among the Gold and the Gilyak on the Amur; and as the common Chinese people are themselves in the habit of wearing old coins as charms, it seems very plausible that the example of the Chinese may have served as an incentive to the Amur tribes, and that Russian traders, familiar with the customs of Siberian peoples, may have suggested the same practice to the tribes of Alaska.

[2] LEGGE, Chinese Classics, Vol. IV, p. 626.

[3] *Li ki*, ed. COUVREUR, Vol. I, p. 52; Vol. II, p. 492.

[4] *Ibid.*, Vol. I, p. 41; Vol. II, pp. 17, 18.

[5] The war-chariot is generally believed to have arisen in Babylonia, and to have spread from this centre to Egypt, Greece, Iran, and India. But the great antiquity which the war-chariot may claim in China prevents us from accepting the conclusion that it was plainly derived there from Babylonia in historical times. Like many other basic factors of ancient Chinese culture, it ranges in the class of those acquisitions which ancient China has in common with western Asia, and which go back to a remote prehistoric age. To these belong the mode of agriculture, the cultivation of

three armored men,— the charioteer in the middle, with a spearman on
his right, and an archer on his left.   There were attached to it seventy-
two foot-soldiers and twenty-five other followers, one hundred men in all;
so that the whole force would amount to a hundred thousand men.
But in actual service, the force of a great state was restricted to three
armies, or three hundred and seventy-five chariots, attended, inclusive
of their armored occupants, by thirty-seven thousand five hundred men,
of whom twenty-seven thousand five hundred were foot-soldiers. [1]   It
seems that body armor was restricted to those fighting from the chariots.
Another safeguard of the warriors was formed by shields decorated
with figures of dragons, or perhaps adorned with feathers. [2]   The latter
affair presents a point of controversy among the commentators: the
one understanding that the feathers were fixed to the shield; the others,
that they were painted on it.   LEGGE adopts the latter view, and translates, "the beautiful feather-figured shield."   Also COUVREUR is
inclined to think that feathers of different kinds were represented on
the shield.   This opinion, however, is not very convincing.   Whereas
it is perfectly plausible that designs of dragons, or, as in recent times, of
tigers were painted on the shields, and doubtless intended to guard the
wearer and to terrify the enemy, it is difficult to see what reasons could
induce man to decorate his shield with a pictorial pattern of feathers.
We are all familiar with the shields of primitive man adorned with real
feathers, particularly among the American Indians; and the primitive
man of the *Shi king* period, in all likelihood, may have done the same. [3]
A document of the Han period brought to light by M. CHAVANNES
(see p. 189), in which pigeon tail-feathers are mentioned in connection
with a buckler, is very apt to corroborate this conclusion.
    The shield was combined with the spear, [4] while later in the Han pe-

---

wheat and barley, tilling of the field by means of the plough drawn by an ox, methods
of artificial irrigation, cattle-breeding, employment of cattle as draught-animals,
the composite bow, the cart based on the principle of the wheel, and the potter's
wheel.

    [1] LEGGE, Chinese Classics, Vol. IV, p. 626; COUVREUR, Cheu King, p. 137.   I
have abandoned Legge's inexact word "mailed" and substituted "armored" for it;
anything like "mail" was unknown in China during the archaic period (compare
Chapter IV).

    [2] LEGGE, *l. c.*, p. 194; COUVREUR, *l. c.*, pp. 135, 136.

    [3] The Tibetans had bucklers ornamented with feathers (see p. 256). An unsophisti-
cated mind may certainly be entitled to raise the question how the Chinese commentators get at the "feathers" in the passage of the *Shi king*, as no direct word to
this effect is employed.   The word *mêng* (No. 7763), into which this meaning is read,
means "to cover, to envelop;" and the term *mêng fa*, after all, may simply mean
"wooden shields covered with hide."   In this sense, the term *mêng tun* ("hide-covered
buckler") is indeed utilized in later literature.

    [4] For instance, BIOT, Chou li, Vol II, p. 223.   In the inscriptions on ancient
bronzes, as reproduced and explained in the *Po ku t'u lu*, the word *sun* ("grand-

riod it was handled together with the sword. The term *kan ko* ("shield and spear") in the *Shi king*[1] is a collective notion comprising defensive and offensive armor, or war-implements. In the administration of the Chou dynasty, there was a special official presiding over the various kinds of spears and bucklers, and commissioned with their distribution.[2] But no contemporaneous description of shields is handed down, from which an exact conception as to their material and form might be gained.

The shields protecting the soldiers in the war-chariots were presumably roof-shaped, as we glean from a text in *Tso chuan*[3] when, in the battle of Ch'ui-pi, fought between the armies of the principalities of Lu and Ts'i, Tse-yüan Tsi of Ts'i pursued Shêng-tse, and shot an arrow at him, hitting the ridge of his shield. In this passage the ridge is designated "roofing-tile" (*wa*), explained by the commentary as the ridge of the shield. This is also the earliest document in which the word *shun* (No. 10,154) appears as a designation for the shield, and, owing to its composition with the classifier 'wood,' leaves no doubt that the shields were wooden.[4] It is worthy of note that during the early period, in the same manner as in armor, no metal was employed for the bucklers; and it is remarkable also that in all later periods of culture when the working of metals was in full swing, none were ever turned to that purpose; wood, rattan, and hide holding their place. The buckler, accordingly, never assumed a vast importance in Chinese warfare.[5]

A fundamental text relating to ancient shields, though dating from the time of the Later Han dynasty, is contained in the dictionary *Shi ming* by Liu Hi. He defines the word *tun* ("shield") as *tung* ("to

---

son") is represented in writing by the rough figure of a youth holding spear and shield, and performing a war-dance.

[1] LEGGE, Chinese Classics, Vol. IV, pp. 484, 578. Likewise in *Li ki* (ed. COU-VREUR, Vol. I, pp. 233, 468).

[2] BIOT, Chou li, Vol. II, p. 238; J. H. PLATH, Das Kriegswesen der alten Chinesen (*Sitzungsberichte der bayerischen Akademie*, 1873, p. 33).

[3] Duke Chao, 26th year, B.C. 516 (compare LEGGE, Chinese Classics, Vol. V, p. 716).

[4] *Shi king*, *Chou li*, and *Shi ki* use the word *tun* (No. 12,223), which is doubtless derived from the verb *tun* (No. 12,225), "to hide away, to conceal one's self." The word *kan* (No. 5814) appears twice in *Shu king*. The commentaries do not interpret the differences between the three words, but explain one by another. The shield, as elsewhere, was occasionally applied also as an offensive weapon. Thus, Fan K'uai, girt with a sword and bearing the buckler on his arm, penetrated into the camp of Hiang Yü, and used the buckler in pushing the guards down, who thus fell to the ground (CHAVANNES, Les Mémoires historiques de Se-ma Ts'ien, Vol. II, p. 279).

[5] Copper shields are mentioned by the Chinese, but refer to foreign tribes; for instance, in the Annals of the Yüan Dynasty under the year 1286, when they were sent from a foreign country called Ma-pa; they are ascribed also to the Shan of Yünnan (see p. 193).

conceal one's self," No. 12,241), and as the object behind which a man hides himself in a kneeling position in order to evade an attack.   Liu Hi enumerates two kinds of foreign shields adopted by the ancient Chinese, — a large and flat one, which originally was indigenous to the country of Wu [1] and peculiar to the generals there, hence styled *Wu k'uei* (No. 6499), "general of Wu;" and a high one, termed *sü tun*, [2] coming from the country of Shu (Sze-ch'uan), but termed by others "shield of the K'iang (Tibetans)" because they asserted that it originated from the K'iang.   Here we notice the ever-recurring Chinese tendency toward imitating and appropriating the armaments of the neighboring tribes.   Liu Hi mentions also the long and narrow shields used by the infantry soldiers in combination with the sword,— styled "foot shields" (*pu* [No. 9485] *tun*); [3] and the short and narrow shields employed on the war-chariots,— styled "small shields" (*kie* [No. 1505] [4] *tun*).   As to the materials chosen for their manufacture, he emphasizes boards and, what is of especial interest, rhinoceros-hide (*si p'i*).   The latter were termed "rhinoceros shields" (*si tun*); the former, "wooden shields" (*mu tun*).   The specimen of a circular buckler of rhinoceros-hide, of Indian manufacture (secured by the writer in Tibet), is illustrated in Plate XXVII.

Culture-objects when once acquired survive through the ages with persistent force, even after the introduction of innovations which seem to be apt to supersede entirely the old material.   We have already referred to the fact that cuirasses have not yet wholly disappeared in modern China.   Indeed, we meet them in all periods of Chinese history, despite new inventions of superior quality.

From the wooden documents found in Turkistan, and recently deciphered with admirable ingenuity by E. CHAVANNES [5] it becomes apparent that hide corselets formed the defensive armor of the Chinese soldiers serving in eastern Turkistan during the Han period.   The contemporaneous texts written out on wooden slips employ either the

---

[1] No. 12,748.   Wu is an ancient kingdom comprising the present province of Kiang-su, the southern part of An-hui, and the northern portions of Chê-kiang and Kiang-si (see Chinese Pottery in the Philippines, p. 42, note 10).

[2] *Sü* (No. 4716) is explained as a war-implement in K'ang-hi's Dictionary, which quotes the passage in question.   This interpretation is not quite satisfactory; for the word *sü* must have a more specific meaning, as shown by the parallelism of the preceding sentence and the following clause, in which it is said that these shields were handled by the *Sü* of the country of Shu.   The word, accordingly, parallel to the preceding generals of Wu, must refer to a military charge or rank in Shu; and it is doubtless derived from a language spoken in Shu, or from a language of the K'iang.

[3] These were actually used in the Han period, as will be noticed in Chapter III.

[4] The word is explained by him in the sense of "small."

[5] Les documents chinois découverts par Aurel Stein dans les sables du Turkestan oriental, p. XVI (Oxford, 1913).

plain word *kia* (No. 187), or the compound *ko kia* (Nos. 393, 569), "hide armor;" and we hear also of an official having charge of armor (No. 758).[1] Simultaneously, another word for body armor, *k'ai*, is twice used in these documents (Nos. 758, 794), and translated by M. CHAVANNES likewise "cuirasse." This seems to be correct only in so far as leather was applied also to this kind of armor, as expressly attested by document No. 794; but it will be seen in the following chapter that the new word *k'ai*, which springs up in the Han period, denotes a new type of armor presenting a combination of hide with metal, and that the rendering by "cuirass" is therefore inadequate. The defensive armor of the Han soldiers was completed by a helmet (No. 794) and a buckler (*tun*), the latter being described as red in the wooden documents (Nos. 75, 77), from which it may be inferred that they were made of wood covered with a red varnish[2] protecting the wood from moisture, red being believed to terrify the enemy; it was the main function of the buckler to ward off the shots of arrows (No. 682). In one case a buckler is especially mentioned as having been made in B.C. 63 by the official Armory of Nan-yang in Ho-nan Province (No. 39); in another case a buckler is on record as having been worked in B.C. 61 by the artisans of the administration (No. 40). Bucklers were decorated with pigeon tail-feathers attached to them (No. 75).[3]

Despite the fact that metal armor, as will be seen in the next chapter, gradually made its way during the period of the two Han dynasties, and was firmly established in the age of the T'ang, mention is still made in the Statutes of the T'ang Dynasty[4] of hide cuirasses (*p'i kia*); rhino-

---

[1] In Ch. 49 of *Hou Han shu* the story is told of how in 75 A.D. General Kêng Kung and his troops, being at war with Kucha, were at the point of starvation, and cooked cuirasses and crossbows so as to feed on the leather and sinews contained in them (CHAVANNES, *T'oung Pao*, 1907, p. 228), — a case sufficiently convincing as to the material of which they were made.

[2] In the same manner as the cuirasses (p. 182).

[3] M. CHAVANNES (*l. c.*, p. 30) thinks that the expression "pigeon-tail" must be a technical term which designates perhaps the leather or hemp handle of the buckler. There is in my opinion no necessity for such a conjecture. "Pigeon-tail," I venture to suggest, is to be understood literally, inasmuch as the buckler, as perhaps in the period of the *Shi king*, was adorned along its edges with feathers; in the document in question the report is made that the soldier so and so has received "a red buckler, the pigeon tail-feathers of which had rotted away." The "rotting-away" sounds plausible with regard to the latter, but much less so if a leather or hemp strap were intended. As to offensive armor, M. CHAVANNES correctly emphasizes the fact that the Chinese soldiers of the Han time availed themselves of crossbows, not of bows; this is confirmed by his documents as well as by the Han sculptures, on which men are usually represented as shooting with crossbows, not, as has been said by some observers, with bows. As to swords, it seems preferable to study them from actual specimens of cast bronze and iron, such as are in our collections, instead of from the bas-reliefs, as M. CHAVANNES recommends us to do (compare Plates XX and XXI).

[4] *P'ei wên yün fu* (Ch. 106, p. 73), and *Ko chi king yüan* (Ch. 41, p. 3). The *T'ang leu tien* ("Six Statutes of the T'ang Dynasty") gives a description of the

ceros-hide (*si se*) being employed for them, and sometimes being supplanted by buffalo-hide.

In the History of the Liao Dynasty [1] rhinoceros-hide armor is still recorded for the year 952 as a tribute of the Nan T'ang dynasty to the Court of the Liao. The captains in the army of the kingdom of Nanchao are reported to have used cuirasses made from rhinoceros-hide. [2] During the middle ages, when the rhinoceros grew scarcer, other hides began to take its place. It has been demonstrated above (p. 162) that under the T'ang the district of Kuang-ling sent to the Court tribute of buffalo-hide armor. [3] MARCO POLO [4] says regarding the Mongols that on their backs they wear armor of cuirbouly (boiled leather), prepared from buffalo and other hides, which is very strong; and all contemporary western writers speak of the leather armor used by the Mongols. [5] This fact is confirmed by the Annals of the Yüan Dynasty. [6]

The type of cuirass styled "hoop armor" has possibly at one time existed in China, though there is no description of it. At the Court of the emperors of the Kin dynasty (1115–1234) in Peking, the guards were all clad with armor. On the left were stationed those with a banded cuirass colored blue (*ts'ing t'ao kia*), holding in their hands a flag on which was represented a yellow dragon. On the right were stationed those with a banded cuirass colored red (*hung t'ao kia*), holding a flag with a red dragon represented on it. [7] The word *kia* used in this connection indicates that it is the question of hide cuirass; and the word *t'ao* ("band") defines the peculiar character of this armor in that it was banded or hooped, the bands being cut out of leather, perhaps in a

---

administrative organization of the period K'ai-yüan (713–741) of the T'ang dynasty, the authorship being ascribed to the Emperor Yüan-tsung (713–755), and Li Lin-fu and others contributing to the interpretation of the work (WYLIE, Notes on Chinese Literature, p. 67; PELLIOT, *Bulletin de l'Ecole française d'Extrême-Orient*, Vol. III, 1903, p. 668).

[1] *Liao shi*, Ch. 6, p. 1.

[2] C. SAINSON, Histoire particulière du Nan-Tchao, p. 19 (Paris, 1904).

[3] In *Yen kien lei han* (Ch. 228, p. 14) a book *Ts'e lin hai ts'o* is quoted to the effect that what is designated "rhinoceros-hide armor" in the T'ang History is at present made from buffalo-hide, but is generally styled *si* ("rhinoceros").

[4] Ed. of YULE and CORDIER, Vol. I, p. 260.

[5] W. W. ROCKHILL, The Journey of William of Rubruck, p. 261 (London, *Hakluyt Society*, 1900), and p. 180.

[6] For instance, *Yüan shi*, Ch. 78, p. 12 (K'ien-lung edition).

[7] This information is contained in the *Pei yüan lu*, the narrative of a journey in 1177 A.D. from Hang chou to Peking, described by CHOU SHAN and translated by CHAVANNES (*T'oung Pao*, 1904, pp. 163–192; the passage indicated is on p. 189). It is quoted, though incompletely, in *P'ei wên yün fu* (Ch. 106, p. 74). CHAVANNES' translation "cuirasses avec des cordons bleus" certainly is all right, as far as the translation is concerned; but I am inclined to think that this term is capable of the interpretation as given above. The word *t'ao* ("band") is in GILES, No. 10,817.

manner similar to that of the corresponding Chukchi armor figured and described by WALTER HOUGH[1] and W. BOGORAS.[2]

Another singular kind of armor is alluded to in the *Lan p'ei lu*[3] under the name *jung kia*. The word *jung* (No. 5736) refers to the soft core of the young antlers of the deer (considered by the Chinese an efficient aphrodisiac); and I am inclined to interpret the term *jung kia* as a cuirass strengthened by horn shavings fastened to the surface, for which there are interesting analogies in other culture areas.[4] In the passage

---

[1] Primitive American Armor (*Report of the U. S. National Museum for 1893*, Plate IV and p. 634). An excellent specimen of this type is in the Field Museum (Cat. No. 34,151).

[2] *Publications du Musée d'Ethnographie et d'Anthropologie de St. Pétersbourg*, II, Plate XII, Fig. 1 (St. Petersburg, 1901). The Chukchi hoop armor, however, is not related to the so-called banded mail of the European middle ages, as asserted by HOUGH (*l. c.*, p. 633) and repeated by BOGORAS (The Chukchee, *Jesup North Pacific Expedition*, Vol. VII, p. 162). In the European types it has been shown that the banded appearance, as it occurs in mediæval illustrations, was produced by thongs of leather which were strung through adjacent rows of chain-links (BASHFORD DEAN, Catalogue of European Arms and Armor, p. 22, New York, 1905),— a feature entirely lacking in the Chukchi armor.

[3] Quoted in *P'ei wên yün fu*, Ch. 106, p. 74. This is a brief work containing likewise the narrative of a mission to the Court of the Kin emperors in 1170 by Fan Ch'êng-ta (1126–1193), and reprinted in *Chi pu tsu chai ts'ung shu*. In the text of this work it is added that the guards had spears with handles inlaid with gold leaf, and flags painted with blue dragons; those in the east had yellow flags, and those in the west white ones.

[4] AMMIANUS MARCELLINUS (XVII, 12) narrates that the armor of the Quadians and Sarmatians consisted of small scales of polished horn arranged on a linen coat like the plumage of a bird (loricae ex cornibus rasis et levigatis, plumarum specie linteis indumentis innexae); and PAUSANIAS (I, 21, 5) relates that a Sarmatian scale armor made of horses' hoofs was preserved as a curiosity in the Temple of Aesculapius at Athens. RATZEL (Über die Stäbchenpanzer und ihre Verbreitung im nordpazifischen Gebiet, *Sitzungsberichte der Bayerischen Akademie der Wissenschaften*, 1886, p. 191) mentions, after a letter received from William H. Dall, an armor made by the Tlingit from slices of deer-hoof fastened to a foundation of elk-skin in the manner of scale armor. In the Philippine collection of the Field Museum (Cat. No. 34,493, gift of Mr. E. E. Ayer), there is a suit of armor composed of rectangular laminæ of buffalo (carabao) horn, mutually connected by means of rows of brass rings. This armor was made by the Moro on Basilan Island. It is identical with the specimen figured by L. SCHERMAN (Berichte des K. Ethnographischen Museums in München IV, 1911, *Münchner Jahrbuch der bildenden Kunst*, 1912, p. 96, Fig. 18), which is stated to hail from the Sulu Archipelago, and to be characteristic of this region. In the Field Museum, however, there is also a suit of armor of exactly the same type, in which the laminæ are entirely wrought from brass, and likewise joined by means of brass rings. This metal suit, according to the traditions of the natives, was captured in 1631 when a Spanish expedition was massacred at Lake Lanao; they assure us also that the suits of carabao horn were turned out in imitation of this Spanish model. It is therefore obvious that the metal harness in question, as moreover attested by the evidence of the object itself, is of Spanish make, and served as model for the Philippine as well as the Sulu horn armors. Suits of armor have always been highly prized articles and carried away to remote corners by barter or capture in war; and it is always necessary to be on one's guard in making correct attributions. We may even go so far as to say that it would be impossible for the natives of the Philippines to construct such a complicated affair from their own inventiveness. Their purely native armor is unpretentious, being made from woven hemp stuffed with matted hemp fibre. This is the national North-Malayan type of body armor, the same as

referred to it is said that in the east and west galleries of the imperial palace the guards were clothed with armor, and that those posted east wore armor of horn dyed red (*hung jung kia*), those posted west wore armor of horn dyed green and blue (*pi* [No. 9009] *jung kia*). It thus seems that the Kin or Niüchi had a predilection for curious armor.

Reference to the cuirass of the Mongols has already been made above (pp. 180, 190).

"They ride long like Frenchmen, and wear armor of boiled leather, and shields and arblasts, and all their quarrels are poisoned,"— thus MARCO POLO [1] describes the equipment of the inhabitants of the kingdom of Nan-chao in Yün-nan called by him Carajan. Yule is inclined to prefer the reading "cuir de bufal" offered by another text, as some of the Miao-tse of Kuei-chou are described as wearing armor of buffalo-leather [2] overlaid with iron plates.

Hide was indeed the chief material utilized for body armor by the aboriginal tribes inhabiting southern China. In this respect we are well informed by several reliable and observant authors of the Sung period. The famous Fan Ch'êng-ta (1126–1193), [3] official, poet, florist, traveller, and ethnographer, has the following description in his valuable account of the regions of southern China, [4] "As regards the armor of the *Man* tribes, harness and helmets are wrought to a large extent only in the kingdom of Ta-li. [5] Elephant-skin is used for this purpose in such

---

we find on Formosa. The aborigines of Formosa, at the time when the Chinese made their first acquaintance in the beginning of the seventh century, were in a transitional stage of life, iron being only sparsely used, while bone and horn took its place; and a hoe with stone blade was employed in tilling the fields. The interesting account given in the Annals of the Sui Dynasty (*Sui shu*, Ch. 81, p. 5) ascribes to them knives, spears, bows and arrows, swords and daggers; and adds that owing to the scarcity of iron in the country the blades are thin and small, being replaced to a great extent by bone and horn, and that "of plaited hemp they make armor, or avail themselves of bear and leopard skins."

[1] Ed. of YULE and CORDIER, Vol. II, p. 78.

[2] According to the *Nan-chao ye shi*, as previously shown, it was rhinoceros-hide; while the text of Fan Ch'êng-ta which follows above speaks of elephant-skin. In all likelihood these three materials, buffalo, rhinoceros, and elephant, were used side by side.

[3] GILES, Biographical Dictionary, p. 242.

[4] The general title of the work is *Kui hai yü hêng chi* (WYLIE, Notes on Chinese Literature, p. 56; BRETSCHNEIDER, Botanicon Sinicum, pt. 1, p. 165). The single chapters have separate headings; the one from which the above extract is given is entitled *Kui hai k'i chi* ("Records of Implements in Southern China"). My quotation refers to the reprint of the text in *T'ang Sung ts'ung shu*.

[5] Name of the country and the capital of the Shan in the present province of Yün-nan, who ruled as the Nan-chao dynasty, and whose kingdom was destroyed by the Mongols in 1252. It still was independent at the time to which our above account refers. The fact that the armor of the *Man* is traced to the kingdom of Ta-li, then inhabited by the *T'ai* or *Shan*, is of some significance. The *T'ai* were a warlike and chivalrous nation like the Tibetans, and had developed a highly advanced culture

a manner that one large piece covers the breast and another the back, looking like the carapace of a turtle, and being as solid and massive as iron.[1] Then small strips of leather are so combined as to form brassards and nape-guards, made like the iron armor-plates of the Chinese,[2] and all colored vermilion. Helmet and harness, both on the interior and exterior side, are all colored vermilion. By means of yellow and black mineral dye-stuffs they paint designs of flowers, small and large animals, such as are now found on girdle-buckles,[3] — of admirable workmanship. They string also small white shells[4] in connected rows, sew them on to the harness, and decorate the helmets with them. Presumably they are survivals of those ancient helmets adorned with shells on vermilion strings mentioned in the *Shi king.*" [5]

---

betraying, in opposition to the Chinese, a keynote of striking individualism. Every adult was a soldier; and it is a surprising fact that there was compulsory military service in the kingdom of Nan-chao, and that the army was highly organized. The History of Nan-chao compiled in 1550 by YANG SHÊN (1488–1559) narrates that the army captains used to wear cuisses, red helmets, and cuirasses of rhinoceros-hide, and carried bucklers of copper; but they marched bare-footed (C. SAINSON, Histoire particulière du Nan-Tchao, p. 19, Paris, 1904). As to its historical relations, the protective armor of the *Man* must therefore be connected with that of the *Shan;* and the *Man* apparently derived it from the superior culture of their neighbors.

[1] Virûdhaka, one of the four guardians of the world (*lokapâla*) in Hindu mythology, wears a helmet from the skin of an elephant's head (GRÜNWEDEL, Buddhist Art in India, p. 138, and Mythologie des Buddhismus, p. 181). An armor of elephant-skin overlaid with gold in the possession of a Mongol prince in 1573 is mentioned by Sanang Setsen (I. J. SCHMIDT, Geschichte der Ost-Mongolen, p. 217). The Jesuit Francisco Combes, in his Historia de Mindanao of 1667 (BLAIR and ROBERTSON, The Philippine Islands, Vol. XL, p. 179), reports that the Joloans on Mindanao in the Philippines are armed from top to toe with helmet, bracelets, coat-of-mail, greaves, with linings of elephant-hide armor so proof that nothing can make a dint on it except fire-arms, for the best sword or cutlass is turned. As the elephant does not occur in the Philippines (its presence on Borneo is presumably due to human agency), these armors, in all likelihood, must have been importations from the Asiatic mainland.

[2] See Chapter V.

[3] The word employed here is *si-pi* (No. 9050), which in this mode of writing, for the first time, appears in Se-ma Ts'ien's *Shi ki* (Ch. 110, p. 6 b) in the sense of a buckle to fasten a girdle. E. H. PARKER (*China Review*, Vol. XX, p. 15), in his translation of this passage, explains *si-pi* as a word of the Sien-pi language. See now R. and H. TORII, Etudes archéologiques (*Journal of the College of Science*, Vol. 36, Tôkyô, 1914, p. 82, and Plate XII). The same word is used again by our author in the description of the swords made in Ta-li; the sheaths are colored vermilion, and painted in their upper part with a design like those occurring on buckles (*si pi hua wên*). Similarly it is employed in the *Ling-wai tai ta* (published by Chou K'ü-fei in 1178) in the description of the saddles of the *Man* (Ch. 6, p. 5), which are varnished red and black like the designs on buckles (*ju si pi wên*). This term is not registered in the *P'ei wên yün fu.*

[4] The *Ling-wai tai ta* (Ch. 7, p. 9), composed by Chou K'ü-fei in 1178, informs us that the shells utilized in the kingdom of Ta-li for the decoration of armor and helmets came from the island of Hainan; they are called "large shells" (*ta pei*), in the works on natural history "purple shells" (*ts'e pei*). They are described as being round on the back, with purple flecks, and with deep cracks on the surface.

[5] See above, p. 185. Such combinations are suggested to the learned Chinese authors by their literary education, but certainly are no evidence for the shell decorations of the *Man* being really due to a stimulus received from ancient China. The

As to the *Li*, the inhabitants of the island of Hainan, the same author states that they make helmets of plaited rattan.

A cuirass of the Lolo is figured and described by F. STARR.[1] It is composed of heavy, moulded plates of thick leather, varnished black and decorated in red and yellow, the shoulders being protected by two projecting wings. From this plastron is suspended an apron of seven horizontal rows of scales, each row overlapping the one above it, and the scales in each row overlapping. The mode of wear of this armor may be seen in the portrait of the Lolo chief Ma-tu figured by CH. FRANÇOIS,[2] who states that these cuirasses are made of buffalo-skin painted with various colors, somewhat similar in shape to the ancient Japanese armor.[3]

Two specimens of Lolo armor are described by HERBERT MUELLER,[4] which are of the same type as the one figured by Starr, only that those have the central breastplate, which is apparently lost in the latter specimen. Neither Starr nor Mueller has recognized what type of armor is here represented. It is not armor of a uniform structure, but one in which two principles are combined, that of sheets, and that of plates or laminæ. The sheets form the body armor proper, ten in number,

---

employment of shells for decorative purposes, on the contrary, is a general characteristic of all cultures in south-eastern Asia and Tibet, where they are employed in a manner foreign to the Chinese. The Tibetan women use large shells as bracelets, and wear girdles, to which rows of shells are attached. It is surprising to find these in the high mountainous regions of Sze-ch'uan (for instance, in Romi-Drango), in isolated spots remote from the sea, whither these shells must have been brought from India via Tibet, or from Burma by way of Yün-nan. The women of the P'u-jên, a tribe of the T'ai or Shan stock formerly inhabiting Yün-nan, used to wear a short skirt, to which ten rows of marine shells were fastened all round (C. SAINSON, Histoire particulière du Nan-Tchao, p. 164). The women of the White Kuo-lo or Lo-lo covered their heads with black cloth adorned with shells (*ibid.*, p. 167); compare also pp. 170, 175, 179, 185, in regard to other tribes who observed the same practice. An interesting study of the Indian shell industry was recently published by J. HORNELL (The Chank Bangle Industry, *Memoirs As. Soc. Bengal*, Vol. III, pp. 407–448, Calcutta, 1913).

[1] Lolo Objects in the Public Museum, Milwaukee (*Bulletin of the Public Museum of the City of Milwaukee*, Vol. I, 1911, p. 216 and Plate III, 8).

[2] Notes sur les Lo-lo du Kien-tchang (*Bulletin de la Société d'Anthropologie*, 1904, p. 640).

[3] The correctness of this comparison seems to me doubtful. PLAYFAIR (*China Review*, Vol. V, p. 93) has drawn from a modern Chinese source the following notes on armor among the Kiu-ku Miao: "The crown of the head is protected by an iron helmet which leaves the back of the head exposed. On the shoulders they wear two pieces of hammered iron armor, of considerable weight, which act as a face-guard. Their body armor covers the whole of the back and the chest. In addition they wear iron chain mail covering the entire body and weighing about thirty catties; they have the appearance of being enclosed in a cage. Their legs are cased in iron greaves of great strength. They carry in their left hand a wooden shield, in their right a sharp-edged spear." Chain mail is discussed in Chapter IV.

[4] *Baessler-Archiv*, Vol. III, 1912, p. 59 and Plate III.

a breast and a back sheet,[1] and eight below these for the protection of the abdomen and loins. Combined with this leather sheet armor are tasses consisting of six or seven horizontal rows, each composed of small rectangular leather laminæ, arranged in vertical position. The leather sheets and plates are varnished red on the outside[2] and yellow on the lower side. Mr. Mueller remarks that parallels to this armor are hardly known, but that, as far as can be judged from the pictures preserved, a certain relationship, however distant, with ancient Chinese armor seems to exist. Unfortunately he does not state to what kind of pictures he refers, nor in what the supposed resemblance should consist. There is hardly any solid foundation for this opinion. This type of armor, on the contrary, although it agrees in some features with one represented on certain Chinese clay figures of the T'ang period (Plate XXXI), does not meet with any exact counterpart among Chinese specimens known to us; nor is such a connection at the outset very probable, since the affinities of *Man* armor, as has been pointed out, go with that of the *Shan*, and are accordingly focussed on another culture-zone.

Besides the word *kia*, another word for armor occurs in the *Shi king*, and this is the word *kiai* (No. 1518). It is once used with reference to great armor donned by a king;[3] and on another occasion it refers to a team of four horses in a war-chariot, clad with armor.[4] LEGGE, following the Chinese comment, is of the opinion that the meaning of *kiai* is identical with that of *kia;* but they are two different words written with two different symbols, and it is therefore justifiable to presume that they denote two different types of armor. As the word *kiai* is used to designate the scales of fishes, turtles, lobsters, and other aquatic scaly animals, it is most likely that it was this notion of the word transferred to a type of body armor, and that it related to scale armor (*lorica squamata*), the scales being cut out of hide or leather.[5] There

---

[1] Plastron and dossière.

[2] In accordance with the ancient Chinese cuirasses, as mentioned in *Tso chuan* (see above, p. 181).

[3] LEGGE, Chinese Classics, Vol. IV, p. 606.

[4] *Ibid.*, p. 131.

[5] LEGGE (*l. c.*, p. 194) states that the armor (not mail) for the horses was made of thin plates of metal, scale-like. It is most improbable that the scales were of metal at the time of the *Shi king*. See Chapter VII. The same semasiological development as in Chinese *kiai* is illustrated in the Tibetan word *k'rab* and the Burmese word *k'yap*, that in the first instance denote scale (scale of a fish), and secondly a body armor, which is now the usual meaning; and it is further interesting that Tibetan *k'rab* has also the meaning of "shield, buckler" (see JÄSCHKE, Tibetan-English Dic-

is unfortunately no description of this armor in any ancient text.   In
the *Li ki* the word occurs several times, the rules of politeness excusing
the warrior clad with a *kiai* from making a bow;[1] but nothing is brought
forward to add to the knowledge of the subject.[2]   I have never seen in
China any suit of armor made of scales of leather; and they are not likely to have been made at later ages when metal was available.   In
Japan, such specimens have fortunately survived; and the one figured
by BASHFORD DEAN[3] may give us an excellent idea of the appearance of
the ancient Chinese scaly leather coats.   It is attributed to the Fujiwara period (around 1000 A.D.), and described as a primitive type of
Japanese harness, the single laminæ being of boiled leather, cut and
beaten into pieces shaped like fish-scales.   A suit of copper scale
armor obtained in Sze-ch'uan (Plate XIV) may be regarded as the
natural continuation of the ancient leather armor of the same type.
The scales are fastened by means of brass wire to a foundation of sackcloth, and overlap one another.   This specimen, weighing 38¼ pounds,
as evidenced by the effects of many blows and bullet-holes visible in the
metal, has actually been employed in warfare.[4]

Scale armor is distinctly mentioned in the *Wan hua ku*, a work
written at the end of the twelfth century; but this passage is taken from
the *T'ang leu tien*, and therefore refers to the T'ang dynasty.[5]   The

---

tionary, p. 49).   In all probability, the Chinese and Tibetan words *kiai* (or *kai*) and
k'rab are anciently related, in the same way as Tib. *k'rag* ("blood") and Chinese
hiuet, Tibetan *skrag-pa* ("to be afraid of") and Chinese *kiü* (W. GRUBE, Die sprach-
geschichtliche Stellung des Chinesischen, p. 16), Tib. *sgrog-pa* ("to tie") and Chin.
kiao (CONRADY, Eine indochinesische Causativ-Denominativ-Bildung, p. VII).   Also
the Chinese word *kia*, "armor" (ancient pronunciation *kiap*, rhyme *hiap*), may be
allied to Tibetan *k'rab*.   It will be seen below (Chapter IV) that scale armor represents the earliest type of armor in Tibet, Persia, and India.

[1] COUVREUR, Li ki, Vol. I, p. 65; Vol. II, p. 13.

[2] The scales of hide armor were called *kia cha* (No. 127).   This may be inferred
from a passage in the *Chan kuo ts'e* (quoted in *P'ei wên yün fu*, Ch. 97, p. 5 b), where
Su Tai (third and fourth century B.C.; GILES, Biographical Dictionary, p. 682)
addresses Yen Wang, and says, "You cut the scales of the buff-coat yourself, and
your wife fastens them together by means of cords."   The word *siao* (No. 4309),
which is here utilized and means "to scrape, pare, trim," indicates that leather is in
question, and that the leather strips were trimmed into a certain shape called *cha*.
Regarding the technical meaning of this word see p. 210, note 3.

[3] Catalogue of the Loan Collection of Japanese Armor, p. 39 (*The Metropolitan
Museum of Art*, Hand-Book No. 14, New York, 1903).

[4] Consul BEDLOE (*Consular Reports on Commerce, Manufactures*, etc., No. 147,
p. 494, Washington, 1892) states, "Scale mail, at an early period, was carried to a
high perfection.   The scales were applied to cloth or leather at first, as spangles are
to gauze, and later as tiles or slates are to the boards of a roof.   They were composed
of iron, pewter, silver, gold, or of various oriental alloys.   In making a suit, scales of
one kind were usually employed, but combinations were frequent, in which metals
of contrasting colors were used.   A good suit of armor can be bought at prices ranging from $10 to $150."

[5] BRETSCHNEIDER, Botanicon Sinicum, pt. 1, p. 160, No. 330.   The above text
will be found in the Chapter on Armor (*kia chou pu*) in *T'u shu tsi ch'êng*. *Ko chi*

third kind of armor known at that time is termed in that book *si lin kia* ("armor of thin scales"), and is classified among iron armor. The very name implies that it is a question of scale armor. The fourth variety of armor is styled *shan wên kia* ("armor with a mountain pattern"); a zigzag design or a continuous row of triangles being understood by the latter name. Also this, likewise made of iron, was perhaps scale armor;[1] as presumably also the fifth, designated "black hammer armor" (*wu chui kia*), likewise of iron. No descriptions of these pieces are furnished in the book mentioned.

Leather scale armor was still used by the Mongols, as attested by Friar WILLIAM OF RUBRUCK (1253), who states, "I saw two who had come to present themselves before Mangu, armed with jackets of convex pieces of hard leather, which were most unfit and unwieldy."[2]

In the Ming period the technical term for armor-scales is "willow-leaf" (*liu ye*). We read in the Statutes of the Ming Dynasty (*Ta Ming hui tien*) that in 1393 six thousand sets of "willow-leaf armor" and helmets of chain mail were ordered for the soldiers of the body-guard serving in the Imperial City.

The great antiquity of hide scale armor is an important fact to us, as there are certain ancient clay figures on which this type of armor is represented. These belong to the earliest that we have, and range in the archaic period;[3] and it will be seen from the notes devoted to their dis-

---

king yüan (Ch. 41, p. 3) and *P'ei wên yün fu* (Ch. 106, p. 73) give exactly the same quotation extracted from the *T'ang leu tien* (the "Six Statutes of the T'ang Dynasty"), drawn up by the Emperor Yüan-tsung in the early part of the eighth century (WYLIE, Notes on Chinese Literature, p. 67; and above, p. 189). The only additional matter prefixed to the latter text is that the thirteen kinds of armor enumerated were ordered to be made by the Imperial Armory (*wu k'u*).

[1] *P'ei wên yün fu* (Ch. 106, p. 74) quotes the *T'ang shi lu* to the effect that the armors called *shan wên kia* were made by the Emperor T'ai-tsung from iron (black metal) dyed in five colors, so that the "mountain pattern" may have been brought out by the color-work. Five-colored armor (*wu ts'ai kia*) is mentioned in *T'ang shu* (Ch. *li yo chi, ibid.*, p. 73). The Pek-tsi, a Korean tribe, brought "varnished armor of metal" (*kin hiu k'ai*) to the Chinese General Li Tsi (GILES, Biographical Dictionary, p. 421), who subjugated Korea between 644 and 658; on these armors, which were used by the Chinese cavalry, five mountain patterns (*shan ngu wên*) were represented by means of iron, which may be understood in the sense that five iron scales were arranged in such a manner as to suggest the design of a mountain. This passage is contained likewise in *T'ang shu* (Ch. 220, p. 3 b).

[2] W. W. ROCKHILL, The Journey of William of Rubruck, p. 261 (London, *Hakluyt Society*, 1900). In the Mongol period, designs of a tiger or lion skin, and the design of metal-armor scales, were also painted on hide armor (*Yüan shi*, Ch. 78, p. 12, K'ien-lung edition).

[3] The clay figures in our collection come down from different periods. A rigid classification coinciding with dynastic periods cannot be established: two large groups may be distinguished,— archaic and mediæval. The two merge into each other. The former may be said to comprise roughly the Chou and Han periods, and to go down perhaps with some types into the fourth and fifth centuries; the latter occupy an epoch from the sixth to the eighth century. The term "archaic" is merely

cussion in the second part of this publication that, according to my interpretation, they are intended for the figure of the ancient shaman[1] (*wu*, or *fang siang shi*).

Among the exorcists of the Chou period, the *Fang siang shi*[2] occupies a prominent place. According to the *Chou li*,[3] he donned a bear-skin decorated with four golden eyes,[4] black trousers, and a red jacket. Armed with a spear and a shield, accompanied by a suite of a hundred attendants, he performed the purifications of every season, searching through the houses and driving out disease. At a great funeral service he strided in front of the coffin, and accompanied it to the grave.

---

intended to convey a chronological notion, but is not applied here with reference to technique or style. The age of the T'ang dynasty may safely be regarded as the *terminus ad quem* for the industry of burial clay figures, for we know surely enough that under the Sung and Ming dynasties the paraphernalia for the grave were carved from wood, but not modelled in clay. This question will be treated fully in Part II.

[1] Our word "shaman" is derived from the Tungusian languages (Manchu *saman*, Gold *šama*). The Mirror of the Manchu Language (*Manju hergen-i buleku bithe*) explains the word *saman* by means of the Chinese phrase *chu shên jên* ("a man who invokes or conjures the spirits"); and it is defined, *enduri weceku-de jalbarime baire nialma* ("a man who prays to and conjures spirits by sacrificing"). It is said in the same Dictionary that the *saman* acts near the sick-bed, and that there are male and female *samasa* (plural of *saman*). The Tungusian word has no connection whatever with Chinese *sha-men* (from Sanskrit *çramaṇa*, Pāli *samana*) denoting a Buddhist ascetic (YULE and BURNELL, Hobson-Jobson, p. 820); a Buddhist monk and a Siberian shaman will always remain two distinct affairs. PELLIOT (*Journal asiatique*, Mars-Avril, 1913, p. 468) has traced the word *šaman* in the language of the Niüchi to a Chinese document of the twelfth century. The identity of the notion conveyed by the Chinese word *wu* ("sorcerer") with the word "shaman" becomes evident from *T'ang shu*, where in the description of the Kirghiz it is remarked, "They call their sorcerers *kan* (*hu wu wei kan*)." The latter word (formerly articulated *kam*) is identical with Turkish *kam*, the general designation for the shaman in all Turkish dialects (compare W. SCHOTT, Über die echten Kirgisen, *Abhandlungen der Berliner Akademie*, 1865, p. 440). While reading the proofs, I receive No. 3 of the *Revue orientale* (Vol. XIV, 1914), in which J. NÉMETH devotes a special investigation to the origin of the word *šaman*: by applying methods of comparative philology, he arrives at the result that the word is an ancient property of the Turkish-Mongol languages.

[2] Chêng K'ang-ch'êng, in his commentary to the *Chou li* (BIOT, Vol. II, p. 150), explains the word *fang siang shi* as "expellers of formidable things," by substituting two other words for *fang siang* yielding this sense; but this conjecture is not adopted by the editors of the *Chou li* under K'ien-lung. BIOT translates the term, much too literally, by *inspecteurs de région*, or by *préservateur universel*. GRUBE (Religion und Kultus der Chinesen, p. 51) renders it "supervisors of the four points of the compass." DE GROOT (The Religious System of China, Vol. VI, p. 974) proposes the translation, "inspectors or rescuers of the country to the four quarters." These translations do not render account of the two words *fang* and *siang*: *fang* (No. 3435) means not only "place, region, quarter," but also "a recipe, a prescription;" and *fang shi*, according to GILES, is "a master of recipes,— a medicine man; a necromancer." The word *siang*[4] (No. 4249) means "to judge of by looks; to practise physiognomy" (hence in Buddhism: the *lakshaṇa* or physical marks of beauty of a Buddha). The *fang siang shi*, accordingly, is a "doctor" who has two functions,— he prescribes medicines, and practises the art of physiognomy (*siang fa*).

[3] BIOT, Vol. II, p. 225.

[4] Apparently a mask, which was worn by the Chinese shamans in all exorcising ceremonies (see DE GROOT, The Religious System of China, Vol. VI, pp. 974–980, 1151, 1187 *et seq.*; also, Vol. I, p. 162).

When the coffin was lowered into the grave, he struck the four corners with the spear, in order to chase away the spirits *wang-liang*.[1] The bear-skin, a Chinese commentator explains, serves the purpose of lending him a formidable appearance; and the four golden eyes testify that he spies in the four regions of the empire all places where contagious diseases are raging. The spear seems to indicate that he combats malignant spirits, and the shield is his means of defence against their attacks.

The two figures of shamans represented on Plates XV–XVII are clad with tight-fitting, sleeveless leather jerkins, the material being cut out in the form of scales arranged in regular horizontal rows. On the front (Plates XV, XVII) the scales are carefully outlined in black ink or varnish over a coating of pipe clay;[2] on the back of one of the figures (Plate XVI) they are impressed in the surface of the clay, presumably by means of a stamp. This process is not applied to the other figure, whose back is plain. In both, the jerkin is held by means of a leather belt tightly drawn around the loins. It does not seem to have a slit in front, and was presumably put over the head. The shaman in Plates XV and XVI wears a hide helmet surmounted by a queer crest, and laid out in vertical grooves; on the back (Plate XVI) coifs of hide scales are attached to it. The other shaman (Plate XVII) is adorned with a snail-like, high tuft of hair held by a hoop. Both are manifestly represented in the attitude of warriors, displaying the same pose of arms and feet. The right arm is raised, the thumb being placed against the second finger: they are apparently in the act of throwing a spear; and the spear, presumably of wood, may have actually been in their hands. The left arm reaching forth with clinched fist, and the feet wide apart, correspond to this action; and the two men naturally concentrate their weight on their right sides. The lively fighting attitude and the body armor show us that the two shamans are engaged in a battle with the demons; and, if the tradition of the Chinese is correct that such clay figures were interred in the graves during the Chou period, we may infer that, as the shaman warded off pestilence and malignant spirits from the grave before the lowering into it of the coffin, he continued in this miniature form to act as the efficient guardian of the occupant of the grave.

Helmets bedecked with scales occur also in Chinese illustrations (Fig. 33), and seem to have remained in the possession of shamans, even though they did not don the scale armor. The clay figure of a magician

---

[1] No. 12,518. These sprites are mentioned among those haunting travellers in the sand deserts of Turkistan (*Pei shi*, Ch. 97, p. 5).

[2] It is impossible to bring these fine lines out in the photographs.

(Plate XVIII), which is much later than the two others shown and presumably no older than the T'ang period, has a helmet with hood, on which rows of scales are outlined in ink.   A cape of tiger-skin envelops his shoulders.   He wears a necklace and jewelry with floral designs on his chest.   His coat is girdled; and a shirt of mail, presumably plate mail, [1] is emerging from beneath it.   In his left hand, which is perforated, he seems to have seized a spear or sword. [2]   A rectangular bag, which possibly serves for the storage of his paraphernalia, is attached to the belt on his left-hand side.   The wearing of a coat over the armor is characteristic of the T'ang period; and the artistic, though conventional, modelling of the face would seem to point to the same epoch.

In general, the conditions of defensive armor, as encountered in the archaic epoch of China, show a striking coincidence with those found in other ancient and primitive culture-groups of Asia, and those still alive in primitive societies.   On the whole, the military equipment of the ancient Chinese in principle agrees, for instance, granted the difference of material, with that of the Scythians as described by STRABO (VII, 3), who states that they used raw ox-hide helmets and cuirasses, wicker shields, spears, bows, and swords.

---

[1] See Chapter V.

[2] Presumably one of wood, which has decayed under ground.

III. DEFENSIVE ARMOR OF THE HAN PERIOD

"Your servant understands that, according to the classics, the perfection of government consists in preventing insurrectionary troubles, and the highest point of military art is to avoid the occasion of war."

YANG HIUNG in *Ts'ien Han shu.*

The sculpture of the Han period unfortunately furnishes no decisive contribution to the question of body armor. While possibly the artists may have intended in some cases to represent armor, as perhaps in some of the fighting horsemen, the stone work does not minutely indicate texture, and the material is such that no positive inferences can be drawn from it.[1] The only piece of defensive armor that is clearly enough outlined on these monuments is the shield or buckler, usually handled in connection with a sword. It is oblong and rectangular in shape with a convex curvature in the centre, causing a hollow on the inner side where the wearer's hand finds its place, and is notched in the middle of the upper and lower ends (Fig. 25). It is a parrying shield easily movable, and sufficient to protect the left arm and to ward off blows struck at it.[2]

It is notable that many soldiers represented on the Han monuments carry their shields also in their right hands, while manipulating the swords in their left; I presume that the fighters, when wearied out, sought relief in this manner by changing weapons from one hand to the other. In Fig. 25 a left-handed, and in Fig. 26 two right-handed shield-bearers have been selected. The same shield is employed also by soldiers fighting from war-chariots.

Another form of shield is much larger, more convex, almost roof-shaped, decorated with what appears like a tree design, and capable of hiding a man's face and the upper part of his trunk (Fig. 27).[3]

---

[1] The difficulty of studying from the bas-reliefs the costume and the ornaments displayed on it, is acknowledged also by M. CHAVANNES in his recent work Mission archéologique dans la Chine septentrionale, Vol. I, part 1: La sculpture à l'époque des Han, p. 39 (Paris, 1913). On a stone of the Hiao-t'ang-shan, M. CHAVANNES (p. 82) has correctly recognized some warriors clad with cuirasses; but hardly any other conclusion than that it is in general the question of hide armor can be drawn from these representations. These warriors are barbarians styled *Hu,* and in all probability Huns (*Hiung-nu,* who are frequently termed also *Hu*). We shall come back to this monument below in speaking of the tactics of the Huns.

[2] See, for example, CHAVANNES, Mission, Nos. 131, 136.

[3] *Ibid.,* No. 190. CHAVANNES (La sculpture à l'époque des Han, p. 251) states that this buckler is of rattan, doubtless for the reason that there are still rattan shields in China; but these are always circular, almost half-spheroidal, and plaited in basketry style. The present specimen is a rectangle, and exhibits no characteristic features of

201

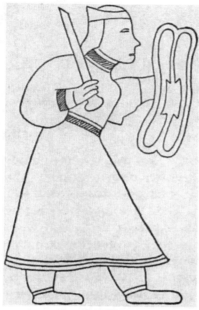

FIG. 25.
Left-handed Shield-Bearer (Sketch from Rubbing of Han Bas-relief).

FIG. 26.
Right-handed Shield-Bearers (Sketch from Rubbing of Han Bas-relief).

In the "Battle on the Bridge," [1] a picture executed with a great deal of life and motion, the manner of handling the buckler in close combat is vividly illustrated. The commander of the force, passing the bridge in his chariot, defends himself with his sword

FIG. 28.
Soldier with Circular Buckler (Sketch from Rubbing of Han Bas-relief representing the Battle on the Bridge).

FIG. 27.
Han Soldier with Rectangular Buckler (Sketch after Chavannes, Mission, No. 190).

against an arbalist whose crossbow he has adroitly overturned with a thrust of his shield, while a footman is attacking his rearing horse

rattan plaiting. It is much more likely to be of wood covered with hide, on which the design is painted. The rattan shields have often been described and illustrated (AMIOT, Art militaire, *Mémoires concernant les Chinois*, Vol. VII, p. 371, and Plate XXX, figs. 10 and 11; DE GUIGNES, Voyages à Peking, Vol. III, p. 20; Atlas of STAUNTON's Embassy, Plates XVII and XIX, No. 5, etc.). In Peking I had occasion in 1901 to see these shields used by fencers, and procured two specimens painted with tiger-heads for the American Museum, New York. The general opinion of the Chinese is that rattan shields are a matter of recent development, and that originally shields were made from a combination of wood and hide (see *Huang ch'ao li k'i t'u shi*, Ch. 15, p. 21, where the earliest relevant text quoted is the *Ki siao sin shu* of 1566 by Tsi Ki-kuang, followed by the *Wu pei chi* of 1621 by Mao Yüan-i). The earliest illustration of the rattan shield I am able to trace is in the *Lien ping shi ki* (Ch. 5, p. 5, ed. of *Shou shan ko ts'ung shu*, Vol. 52), written in 1568 (WYLIE, Notes, p. 91). Merely judging from its circular shape, the round shield above referred to, in the hand of the soldier at the foot of the bridge, might be a rattan shield; but I venture to doubt that the latter was in existence during the Han period. The shield in question may as well be of wood or hide (compare Figs. 28, 30). The rattan shield painted with a tiger's head was officially introduced into the army under the Manchu. This troop was uniformed with a short jacket of yellow cotton stuff on which tiger stripes were represented in black, a pair of leggings and boots with the same design, and a hood in the shape of a tiger-head (see *Huang ch'ao li k'i t'u shi*, Ch. 13, pp. 49–50; the shield is figured and described in Ch. 15, p. 21).

[1] CHAVANNES, Mission, No. 136.

with a spear.   On this representation we notice another type of shield of circular shape (Fig. 28) on the arm of a warrior who is posted on the left-hand side at the foot of the bridge.   The question as to the material from which this shield may have been wrought is not susceptible of positive decision.   Certain it is, however, that three distinct types of buckler are depicted on the monuments of the Han.[1]

Of the three types of Han bucklers, the first may be ascribed as peculiar to the period, in so far as it does not seem to have survived in later ages; it is not alluded to in military literature, nor is it traceable among the specimens of shields in vogue during the Ming and Manchu dynasties.   The case is different with regard to the two remaining types. The greatest authority on military matters is Mao Yüan-i, who published his work *Wu pei chi* (not mentioned by Wylie) in 1621 (80 volumes). It is the most comprehensive work of this class, and the one best illustrated.   All relevant illustrations of the *T'u shu tsi ch'êng*, which quotes this author as Mao-tse, are derived from his work.   In accordance with an older work *Wu king* ("Canon of Military Matters"), he discriminates between two main types of shields, the long shield of the footmen (Fig. 29), and the round shield of the horsemen (Fig. 30). The former is entirely made of wood, and, being as tall as a man, completely screens his body.   It rests on the ground, and is a veritable fence or bulwark.[2]   The latter, of wood covered with hide, is carried by the cavalier on his left arm, which is passed through the two straps in order to protect his left shoulder against arrow-shots, while he brandishes in his right hand the short sword.[3]   Mao admits that it offers no advantages, and it certainly was more an encumbrance than a safeguard.   As the round buckler is peculiar to the horsemen, we may suppose that the Han soldier armed with it is an equestrian engaged in a dismounted combat.   There are instances on record to the effect that the soldiers, especially when the decisive moment approached, dismounted from their horses, marched on foot, sword in hand, and engaged in close combat.[4]

From the wooden documents of Turkistan recently edited and translated by M. CHAVANNES we learn that the shields used by the soldiers of the Han period were red; that is to say, they were made of wood, and

---

[1] Thus likewise CHAVANNES, La sculpture, p. 37.

[2] This is the same type of shield as that figured and described by PH. F. v. SIEBOLD (Nippon, 2d ed., Vol. I, pp. 336, 337).

[3] The horsemen of the Kirgiz, who wore wooden cuishes, fastened a round shield to their left shoulder to ward off arrow-shots and sword-cuts (*T'ang shu*, Ch. 217 B, p. 8).

[4] Compare the battle deciding the fate of Hiang Yü in *Shi ki*, Ch. 7 (CHAVANNES, Les Mémoires historiques de Se-ma Ts'ien, Vol. II, pp. 318–320).

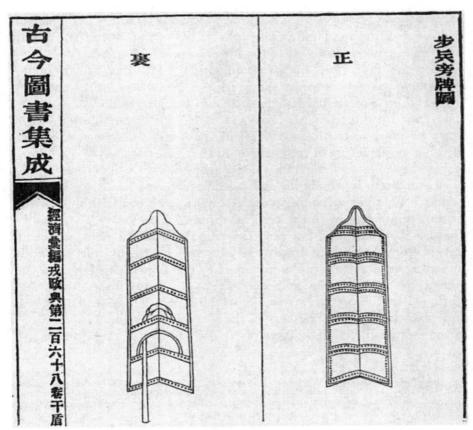

FIG. 29.
Shield of Foot-Soldiers, Exterior (to the right) and Interior (to the left).   From *T'u shu tsi ch'êng*.

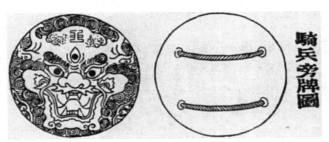

FIG. 30.
Round Shield of Equestrian Soldiers, Exterior and Interior.

coated with a red varnish to protect the material from the influences of the weather.[1] They were turned out in the official armory of Nan-yang in Ho-nan Province,[2] and in all probability were adorned with the tail-feathers of pigeons fastened to the lower edge. The wooden documents employ the word *tun*,[3] once formed with the classifier 'spear' (*mao*);[4] and in one passage[5] appears the word *p'ai* (No. 8574), which, as far as I know, is thus attested for the first time in the Han period.[6]

In his Introduction M. CHAVANNES has given an admirable summary of the information garnered in these early documents, and has drawn a vivid picture of the garrison life in those outposts of the Chinese empire.[7] He has sounded also the sentiments by which those soldiers were animated, by rendering several fine pieces of poetry of the T'ang period. There is still another, contemporaneous source which permits us some inferences as to the emotional life of those brave Han frontier-guards. CHAVANNES[8] has ably described the function of the signal-towers erected along the frontier at intervals averaging thirty *li*, which served as optical telegraphs announcing the approach of hostile van-guards by means of huge beacon-fires. In many cases the guards stationed in these towers were kept alert in repelling undesirable invaders.[9] In the burial pottery of the Han period, which is a microcosm of the culture life of those days, we find a number of miniature models

---

[1] Compare above, p. 189.

[2] It seems to have been customary in the Han period to occasionally inter armor and shield with a general. We learn that the son of the marshal Chou Ya-fu purchased from an officer of the Imperial Armory a cuirass and buckler intended for the funeral of his father (L. WIEGER, Textes historiques, p. 448). This act led to an accusation against the old general, which resulted in his suicide; the illegal point of the case, however, was sought in the step of purchasing imperial property, not in the intended burial; and the charge was forced, as the Emperor was intent on causing the downfall of the old officer. The *Ku kin chu* by Ts'ui Pao of the middle of the fourth century relates that in the third year of the reign of the Emperor Chang (78 A.D.) people dug up the ground of a burial-place at Yüan in Tan-yang (An-hui Province) and found in it a piece of armor. It was a cuirass (*kia*).

[3] CHAVANNES, *l. c.*, Nos. 77, 763.

[4] No. 75.

[5] No. 682.

[6] The Annals of the Han Dynasty employ neither of these words, but the word *shun*.

[7] I can only join Mr. L. C. HOPKINS (*Journal Royal As. Soc.*, 1914, p. 475) in the wish that the substance of this essay may be made more generally accessible. Perhaps the Royal Asiatic Society itself might undertake to publish an English translation of it in a separate issue.

[8] *L. c.*, pp. XI–XIII.

[9] To quote one example, in 108 A.D., the K'iang (Tibetans) with a force of over ten thousand men attacked the watch-towers near Kan-chou fu in Kan-su Province, and killed or captured the officers and privates occupying them (CHAVANNES, *T'oung Pao*, 1906, p. 257). Beacon-towers in which lookout soldiers were kept, *tun t'ai* (No. 12,205), were still in existence under the Ming dynasty, and are well described by Persian travellers in the fifteenth century (see BRETSCHNEIDER, *China Review*, Vol. V, p. 34). Compare Fig. 31.

兵雜集卷六

敵臺圖 一

FIG. 31.

Military Watch-Towers of the Ming Period (from *Lien ping shi hi* of 1568).

representing such watch-towers; and all these, according to the unanimous testimony of the Chinese, have been found in graves of Kan-su Province. The conclusion would seem justified that pottery of this type was interred, as worthy emblems of their martial calling, with renowned officers who had deserved well of their country in the frontier wars and had died the honorable death of the soldier. On Plate XIX is illustrated a green-glazed model of a three-storied watch-tower rising from the bottom of a round bowl: on the two parapets and roofs the sentinels are engaged in showering from their crossbows a volley of darts on an advancing column of scouts.[1]   Here we enjoy seeing before us in action the undaunted heroes of the Hunnic wars whose sentiments were immortalized by Li Po. The imposing loftiness of the structure standing with the force of a pyramid, the beautiful architectural forms, the jutting wooden beams supporting the corners of the parapets, are notable features making this bit of clay a live and unique document of the culture of the Han period.

There are also less elaborate pottery models of such watch-towers. One in the Museum collection[2] shows a single story with windows on three sides and a door ajar in the front wall; the windows are provided with elegant lattice-work. Another specimen[3] represents the section of a city-wall with a roofed, square tower in the corner, to which a staircase leads up.

The most signal fact about defensive armor under the Han is that metal suits gradually made their way during this period. We meet, for the designation of it, a new word k'ai (No. 5798), written with a character in which the classifier kin ("copper"[4] or "metal") enters, and which does not occur in the ancient canonical texts. From the terminology of the dictionary Shuo wên (around 100 A.D.) we gather that armature had then grown more complete, that there were metal helmets (tou mou), brassards (han),[5] and metal protectors for the nape (ya-hia).[6]   The old

---

[1] This beautiful piece of Han pottery is in the collection of Mr. Charles L. Freer of Detroit, to whom I am greatly indebted for the photograph and his kind permission to publish it. The object was acquired by Mr. Freer as early as in the seventies, and is the first specimen of Han pottery that came to America; presumably it was even the first to come out of China.

[2] Cat. No. 118,489; 27.5 cm high, green glaze decomposed into silver oxidation.

[3] Cat. No. 120,901; gray clay, unglazed; excavated by Dr. Buckens, physician in the service of the Peking-Hankow Railway, near Chêng-chou, Ho-nan Province.

[4] "Copper" is probably the original meaning, but not, as supposed formerly, "gold." In the Chou li gold is always designated huang kin ("yellow metal").

[5] GILES (No. 3791) translates "greaves; leg-guards for soldiers," which is doubtless also correct; but the definition of this word in the Shuo wên is pei k'ai; that is, arm-guards.

[6] See COUVREUR, Dictionnaire chinois-français, p. 115 b (also in PALLADIUS, Chinese-Russian Dictionary). Compare Chinese text opposite.

word *han* (p. 175) was now likewise connected with the classifier "metal" (No. 3816); and an entirely new word *ye* (No. 12,996), composed of the phonetic element *ye* ("leaf") and the same classifier, springs up to denote a new contrivance in the structure of protective armor,—a metal lamina (literally "metal leaf"). These facts combined go to prove that far-reaching innovations had set in after the close of the

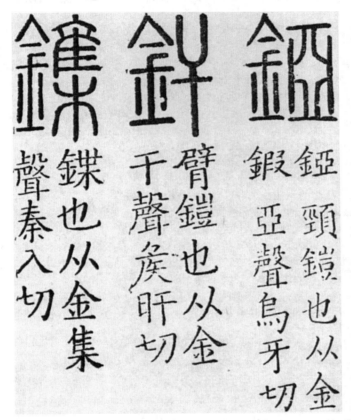

Chou dynasty, and that the Han period must have revolutionized the entire method and technics of armature. Chêng K'ang-ch'êng, the famous commentator of the *Chou li*, who lived in the second century A.D., says anent the armorers of the Chou time[1] that the ancients employed hide in the manufacture of corselets (*kia*), but that now (in the author's time) metal (*kin*) was utilized for the same purpose, and that this product is designated *k'ai*. Of what metal was this new armor made? And what type of armor was represented by it? The most interesting contribution to this question is made by Chung Ch'ang-t'ung,

---

[1] BIOT, Chou li, Vol. II, p. 152.

an author who lived in the beginning of the third century A.D., and who is known as the editor of the Taoist writer Yin Wên-tse.[1]   He is quoted as follows in the *Yen fan lu*:[2] "In days of old, war-chariots were employed in warfare, and the fashion of iron plates was not yet in use for armor; at the present time, hide armor, though it can still offer sufficient resistance to a crossbow, will needs lead to the loss of the army and the destruction of the empire.   Regarding this matter, it was at the time of the Posterior Han (25–220 A.D.) that armor received iron laminæ, but it is not known what the state of affairs was at the time of the Anterior Han (B.C. 206–23)."   Here it is plainly expressed that iron armor came up under the Later Han dynasty, and the expression *t'ie cha*[3] leaves no doubt that it was armor composed of iron laminæ.

In this connection another notice incorporated in *Ko chi king yüan* (Ch. 41, p. 1 b) would be of interest, if any dependence could be placed as to the value and the time of the source from which it is quoted. This is a work called "Dissertation on Corporal Punishments" (*Jou hing lun*) by K'ung Jung, a descendant of Confucius in the twentieth degree, who, according to GILES,[4] died in 208 A.D.   Nothing is known to me regarding this work; M. PELLIOT, in his careful bibliographical study of Chinese law,[5] does not mention it.   In the present case, it would be indispensable to know exactly when that work was composed, as the author lays stress on a contemporaneous event, and to ascertain whether the incriminated passage was really contained in the original

---

[1] WYLIE, Notes on Chinese Literature, p. 156; L. WIEGER, Taoisme, Vol. I, Le canon, p. 184, No. 1159.

[2] Completed in 1175 by Ch'êng Ta-ch'ang (WYLIE, Notes on Chinese Literature, p. 160) and reprinted in the *T'ang Sung ts'ung shu*.

[3] The word *cha* (No. 127) refers to the wooden or bamboo tablets used for writing and united into bundles of books before the invention of paper.   The discoveries in Central Asia have rendered us familiar with the form of these wooden documents. The plates, as used in the manufacture of armor, have indeed a very similar shape; and hence the transfer of the name of the latter is easy to understand.   COUVREUR (p. 736 b) translates *cha* by "les couches de cuir ou les plaques de métal qui composent une armure;" PALLADIUS in his Chinese-Russian Dictionary (Vol. II, p. 379) by "fish-scale, armor;" GILES gives the meaning "a layer" and "numerative of *kia*, armor."   There are some passages in the *Tso chuan* and *Han shi wai chuan* (see *P'ei wên yün fu*, Ch. 97, p. 6) where *cha* doubtless relates to the different layers of a hide armor; but as a rule it originally refers, as stated above (p. 196), to the scales of a hide scale armor.   This is also the opinion of K'ung Ying-ta (574–648), who, in his work *Shang shu chêng i*, gives the following definition of the word *ye* (No. 12,996),— "metal lamina or plate in armor; the metal lamina of armor is the same as that is called *cha* in the *K'ao kung chi* (in the *Chou li*)."   The word *cha*, however, does not occur in the text of the *Chou li*, but only in the commentaries.   In the same sense, the K'ang-hi Dictionary defines the word *cha* as *kia ye*, "armor leaves," that is, plates or laminæ covering the armor.

[4] Biographical Dictionary, p. 401.

[5] Le droit chinois (*Bulletin de l'École française d'Extrême-Orient*, Vol. IX, 1909, pp. 27–56).

edition.   Not being able to do so, I can give it only with all reserve: "The holy men of antiquity made armor of rhinoceros-hide; now the *pên ling*[1] have iron armor."

The fact that the word *k'ai*, and the new type of body armor understood by it, were actually employed during the Han period, is now obviously brought out by the contemporaneous wooden slips discovered in eastern Turkistan, and which have been edited and translated by E. CHAVANNES.[2]   As already mentioned, the word *k'ai* occurs there on two of the wooden documents (Nos. 758, 794); while the ancient word *kia* is preserved in three other cases.   Both types, *kia* and *k'ai*, accordingly, were in use among the outlying Chinese garrisons of the Han period; and as explicitly recognized by Chinese authors, the *k'ai* differed from the *kia* in the essential point that they were reinforced by metal pieces. The foundation of the armor *k'ai* consisted likewise of leather or hide; and in CHAVANNES' document No. 794 the question is of "four pieces of hide, two halves being so connected as to make two suits of armor." The "halves" seem to refer to two large pieces of hide covering chest and back.

The metal helmet appearing under the Han and perhaps under the Ts'in dynasty (p. 175) is the natural accompaniment of metal armor; the *galea* of ancient times gives way to the *cassis* (Figs. 32, 33).   The word *tou mou* for the metal helmet mentioned above appears, indeed, on one of the contemporaneous wooden slips of the Tsin dynasty (265–313).[3]

If the metal of the Later Han dynasty was iron, — what was the metal employed during the Former Han dynasty?   And what was the shape of the metal pieces attached to the hide foundation?

It is not very likely, for technical reasons, that hide armor was immediately followed by armor consisting of iron laminæ.   The latter denotes a much more advanced stage of civilization, and presupposes acquaintance with the art of forging iron; it is also a much more complicated structure, its manufacture requiring a skill far superior to the more mechanical mode of preparing a coat of hide.   We are fortunately in a position to show from both literary and archæological evidence that iron hide armor was preceded by copper hide armor.   In the work *Yen fan lu* quoted above, the observation is made that "in the times of remote antiquity and in the period anterior to the Ts'in and the Han leather armor named after the rhinoceros was much used in the army, but that in the records of Se-ma Ts'ien's *Shi ki* mention

---

[1] Apparently the title of a military office at the time of the Han dynasty.

[2] Les documents chinois découverts par Aurel Stein dans les sables du Turkestan oriental (Oxford, 1913).

[3] CHAVANNES, *l. c.*, No. 794.

is made of armor fabricated from forged copper (*tuan kin wei kia*); that, however, on close examination, the employment of the latter is still much restricted."[1]

We shall not be far wrong in concluding that the metal pieces employed for the reinforcement of armor in the period of the Anterior

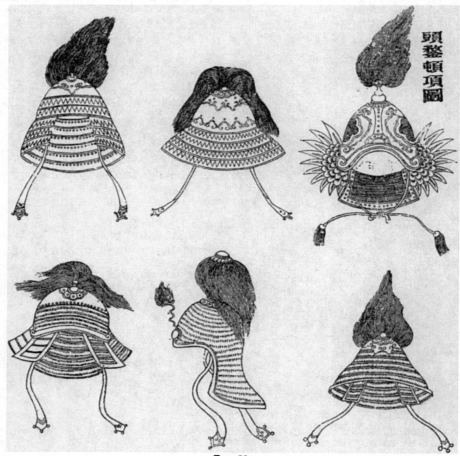

FIG. 32.
Sketches of Helmets (from *T'u shu tsi ch'êng* which reproduced them from *Wu pei chi*), representing the Tradition of the Ming Period.

---

[1] The expression "to forge defensive armor" (*tuan kia*) occurs in *Shi ki*, Ch. 112, in the biography of Chu-fu Yen (compare *P'ei wên yün fu*, Ch. 106, p. 56 b). In the age of the Three Kingdoms (221–277) metal armor, for which copper or iron was utilized, was firmly established, as we see from the life of the famous General Chu-ko Liang (*San kuo chi, Wu chi*, Ch. 19, p. 1 b), who lived from 181 to 234 (see GILES, Biographical Dictionary, p. 180). In *Tsin shu* and *Sung shu*, metal armor is frequently mentioned. An iron mask (*t'ie mien*) for the protection of the face is first mentioned as being employed in the period Yung-kia (307–313 A.D.) by General Chu Ts'e (styled Chung-wên) in the battle of Hia-k'ou, in Han-yang fu, Hu-pei Province (*Tsin shu*, Ch. 81, p. 6).

Han were of that metal then most generally employed,—copper. And a number of perforated, thin copper plates exhumed in the environment of Si-ngan fu from a grave of that epoch tends to confirm this opinion. These laminæ, some of which are sketched in Fig. 34, can but have served the purpose of being sewed on to the surface of a cuirass. They were employed for the making of a *k'ai*, and formed the natural continua-

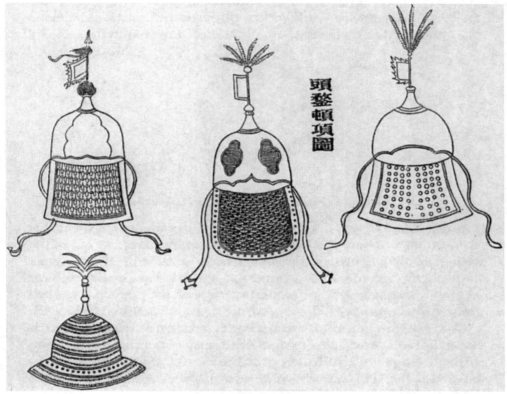

FIG. 33.
Sketches of Helmets (from *T'u shu tsi ch'êng* which reproduced them from *Wu pei chi*), representing the Tradition of the Ming Period.

tion of the ancient scale armor *kiai* discussed at the end of the previous chapter. The scales in the latter were cut out of leather: in the third and second centuries B.C., the Han made a decided advance by gradually transforming these leather into copper scales; and the Posterior Han, in the first centuries of our era, went a step farther in substituting iron for copper. The specimens in Fig. 34 demonstrate that the copper pieces leaned in their forms toward scales, though they approach to a higher degree the shape of a leaf (hence the term "leaf" which we meet in the Han authors). A slow and gradual development must have been in operation toward effecting that uniform oblong, rectangular shape which we are wont to designate as "plate." There is, for lack of monuments, as yet no means of exactly ascertaining the date when this type of regular iron plate armor sprang up in China. The term *t'ie cha* employed by Chung Ch'ang-t'ung, discussed above, is very tempting in leading us to assume that it existed at least toward the end of the Posterior Han period in the third century A.D.; the word *cha* relates to the rectangular wooden writing-slips still prominent in the administrative system of the Han, and the application of this word to the plates of

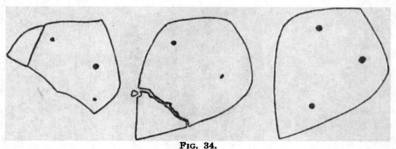

FIG. 34.
Bronze Scales of Armor of Han Period (half of actual size).

an armor is most happy. As these wooden slips possessed regular forms, we are allowed to infer that also the iron plates in the armor of the Han were gradually adapted to the same uniform standard. In the age of the T'ang (618–906) iron plate armor presents itself as an accomplished fact, and was made with a technical perfection which must have been preceded by centuries of diligent and intelligent practice (see Chapter V).

The existence of protective laminæ of rectangular shape under the Han may be inferred also from another matter peculiar to that age.[1] In the biography of Ho Kuang, who died in B.C. 68, the great "king-maker" of the Han dynasty, as Mayers calls him, mention is made of "jade clothes" (*yü i*). Yen Shi-ku (579–645), the famous commentator of the Han Annals, explains this term as denoting a coat of the form of an armor (*k'ai*), consisting of jade slabs joined together by means of gold threads; these jade slabs were shaped into regular plates (*cha*), one foot long and two inches and a half wide; they formed a perfect enclosure, and reached down to the feet. Another style of this garment, compared likewise with armor by Yen Shi-ku, was composed of strung pearls or

---

[1] The following information is drawn from the *Han tsien* (No. 1648) of Kua Ts'ang-lin of the Sung; the edition before me is by Wu Ki-ngan of the Ming, and was published in 1600. This is a most valuable work for the study of Han culture, being arranged in the form of a glossary of subject-matters (corresponding to our archæological dictionaries) extracted from the Han Annals together with the commentaries; it allows us to ascertain at a glance what objects of culture existed under the Han.

beads in the upper part, while only the skirt was formed by jade plates. It is self-evident that these jade plates, of which we hear nothing at any earlier period, were produced in imitation of metal armor-plates; and Yen Shi-ku's simile with an armor strongly supports this opinion.

By what factor was the innovation and progress of the Han in matters of defensive armor caused? The development of the defence of the body moves along as the natural consequence of the advance in weapons of offense. "The history of invention as applied to war has been the record of alternate advances in this line, and in overcoming defence." [1] The steadily growing perfection of weapons necessitated a corresponding increase in the efficiency and power of resistance of body armor. The chief weapons of the Chou period were spear and bow; and the armor of rhinoceros-hide offered to them adequate opposition. In the age of the Han we meet the more effectual crossbow and the two-edged sword; and Chung Ch'ang-t'ung justly says that hide armor then was no longer a suitable shelter for the arrows shot from crossbows, if the interests of an army were to be maintained. The copper or bronze swords in vogue among the Former Han dynasty gradually gave way to iron swords under the Later Han dynasty; and parallel with this movement, we notice a logical development from plain hide and hide scale armor to copper scale and iron scale, and ultimately to iron plate armor. Thus, judging from appearances, it may be conceived that this sequence in the gradual perfection of armor might have been evolved from purely inward causes and necessities, and that no factors of any outward influence need be invoked in order to account for it; but such a conclusion hazarded without any regard to historical agencies would be plainly illusory.

It cannot be denied that an entirely different point of view may be pursued in this problem. It may be argued that the Chinese, despite the numerous aggressive and defensive wars which they have made on the adjoining tribes, cannot be called, in the strict sense of the word, a warlike nation, and that they were always deficient in inventions of military implements. At all times they were ready to adopt any superior arms from their more belligerent neighbors, and to vanquish their enemies with their enemies' devices. The crossbow is properly claimed as a contrivance of the aboriginal tribes of southern China; and the type of the short bronze sword of the early Han (see Plate XX) bears such a striking similarity to that of the Siberian bronze age, that imitation due to historical contact may justly be suspected. Under the Han, cast-bronze swords (Plate XX) gradually gave way to

---

[1] O. T. MASON, The Origins of Invention, p. 389.

cast-iron swords (Plate XXI), the latter being cast in the same shape as the former. The process of transformation is identical with the one that we observe in the antiquities of Siberia. The excellent plates of ancient Siberian bronze and iron swords published by W. RADLOFF,[1] in which bronze is colored green and iron brown, afford a good object-lesson for the study of the gradual transition from bronze to iron: here, for instance, we note that the hilt is changed into iron, whereas bronze is retained for the blade (Plate XII, No. 4); or that the blades become iron, and the hilts remain of bronze (Plate XIII, Nos. 1–3), until ultimately there spring up types purely of iron which faithfully preserve the forms and ornaments of the more ancient bronze swords. We know from literary documents that the Han still turned out weapons of bronze, that under the Former Han the latter were gradually superseded by iron weapons, and that these were definitely established under the Later Han: the year 219 may safely be regarded as the term when weapons were made exclusively from iron, and when bronze was discarded for this purpose.[2] It will therefore be in general correct to assume for archæological purposes that bronze swords bearing the characteristics of the Han, with greater probability belong to the period of the Former Han dynasty (B.C. 206–23), while cast-iron swords of the same features most probably range in the period of the Later Han dynasty (25–220 A.D.). The casting of iron for implements of every-day use is peculiar to that age: the Chinese then ingeniously applied to iron the same process as formerly to bronze, casting it in sand moulds, and perpetuating in the new material their ancient bronze forms. Thus we have large bulging vases (of the type styled *hu*) with movable lateral rings and inscriptions in Han style cast in high relief on the exterior of the bottom,[3] — of the same shape as the corresponding vases in bronze and pottery. There are, further, stoves, large cooking-kettles, cooking-pans, coin-moulds, bells, lamps, chisels, knives, and mountings for chariot wheel-naves, — in style and decoration breathing the spirit of Han culture, and the complete decomposition of the thick iron core testifying to their great antiquity. The cast-iron spears shown on Plate XXI, owing to the decay of the iron substance underground, have almost lost their original forms. The swords are in a somewhat better state of preservation. They are two-edged, like the older bronze prototypes,

---

[1] Siberian Antiquities (*Materials toward the Archæology of Russia*, No. 5, in Russian, St. Petersburg, 1891).

[2] See the interesting observations of F. HIRTH (Chinesische Ansichten über Bronzetrommeln, pp. 18–22, and The Ancient History of China, pp. 234–237).

[3] It is the well-known formula *i hou wang* ("may it be serviceable to the lords!").

with massive iron hilts, but with lozenge-shaped guards of bronze coated with a dark and polished patina.

We are now confronted with the fact that the Han period has run through the same phase of development with regard to offensive and defensive armor. It is therefore inevitable to conclude that a correlation exists between these two developments, and that the production of defensive iron armor under the Posterior Han is prompted by the coeval coming into existence of iron weapons. The two phenomena are in mutual proportions. In the same manner, the perfection of bronze arms under the Anterior Han must have resulted in the machination of bronze protective armor. The same causes bring about the same effects; and if the agencies of the cause, the weapons, are suspected with good evidence of foreign origin, the same suspicion is equally ripe for the effects — defensive armor. The one is inconceivable without the other. In the ancient Siberian swords we meet the same process of development from bronze to iron as in ancient China, and this parallelism plainly reveals the historical interrelation of the two culture groups. This being the case, the further supposition is justifiable that also the progress made under the Han in body armor might be due to an impetus received from the same quarter. At this point due attention must be paid to the great historical connections linking all Asia in matters of military art. No human invention or activity can be properly understood if viewed merely as an isolated phenomenon, with utter disregard of the causal factors to which it is inextricably chained. Every cultural idea bears its distinct relation to a series of others, and this reciprocity and interdependence of phenomena must be visualized in determining its historical position. The development of harness must be viewed in close connection with the mode of military tactics, the science of warfare: every progressive step advanced in the latter draws a natural reaction on the form of armament, and a transformation of the latter is a sure sign of the fact that a considerable change in tactical conduct has preceded it. It is therefore from the history of tactics that we must derive our understanding of the technique of armor. The problem now set before us is,— What great movement in military tactics caused the radical transformation of arms experienced by the peoples of China, Central Asia, and Siberia around the centuries of our era? This movement, in my opinion, proceeded from ancient Irān. I shall endeavor to demonstrate that far-reaching tactical reforms were launched in Irān and deeply affected the entire ancient world, and that these innovations spread from Irān to the Turkish tribes of Central Asia, and were handed on by the latter to the Chinese. Developments of tactics and armature moved along very similar lines in the three groups.

First of all, attention should be called to the fact (and this cannot be an accident) that the new parts of the armor added in China during the Han period are exactly those which we find in ancient Persia. The nape-guard (*ya-hia*)[1] meets its counterpart in the *kūiris* named in the Avesta, rendered in the Pahlavī version *grīvpān* ("neck-guard") and explained by the gloss, "attached behind from the helmet to the corselet."[2] The Avesta mentions also leg-guards, *rānapānō* ("thigh-protector") which are interpreted as greaves; and according to Jackson, the helmet is described in the Avesta as made of iron, brass, or gold.[3] Likewise the new mode of fighting prevailing in the Han period — the use of the sword in connection with shield and armor — is paralleled in Persia when we read in Xenophon's *Cyropædia* (II, 1, 21) that Cyrus, in training his men, relieved them from practice with the bow and the javelin, and exercised them in but one direction, to fight with sword, shield, and armor.[4]

Further, it is essential to grasp the fundamental fact of the difference between mounted archers and true cavalry, and the development of these two different arms and means of tactics among the Iranians. Herodotus (VII, 84) states that the Persian horsemen were equipped in the same manner as the infantry, except that some of them wore upon their heads devices wrought of brass and steel. Accordingly, the Persian cavalrymen of that time must be credited with the wearing of sleeved tunics of diverse colors, bedecked with breastplates of iron scales like fish-scales, as attributed by Herodotus (VII, 61) to the infantry. The description of Herodotus (IX, 49) leaves no doubt that the Persian horsemen fighting the Greeks were only a body of infantry mounted on horses and chiefly depending upon their bows, at which Herodotus expresses astonishment by remarking that, though horsemen, they used the bow; they were, accordingly, mounted archers.

This mode of fighting was spread over the entire Scythian and Iranian world. The Scythians shot with bow and arrow from horseback (Herodotus, IV, 131), and singly skirmished in open order against their opponents, attacking them here and there where chance or advantage offered; they were at the same time nowhere and ubiquitous, effectually screening their operations. The Massagetæ (Herodotus, I,

---

[1] A Chinese word suspicious of foreign origin.

[2] A. V. W. Jackson, Ancient Persian Armor (in *Classical Studies in Honor of Henry Drisler*, p. 118, New York, 1894).

[3] *Ibid.*, p. 119. The greaves are mentioned also by Xenophon (*Anabasis*, VIII, 6); Herodotus (VII, 84) ascribes brass and steel helmets to the Persian cavalry men; Xenophon (*Cyropædia*, VI, 1, 2) speaks of brazen helmets, and in one case (VI, 4, 2) of a golden helmet.

[4] Compare also *Cyropædia*, I, 2, 12.

215) were familiar with the mode of fighting both on horseback and on foot, which indicates that when in the saddle they were mounted foot-men. The Parthian mounted archers were dreaded and detested by the Romans, chiefly because in taking to flight they shot their arrows back-ward at the pursuing enemy.[1] The Mongols, during their invasions, availed themselves of the same mode of tactics. "In battle they with-draw in good order, as soon as they are at a disadvantage," says the Ar-menian historian Haithon, "but it is very dangerous to pursue them, as, though turning back, they are able to shoot during the flight, and thus wound men and horses."

According to XENOPHON (Anabasis, VIII, 6, 7), there were around Cyrus about six hundred cavalry, the men all armed with breastplates, greaves, and helmets, except Cyrus, who presented himself for battle with his head unprotected;[2] and all the horses of the cavalry that were with Cyrus had defensive armor on the forehead and breast. Here, then, for the first time is the question of real cavalry; horse and man being completely armored, and this new equipment being a sign of a new mode of tactics, while in the age of Herodotus the horse of the Persians was not yet caparisoned.[3] Though the term "cataphracti" is not used by Xenophon, the institution described by him is either the forerunner of the latter or identical with them.

In Cyropædia (VI, 4, 1), besides the frontlets and breastplates of the horses, single horses with greaves, and chariot horses with plates upon their sides are mentioned; so that the whole army glittered with brass, and shone with purple garments. Abradatas equipped the horses of his chariot with brazen mail (ibid., VI, 1, 51).[4] In the same work (VII, 1, 2) it is on record that all those who were with Cyrus were fur-nished with the same equipment as himself; purple coats, brazen armor, brazen helmets, white crests, short swords, and each with a spear made of the timber of the corneil-tree. Their horses were armed with brazen forehead-pieces, breastplates, and shoulder-pieces which simultaneously served as thigh-protectors to the rider. The rider allowed his feet to hang down behind these flank-pieces which safeguarded his thighs.

---

[1] E. BULANDA, Bogen und Pfeil bei den Völkern des Altertums, p. 61 (Wien, 1913).

[2] On the armor of Cyrus see XENOPHON (Cyropædia, I, 4, 18; VII, 1, 2).

[3] The Massagetæ (HERODOTUS, I, 215), who in their costume and mode of living resembled the Scythians, had their horses caparisoned with breastplates of bronze, while gold was utilized for the bridles, the bit, and the cheekplates. The fact that the horses in the army of Xerxes were not caparisoned is practically demonstrated by the Nisæan charger of the Persian noble Masistius, which received an arrow in its flank (HERODOTUS, IX, 22). Neither were the horses of the Assyrians caparisoned, who possessed only mounted infantry, not cavalry in the strict sense.

[4] Compare also VI, 2, 17.

Finally, in his concluding chapter (VIII, 8, 22), in which Xenophon laments the gradual degeneracy of the Persians after the death of Cyrus, he sums up again by saying that Cyrus, after breaking them of the habit of skirmishing at a distance, armed with breastplates both men and their horses, gave every one a javelin in his hand, and trained them to close fighting; but now, the historian complains, they neither skirmish from a distance, nor do they engage hand in hand.   In this passage it is clearly stated that Cyrus was the father of a new mode of tactics, and that this method was exactly what we understand by regular cavalry in the modern sense,— horsemen engaging in close combat, and charging their opponents with all possible speed by means of javelin, spear, lance, or sabre.   The *Cyropædia*, of course, is nothing more than an historical romance, and the attribution to the elder Cyrus of the new tactical principle is plainly an anachronism; it must, however, have been in full operation among the Persians in Xenophon's time.   It cannot have existed under Cyrus, as we do not find it in the army of Xerxes invading Greece.

The mail-clad warriors of the Persians and related nations became known in the antique world under the name cataphracti (κατάφρακτοι) or *catafractarii*, derived from *cataphracta*, the designation of their defensive armor.   Sarmatians clad with such armor are represented on the Column of Trajan; actual fragments of armor of this sort discovered in graves of southern Russia, and, further, the notices of classical authors, enable us to form some idea of the appearance of these suits of armor.[1] They consisted of a foundation of cloth or leather, to which scales or laminæ of metal (copper or iron), more rarely of horn or bone, were sewed on in such a manner that the single rows overlapped, each row covering the upper part of the row immediately below.   The result, accordingly, was a type of scale armor (φολιδωτός), the details in the arrangement of which naturally escape us.   It was singularly flexible, provided with sleeves, and enveloping the entire body except that portion of the thighs which grips the horse.   It was well adapted to the form of the trunk, and permitted the soldier ample freedom of motion. The horses likewise were completely armored with the same kind of scales, though they were frequently caparisoned with leather only (AMMIANUS, XXIV, 6),[2] as they were handicapped by the weight of the metal.   The man had to be lifted on his horse.   He was equipped with a long spear, which was supported by a chain attached to the horse's neck, and at the end by a fastening attached to the horse's thigh, so as

---

[1] Compare the excellent article of E. SAGLIO in *Dictionnaire des antiquités grecs et romains*, Vol. I, p. 966.

[2] Operimentis scorteis equorum multitudine omni defensa.

to get the full force of the animal's weight into the spear-thrust.[1]   At a
given signal, the squadron composed of such horsemen dashed forth for
the assault of the enemy, and was a formidable weapon against the
infantry armed with bows, as the body protection rendered the horsemen
arrow-proof.   There were also cataphracti armed with bows, as follows
from the figure of such a cavalier represented on the Column of Trajan,
and shooting backward.   It is clear that this troop could be efficient
only as a united body and for the purpose of a surprise charge; when
successfully repelled, the result must have been disastrous to the clumsy
horsemen.   The single ones were incapable of defending themselves;
and we hear that the Gauls who accompanied the army of Crassus
practised the stratagem of seizing their lances and pulling them off the
horses.   The difference in principle between the former mounted
bodies of archers and this new system of cavalry is obvious: the mounted
infantry soldier was an individual, and as such an independent fighting-
unit, able and mobile on any occasion, be it charge, enduring battle, or
pursuit; this troop did not advance at command in any regular alignments, but dispersed in open order, small bands suddenly sallying forth
here and there, and as swiftly turning round, now attacking, then
feigning flight, exhausting their opponents in pursuit, then rallying and
pushing forward again till the contest was decided.   The new cavalry
troop was a machine set in motion by the will and word of a single commander.   It was effective as long as the body preserved the agility of its
members and worked with collective action as an undivided unit.   Its
success was bound up with the speed, security, and force of its assault;
when the charge failed, its case was lost.

When and by whom this new mode of tactics was invented is unknown. We have seen that it existed in Persia at the time of Xenophon,
and the idea seems to have indeed originated among Iranians.   Subsequently we find it in the army of Antiochus Epiphanes; and from the
time of Antoninus Pius it became common in the armies of the Romans,
soldiers of this description being frequently mentioned in inscriptions
of that period.   Thus we see the Romans adopt the strategy of their
adversaries,— a bit of history which, as we shall see presently, repeats
itself in China.   The Iranian mode of strategy with the peculiar body
armor for man and horse spread likewise to the Scythians (see p. 220),
and to Siberia as far as the Yenisei, as witnessed by the famed petroglyph of a mounted lancer equipped with plate mail.   This horseman indeed represents a *cataphractus* (Fig. 35).   This monument may be

---

[1] SMITH, WAYTE, and MARINDIN, Dictionary of Greek and Roman Antiquities,
3d ed. (Vol. I, p. 384).

roughly dated in the time of the Siberian iron age, and is surely coeval with the period of Chinese-Turkish relations in the epoch of the Han.

In fact, the Turkish tribes who fought the Chinese at that time had undergone a similar development from the primitive and crude warfare of mounted archers to the principle of organized cavalry, like their Iranian neighbors; and the Turks, on their part, were duly seconded in this respect by the Chinese. We know surely enough that the pri-

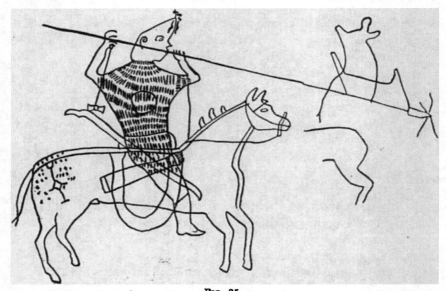

FIG. 35.
Mounted Lancer Clad with Plate Mail, Rock-Carving on the Yenisei, Siberia (from Inscriptions de l'Iénissei, Helsingfors, 1889).

meval Chinese did not possess cavalry, and that their battles were fought by soldiers on foot or in war-chariots (p. 185). We know, further, that the tactics of mounted infantry archers, in imitation of Turkish practice, were first organized in China by King Wu-ling (B.C. 325–299) of Chao; that he introduced the narrow-waisted and tight-fitting barbaric costume among his subjects, and taught them shooting with the bow while on horseback.[1] Regular cavalry, we see, came up in China from under the Anterior Han, and this was still less a truly Chinese idea than the mounted infantry. It was adopted from the Huns; and the Huns, I venture to assert, — though this impression cannot be supported at present by a literary document, — had learned this lesson from Iranians. There is no escape from the conclusion that historical contact and derivation must have been in operation, for it would be against all

---

[1] See the writer's Chinese Pottery, p. 216.

reason to assume that both the Huns and the Chinese should independently have run through the same stages of development of a complex series of phenomena as the Iranians did several centuries before this period. The inward identity of these developments on the three sides, resulting in the same styles of body armor improved by the utilization of metal, and the same manner of fighting, is sufficient proof for the fact that the one nation successively adopted the new practice from the other.

It would be beyond the scope of the present investigation to enter into the details of the history of this military institution in China. China's military history has been much neglected, though it offers a wide field for studies of great culture-historical interest. Among these, a research into the subject of cavalry is worthy of special consideration. A few suggestive remarks may here be offered.[1]

The Huns, the Hiung-nu of the Chinese Annals, were born fighters, tribes of horsemen, and expert archers. According to the picture of their life drawn by Se-ma Ts'ien,[2] they taught their children to practise riding on the backs of sheep, and to shoot birds and rodents with bow and arrow. Qualification in archery made the soldier, "and every soldier strong enough to bend a bow was a cuirassed horseman."[3] This plainly indicates that the soldiery of the Huns consisted of mounted archers fighting in open order and individually, like the Scythians; and the historian further adds that their offensive weapon for distant fighting was the bow and arrow,[4] while in close combat they employed swords and short spears. Whether they engaged also in dismounted combat, we do not know. When Se-ma Ts'ien adds that they were not ashamed of flight, this is duly connected with their mode of fighting, as set forth above (p. 218) in regard to Iranians and Scythians:[5] their flight was a

---

[1] An interesting work giving a digest of the military affairs of the Han dynasty is the *Pu Han ping chi* (reprinted in *Chi pu tsu chai ts'ung shu*).

[2] *Shi ki*, Ch. 110, p. 1 b.

[3] Thus in the translation of E. H. PARKER (*China Review*, Vol. XX, p. 1), which seems to me exact. HIRTH (Ancient History of China, p. 168) translates, "Having grown to become soldiers, they would thus become excellent archers, when they were all supplied with armor on horseback." This, though generally rendering the sense of the passage, is hardly in Se-ma Ts'ien's text; at any rate, the words *kia ki* cannot be separated, but form a technical term, "a horseman clad with hide armor." The word *kia* in Se-ma Ts'ien invariably refers to hide armor or cuirass, not to metal armor, which is *k'ai*.

[4] As swift and mounted archers the Huns appeared in Europe (motibus expediti, et ad equitandum promptissimi: scapulis latis, et ad arcus sagittasque parati. JORNANDES, XXIV), as did the Mongols at a later date.

[5] MARCO POLO (ed. of YULE and CORDIER, Vol. I, p. 262) very aptly says in regard to the Mongols, "As they do not count it any shame to run away in battle, they will sometimes pretend to do so, and in running away they turn in the saddle and shoot hard and strong at the foe, and in this way make great havoc. Their horses are trained so perfectly that they will double hither and thither, just like a dog, in

sham-flight to deceive and exhaust their opponents, and they did not fail during this manœuvre of retreat to send their arrows backward. Their cuirass (*kia*) was of leather obtained from the skins of their domestic animals, from which also their ordinary clothing was prepared; in addition to leather garments, they had coats of felt.

The re-organizer of the military power of the Huns was the famed Moduk[1] (Mau-tun), who at the end of the third century B.C. welded the scattered tribes into a compact unit. Moduk was the son of the Shan-yü[2] T'ou-man, who afterwards had a younger son by a favorite consort. Wishing to disinherit Moduk, and to place this younger son on the throne, he sent Moduk as hostage to the old enemies of the Huns, the Yüe-chi (Indoscythians), and then went on the war-path against the latter. Moduk, his life being thus imperilled, thought of his safety, and, stealing one of the swiftest horses of the Yüe-chi, fled homeward. His father, who thought this was an heroic deed, placed him in command of ten thousand horsemen. The ambitious Moduk then plotted against his father's life and throne. The Chinese historian Se-ma Ts'ien[3] narrates the story of how he achieved his scheme, in a highly anecdotal form, from which important events are apparently omitted. The story is that Moduk, making sounding arrows,[4] trained his equestrian

---

a way that is quite astonishing. Thus they fight to as good purpose in running away as if they stood and faced the enemy, because of the vast volleys of arrows that they shoot in this way, turning round upon their pursuers, who are fancying that they have won the battle. But when the Tartars see that they have killed and wounded a good many horses and men, they wheel round bodily, and return to the charge in perfect order and with loud cries, and in a very short time the enemy are routed. . . . And you perceive that it is just when the enemy sees them run, and imagines that he has gained the battle, that he has in reality lost it, for the Tartars wheel round in a moment when they judge the right time has come. And after this fashion they have won many a fight." This picture holds good as well of the Scythians, Huns, and T'u-küe. From the numerous representations of the mounted archer shooting backward on the relief bands of the Han pottery we see how deeply impressed the Chinese were by this feat of military skill.

[1] This is the correct Turkish restoration of the name, as based on the data of the Chinese commentators, according to O. FRANKE (Beiträge aus chinesischen Quellen zur Kenntnis der Türkvölker und Skythen Zentralasiens, *Abhandlungen der preussischen Akademie*, 1904, p. 10). He reigned B.C. 201 to 177.

[2] Title of the sovereigns of the Huns. Compare Plate XXII for a Chinese pictorial representation of one of the Shan-yü.

[3] *Shi ki*, Ch. 110, p. 3 b. Compare A. WYLIE, History of the Heung-noo in their Relations with China (*Journal of the Anthropological Institute*, Vol. III, 1874, p. 408); E. H. PARKER, The Turco-Scythian Tribes (*China Review*, Vol. XX, p. 7); and F. HIRTH (Sinologische Beiträge zur Geschichte der Türk-Völker, p. 254, St. Petersburg, 1900), who very well characterizes Moduk as a hero.

[4] He did not invent them, as Wylie translates. Also GILES (No. 10,928; *ming ti*) states that the sounding arrows were "invented by Mao-tun or Meghder" (similarly PALLADIUS, Vol. I, p. 174). ASTON (Nihongi, Vol. I, p. 87) makes Parker say that the sounding arrows are not Chinese, but an invention of the Huns; but PARKER (*China Review*, Vol. XX, p. 7), referring to the *nari-kabura* of the ancient Japanese, observes only that the latter seem to have imitated the Huns. In my opinion it is

archers in shooting with them. An order was issued by him to the effect that all his men, at whatever goal he should discharge a sounding

---

begging the question to speak in this case of an invention of Moduk, or of a Hunnic invention, or of invention at all; for such a contrivance is not an invention creditable to an individual or a single tribe. It represents the result of a gradual finding and experimenting, the how, when, and where of which is lost. All we may safely assert is that chronologically we first meet these buzzing arrows among the Huns, — and the text of the *Shi ki* contains the oldest record of them, — and that numerous archæological finds made in central and western Siberia testify to the fact that this type of arrow was formerly generally diffused among the Turkish stock of peoples (compare B. ADLER, Pfeifende Pfeile und Pfeilspitzen in Sibirien, *Globus*, Vol. 81, 1902, pp. 94–96; this brief notice is purely descriptive, without an historical point of view). Moduk did not invent the sounding arrow, which surely existed before his time, and which was used by his countrymen for hunting purposes; but he turned it to a novel use by availing himself of the whizzing noise as a signal for a cavalry attack. With this specific end in view he had such arrows "made," as the Chinese text says, which implies that they were previously known. HIRTH (*l. c.*, p. 254, note) has justly doubted whether Moduk may be regarded as the "inventor" of the sounding arrow, since a similar expression (*hao shi*, No. 3872, "sounding arrows, discharged by bandits as a signal to begin the attack") is metaphorically employed by the philosopher Chuang-tse of the fourth century B.C. But the *ming ti* of Moduk must have been affairs somewhat different from the latter, otherwise we should not have the two different terms. There are indeed (and the ethnographical point of view should never be neglected) diverse types of sounding arrows in our collections. An arrow can be made "sounding" by merely having one or several perforations in the iron blade; and the humming is essentially intensified by a special whistling apparatus inserted between shaft and head. This device is an oval-shaped knob of wood or bone, perforated like a whistle with two, four, or more holes, on which the wind plays when the arrow sharply cuts the air. I venture to presume that the sounding arrow mentioned by Chuang-tse belonged to the first of these types, and that of Moduk to the second; the interpretation given by Ying Shao (*Shi ki*, Ch. 110, p. 3 b) of the term *ming ti* leaves no doubt as to this fact. Again in the Chinese Annals we hear of sounding arrows being in the possession of the T'u-küe or Turks (for instance, *Chou shu*, Ch. 50, p. 3; *Pei shi*, Ch. 99, p. 2; and JULIEN, Documents historiques sur les Toukioue, p. 9). A new term appears in the Annals of the T'ang Dynasty (*T'ang shu*, Ch. 39, p. 9), — *hiao* arrows (*hiao shi*). The word *hiao*, not listed in any of our dictionaries, is written with a character composed of the classifier 'bone' (*ku*) and the phonetic element *hiao* ('filial piety'). This reading is indicated in the Glossary of the T'ang Annals (Ch. 4, p. 2 b) where the word is explained by the older term *ming ti* ("sounding arrow"). The manner of writing the word indicates that the question is here of arrows with a whistling contrivance carved from bone. These arrows, according to *T'ang shu*, were sent as tribute from the district Kuei-ch'uan in Kuei chou, now the prefecture of Süan-hua in Chi-li Province (PLAYFAIR, Cities and Towns of China, 2d ed., No. 7363). Sounding bone arrows, accordingly, were made and used in China during the T'ang period; and in coming to Japan, we need not invoke the Huns, but are confronted with the plain fact of an idea directly imported from China. The *Kōjiki* of 712 A.D. (B. H. CHAMBERLAIN's translation, p. 72) relates that "the Impetuous-Male-Deity shot a whizzing barb into the middle of a large moor, and sent him [the Great Deity] to fetch the arrow, and when he had entered the moor, at once set fire to the moor all round." The text employs the same characters for the word as *Shi ki* and *Ts'ien Han shu* (Ch. 94 A, p. 2 b: *ming ti*), but they receive the Japanese reading *nari-kabura* (literally, 'singing turnip'). CHAMBERLAIN, in the introduction to his translation of the Kōjiki (p. LXIX), justly emphasizes that this peculiar kind of arrow belongs to the traces of Chinese influence on the material culture of old Japan (Japanese illustrations in PH. F. v. SIEBOLD, Nippon, 2d ed., Vol. I, p. 342, and G. MUELLER-BEECK, *Mitteilungen der deutschen Ges. Ostasiens*, Vol. IV, p. 3, Plates 5 and 6). In the *Nihongi* of 720, a sounding arrow with eight eyes or holes is mentioned (ASTON, Nihongi, Vol. I, p. 87; K. FLORENZ, Japanische Mythologie, p. 206). Reverting to China, we have for the Mongol period Rubruck's account to the effect that Mangu made a very strong bow which two men could

arrow, should aim at the same, under penalty of decapitation. To ascertain how far his followers might be relied upon, he speedily put them to the test. Taking the sounding arrow, he aimed at his favorite horse, when some of his attendants hesitated to follow his example, and were decapitated on the spot. A sterner test was soon in store: his attendants stood aghast at seeing the sounding arrow fly at his cherished wife; those fearing to comply with the order were at once beheaded. Afterwards he went ahunting and discharged the sounding arrow at King T'ou-man's favorite horse; his men without exception duly followed suit: thus Moduk knew that his adherents could be trusted, and finally resolved on the accomplishment of his grand *coup d'état*. While on a hunting-expedition with his father, he seized a favorable opportunity to let a sounding arrow fly at the Shan-yü, whereupon a volley was fired at him by his adherents. The king fell; and his death was followed by the massacre of his wives (except Moduk's own mother), his youngest son, and all officers of state who refused allegiance to the victor. Moduk set himself up as Shan-yü in B.C. 201.[1]

There is assuredly the fact of a large political movement at the bottom of this narrative. Certainly, there was no need of a brigade or two of cavalry to eliminate the person of the king; it was a wrestle for the kingdom which involved a contest with a huge army. The problem confronting Moduk was how to overrun the king's powerful host. At this point his reform set in: he became the drill-master of his equestrian archers and a prominent cavalry tactician. His task was beset with

---

hardly string, and two arrows with silver heads full of holes, which whistled like a pipe when they were shot; Mangu sent these as a symbolic gift to the King of the Franks (W. W. ROCKHILL, The Journey of William of Rubruck, p. 180). As to the Ming period, these arrows are figured in the *Wu pei chi* of Mao Yüan-i of 1621 (Ch. 102, p. 10). Those used in the army under the Manchu dynasty are illustrated and described in the *Huang ch'ao li k'i t'u shi* (Ch. 14). They exhibit a great number of types and varieties which require a special study; in principle, there are two chief classes, — arrows with sharp iron points stuck into the whistle; and arrows with whistle, but without any iron point. The latter do not serve the purpose of killing, but of making only a certain impression. The Kalmuk of the eighteenth century availed themselves of whizzing arrows in hawk-hunting. When the water-fowl frightened by birds of prey would not rise, it was roused by means of such arrows provided with a bone knob, but without iron; for the fowl should not be slain while in the water (P. S. PALLAS, Sammlungen, Vol. I, p. 147). Such blunt sounding arrows were used till the end of the Manchu dynasty by the imperial body-guards to frighten obtrusive people when the emperor was driving out. Wounds from this weapon, if any, were of course harmless. This type of arrow is styled *pao* (E. v. ZACH, Lexicographische Beiträge, Vol. I, p. 50); it is not, however, as v. Zach explains, merely the bone knob which is so called, but the entire implement. The bone knob is termed *ku pao*. The word *pao* first appears in the *T'ang leu tien* (the "Six Statutes of the T'ang Dynasty") in the sense of a bone arrowhead. At one time, sounding arrows were used in old England, the arrowheads being perforated (J. STRUTT, Sports and Pastimes of the People of England, p. 127).

[1] This is the date given by M. TCHANG (Synchronismes chinois, p. 118). WYLIE gives the date as B.C. 209.

grave difficulties; to break the former deep-rooted habit of irregular fighting on the part of these wild hordes, and to train them to the word of one chief commander, required a master's mind and an iron will-power. Men always wont to unrestricted freedom in the discharge of their weapons, and almost unconstrained as to their movements and operations on the battle-field, were now forced to absolute subjection under the command of the chief, and compelled to fire volleys strictly at his signal, — a genuine cavalry feat.

Speaking *cum grano salis*, Moduk did the same as Cyrus in Xeno-phon's *Cyropædia*, or Maurice of Nassau when in the war of inde-pendence of the Netherlands (1568–1609) he drilled his German mer-cenaries, who were more lightly armed and mounted than their Spanish opponents, to form in two or three lines, to move rapidly, and to make direct charges while firing their pistols at the enemy. Moduk's method of drilling naturally presupposes an orderly array of his troops in rigor-ous alignments. The revolutionary character of his innovation, which was a source of amazement to his countrymen, is indicated by the grad-ual exercises and tests, and the severe punishments meted out to the negligent ones. His military genius is illustrated by the fact that he conceived the bold plan of introducing a radically new mode of tactics, that of organized and compact cavalry, in order to overthrow his father's irregular horsemen. He opposed the art and strategy of war to natural belligerents, the principles of cavalry attacks to unprincipled savage warfare. Was Moduk himself the inventor of this new science of tactics? This can hardly be presumed. We remember that he lived as a hostage among the Yüe-chi. This, of course, was at a time when the Yüe-chi still occupied their seats in the northern part of Kan-su; their westerly migration took place in B.C. 165. Maybe he learned military lessons from the Yüe-chi. The facts, at all events, prove that he had the spirit and nerve of Cyrus in him. The Iranian standard is clearly demonstrated in his doings. In the same manner as Iranian cavalry practice was adopted by the Romans, it deeply influenced the Turkish tribes; and Moduk was the prominent leader and organizer of this reform.

In reading carefully the battles fought by the Huns against the Chinese, we recognize, despite their meagre and incomplete descriptions, that the Huns were most expert cavalry tacticians, who fully practised the rules laid down by Frederick the Great after the lesson which he received from the Austrians at the battle of Mollwitz,— "Every officer of cavalry must ever bear in mind that there are but two things required to beat the enemy: first, to charge him with the greatest possible speed and force; and second, to outflank him." Hunnic skill in manœuvres of the latter sort[1] and their ability for making the best of the field of operations or any accident of territory, are especially notable in the fierce struggle against the army of Li Ling. On outpost and scouting duty they were unsurpassed. The manner in which Moduk in an unusually cold winter forced the army of the first Han Emperor, 320,000 men, mostly infantry, into a siege, enticing it on by feigning defeat and flight and keeping his best forces in ambush, is a feat worthy of this military genius. It is a deplorable loss that the details of this unique campaign have not been recorded accurately.[2]

A "battle of the Huns" is preserved on the stone monuments of the Hiao-t'ang-shan.[3] There we see them galloping on their sturdy ponies, and shooting with bow and arrow. Others are equipped with long halberds, and show us that the Huns charged in the same manner as the cataphracti. One horseman makes an attempt to drag another out of the saddle by means of a long lance with presumably hooked point.[4] A dismounted warrior, clad with a cuirass and with sword in hand, is engaged in cutting off heads. Also some of the mounted archers have donned an armor. Reserves waiting in ambush are kept in the background, shielded behind hilly ground or artificially thrown-up intrenchments.[5] The king of the barbarians is seated in front, giving instructions to a man kneeling before him.

---

[1] It is interesting that there is a Turkish word for this manœuvre, *tulghama*. This practice was introduced by Baber into India, and is described in his Memoirs (PAVET DE COURTEILLE, Baber nameh, Vol. I, p. 194, and P. HORN, Das Heer- und Kriegswesen der Grossmoghuls, p. 22, Leiden, 1894). The cavalry of the Moghuls, consisting of armored lancers mounted on caparisoned horses, certainly is an offshoot of the ancient cataphracti.

[2] A great setback to the study of military matters is the lack in the Chinese annals of any descriptions of battles, such as we have in the classical authors. The annalists are usually content to state the figures of the respective armies, the names of the commanders, date and locality of the battle, and its final dry net result with the quota of the slain and captives; but nothing, as a rule, is given out concerning the military operations in the course of the battle. Only in the biographies of the prominent generals of the Han period do we occasionally encounter a somewhat detailed record of the military evolutions of a combat, though these also are sadly deficient and pass over in silence what we are most anxious to learn. The Confucian scholar never was interested in the military side of the events.

[3] CHAVANNES, Mission, No. 47, and La sculpture, p. 82. In a poem of the first century A.D. by Wang Yen-shên, descriptive of a palace in K'ü-fu, the home of Confucius, are mentioned representations of people from Central Asia (*Hu jên*) depicted in a group on the upper parts of the pillars. They were outlined kneeling in a reverential attitude opposite one another. "There they remained unmoved with their long and narrow heads and their eyes in a fixed gaze like that of a bustard (*tiao*). Over their lofty noses and deep eyes they lifted their highly arched eyebrows. They looked sad as if in danger" (J. EDKINS, in *Chinese Recorder*, Vol. XV, 1884, p. 345).

[4] Such lances are illustrated in *Wu pei chi* and other Chinese works concerning military matters.

[5] M. CHAVANNES (*l. c.*) conceives them as going out of tents. This point of view is possible, but the opinion as given above seems to be preferable. The outlines here in question have hardly any resemblance to tents.

It must certainly be granted, as justly emphasized by CHAVANNES,[1] that the Huns were initiated also into the more "scientific" strategy of the Chinese by those Chinese generals who, from fear of being cashiered and court-martialled at home as a sequel of their defeats, preferred surrender to the enemy. The brave General Li Ling, who was forced to surrender to the Huns, is reported to have trained their soldiers in the art of war as then practised by the Chinese; the Emperor, on hearing these tidings, condemned him as a traitor, and caused his mother, wife, and children to be put to death.[2]

HIRTH,[3] in balancing the advantages and shortcomings of Hunnic and Chinese warfare, thinks that the Chinese have had on their side greatly superior armament and a certain uniformity of organization. The latter observation is doubtless to the point, but I hardly believe that Chinese arms were superior in technique to those of the Huns: the ancient bronze and iron arms discovered in Siberian soil are surely as good as any of ancient China. Possibly the crossbow, which was foreign to the Huns, rendered the Chinese superior in some respect.

The military equipment and organization of the Han, compared with that of the Chou, show a number of fundamental changes which are simultaneously symptoms of radical reforms in the manner of tactics and strategy. The main features of these innovations are the great importance attributed to the horse, — as the renowned General Ma Yüan put it, "the horse is the foundation of all military operations," [4] — the preponderance of horsemen over infantry, the prevalence of the crossbow over the bow, the use of body armor on the part of the horsemen, and the gradual development of a genuine and regular cavalry. The immediate cause of these military reforms was brought about by the endless struggles with the ever-restless nomadic hordes threatening the north-western outskirts of the empire; and imitation of their mode of warfare consequently became imperative. The wearing of armor by the horsemen, as we noticed, was a custom of the Huns; and if the Chinese followed suit, we may well lay it down as an adoption of Hunnic practice. This is not merely an impression in the matter, but a fact confirmed by the report of Ch'ao Ts'o presented to the throne in B.C. 169.[5] In this lengthy memorial the diversity of Hunnic and Chinese warfare is set forth in detail; and for the first time the formation of a

---

[1] Les Mémoires historiques de Se-ma Ts'ien, Vol. I, p. LXIX.
[2] GILES, Biographical Dictionary, p. 450.
[3] Ancient History of China, p. 166.
[4] Hou Han shu, Ch. 54, p. 9.
[5] L. WIEGER, Textes historiques, p. 414.

corps of chevaulégers (*king ki*)[1] is recommended, as the heavy infantry and war-chariots of the Chinese were powerless against the Huns. He further advised employing the tactics of the Huns against the Huns, and hiring mercenaries of the horde I-k'ü for this purpose; while within the boundaries of the empire the Chinese army should continue with the Chinese mode of tactics. This suggestion was not carried out immediately, but we see it brought into effect under the Emperor Wu (B.C. 140–87), who may be regarded as the reformer of Chinese cavalry. The man who really achieved the work and infused new life into the cavalry arm was General Ho K'iü-ping, who completely abandoned the traditional ground of Chinese tactics, and put the institution of chevaulégers into practice.[2] As a youth of eighteen he was an accomplished horseman and archer, and at the head of a squadron of eight hundred chevaulégers, forming the advance-guard of the army, gained laurels against the Huns. In B.C. 121, when only twenty years of age, he was appointed commander-in-chief of the entire force of chevaulégers, and defeated the Huns in six consecutive battles.[3] His common sense is shown by the fact that he positively refused to study Sun Wu's "Art of War," and preferred to trust to his own judgment. This doubtless means that he was a practical man who rejected theories, and by long experience had grasped the warfare of his adversary and appropriated the latter's method as the most promising one. His victories over the Huns are due to the tactics of cavalry which he adopted, while his predecessors under the early Han emperors prior to Wu met with disastrous failures by opposing infantry to the horses of the enemy. Surely the Chinese had bought their experience at a high price.

Cavalry thus grew during the Han period into an independent arm, and finally was the most important one in the wars against the roving tribes of Central Asia. The cavalry had its own organization and administrative powers. As shown by a passage in a memorial

---

[1] Or *p'iao ki* (No. 9134), "fleet cavaliers" (see CHAVANNES, Les Mémoires historiques de Se-ma Ts'ien, Vol. III, p. 559), apparently translation of Turkish *čapkunči* (P. HORN, Das Heer- und Kriegswesen der Grossmoghuls, p. 21, and W. RADLOFF, Wörterbuch der Türk-Dialecte, Vol. III, col. 1922).

[2] A repetition of this spectacle took place in Europe when it suffered in the tenth century from the inroads of the Hungarians, until Henry I of Germany, by adopting the cavalry methods of the enemy, finally succeeded in repelling him. Again, in the thirteenth century, the light horsemen of the Mongols and Saracens got the better of the iron-clad cavalry of central Europe. Only the German Order of Prussia then possessed enough military acumen to form an excellent light cavalry under the designation "Turcopoles" placed at the command of a "Turcopole," which rendered good services against Lithuanians and Poles (M. JÄHNS, Ross und Reiter, Vol. II, p. 86).

[3] His biography is in *Shi ki* (Ch. 111) and *Ts'ien Han shu* (Ch. 50). It has been translated by A. PFIZMAIER (*Sitzungsberichte Wiener Akademie*, 1864, pp. 152–170); see also GILES, Biographical Dictionary, p. 260.

presented by Huai-nan-tse to the Han Emperor Wu, there were then four officially recognized main bodies of troops, — war-chariots, cavalry, archers, and arbalists.[1]

The new order of military affairs is especially expressed by the new military offices instituted by the same Emperor. The high significance which the tactics of cavalry must have reached in his time is very conspicuous in these functions. He established a commander of cavalry (*tun ki hiao wei*), a commander of the squadrons of foreign cavalry (*yüe ki hiao wei*) formed by the men of the country of Yüe subjected to China, a commander of the squadrons of foreign cavalry (*ch'ang shui hiao wei*) formed by the Turks or Huns (*Hu*) of Ch'ang-shui and Süan-ho, and a commander of the Turkish or Hunnic cavalry (*hu ki hiao wei*) stationed at Ch'i-yang.[2] In this institution of Turkish cavalry[3] incorporated with the Chinese army we may recognize a positive sign of the fact that the Chinese had borrowed the whole affair from their Turkish neighbors, and utilized against them their own tactical stratagems. Also in the military colonies founded by the Emperor Wu in Turkistan to break the power of the Turks, detachments of cavalry were established.[4]

The perpetual wars with the turbulent nomads required an immense number of horses. "In view of his campaigns against the barbarians of the north, the Son of Heaven maintained a large number of horses, several myriads of which were reared in the capital Ch'ang-ngan," relates Se-ma Ts'ien.[5] "In B.C. 119, the commander-in-chief and the general of the chevaulégers made a great incursion to attack the barbarians of the north; they took from eighty to ninety thousand captives. Five hundred thousand pounds of gold were distributed as reward. The Chinese army had lost over a hundred thousand horses. We do not here render an account of the expenses incurred by the land and water transportation, the chariots and cuirasses."[6] Here, accordingly, is the question of cavaliers wearing cuirasses.

The generals of the Han dynasty were all clad with armor and mounted on horseback. When in 48 A.D. General Liu Shang was badly defeated by the *Man* barbarians, General Ma Yüan, who had formerly

---

[1] L. WIEGER, Textes historiques, p. 506.

[2] Compare CHAVANNES, Les Mémoires historiques de Se-ma Ts'ien, Vol. II, pp. 525, 526.

[3] The Tibetans (K'iang) also were recruited by the Chinese to form regiments of cavalry (CHAVANNES, *T'oung Pao*, 1906, p. 256).

[4] See E. BIOT, Mémoire sur les colonies militaires et agricoles des Chinois (*Journal asiatique*, 1850, pp. 342, 344, 345).

[5] CHAVANNES, Les Mémoires historiques de Se-ma Ts'ien, Vol. III, p. 561.

[6] *Ibid.*, p. 569.

gained laurels in their pacification, turned in a petition asking to be placed in service again. As he was in his sixty-second year, however, the Emperor declined his offer in view of his advanced age. Ma Yüan then made a personal appeal to him, saying, "Your servant is still able to sit in the saddle with the armor on his body." The Emperor demanded the experiment, whereupon the aged soldier flung himself into the saddle and daringly looked around, in order to demonstrate that he was still of use. The Emperor, filled with admiration, entrusted him with the command.[1] It is on record that General Kêng Ping, who died in 91 A.D., was always at the head of his troops, enveloped with his armor and mounted on horseback.[2] There is thus sufficient evidence at hand that the Chinese derived their whole system of cavalry from the Huns, both cavalry tactics and cavalry equipment; and there can be no doubt of the fact that the Chinese made exactly the same use of cavalry as the Huns.[3] Thus the Iranian ideas have filtered through the Huns into the Chinese. For this reason it is most likely also that the new cuirasses bedecked with copper and iron laminæ, coming up in China during the epoch of the Han, received their impetus from the west, more specifically from the metal scale and plate armors worn by the Iranian and Scythian cataphracti.

As said before, the history of cavalry development in China (and that of military art in general) remains to be written. An interesting observation may still be added here. Under the Sui and T'ang, the light cavalry, apparently the inheritance of the institution of the Han, was in full operation, particularly in the campaigns against the Turkish tribes. It seems, however, that the method of cavalry charges, as established by the Han after Hunnic example, had subsequently fallen into oblivion; for we are informed from the interesting biography of Yang Su inserted in the Annals of the Sui Dynasty[4] that this daring

---

[1] *Hou Han shu*, Ch. 54, p. 12 b; HIRTH, Chinesische Ansichten über Bronze-trommeln, p. 60.

[2] CHAVANNES, *T'oung Pao*, 1907, pp. 223, 224.

[3] A good example of the employment of cavalry for reconnoitring is furnished in B.C. 152 by the feat of Li Kuang, who went out with a guard of a hundred horsemen and suddenly saw himself confronted by a cavalry corps of several thousand Huns. He advanced to make them believe that he represented the vanguard of a large force following. At a short distance from the enemy he gave orders to dismount and to unsaddle, in order to show that he had no mind to retreat. A captain of the Huns sallies out; Li Kuang and ten of his men jump on their horses, and fell him with an arrow-shot. He turns back, unsaddles again, and orders his soldiers to graze the horses, and to take a rest. Until the evening the distrustful Huns durst make no charge. Under cover of night, the Chinese retreated in good order. The interesting biography of Li Kuang has been translated by A. PFIZMAIER (*Sitzungsberichte Wiener Akademie*, 1863, pp. 512–528).

[4] *Sui shu*, Ch. 48, pp. 1–6. According to GILES (Biographical Dictionary, p. 914) Yang Su died in 606 A.D.

commander was obliged to inaugurate again a reform of cavalry tactics. In 598 A.D. the Turkish Khan Ta-t'ou, the Tardu of the Byzantine historians, made an inroad into China; and Yang Su, appointed generalissimo against him, met with unusual success. Formerly, the Chinese annalist tells us on this occasion, the generals in their battles with the Turkish hordes were chiefly concerned about the cavalry of the enemy, and merely observed an attitude of defence by forming a carré of chariots, infantry and riders, the latter being posted in the centre surrounded by the other troops, and the carré being encircled by an abatis.[1] Yang Su held that this means of defence was merely an act of fortifying one's self, but could never lead to a victory; and he entirely abandoned this old-fashioned practice. He formed his troops solely into squadrons of horsemen ready for immediate attack. On learning these tidings, the Khan was overjoyed, exclaiming, "Heaven has accorded me this favor!" Dismounting from his horse, he looked up to Heaven

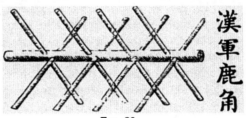

FIG. 36.
Abatis (from *Huang ch'ao li k'i t'u shi*).

and worshipped. At the head of a hundred thousand picked equestrians he advanced, and suffered a distressing defeat from the hand of Yang Su, who charged him with all vehemence. Fortunately we are told also some details as to the method of Yang Su's offensive procedure. He was a harsh warrior, enforcing martial laws with Spartan severity: capital punishment was meted out to whomever infringed the articles of war. In open battle he began operations by rushing one or two hundred riders against the position of the enemy. Did they succeed in breaking him, it was all right; did they fail and retreat, he had all of them, irrespective of their number, beheaded on the spot. Then he proceeded to send forth a squadron of two to three hundred men, until the enemy was beaten. Thus his officers and men were overwhelmed with awe, and "possessed of a heart ready to die." From this time, Yang Su remained victorious in every combat, and reaped the fame of a remarkable commander.[2]

When I make the armament of the Iranian and Scythian cata-

---

[1] In Chinese *lu kio* ("stag horns"). Every visitor to China has seen these affairs in front of Yamen and police stations. The illustration (Fig. 36) is derived from *Huang ch'ao li k'i t'u shi* (Ch. 15, p. 26). These abatis are first mentioned in the life of Sü Huang (*San kuo chi, Wei chi*, Ch. 17, p. 6), then in the life of Ma Lung (*Tsin shu*, Ch. 57, p. 2 b), who made extensive use of this means of defence in open territory.

[2] *Sui shu*, Ch. 48, p. 3.

phracti responsible for the appearance of metal armor in China, and when I am inclined to trace the perfection in the organization of the cavalry among the Huns and Chinese to a movement issuing from Iranian quarters, it should be pointed out, on the other hand, that the cataphracti do not seem to have exerted any directly imitative influence on Huns and Chinese, or that these two nations did not absolutely copy or adopt in all particulars this peculiar mode of warfare. At least, there is no direct documentary testimony to this effect, save the rock-carved lancer on the Yenisei (Fig. 35), which thus far represents an isolated case. The "battle of the Huns" above referred to displays Central-Asiatic horsemen armed with long halberds amidst equestrian archers, and could possibly be invoked as attesting, on the part of the Huns, cavalry charges in the manner of the cataphracti. In the Chinese Annals, however, as far as I know, no instance of a charge of horsemen with spears,[1] on the part of either the Chinese or the Huns, is on record; nor do I find any mention of armored horses in the Han period. The earliest palpable evidence for an armored warrior astride a caparisoned horse is represented by a clay figure pointing to the T'ang epoch.[2] Several references in the Annals allude to such caparison in the sixth and seventh centuries of our era. As the facts are, neither the Huns nor the Chinese could have had any use for the more specific tactics of the cataphracti. These were directed against heavy-armed infantry lined up in regular files. The Huns did not possess any infantry; and the Chinese employed theirs against the Huns only in the experimental stage of their operations, and with such disastrous results that it deterred them from further experiments. On the whole, Hunnic-Chinese expeditions were cavalry wars conducted with light brigades. The long marches, the wretched roads, the difficulty of the field of operations, the uncertainty of supplies and forage, and the exhausting Central-Asiatic climate, formed a serious handicap in the equipment of troops, man and horse, with heavy armament; so that a selective method in what western progress in the art of war had to offer became indispensable.

In the Ming period mail-clad cavaliers managing lances and war-clubs

---

[1] Spears are not mentioned in the Han documents translated by M. Chavannes, but the conclusion would not be warranted that they were then not used by the Chinese army. The renowned General Li Ling, who in B.C. 99 advanced into the territory of the Huns with a small army of five thousand foot soldiers, in the first encounter with the enemy, arrayed his ranks in such a manner that the front line was formed by those armed with spears and bucklers, while the archers and arbalists occupied the rear. The Huns, as well as the T'u-küe and Uigur of later date, according to the Chinese records (*Pei shi*, Chs. 97, p. 5; 99, p. 2), had spears.

[2] See Chapter VII and Fig. 51.

were in existence, as attested by an illustration in the *Lien ping shi ki*[1] (Fig. 37). As this recent epoch lacked any inventiveness in military matters and merely continued the institutions of the T'ang, Sung, and Yüan, it can hardly be credited with the feat of having originated

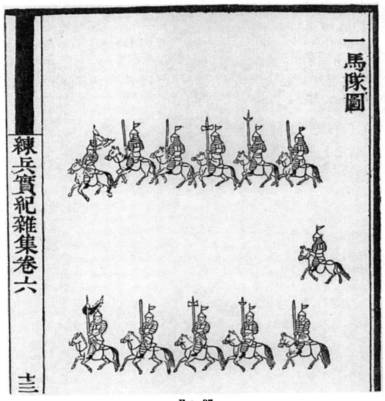

FIG. 37.
Detachment of Mail-clad Cavalry (from *Lien ping shi ki* of 1568).

mounted lancers; for the present, however, I am unable to say exactly at what date this arm sprang up in China.

In YULE's edition of Marco Polo (Vol. II, p. 501) is figured an interesting sketch from a Persian miniature of the thirteenth century, representing two mounted soldiers. They are styled by Yule "Asiatic warriors," and in all probability are intended for Mongols. The one of the two encased with a plate mail is charging with a lance; while his

---

[1] A work on military art by Ts'i Ki-kuang, written in 1568 (WYLIE, Notes, p. 91). It is reprinted in *Shou shan ko ts'ung shu*, Vols. 51 and 52.

opponent is equipped with club and circular shield, a bow-case being suspended from his girdle.

We hear of lancers in the history of the Sui dynasty, particularly in the insurrectionary wars leading to its downfall. Yang Hüan-kan, who died in 613,[1] revolted against the Emperor Yang of the house of Sui; his fortitude and audacity are emphasized in his biography, and it is recorded that in battle he brandished a long lance, while rushing at the head of his troops with loud war-cries.[2] Li Mi (582—618),[3] in his struggle against Wang Shi-ch'ung, availed himself of a cavalry troop equipped with long lances, who, enclosed in a narrow pass, were helpless against the riders of Wang Shi-ch'ung armed with short swords and bucklers.[4]

---

[1] GILES, Biographical Dictionary, p. 903.
[2] *Sui shu*, Ch. 70, p. 2.
[3] GILES, *l. c.*, p. 453.
[4] *T'ang shu*, Ch. 84, p. 3.

IV. HISTORY OF CHAIN MAIL AND RING MAIL

> Steed threatens steed, in high and boastful neighs
> Piercing the night's dull ear, and from the tents
> The armourers, accomplishing the knights,
> With busy hammers closing rivets up,
> Give dreadful note of preparation.
>
> —SHAKESPEARE (*King Henry V*).

In the preceding notes we attempted to establish on the basis of inward evidence a progressive historical sequence indicating a connection which linked Irān, Turān, and China in matters of warfare and armament about the first centuries before our era. We now propose to subject to an investigation a specific case revealing in the time of the early middle ages the transmission of a well-defined type of body armor from Persia to China and other countries.

At the present time we find widely distributed over Asia an interesting type of defensive armor occurring in the two variations of chain mail and ring mail. The word "mail" is derived from French *maille* (Latin *macula*), and originally designates the mesh of a net. Chain mail consists of interwoven links of iron or steel so joined together that the whole affair in itself forms a shirt or coat. Ring mail is composed of rows of overlapping iron or steel rings fastened upon a heavy background of cloth or leather forming a jerkin. Chain mail was a favorite means of defence in the chivalrous age of Europe, during the twelfth and thirteenth centuries. At present specimens are still encountered in Persia, among the tribes of the Caucasus, in India, Tibet, Mongolia, Siberia, and China.[1] Tibet is probably now the only country in the world where chain mail is still donned in actual military service; while all other peoples simply keep it as an heirloom or relic of the past, or, like the chieftains of some Caucasian tribes, may sometimes parade it on ceremonial occasions.

The origin of chain mail, as will be seen from the following notes, is to be sought in Irān. The Persian chain mail is an astounding example of the migration and wide distribution of a cultural object over a vast area. Not only is it diffused over India, Tibet, and China, but also over the whole of Siberia; and it is interesting to note that nearly all observers

---

[1] Reference to the use of chain mail among the Kiu-ku Miao has been made above (p. 194).

237

in those regions are agreed as to its foreign origin.[1] Old PALLAS[2] describes it as existing among the Kalmuk on the Volga, and "consisting in Oriental fashion of a net-work of iron or steel rings." According to his investigations, "it arrived there through commerce with the Truchmen and Usbek, likewise through wars with China; the finest is of *Persian* workmanship, wholly from polished steel, and is valued at fifty horses and even more. Such precious armor as well as fine swords and horses receive individual names among the Kalmuk and Tatar tribes. Armor of brass scales is the most common among the Mongols and in China." In various regions of the Altai, chain mail has been discovered which, according to W. RADLOFF,[3] does not come down from the so-called Siberian iron period, but was imported at much later times from other countries, perfectly agreeing in its form, as it does, with chain mail wrought in the southern part of Asia. A. v. MIDDENDORFF[4] states that shirts of chain mail are still found in the possession of some Tungusians, reminding them of the valiant deeds of their ancestors. But J. GMELIN[5] in the eighteenth century had already observed that they had fallen into disuse among them, and were shown as mere curiosities. They are now alive only in their heroic tales; nor did I encounter any, despite repeated inquiry, among the Tungusian tribes with which I came in contact in eastern Siberia. The same is the case with the Irtysh-Ostyak, a tribe of the Ugrian stock of peoples, whose princes, judging from the references in their epic songs, were formerly in possession of chain mail. S. PATKANOV,[6] to whom this observation is due, comments that chain mail was previously known to almost all nations of western, and partially of middle and eastern Siberia, and that it presupposes a culture and manual dexterity superior to any that could be expected from most of these. Although the former inhabitants of those regions were rather well versed in the art of forging iron and weapons, he inclines toward the opinion that the shirts of mail formerly found among them originated from countries whose peoples were further advanced in culture, and that they were imported from the Orient through the medium of the

---

[1] It is widely spread also over northern Africa (*Zeitschrift für Ethnologie*, Vol. XI, 1879, Verhandlungen, p. 34).

[2] Sammlungen historischer Nachrichten über die mongolischen Völkerschaften, Vol. I, p. 145 (St. Petersburg, 1776).

[3] Aus Sibirien, Vol. II, p. 130 (Leipzig, 1884).

[4] Reise in den äussersten Norden und Osten Sibiriens, Vol. IV, p. 1516 (St. Petersburg, 1875).

[5] Reise durch Sibirien, Vol. II, p. 644; and C. HIEKISCH, Die Tungusen, p. 73 (Dorpat, 1882).

[6] Die Irtysch-Ostjaken und ihre Volkspoesie, Vol. II, p. 014 (St. Petersburg, 1900). In the Turkish epic poetry these iron armors are likewise mentioned (A. SCHIEFNER, Heldensagen der Minussinschen Tataren, p. XVI, St. Petersburg, 1859).

Volga and Kama peoples, or rather from the southern Turko-Tatar tribes who seem to be very familiar with this kind of defensive armor. The representation of chain mail on figures in the cave-temples of Turkistan[1] might be directly traceable to Iranian influence, which is overwhelmingly manifest in those monuments.   But let us first examine the state of affairs in regard to ancient Persia.

Specimens of Persian armor of very ancient date, unfortunately, seem not to have survived; and our knowledge of the subject is largely founded upon literary records, and on reconstructions based on the appearance of warriors as often represented in the stone sculpture of the Sassanian period.   In regard to the armor of the ancient eastern Iranian tribes, W. GEIGER[2] remarks that it possibly consisted of metal scales or of a texture of brazen rings.   The fundamental passage for our knowledge of ancient Persian armor remains HERODOTUS (VII, 61); and A. V. W. JACKSON,[3] taking it as the starting-point of his study, has made a very valuable contribution to the subject.   According to the statement of Herodotus, the ancient Persians wore tunics with sleeves of diverse colors, having upon them iron scales of the shape of fish-scales; and this comparison leaves no doubt that scale armor, and not chain mail, is meant.[4]   The nobles and commanders seem to have worn breastplates of golden scales, bedecked with a purple tunic (HERODOTUS, IX, 22).   This passage shows that Persian armor was solid enough to

---

[1] A. GRÜNWEDEL, Altbuddhistische Kultstätten in Chinesisch-Turkistan, pp. 8, 25 (Berlin, 1912).

[2] Ostiranische Kultur im Altertum, p. 444 (Erlangen, 1882).

[3] Herodotus VII, 61, or the Arms of the Ancient Persians illustrated from Iranian Sources (Classical Studies in Honor of Henry Drisler, pp. 95–125, 6 figs. and 1 plate, New York, 1894).

[4] According to O. SCHRADER (Reallexikon, p. 611), chain mail then became known in Europe for the first time.—The Persian shield mentioned by Herodotus under the name gerron, and contrasted with the Greek aspis, in my opinion, has not received full justice from the hands of Professor JACKSON (l. c., p. 99).   The additional note of Prof. Merriam (p. 124) is very ingenious, but it should not be forgotten that AMMIANUS MARCELLINUS (XXIV, 6, 8) describes the Persian shields as oblong and curved (convex), of plaited willow, and covered with rawhide, and as used by the infantry composed of the rural population (quorum in subsidiis manipuli locati sunt peditum, contecti scutis oblongis et curvis, quae texta vimine et coriis crudis gestantes, densius se commovebant).   Similar types of shields, in which wood and skin were combined, occurred among the Arabs (G. JACOB, Altarabisches Beduinenleben, p. 136; G. MIGEON, Manuel d'art musulman, Vol. II, p. 246, Paris, 1907).   Typologically, they correspond to the circular Chinese shields plaited from cane or rattan, and painted with the head of a tiger (p. 203). The gerra alluded to by Herodotus were, I am inclined to think, likewise devices of plaited willow. G. RAWLINSON translates, "They bore wicker shields for bucklers." Also XENOPHON (Anabasis, I, 8) speaks of Persian troops with wicker shields, and next to them heavy-armed soldiers with long wooden shields reaching down to their feet (the latter were said to be Egyptians). The ancients, according to the testimony of VEGETIUS (Instituta rei militaris, I, 11), who lived at the end of the fourth century A.D., availed themselves of round shields, likewise plaited from willow twigs (scuta de vimine in modum cratium corrotundata).

resist the blows of the Greeks, as the blows falling upon the breastplate of Masistius had no effect. Only a certain portion of the Persian army was shielded by armor, for in the battle of Plataea they perished in great numbers owing to their light clothing, contending against the heavily armed Greeks (HERODOTUS, IX, 63). AMMIANUS MARCELLINUS (XXIV, 6; XXV, 1) informs us that the Persians opposed the Romans with such masses of mailed cavalrymen, that the iron scales of their armor suits, following the movements of the body, reflected a glaring splendor, and that their helmets, representing in front a human face, covered their heads completely, openings being left only for the eyes and nostrils, — the only spots where they were vulnerable. [1]

The iron scale armor of early times was retained in the age of the Arsacides and Sassanians. Then, also, the force of the Persian army was the cavalry, consisting of the nobles. The horsemen occupied the first place in the order of battle, and success depended chiefly on their strength and bravery. On the Sassanian rock-carvings, chain mail appears beside scale armor. A bas-relief, probably from early Sassanian times, represents such a Persian horseman clad with chain armor reaching almost down to his knees, and provided with sleeves; his neck-guard is so high as to envelop his head completely; he wears a helmet with floating ribbons, and carries a lance nearly two metres long in his right hand and a small shield in his left, a quiver being attached to his belt. Head, nape, and chest of the horse are likewise protected by chain armor. [2] At the time of the Khusrau, the complete

---

[1] Contra haec Persae objecerunt instructas cataphractorum equitum turmas sic confertas, ut laminis coaptati corporum flexus splendore praestringerent occursantes obtutus.—Ubi vero primum dies inclaruit, radiantes loricae limbis circumdatae ferreis, et corusci thoraces longe prospecti, adesse regis copias indicabant.—Erant autem omnes catervae ferratae, ita per singula membra densis laminis tectae, ut juncturae rigentes compagibus artuum convenirent: humanorumque vultuum simulacra ita capitibus diligenter apta, ut imbracteatis corporibus solidis, ibi tantum incidentia tela possint. haerere, qua per cavernas minutas et orbibus oculorum adfixas parcius visitur, vel per supremitates narium angusti spiritus emittuntur.

[2] CHRISTENSEN (L'empire des Sassanides, p. 60, Copenhague, 1907), who describes this armor, says that it is scale armor. The monument to which he refers seems to be identical with the one illustrated by J. DE MORGAN (Mission scientifique en Perse, Vol. IV, p. 319) after a bas-relief of Takht-i-Bostān, and identified with Khosrau II Purwēz (591–628). DE MORGAN, however, interprets this armor as chain mail, which plainly appears on the helmet as reconstructed by him, enveloping the entire face and neck, two almond-shaped openings being left for the eyes; this coif of mail attached to the iron calotte of the helmet, according to DE MORGAN, is joined to the mail of the armor. SARRE and HERZFELD (Iranische Felsreliefs, p. 203, Berlin, 1910), in their description of this bas-relief, give the same interpretation of chain mail. According to the same authors (p. 74), the costume of a king on a Sassanian relief of Naqsh-i-Rustam consists of scale armor, and ring mail for the protection of arms and legs. On another relief (p. 83) the same kind of armature is pointed out, scale armor reaching down to the hips, while arms and legs seem to be enveloped with ring mail. In two other places (pp. 203, 249), however, chain mail reaching down to the knees is pointed out. I am under the impression that DE MORGAN and SARRE,

outfit of the horsemen consisted of horse mail, a shirt of mail, a breast-plate, cuishes, a sword, lance, shield, a club attached to the belt, a hatchet, a quiver containing two stringed bows and thirty arrows, and two twisted strings in reserve fastened to the helmet.[1] The manufacture of armor was at the height of perfection in the Sassanian epoch. When the Arabs overran the Persian Empire and conquered Ktesiphon, they found in the well-equipped arsenals the king's cuirass with brassards, cuishes, and helmet, the whole wrought in pure gold.[2]

Chain mail, which doubtless existed under the Sassanians, is distinctly mentioned in the Avesta (*Vendidad*, XIV, 9) under the name *zrādha*. According to JACKSON,[3] this word is presumed to designate the ringed mail-coat; so called, it is thought, from its rattling. The word is derived from the root *zrād* (corresponding to Sanskrit *hrād*), which means "to rattle." The Pahlavī version of the Vendidad passage renders the word *zrādha* by *zrāī*, which answers to Firdausī's[4] Persian word *zirih*, already explained by VULLERS in his *Lexicon Persico-Latinum* as "vestis militaris ex anulis fereis conserta." The identification of *zirih* or *zireh* with chain mail seems to be certain, for under the

---

in their interpretations of armor on the bas-reliefs, are somewhat influenced by the statement of Herodotus. There can be no doubt, however, that chain mail was known in Persia during the Sassanian epoch, and at the much earlier age of the Avesta (see above).

[1] Compare A. CHRISTENSEN (*l. c.*, p. 60); C. INOSTRANTSEV, Sassanidian Studies, p. 80 (in Russian, St. Petersburg, 1909).

[2] CHRISTENSEN (*l. c.*, p. 106).

[3] *L. c.*, p. 117. BARTHOLOMAE (Altiranisches Wörterbuch, p. 1703) renders the word only by "Panzerkoller, Panzer."

[4] Compare the passage from the *Shāh-nāmeh* quoted by JACKSON (*l. c.*, p. 107). O. SCHRADER (Sprachvergleichung und Urgeschichte, p. 103; and Reallexikon, p. 611) assumes that Avestan *zrādha* had the meaning "scale armor," and is identical with the one described by Herodotus. This opinion seems to me unfounded; Persian *zirih*, which is derived from that word, and the same transmitted to India, have the significance "chain mail;" so that also *zrādha* is most likely to have had the same meaning. Schrader's point of view is merely prompted by the desire to make the interpretation of the word conform with the passage of Herodotus. This is naturally one-sided: Irān must have possessed various types of armor from ancient times, and chain mail must have pre-existed there before it was propagated from this centre to all parts of the world. From the Chinese account given below, it follows that chain mail held its ground in Sogdiana in the beginning of the eighth century; and if Jackson's identification of the Sino-Persian term *ket-li-dang* occurring in the Annals of the Sui Dynasty (see this volume, p. 28, note 1) is correct, we should have additional evidence for the employment of chain mail in Sassanian Persia. Of course, I do not mean to say that scale armor was out of commission during the Sassanian period; it may very well have persisted during that time, together with a variety of other kinds of armor. The fact that such were then in existence is brought out by the figure of the Persian grandee hunting a boar and a lion on the famous silver bowl in the Eremitage of St. Petersburg (A. RIEDL, Ein orientalischer Teppich vom Jahre 1202, p. 28; and reproduced in many other books). A real history of Persian armor remains to be written.

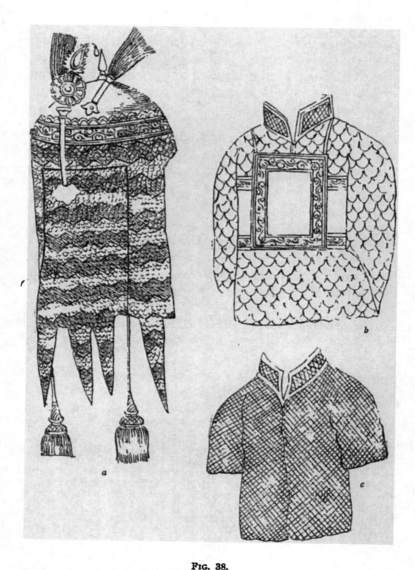

FIG. 38.

Helmet and Chain Mail from *Ain I Akbari* (Blochmann's translation, Vol. I, Plate XIII): (*a*) Helmet with Nasal and Coif of Mail; (*b*) Chain Mail with Breastplate (*bagtar*); (*c*) Chain Mail composed of Steel Links (*sireh*).

same name we meet this armor in the soldiery of the Indian Moghuls.[1]
It is figured among the sketches of the *Ain I Akbari*, a history of the
Emperor Akbar, written in 1597 by Abul Fazl Allami (1551–1602).[2]
As this work has now become exceedingly rare, three illustrations from
it are here reproduced from a copy in the writer's possession (Fig. 38).
They are instructive from more than one point of view. First, they
furnish actual proof of Persian chain mail, as well as helmet, having
been transmitted from Persia into India. Second, as regards the
manner of drawing, it will be noticed that the coat in Fig. 38 *b* is strikingly similar to the Chinese sketch of ring mail in Fig. 41. Both convey
the impression of scale armor, but are explained as, and intended for,
chain mail and ring mail respectively.[3] It is exceedingly difficult to
produce a good sketch of either; and it is interesting to note that two
draughtsmen, independent of each other, have had recourse to the
same mechanical means of representing them. They teach, as many
other cases, that caution and criticism are necessary in diagnosing
types of armor after pictorial or other designs.[4] The helmet (Fig. 38 *a*)
with nasal and coif of mail (*mighfar*) is the same as that still extant in
India, and from there conveyed to Tibet (Plate XXVIII). IRVINE
(p. 565) describes the *zirih* as a coat of mail with mail sleeves, composed
of steel links, the coat reaching to the knees. There are six specimens in
the Indian Museum. Armor in the collection of the Nawāb Wazīr at
Lakhnau is described in 1785 as follows: "The armor is of two kinds,
either of helmets and plates of steel to secure the head, back, breast,
and arms, or of steel network, put on like a shirt, to which is attached a

---

[1] W. IRVINE, The Army of the Indian Moghuls (*Journal Royal As. Soc.*, 1896,
p. 565).

[2] Translation of H. BLOCHMANN, Vol. I, Plate XIII (Calcutta, 1873).

[3] IRVINE (*l. c.*, p. 564) remarks that from this figure it may be inferred that, in
a more specific sense, *baktar* or *bagtar* was the name for fish-scale armor. Yet BLOCH-
MANN's explanation of this figure, according to the *Ain I Akbari*, is "chain mail with
breastplate (*bagtar*)."

[4] Chinese sketches of defensive armor certainly are far from being good or accurate; on the contrary, they are purely conventional in style, a fixed and ready-made
motive or model being employed for each type of armor. Yet they are not much
worse than corresponding designs from India, Persia, and mediæval Europe. At all
events, they are interesting, and in many respects even instructive. Whatever their
defects may be, if we are willing to understand the symbolic language of the draughtsmen, their productions allow us in the majority of cases to recognize what type of
armor is intended by them, in the same manner as inferences as to the type of armor
intended may be deduced from the terminology of the language. In cases where no
actual specimens are at our disposal, the Chinese illustrations may still claim a primary importance; where we have specimens to study, as in the case of chain mail and
plate armor, the sketches of the Chinese afford opportunity for an instructive comparison; and for this reason I have drawn upon these sources also. They may render
us essential assistance in interpreting the types of armor represented in statuary
and painting.

netted hood of the same metal to protect the head, neck, and face.   Under the network are worn linen garments quilted thick enough to resist a sword.   The steel plates are handsomely decorated with gold wreaths and borders, and the network fancifully braided."

Thus Persian chain mail spread to India in the Moghul period. W. EGERTON[1] observes that Persian arms were generally worn by the upper classes in India, and that the blades of swords were often Persian, even though mounted in India; in fact, as Persian artificers were frequently employed at the principal native courts, it is difficult sometimes to say whether a piece of armor is Persian or Indian.

Whether ancient pieces of chain mail are still preserved in Persia, I am unable to say.[2]   Plates XXIII—XXV illustrate a piece of mail complete with all paraphernalia, the shirt with long sleeves being open in front.   It was obtained at Tiflis by Mr. Charles R. Crane of this city, and is said to have served as the parade armor of a chieftain of the Khewsur.[3]   It is doubtless of Persian manufacture, as proved principally by the Persian designs on the arm-guard (Plate XXV, Fig. 2).   J. MOURIER[4] has already observed that the helmets with coifs of mail and the suits of chain mail found among the tribes of the Caucasus seem to be of Persian origin.   The rings forming the texture of that mail consist of thin iron wire loosely twisted together, being neither welded nor riveted.   This rather degenerate style of workmanship testifies to the fact that the suit in question was merely intended for ceremonial or pageant purposes: an energetic sword-blow would probably shatter the whole outfit.   The iron casque of the well-known Persian form, called in Persian *zirih-kulāh*, is provided with a sliding nasal (noseguard), and with a couvre-nuque consisting of a long coif of mail guarding forehead, cheeks, neck, and shoulders.   On Plate XXV the two-edged sword, arm-guard, hauberk, and gauntlet, completing the set, are shown.

The Arabs have undoubtedly derived chain mail from the Persians. All the available historical evidence is decidedly in favor of Persian prior-

---

[1] An Illustrated Hand-Book of Indian Arms, p. 142 (London, 1880).

[2] According to EGERTON (*l. c.*, p. 141), armor is now no longer worn in Persia, except to add to the pageant of their religious processions, held annually in the month of Muharram, to commemorate the death of Hassan and Hussain, the Shiah martyrs. Many that are of modern manufacture have been made for ornament rather than use, and betray in their style the decline of the art.   The best period, judging from the examples preserved, seems to have extended from the time of Shah Abbas to that of Nadir Shah.   The armor of Shah Abbas is in the British Museum; it is figured in G. MIGEON (Manuel d'art musulman, Vol. II, p. 251, Paris, 1907).

[3] I am under obligation to Dr. Charles B. Cory, the present owner of the armor, for his courtesy in placing it at my disposal.

[4] L'art au Caucase, pp. 156, 157 (Paris, 1907).

ity.[1]   Among the ancient Arabs of the pre-Islamic epoch we meet with leather and iron armor,[2] without any clear description of their appearan.ce.   The latter seem previously not to have consisted of mail, though th[1]s cannot be stated positively; but according to the descriptions of the poets, chain mail comes into question in the majority of cases.[3]   Tradition ascribed its invention to King David, and the Korān (Sūre XXI, 80; XXXIV, 10) sets forth that God himself taught David how to smelt iron, with which to make the rings, and to join them into a solid armor. This story certainly is devoid of historical value.   The place Salūk in Yemen was of old renowned for its armor consisting of a double row of rings.   Also "Persian armor" is mentioned in Arabic records, whereby garments lined with silk and cotton were understood.   "Armor from Sogd" (Sogdiana) became known after the foreign conquests of the Arabs.[4]   Possibly also scale armor was worn.[5]

Chao Ju-kua narrates that the ruler of Basra, when he shows himself in public, is accompanied by more than a thousand mounted retainers in full iron armor, the officers wearing chain mail.[6]

During the early middle ages of Europe, the horses of armies were not caparisoned.   Only from the beginning of the thirteenth century, probably under the influence of the Crusades, were they protected by chain-mail covers.[7]

According to MAX JÄHNS,[8] the chain mail (*Parsen, Barschen*), as it first appears during that time in the armature of the horse, is probably of oriental, and more specifically of Persian origin.   Dr. BASHFORD DEAN,[9] the great authority on armor in this country, offers the following suggestive summary of this subject: "Chain mail marked a distinct epoch in the development of arms and armor: for it was light, flexible, and extremely strong.   And it soon, therefore, came to supplant the

---

[1] Compare the notes of C. H. BECKER (*Der Islam*, Vol. IV, 1913, pp. 310–311).

[2] F. W. SCHWARZLOSE, Die Waffen der alten Araber aus ihren Dichtern dargestellt, pp. 325, 328 (Leipzig, 1886).

[3] *Ibid.*, p. 331.

[4] *Ibid.*, p. 334.

[5] G. JACOB, Altarabisches Beduinenleben, p. 136 (Berlin, 1897).   BECKER (*l. c.*) mentions also Arabic cotton armor (*lubbāda*); what he calls ring mail (*Ringpanzer*), I believe, strictly speaking, is chain mail.   In the age of the T'ang (618–906) the soldiers of the Arabs were equipped with bow, arrows, long spears, and metal armor (*T'ang shu*, Ch. 221 B, p. 8 b).

[6] *Lien huan so-tse kia*, literally, "armor of chains, the links of which are mutually connected" (see HIRTH and ROCKHILL, Chau Ju-kua, p. 137).

[7] G. STEINHAUSEN, Geschichte der deutschen Kultur, p. 247; L. BECK, Geschichte des Eisens, Vol. I, p. 863.

[8] Ross und Reiter, Vol. II, p. 137.

[9] Catalogue of European Arms and Armor (*The Metropolitan Museum of Art*, Hand-Book No. 15, p. 21, New York, 1905).

cruder defences of Carolingian times. Some authorities maintain that this form of armor was borrowed from the Orient; and certain it is that its development in the twelfth and thirteenth centuries was largely influenced by oriental models. If, however, this form of armor were derived originally from the East, it is a rather remarkable fact that its early appearance in Europe should be traced so clearly to the northern peoples, and that the 'byrnie' (*brünne*), or shirt of mail, should have become a characteristic part of the equipment of a Norseman. Nevertheless it may still have been derived primitively from the East, since it is well known that the early excursions of the Viking carried them well into the Mediterranean, and that even by the eighth century they were well acquainted with many objects of oriental origin." The Arabs and Byzantines have transmitted chain mail to Europe; and a share in this movement may be attributed to the cultural exchanges between East and West during the crusades.

At the time of Mohammed the Arabs had already adopted the Persian practice of protecting horse and man with armor, the armored horsemen and horses being designated *mudjaffaf;* that is, clad with the *tidjfāf*, the Persian felt armor.[1]

When we come to China, the situation is the same as in Europe and in India. Historical evidence is not lacking for the foreign origin of Chinese chain mail. Indeed, the first record alluding to it, the *T'ang shu*,[2] in its account of *K'ang* (Sogdiana, Samarkand), states that in the beginning of the period K'ai-yüan (713–741), Samarkand sent to China chain armor (*so-tse k'ai*) as tribute.[3] The famous poet Tu Fu, who

---

[1] Compare C. H. BECKER (*Der Islam*, Vol. IV, 1913, p. 311). BECKER states that the history of defensive armor in the Islamic world still remains to be written; but his remarks render it sufficiently clear that the origin of these things is to be sought in Persia, and that they were transferred to Europe through the medium of the Arabs and Byzantines. The soldiers of the Byzantine army were protected for the most part by scale armor, though, judging from quite early monuments, ring or chain mail was sometimes used (O. M. DALTON, Byzantine Art and Archæology, p. 684, Oxford, 1911).

[2] Ch. 221 B, p. 1 b.

[3] A tribute of armor from Samarkand is still recorded in the *Ming shi* under the year 1392 (see BRETSCHNEIDER, *China Review*, Vol. V, p. 123). It can of course be presumed only that the chain mail sent by Samarkand was of Persian origin; but this conclusion is most probable, as the culture of Sogdiana, the capital of which was Samarkand, was thoroughly Iranian. From what was said above on "armor from Sogd" it seems that among the Arabs Sogdiana was regarded as a famous seat of the manufacture of armor. In view of the fact that chain mail is an Iranian import in China it is curious that in the Persian legend of Alexander's expedition to China, the King of China presents to him among many other things a hundred long coats of mail (H. ZOTENBERG, Histoire des rois des Perses, p. 440). In *T'ang shu* (Ch. 220, p. 3 b), where an account of the foreign tribes of the east, including Koreans and Tungusians, is given, mention is made of a *so kia* ("chain cuirass"); the word *k'ai* is not used, and the question is probably of a leather corselet with rings attached to its surface.

lived about this time (712–770), alludes in a verse to a "metal-chain cuirass" (*kin so kia*).[1] Chain armor (*so-tse kia*)[2] is distinctly mentioned in the *Wan hua ku*, a work written at the end of the twelfth century,[3] in which are enumerated the designations for thirteen kinds of armor known at that period. Chain armor is there listed as the twelfth in the series; and it is expressly stated that it ranges in the class of iron armor (*t'ie kia*). In all probability, however, this passage is taken from the *T'ang leu tien* (the "Six Statutes of the T'ang Dynasty") drawn up by the Emperor Yüan-tsung in the first part of the eighth century (p. 189); and as the thirteen kinds of armor on record are said to have been made at that time in the Imperial Armory, we may presume that chain mail was turned out by the Chinese as early as the T'ang period, after models first introduced from Samarkand.

In the Biography of Han Shi-chung, who died in 1151,[4] a "chain connected armor" (*lien so kia*) capable of resisting bows is credited to this general;[5] but it would seem that this newly-coined term does not refer to a real chain mail, but rather to ring mail, in which rows of iron rings are fastened to a foundation of leather (see p. 252).

According to the testimony of WILLIAM OF RUBRUCK, chain mail, which he styles haubergeon, was known to the Mongols.[6] In the year 1345, during the reign of the Emperor Shun, Djanibeg (1342–1356), son of Uzbeg,[7] sent to China, among other products, swords, bows, and chain mail coming from Egypt (Mi-si-rh).[8]

Chain armor had no official recognition in China, and was never introduced into the army. It is conspicuously absent in the military regulations of the Ming dynasty, nor is it mentioned in the well-informed military work *Wu pei chi*. We have to go as far down as in the K'ien-lung period to renew its acquaintance. We meet it there again as a foreign import. In the Imperial State Handbook of the Manchu

---

[1] *P'ei wên yün fu*, Ch. 50, p. 70 (under *so*), or Ch. 106, p. 74 (under *kia*). There is also a quotation given there to the effect that "the finest of armors are designated chain mail," derived from a poetical work *Erh lao t'ang shi hua*, the date of which is unknown to me.

[2] Entered in GILES's Dictionary, p. 1264 c, with the same translation.

[3] BRETSCHNEIDER, Botanicon Sinicum, pt. 1, p. 160, No. 330.

[4] GILES, Biographical Dictionary, p. 251. His biography is in *Sung shi* (Ch. 364, p. 1).

[5] *Sung shi*, Ch. 364, p. 6 b.

[6] W. W. ROCKHILL, The Journey of William of Rubruck, p. 261 (London, 1900). Rubruck reports that he once met two Mongol soldiers out of twenty, who wore haubergeons. He asked them how they had got hold of them; and they replied that they had received them from the Alans, who are good makers of such things, and excellent artisans.

[7] BRETSCHNEIDER, Mediæval Researches, Vol. II, p. 15.

[8] *Yüan shi*, Ch. 43, p. 5 b (K'ien-lung edition).

Dynasty (*Huang ch'ao li k'i t'u shi*, Ch. 13, p. 53) a piece of chain mail is illustrated (reproduced in Fig. 39) under the name *so-tse kia*. It is recorded that in 1759, after the subjugation of Turkistan, numerous captives were made, and innumerable spoils of arms obtained which were hoarded by imperial command in a building of the palace, the *Tz'e kuang ko*. Among these trophies were several pieces of chain armor; and

FIG. 39.
Iron Chain Mail from Turkistan (from *Huang ch'ao li k'i t'u shi*).

a document recording this event was draughted, and deposited beneath those objects in the treasury. This shows that in the K'ien-lung period chain armor was foreign to the Chinese and considered an object of curiosity and rarity. The specimen consists of a jacket and trousers. The rings are said to be iron; but it is not stated whether they are riveted, nor can this be gathered from the illustration. The shirt of mail is closed in front, and put on over the head. The collar, as explained in the text, is made of white cotton and tied up by means of a cord.

Two specimens of chain mail secured in China are represented on Plate XXVI. Both are jackets with sleeves, having a short slit underneath the neck, and being tied up by means of a leather band. Though identical in appearance, they are of different technique. The shirt of mail shown in Fig. 1 of the Plate consists of riveted steel rings; the one in Fig. 2, of welded iron rings. The former was obtained at Si-ning, Kan-su Province, with the information that it had previously hailed from Tibet; the latter, at Si-ngan, Shen-si Province. These two coats, accordingly, are technically much superior to the one from the Caucasus, in which the rings are merely of twisted iron wire not welded. It is thus clear that there are coats of mail widely varying in the technical process and in quality. To decide the question as to the locality where the two specimens were manufactured would require a larger comparative material than is at my disposal. The Tibetans, as will be seen presently, must be discarded as being unable to produce chain mail. The Chinese, as we noticed, may have themselves made it in the T'ang period; it is certain, however, that none is turned out in China at the present time. Altogether, these specimens are scarce; and modern Chinese accomplishments in iron and steel are so crude and inferior, that it is difficult to believe in the Chinese origin of the two pieces of mail. Particularly the mail in Fig. 1 of Plate XXVI represents such a complex and toilsome technicality, involving so great an amount of time and patience as can be credited only to a highly professional and skilful armorer, who was a specialist in this line; the process of riveting steel rings, moreover, is not practised by the Chinese. My personal impression in the matter, therefore, is that the two mails were fabricated in Persia or Turkistan, and thereupon traded to China.

An offensive weapon deserves attention in this connection, because a chain is utilized in it, and its invention is ascribed by the Chinese to a foreign tribe. This is the t'ie lien kia (No. 1132) pang, a weapon consisting of two wooden cudgels, the one nearly three times the length of the other, their upper ends being connected by an iron chain (Fig. 40). The longer cudgel is round, and is held by its lower end in the hands of the soldier; the shorter one is square in cut, and provided at the end with a sharp iron point intended to hit the enemy's head. The chain allowing it ample freedom of motion, it is swung around in a wide circle, thus making it a fierce and powerful weapon. The Wu pei chi, illustrating and describing this instrument (Ch. 104, p. 14), states that its original home was among the Si Jung (the Western Jung), one of the general designations for the Turkish and Tibetan tribes living north-west

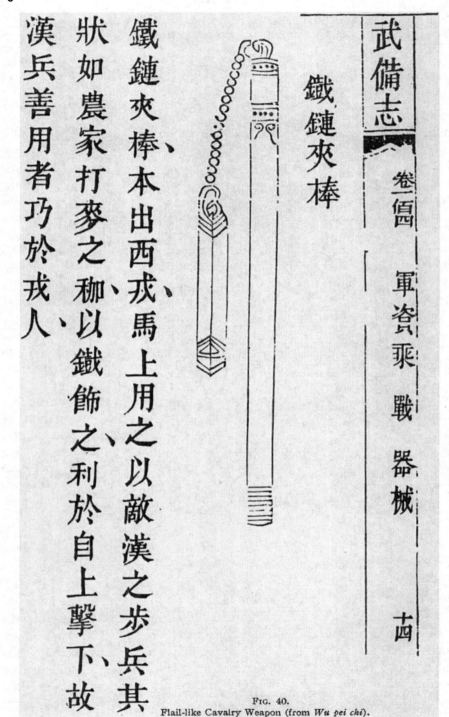

武備志　卷佰四　軍瓷乘戰器械　十四

鐵鏈夾棒

鐵鏈夾棒、本出西戎、馬上用之以敵漢之步兵其狀如農家打麥之耞以鐵飾之利於自上擊下、故漢兵善用者乃於戎人、

FIG. 40.
Flail-like Cavalry Weapon (from *Wu pei chi*).

from China; that they made use of it, while riding on horseback, in fighting Chinese infantry; and that the Chinese soldiers learned to handle it, and are more clever at it than the Jung. Its shape is compared to a threshing-flail; and it may even have been derived from this

FIG. 41.
Ring Mail of Steel Wire (from *Wu pei chi* of 1621).

implement, with which it agrees in mechanical principle. It is still known in Peking under the name of "threshing-flail," and is used in fencing. I saw this sport practised in 1902, and at that time secured a specimen for the American Museum, New York. In the time of the Emperor K'ien-lung it was still employed in the Chinese army.[1]

---

[1] *Huang ch'ao li k'i t'u shi*, Ch. 15, p. 25 b. According to this work, the weapon is first mentioned in the *T'ung tien* of Tu Yu, who died in 812, where it is said that it was manipulated by women on the walls to resist invaders. Ti Ts'ing, the famed general in the wars against the western Liao (biography in *Sung shi*, Ch. 290), who died in 1057, employed it on horseback.

Different from chain mail, though allied to it, is the ring mail. The *Wu pei chi*, as far as I know, is the only source to inform us of the existence of this type of armor in China (Fig. 41). The cut of this book is here reproduced, not only because it is unique in the representation of this specimen, but also because it is very instructive in showing us again how difficult it is to draw inferences from oriental illustrations as to the real type of armor intended by the artist. Any expert in armor, casting a glance at this sketch furnished by the Ming edition of the *Wu pei chi*, could voice no other opinion than that it is meant to represent a type of scale armor. But the author, as plainly stated in the heading, means to represent a ring armor made of steel wire; and the description added by him leaves no doubt of this intention. He states that "armor of connected rings wrought from steel wire was formerly made by the Si K'iang, and that the structure of the rings is identical with the large iron wire rings of his time, with openings as big as in a coin; in shape, it is like a sort of shirt, and it is held together above by a collar; it is not open in front, but put on over the head; spears and arrows can hardly ever pierce it and cause wounds." Unfortunately he omits to state what the foundation is to which the rings are fastened; but from the drawing, in which the rings are arranged in overlapping rows, it is necessary to conclude that they were attached to a solid garment, in the same manner as our ring mail, which consisted of steel rings sewed edgewise upon leather or strong quilted cloth.

The name K'iang (No. 1264) mentioned in this text, as is well known, is a general designation for the multitude of ancient Tibetan tribes, at a time when they were still settled in the western parts of Chinese territory. A. WYLIE[1] has translated from the Annals of the Later Han Dynasty the records pertaining to them. They were exterminated by the Han dynasty.[2] The Chinese tradition tracing ring mail to Tibetan tribes is significant, though it is not necessary to adopt the opinion that the latter ever really made it. Yet the fact remains that ring mail still occurs among the Tibetans. There is even a Chinese source of the middle of the eighteenth century alluding to it. In the *Si-tsang ki* ("Records of Tibet"), a small but interesting work on Tibet in two volumes, published in 1751 by Chu K'i-tang (Ch. 1, p. 23), three kinds of armor in use among the Tibetan soldiers are enumerated,—the scale armor (*liu ye*, "willow-leaves"), the ring armor (*lien huan*, "connected

---

[1] History of the Western Kiang (*Revue de l'Extrême-Orient*, Vol. I, 1883, pp. 424–478).

[2] CHAVANNES, Les Mémoires historiques de Se-ma Ts'ien, Vol. III, pp. 591, 595; and Trois généraux chinois (*T'oung Pao*, 1906, pp. 256–258).

rings"), and the chain armor (*so-tse*).[1] This naturally carries us to Tibet and its relations to Persia in the matter of chain mail; but before taking leave of China, it should be emphasized that chain mail remains the only type of armor borrowed and imported by her directly from a foreign country. With this exception, the making of armor, though foreign impulses cannot be denied, is purely indigenous, and also Chinese in its essential characteristics. From a negative point of view, its independence from the west is exhibited by several features that are lacking in Chinese, but which occur in western armor: as, for instance, the curious nasal (or nose-guard), characteristic of Persian, Indian, and Turkish helmets (Plates XXV and XXVIII); and gauntlets, absent in China, but met in Persia, India, and Japan.

The Persians seem to have had relations with Tibet at an early date. In the "Histoire des Rois des Perses," translated (from an Arabic source composed between 1017 and 1021) by H. ZOTENBERG (p. 434), Alexander the Great is made to undertake an expedition into Tibet, whose king offers him submission and a tribute of a hundred loads of gold and a thousand ounces of musk. The two products of Tibet most eagerly solicited by the Persians are clearly emphasized in this legend. Among the wonders possessed by King Abarwīz figured the "malleable gold" extracted for him from a mine of Tibet (*ibid.*, p. 700); this was a block of gold five hundred grains in weight, flexible like wax; when pressed in one's hand, it passed through the fingers and could be modelled; figures were fashioned from it, and it would then assume its former shape again.

The Annals of the Sui Dynasty[2] have preserved a most interesting account of a country styled Fu, situated over two thousand *li* north-west of Sze-ch'uan. As I hope to show in detail on a future occasion, the question here is of a Tibetan tribe with a thoroughly Tibetan culture. The particular point that interests us in this connection is that this tribe of Fu possessed helmets and body armors of varnished hide, and that armor played a significant part in its funeral ceremonies. The corpse was placed on a high couch; it was washed, and dressed with helmet and cuirass; and furs were piled upon it. The sons and grandsons of the dead man, without wailing, donned their cuirasses, and performed a sword-dance, while exclaiming, "Our father has been carried away by a demon! Let us avenge this wrong and slay the demon!"

---

[1] As the Tibetans, even less than the Chinese, can be credited with the manufacture of chain mail, and as Tibetan chain mail is plainly stamped as a Persian import, suspicion is ripe that also Tibetan (and consequently Chinese) ring mails are derived from the same source; but strict evidence for the antiquity of ring mail in Irān yet remains to be brought forward.

[2] *Sui shu*, Ch. 83, p. 8.

This truly was the burial rite of a militant and valiant people, the dead being believed to continue their lives as warriors, and the survivors combating with their arms the demon who was supposed to have swept him away.   A similar idea was symbolically expressed on the burial-places of the Tibetan heroes, who during the age of the T'ang had fallen in their bitter strifes with the Chinese.   As related in the T'ang Annals, white tigers were painted on the red-plastered walls of the buildings belonging to their sepulchral mounds scattered along the upper course of the Yellow River: when alive, they donned a tiger-skin in battle, so the tiger was the emblem of their bravery after death.[1]

The Tibetans were a warlike nation in the early period of their history, and at times the terror of their neighbors, even of China.   The Annals of the T'ang Dynasty,[2] which call them T'u-po (Tibetan *Bod*), and describe at length their relations with the empire from the seventh to the ninth century, praise their armor and helmets as excellent, covering the entire body, and leaving openings for the eyes only;[3] so that power-ful bows and sharp swords cannot wound them very much.   This pass-age, however brief, allows the inference that Tibetan armor of that period was of iron (for it is designated with the word *k'ai*, No. 5798); that it was a complete armor with brassards, cuishes, and greaves; and that the helmet was provided with a visor.[4]   The "gold" armor,[5] which King Srong-btsan sgam-po, according to *T'ang shu*, is said to have transmitted as a gift to the Emperor T'ai-tsung when he wooed the hand of a Chinese princess, is perhaps not to be taken too literally; the word *kin* may simply mean "metal."[6]

Among the eastern Tibetan tribes we have proof for the existence of iron armor as early as the sixth century.   The *Pei shi*[7] imparts the interesting news that in the first year of the period Pao-ting of the Pei Chou dynasty (561 A.D.) the Pai-lan, a tribe of the K'iang, who in matters of customs and products agreed with the Tang-ch'ang,[8] sent

---

[1] *T'oung Pao*, 1914, p. 77.

[2] *T'ang shu*, Ch. 216 A, p. 1 b.

[3] A striking analogy with the Persian helmet as described by Ammianus Mar-cellinus (above, p. 240).

[4] Presumably of a similar type as the royal Persian helmet figured by J. DE MORGAN (Mission scientifique en Perse, Vol. IV, p. 320, Paris, 1897).

[5] Thus translated by S. W. BUSHELL, The Early History of Tibet, p. 10 (reprint from *Journal Royal Asiatic Society*, 1880).

[6] A golden (*huang kin*) armor, referring to the T'ang period, is mentioned in *Ming huang tsa lu* (Ch. B, p. 2).

[7] Ch. 96, p. 9 b.

[8] Regarding these tribes compare S. W. BUSHELL (The Early History of Tibet, p. 94), and W. W. ROCKHILL (The Land of the Lamas, p. 337).   Tibetan armor has not infrequently been sent to China; specimens are preserved, and may still be seen

envoys with a tribute of cuirasses made from rhinoceros-hide (*si kia*) and iron armor (*t'ie k'ai*).

There is a somewhat vague Tibetan tradition relative to the period of the early legendary kings, to the effect that armor was first introduced into central Tibet from Lower K'ams (*Mar K'ams*) in the eastern part of the country.[1] It is difficult to decide as to what type of armor is to be understood in this passage, in which occurs the general word *k'rab*, the original meaning of which, as we tried to show (p. 195),[2] must have been "scale armor." It may be permissible to think, in this case, of a style of hide armor, as it was in vogue among the Fu and the neighboring Shan and Man; but the tradition which here crops out is somewhat weak and hazy.

Coats of mail are frequently alluded to in Tibetan epic literature and historical records. In the History of the Kings of Ladākh they are mentioned under the reign of the seventeenth king, bLo-gros C'og-ldan, as being brought from Guge, eighteen in number; the most excellent of them receiving individual names, as was the case also with swords, saddles, turquoises, and other precious objects.[3] The usual types of armor in Ladākh were chain or scale armor. The fact that they are recorded as coming from Guge is significant, for Guge must have had ancient relations with Persia;[4] and the chain mail of Guge was most probably of Persian origin. The plain fact remains that the Tibetan blacksmiths do not turn out iron chain mail, nor are they capable of making it; so that they are most unlikely ever to have made it at any earlier time. The supposition of an import is therefore the only solution of the problem.

The *Wei Tsang t'u chi*, a description of Tibet by Ma Shao-yün and Mei Si-shêng written in 1792, has the following note on the outfits of

---

in many Lama temples. The *Ming shi* tells of a tribute of armor, swords, and products sent in 1374 by the country of Ngan-ting in the territory of the Kuku-nōr, which was classified among the Si Fan (BRETSCHNEIDER, *China Review*, Vol. V, p. 32).

[1] CHANDRA DAS, in *Journal Asiatic Society of Bengal*, 1881, pt. I, p. 214.

[2] B. HOUGHTON (Outlines of Tibeto-Burman Linguistic Palæontology, *Journal Royal Asiatic Society*, 1896, p. 41), in pointing out the coincidence of Tibetan *k'rab* and Burmese *k'yap*, remarks that each word denotes originally a flat, thin thing or scale, and that hence they come to mean scale armor. "It is, of course, possible," he adds, "that this was possessed by the Burmans in Tibet, but on the other hand it is equally probable that the words have been applied independently on the introduction of this particular kind of armor, (? from China)." This view seems forced. The words *k'rab* and *k'yap* are not loan-words from Chinese, but on equal footing with Chinese *kia* and *kiai*, and speak in favor of scale armor having been a very ancient means of defence in the Indo-Chinese group of peoples.

[3] Compare MARX, in *Journal Asiatic Society of Bengal*, Vol. LX, pt. I, 1891, pp. 122, 123. Also among the ancient Arabs, excellent armors were named (SCHWARZLOSE, *Die Waffen der alten Araber*, p. 69).

[4] LAUFER, *T'oung Pao*, 1908, p. 13.

the Tibetan army of that time:[1] "When the troops go on an expedition, they wear armor consisting of helmets and cuirasses. The latter are made of iron scales[2] or of chains. On the helmet of the cavalry is attached a red crest or a peacock-feather. From their waist hangs a sword, on their back is slung a gun, and in their hand they carry a pike. On the infantry helmet is a cock's feather. They have hanging to their waist a sword, without counting a dirk. Under their arm is a bow and arrow, and in their hand a buckler of rattan or wood. Some also bear a pike in their hand. Their wooden bucklers measure one foot six inches across, and three feet one or two inches in length, and are painted with pictures of tigers, and ornamented with different-colored feathers;[3] outside they are covered with sheet iron."

If the assumption is correct that Tibetan chain mail is Persian in origin, the scale armor would remain to be looked upon as the national body armor of Tibet, at least as the older type which preceded the introduction of chain mail.[4] In former times, it seems to me, the latter was traded over a direct route from Persia into Guge in western Tibet, on the same path along which religious ideas of the Zoroastrians poured in and exerted a deep influence on the shaping of the Tibetan Bon religion, while during the last centuries northern India became the mart which supplied Tibet with this much-craved article.

The Tibetan and Persian relations in matters of arms are expressed also by the identity of the Tibetan and old-Persian sword. Indeed, the Tibetan sword, as still in use at present, is the same as that re-

---

[1] ROCKHILL, *Journal Royal Asiatic Society*, 1891, p. 215.

[2] Mr. ROCKHILL has, "made of linked willow-leaf (shaped iron plates)." But the expression *liu ye* ("willow-leaf"), as we see from the regulations of the Ming dynasty, refers to scale armor, not to plate armor. Mr. WADDELL (Lhasa and its Mysteries, p. 168) speaks of cuirasses consisting of small, narrow, willow-like leaves about an inch and a half long, threaded with leather thongs, still worn by Tibetan soldiers, a few of whom also wear coats of chain mail. The Chinese physician Dr. Shaoching H. Chuan, who visited Lhasa with the Chinese Mission to Tibet in 1906–1907 has written a very interesting and well-illustrated article on Lhasa under the title The Most Extraordinary City in the World (*Nat. Geogr. Mag.*, 1912, pp. 959–995); on pp. 978 and 980 are good illustrations of Tibetan soldiers wearing chain mail.

[3] In the Tower Armory there is a shield of the Angami-Naga, faced with bearskin, the side ornamented with tufts of feathers (HEWITT, Official Catalogue of the Tower Armories, p. 100). Compare p. 210.

[4] In ancient India, likewise, scale armor seems to represent the older type. The *Çukranīti* describes solely this type of armor by saying that "armor consists of scales of the breadth of a grain of wheat, is of metal and firm, has a protection for the head, and is ornamented on the upper part of the body" (G. OPPERT, On the Weapons, Army Organization, and Political Maxims of the Ancient Hindus, p. 109, Madras, 1880). A suit of Tibetan scale armor is illustrated by A. GEORGI (Alphabetum Tibetanum, Rome, 1762, Plate IV) in the figure of a shaman, entitled *cio kion* (that is, *c'os skyong*, "protector of religion").

constructed by J. DE MORGAN[1] after a bas-relief of Takht-i-Bostān, both in its shape and in the style of its decoration, for which inlaid stones were employed. The history of the sword, however, is somewhat different from that of chain armor, and is not connected with an importation of swords from Persia into Tibet. The swords of the Turkish tribes of Central Asia, to which the Tibetan swords are related, must be taken equally into consideration; and it seems that this type of sword is a common property of the whole group, of such great antiquity that the accurate history of its distribution can no longer be traced.[2]

The Tibetans make (or rather, made) use also of the circular and convex rhinoceros-hide shield of Indian manufacture, ornamented with four brass bosses (Plate XXVII, Fig. 1).[3] This shield is employed likewise in Burma and Siam. The national Tibetan shield is made from rattan plaited in the basketry style of circular coils (Plate XXVII, Fig. 2). Of what type the shield of the ancient Tibetans (K'iang), adopted by the Chinese, was (p. 188), we do not know.

Also the Tibetan helmet (Plate XXVIII), composed of steel sheets incrusted with gold and silver wire, forming floral designs, and with attached coif of mail and sliding nasal, is of Indo-Persian origin (compare Plate XXV).

---

[1] Mission scientifique en Perse, Vol. IV, p. 321 (Paris, 1897). Compare this volume, p. 15.

[2] The swords represented on the monuments of Turkistan belong to the same type (see A. GRÜNWEDEL, Altbuddhistische Kultstätten, pp. 26, 27, and many other examples).

[3] For Indian specimens see W. EGERTON, An Illustrated Handbook of Indian Arms, pp. 95, 111, 118, 134 (London, 1880). Rhinoceros-hide shields are mentioned in the *Ain I Akbari* of Abul Fazl Allami (translation of H. S. JARRETT, Vol. II, p. 281, Calcutta, 1891).

V. THE PROBLEM OF PLATE ARMOR

> "The skilful leader subdues the enemy's troops without
> any fighting; he captures their cities without laying siege
> to them; he overthrows their kingdom without lengthy
> operations in the field.   With his forces intact he will dis-
> pute the mastery of the Empire, and thus, without losing
> a man, his triumph will be complete."
>
> SUN-TSE, *Art of War* (translation of LIONEL GILES).

We had occasion to allude to plate armor[1] in the chapter on defensive
armor of the Han period, stating that in all probability it existed in the
China of those days; we referred also to its possible occurrence among the
armor worn by the *cataphracti* of the ancients, and figured a Siberian
petroglyph from the Yenisei representing a mounted lancer clad with
such mail.   We now propose to discuss this problem in detail,—a problem
of fundamental historical importance, as it reveals ancient relations
between many peoples of Asia, and touches also the question as to the
connection of Asiatic with American cultures.   Classical and other
archæologists have not yet ventilated this problem, apparently for the
only reason that they did not sharply enough discriminate between
the various types of body armor.   "Scale armor" was the catchword
under which everything of this sort was pressed together.[2]   But plate
armor must be strictly differentiated from scale armor as a special type,
which sprang up independently.   The laminæ forming plate armor
are rectangular and flat, and mutually lashed together; and in the same
manner the parallel horizontal rows are connected one with another.
Such connection is absent in scale armor, in which each scale is individually treated and attached to a background; the background is in this
case a necessity, while in plate armor it is dispensable.   The laminæ
of scale armor are arranged like roofing-tiles or the scales of a fish,
one placed above another; while in plate armor the laminæ, as a rule,
are disposed one beside another, or but slightly overlapping.   Plate

---

[1] The word "plate armor" is used here throughout in the sense adopted by American
ethnologists, — armor consisting of horizontal rows of narrow, rectangular laminæ
(regardless of the material), the single laminæ or plates being mutually lashed together by means of thongs, and the various rows being connected in a similar manner.   Students of European armor usually take the term "plate armor" to designate
armor composed of large sheets of metal closely enveloping chest and back.   This
type is here styled "sheet armor."

[2] In England, plate armor is usually styled "scale armor." E. H. MINNS (Scythians
and Greeks, p. 74, Cambridge, 1913), for instance, speaks of "a system of thongs
plaited and intertwined as in Japanese and Tibetan scale armor."   This, of course, is
plate armor; scales are never intertwined.

armor is more flexible and lighter in weight, and hence recommended itself to all nations who became acquainted with it. Plate armor can be easily donned over or beneath any garment, and does away with the uncomfortable leather jerkin. For this reason it proved the most favorite and enduring type of armor in China. It was capable of development and refinement, while scale armor always remained stationary.

It is the ethnologists who were the first to place us on the track of this subject; and there are chiefly two scholars, Friedrich Ratzel and Walter Hough, who took the leadership in this research. Our best course will therefore be to begin by reviewing their studies of the subject, and then to see how their results compare with the new material now at our disposal.

FRIEDRICH RATZEL[1] was the first to make a thorough investigation of the geographical dissemination of plate armor, as far as the material was accessible in his time (1886), among the tribes of north-western America and the Chukchi, also on the Society, Austral, and Gilbert Islands in the South Sea. He was particularly struck by the observation that such armor was lacking in other parts of the world, and that its appearance in the Arctic regions was out of proportion to the general poverty of culture there prevailing. The belief in its independent existence among these peoples conflicted with his axiom that the indolence of inventive power is a fundamental law of the primitive stages of ethnic life. In order to explain the phenomenon of plate armor, Ratzel had recourse to Japan, where he deemed armor had reached its greatest development,[2] and where the threads of ancient tribal connections indicated by these peculiar productions ran together; and he believed in a direct contact between Japan and the north-west coast of America in the distribution of plate armor, to the exclusion of the Asiatic Continent. Although the result of this investigation is seemingly historical, the methods and the point of view pursued are purely geographical; and an historical mind cannot fail to notice the weak points of this argumentation. The existence of plate armor in Japan, for instance, is merely accepted as a fact given in space, without inquiry into its historical foundation and development, and without the knowledge of corresponding objects in China and other parts of Asia being much older.

---

[1] Über die Stäbchenpanzer und ihre Verbreitung im nordpazifischen Gebiet (*Sitzungsberichte der Bayerischen Akademie der Wissenschaften*, 1886, pp. 181–216; 3 plates).

[2] H. SCHURTZ (Urgeschichte der Kultur, p. 355) has adopted the opposite point of view, and interprets that the curious plate armor characteristic of the peoples of the Bering Sea has served as model for the Japanese armor made from lacquered pieces of leather, as certain traditional decorations in the former also seem to prove. This opinion is out of the question, for technical and historical reasons.

WALTER HOUGH, in his intensely interesting and valuable study "Primitive American Armor,"[1] arrives, after a careful survey of the subject, at the conclusion that "plate armor in America is a clear case of the migration of invention, its congeners having been traced from Japan northeastward through the Ainu, Gilyak,[2] and Chukchi, across Bering Strait by the intervening islands to the western Eskimo. Here the armor spread southward from the narrowest part of the strait, passing into the slat armor of the North-west Coast, which is possibly a development of the plate idea. The plate armor also may have spread to the eastern coast of North America. Hence, it appears to be con-

---

[1] *Report of the U. S. National Museum for 1893*, pp. 625–651 (Washington, 1895; 22 plates).

[2] This is a debatable point. J. BATCHELOR (The Ainu of Japan, p. 287, London, 1892) says, "The Ainu also wore armor in their wars; but it was of a very light kind, consisting entirely of leather. Some of them, however, wore Japanese armor which they took from the dead in warfare. This is also one way in which they came by their swords and spears." It seems quite certain that the Ainu have never made any plate armor; and what is found among them of this class is plainly derived from the Japanese. Nor can the Gilyak be credited with plate armor. The only specimen of iron plate armor ever discovered in this tribe, and figured and described by L. v. SCHRENCK (Reisen und Forschungen im Amur-Lande, Vol. III, p. 573), is, as SCHRENCK says, of Manchu origin; and he adds expressly that the iron armors, according to the *unanimous statement of the Gilyak*, originate from the Manchu. Dr. HOUGH, who has reproduced Schrenck's drawing of the helmet and of a piece of the armor, seems to have overlooked the description in Schrenck's text, though also on the plate the attribute "old Manchu" is added to both specimens, in contradistinction to the indigenous real Gilyak armor coat plaited from fibre. The Gilyak, therefore, cannot be cited, as Dr. HOUGH has done, as a stepping-stone in the migration of plate armor from Japan to the Eskimo. Also Mr. BOGORAS (The Chukchee, *Jesup North Pacific Expedition*, Vol. VII, p. 164), whose exactness and carefulness is otherwise deserving of the highest praise, has fallen into the same error by reproducing and describing Schrenck's drawing as "Gilyak armor," without paying attention to Schrenck's text. If, therefore, the statement of Bogoras should be correct, — that the shape of the plates, and the manner of connecting them, in an iron armor of the Chukchi, are quite similar to those observed on the remnants of this "Gilyak armor," — this would seem to say that the Chukchi armor in question would have to be connected with Chinese, and not with Japanese culture, as Mr. BOGORAS is tempted to believe; it will be seen on the following pages that other weighty reasons militate strongly against this Japanese theory. SCHRENCK, beyond any doubt, is correct in his statement; and his result agrees with my own inquiries among the Gilyak for armor, and also with my study of Chinese armor. Only SCHRENCK'S definition of "Manchu" must be modified into "Chinese." This error is excusable, as any investigation of Chinese armor had not been made in his time. The Manchu cannot be credited with any original invention in the matter of armor: they adopted it, like so many other things, from the Chinese; and it can be shown step by step, substantiated by official documents, that the Manchu, as in numerous other matters, have also faithfully copied the military equipment established by the Ming dynasty. There is no Manchu type of armor which has not yet existed in, and could not be derived from, the Ming period. SCHRENCK'S Gilyak armor, accordingly, is plainly a modern Chinese specimen, that must forfeit any claim to the historical utilization, to which it has been submitted; it cannot be brought into relation with Japan, nor with the Chukchi, nor with the Eskimo. This ethnographical continuity asserted by HOUGH cannot be proved, nor does it in fact exist. RATZEL (*l. c.*, p. 214) had justly emphasized the entire lack of plate armor among the peoples of Yezo, Saghalin, and the adjacent mainland. Thus the Japanese theories of Ratzel and Hough, though reaching the same end, materially differ in point of construction.

clusive that plate armor in America had Asiatic origin." On p. 633 Dr. Hough states as follows: "The hoop or band armor mentioned as type 4 is found only on the Siberian side of this area and, as well as the plate armor, recalls well-known forms in Japan. This hoop armor is interesting as showing the reproduction of plate armor types in skin, being made of horizontal bands of sealskin instead of rows of ivory plates, the rings telescoping together when the armor is not in use." In describing Eskimo armor made of five imbricating rows of plates of walrus ivory, Dr. Hough observes that in the form, lashing, and adjustment of the plates it is identical with certain types of Japanese armor.[1] His conclusions are the more remarkable, as the previous investigation of Ratzel was unknown to him, and his result has apparently been attained independently. We are here confronted with the interesting case that two ethnographers of high standing have made a notable and praiseworthy attempt to apply an historical point of view to a purely ethnographical situation, with a result so tempting and seemingly convincing that some of the best representatives of our science have readily accepted it.[2] But in the light of a plain historical fact, the position taken by Ratzel and Hough in this question becomes untenable.[3]

---

[1] Compare also HOUGH (*American Anthropologist*, Vol. XIV, 1912, p. 40).

[2] BOGORAS (*l. c.*, p. 162), for instance, seems to accept Hough's results; the Chukchi hoop armor is, to him, "evidently an imitation in skin of plate armor" (repeated after HOUGH, p. 633). R. ANDREE (*Globus*, Vol. 69, 1896, p. 82) acceded to the theory of Hough.

[3] This case well illustrates the difficulty of historical reconstructions built exclusively on the basis of observed data of purely geographical and ethnographical character. As soon as the light of authenticated historical facts is obtained, our preconceived assumptions and conclusions will always be subject to considerable modifications. In my opinion it is therefore impossible to elaborate with assured results historical reconstructions founded on purely ethnological data. Our mind, owing to our scientific training, can evolve only a logical sequence of thoughts, and interpret given data in a highly logical manner only; but history itself is not logical; on the contrary, it is irrational and erratic, moving in zigzag lines, like lightning; it is a labyrinth of dark passages running in all directions; and, above all, it is more imaginative than the boldest flight of our fancy could possibly be. The unexpected, the unforeseen, has always happened; and this is what cannot be supplied or supplemented by the logic of our rational mind. Reconstructions certainly are justifiable and should be attempted, but must never be taken as a substitute for history, or even as real history; they will always remain more or less subjective and problematical, and may be of value as a working hypothesis. It should never be forgotten, however, that the subjective criterion of conceivableness or plausibility, or of an appeal to our common sense, will but seldom prove before historical facts. The rule may even be laid down that whatever may appear to our conception as quite natural, self-evident, or logical, may hardly ever have happened that way, or need not have happened that way, but otherwise. Our knowledge of most subjects is still too meagre to allow at the present time of culture-historical reconstructions embracing a wide area of the globe. To these belongs also the theme of plate armor, the specific history of which must first be traced in the single culture zones where it occurs, before its general history can be built up with any encouraging result. Plate armor

In the north-east of China, beyond the boundaries of Korea, in the east conterminous with the ocean, the northern limit being unknown, we find from very remote ages the habitat of a most interesting people, the Su-shên, who have greatly stirred the imagination of Chinese and Japanese chroniclers. They were the Vikings of the East, raiding on several occasions the coasts of northern Japan, and fighting many a sea-battle with the Japanese in the seventh century.[1] For a thousand years prior to that time, the Chinese were acquainted with this tribe and its peculiar culture: even Confucius is said to have been posted in regard to them, and to have been aware of the fact that they availed themselves of flint arrowheads, usually poisoned, which were then preserved as curiosities in the royal treasury of China. From Chinese records we can establish the fact that the Su-shên lived through a stone age for at least fifteen hundred years down to the middle ages, when they became merged in the great flood of roaming Tungusian tribes. They had also stone axes, which played a rôle in their religious worship. A mere supposition is that they belonged to the Tungusian stock of peoples; yet this remains to be ascertained. They may as well have been related to one of the numerous groups of tribes occupying ancient Korea, or, which is still more likely, to the so-called Palæ-Asiatic tribes of the North-Pacific region; but the whole ancient ethnology of north-eastern Asia remains as yet to be investigated.

Under the year 262 A.D. it is on record in the Annals of the Three Kingdoms[2] that the Su-shên presented to the Court of China a tribute of a mixed lot of harness, altogether twenty pieces, including armor made of leather or hide, of bone, and of iron, with the addition of four hundred sable-skins.[3] On the iron armor, which was foreign to the culture of the

---

certainly is not by any means so rigidly restricted as assumed by Ratzel and Hough; it will be seen that it takes its place in China, western Asia, ancient Siberia and Turkistan, where it is assuredly much older than in Japan.

[1] Compare Jade, p. 59. The Han Annals state that the Yi-lou, another name for the Su-shên, were fond of making piratical raids in boats; the Wo-tsŭ settled in the north-eastern part of Korea, and bordering in the south on that tribe, "dreaded it so much that every summer they were wont to hide in the precipitous caves until winter, when navigation was impossible, at which time they came down to occupy their settlements" (E. H. PARKER, *Transactions Asiatic Society of Japan*, Vol. XVIII, 1890, p. 201). In the same study of Parker (pp. 173 *et seq.*) a history of the Su-shên will be found.

[2] *San kuo chi, Wei chi*, Ch. 4, p. 13 a (compare *T'oung Pao*, 1913, p. 347).

[3] I am inclined to understand this passage in the sense that there were three distinct kinds of armor, made entirely either of leather, or of bone, or of iron. It is impossible to presume that bone was used in connection with iron in the make-up of one and the same suit of armor. The iron armor, we are forced to conclude, must have formed an individual type in itself, and assuredly one alien to the culture of the Su-shên, who, we know with certainty, were not acquainted with the technique of metals for an extended period, and availed themselves of flint arrowheads. Before going to press, I notice from the work of R. and K. TORII (Études archéologiques,

Su-shên, I shall comment later.   Hide armor and bone armor formed
the national harness of the Su-shên, as we may infer from another
memorable passage in the Annals of the Tsin Dynasty[1] relating to the
period 265–419 A.D., where the characteristic arms of the tribe are
enumerated as wooden bows, stone crossbows, hide and bone armor.[2]
It is remarkable that the Chinese do not ascribe bone armor to any
other of the numerous tribes, with whom they became familiar during
their long history, and whose culture they have described to us.   In all
likelihood, the term "bone armor" occurs in their records only in those
two passages; and it is not at all ambiguous.   There is but one thing
that can be understood by it, — the well-known type of bone armor, as it
still occurs among the tribes occupying the northern shores of the Pacific
on the Asiatic and American sides, particularly among the Chukchi and
Eskimo, and in that region exclusively.[3]   The Eskimo ivory plate armor
represented on Plate XXIX will give some idea of what the Su-shên

---

Journal of the College of Science, Vol. 36, Tōkyō, March 29, 1914, p. 73), which has
just reached me, that the two Japanese authors understand this passage in exactly
the same sense.

[1] *Tsin shu* (compiled under the T'ang dynasty by Fang K'iao and others),
Ch. 97, p. 2 b.

[2] The question in this passage, accordingly, is of the armor, offensive and defensive, possessed and made by the Su-shên in the beginning of the middle ages.   Hide
and bone armor are attributed to them, while iron armor is not mentioned.   The
text might be construed to mean that the Su-shên possessed but a single type of
armor, composed of both bone and leather; that is, plates of bone lashed together by
means of hide thongs; bone armor is unthinkable without such a ligament, but this
consideration need not preclude the assumption that the Su-shên fabricated also pure
hide armor.   The ethnographical fact that in the culture-area to which this tribe
belonged hide and bone armor still occur side by side, must be equally considered in
this question; and for this reason we may well understand the passage of the Tsin
Annals in the sense that the Su-shên had hide or leather armor, and bone armor.   But
this point of view is of minor importance.   The same passage in the *Tsin shu* indicates
a tribute sent by the Su-shên toward the end of the period King-yŭan (260–264) and
consisting of arrows, stone crossbows, armor, and sable-skins.   What kind of armor
it was on this occasion is not specified; but the general word *kia* refers to a hide armor
or cuirass.   J. KLAPROTH (Tableaux historiques de l'Asie, p. 85) attributes "cuirasses
made from skin and covered with bone" to the Yi-lou; the latter are identical with
the Su-shên, and the text from which Klaproth translated must be the same as that
of the *Tsin shu* referred to above.   The text relative to the Yi-lou inserted in
Hou Han shu (Ch. 115, p. 2 b) makes no allusion whatever to armor, but I am not
inclined to infer from this silence that the Yi-lou or Su-shên lacked armor in the Han
period.

[3] As stated by me in *T'oung Pao* (1913, p. 349), the plates of this bone armor
were presumably carved from walrus ivory, in the same manner as in the present
Eskimo and Chukchi plate armor.   Dr. W. HOUGH of the U. S. National Museum in
Washington, to whom I addressed the question as to whether ivory or ordinary bone
was utilized to a larger extent in these pieces has been good enough to write me as
follows: "The Eskimo armor in the Museum and such suits as I have seen are
mostly made of walrus ivory, and so far as I can remember, there are no combinations
of ivory and bone in the same piece.   On the other hand, there are fragmentary parts
of armor from St. Lawrence Island and from the Alaskan mainland which are made
of bone; just what bone I cannot say, probably the whale."

tribute armor was like.[1]   The point here at issue, then, is the fact that
the entry of the Chinese annalist, under the year 262, regarding the
presentation of bone armor on the part of the Su-shên, is the earliest
recorded reference to bone armor in history, capable of throwing a
flashlight on events in the North-Pacific culture area, so glaringly devoid
of any records.

The date 262 is of far-reaching consequence.   Certainly, like all
dates where inventions or culture ideas are involved, it is a mere symbol,
that requires a certain latitude in its translation.   The tribute of 262
indicates that bone armor had been made prior to that date by the
Su-shên, or generally within the culture-zone to which they belonged;
and since complex inventions of such character require time to mature,
and the laborious efforts of several generations, it is justifiable and
reasonable to conclude that the beginnings of the invention go back to a
far earlier period.   Plate armor of bone must therefore be infinitely
older than could heretofore be supposed from the mere circumstantial
evidence of present geographical distribution; and it follows also that
the geographic area of bone armor must have been much more extended
in ancient times, and reached farther south along the shores of Asia.   In
other words, the culture area under consideration, as it now presents
itself to our eyes, must have occupied a larger territory in the times of
which we speak, — a conclusion confirmed to me also by other reasons;
and the Su-shên must have either ranged among the representatives of
North-Pacific culture, or have been strongly influenced by it.   If as
early as 262 the Su-shên were in possession of bone plate armor, this
type of harness cannot be explained as having been made in imitation
of Japanese plate armor — for the plain reason that Japanese plate
armor was at that time not in existence.   Metal armor in Japan cannot
be pointed out before the close of the eighth century.   Fragments of
armor consisting of scales of bronze incrusted with gold, and preserved
in the Museum of Tōkyō, are assigned to about the year 800 A.D. by
BASHFORD DEAN,[2] our great authority on Japanese armor; while fragments of iron plate armor are not older than about 1050 and 1100; that

---

[1] The number of perforations in the plates is not always six, as in the specimen
illustrated.   A large number of detached Eskimo ivory plates in the Field Museum
(Cat. No. 34,154) exhibits on an average twelve perforations, two and two being close
together.   Sometimes a third perforation is added to the two in the corners, and sometimes an additional perforation is drilled through the centre of the upper or lower side.
A very interesting specimen in our collection (Cat. No. 34,153) is a pair of Eskimo
cuisses (leg-guards) of mastodon ivory, 16.5 cm long, with rows of perforations
along the top and bottom edges.   These objects were obtained by A. M. Baber from
the Asiatic Eskimo on the Tchukotsk Peninsula.

[2] Catalogue of the Loan Collection of Japanese Armor, pp. 20, 28 (New York,
1903).

is, they belong to the latter part of the Fujiwara period (900–1100). Before this time, padded coats and hide cuirasses were the usual means of body protection; the latter sometimes assumed the form of scale armor, the scales being cut out of pieces of boiled leather.[1]

The Chinese Annals of the Sui Dynasty,[2] in the interesting account on Japan, state that the Japanese (Wo) make armor of varnished leather (*tsi p'i wei kia*) and arrows of bone. At that time, which, from the standpoint of Japanese development, is designated as the protohistoric or semihistoric period, defensive armor cannot have played any significant rôle in ancient Japan, as it is conspicuously absent in her two oldest records, the Kōjiki (composed in 712 A.D.) and the Nihongi (720 A.D.).[3] In the year 780 an order was issued by the government that leather armor should be used, because the kind hitherto worn (that is, padded coats) was continually requiring repair. This order permitted, further, the use of iron instead of leather, and advised that all armor should be gradually changed to metal.[4] It is therefore clear that at the time, when our Su-shên account of bone armor is at stake, the Japanese did not possess any metal or any plate armor, and that it is even questionable whether they then availed themselves of defensive armor at all. We are hence prompted to the conclusion that bone plate armor, being at least from six to eight hundred years older than Japanese plate armor, cannot have been made as a reproduction of the latter, and that Japan cannot be made responsible for it. Thus the whole theory of a connection of American and Northeast-Asiatic plate armor with Japan must naturally collapse.

If the opinion should be correct of those who believe that American-Asiatic plate armor must have been made in imitation of a form of iron

---

[1] Catalogue of the Loan Collection of Japanese Armor, p. 38 (New York, 1903). According to W. GOWLAND (The Dolmens and Burial Mounds in Japan, p. 47, Westminster, 1897), no bronze armor has as yet been found in the dolmens of Japan; and iron armor, too, is by no means of very common occurrence.

[2] *Sui shu*, Ch. 81, p. 6 b (also *Pei shi*, Ch. 94, p. 72). It is notable that the account of Japan in the Annals of the Later Han Dynasty (Ch. 115, p. 5 b) makes no mention of body armor, but points out only the shield and the use of offensive weapons, such as spear, wooden bow, and arrows with bamboo shafts and bone heads. Arrows with iron heads employed in Japan are first reported in *Tsin shu* (Ch. 97, p. 3).

[3] O. NACHOD, Geschichte von Japan, Vol. I, p. 155 (Gotha, 1906). But shields are several times mentioned as offerings. The Annals of the Later Han Dynasty, as pointed out, confirm the existence of shields. The idea generally entertained that Japan has had a bronze and an iron age, in my opinion, is erroneous. The bronze and iron objects found in the ancient graves have simply been imported from the mainland, and plainly are, in the majority of cases, of Chinese manufacture. Many of these, like metal mirrors, certain helmets and others, have been recognized as such; but through comparison with corresponding Chinese material, the same can be proved for the rest. Ancient bronze objects are so scarce in Japan that, even granted they were indigenous, the establishment of a "bronze age" would not be justified, nor is there in the ancient records any positive evidence of the use of bronze.

[4] BASHFORD DEAN, *l. c.*, p. 27.

armor, two other theoretical considerations could be advanced. There remain the Chinese and the ancient Turks of Siberia and Central Asia; and it might be argued that Chinese or Siberian harness of iron plate could have furnished a suitable model for the Arctic harness-maker. To such a point of view, however, serious objections could be raised; and here again, first of all, on purely historical grounds. The utilization of iron in the making of armor, as we noticed in Chapter III, does not become apparent in ancient China till as late as the first centuries of our era, its beginnings being justly laid by the Chinese in the period of the Later Han dynasty (25–220 A.D.; see p. 210), and thus it appears from inward evidence. This primeval iron armor, in all likelihood, was not yet a true iron plate armor, but merely a hide cuirass reinforced by iron laminæ; rectangular iron plates may have then existed, but the matter is still problematical. Even presuming that iron plate armor might have obtained during the epoch of the Later Han, for which there is as yet no positive evidence, we should be forced to infer that the developments of the ancient Chinese iron armor and the northern bone armor, in this case, have necessarily been contemporaneous events. The tribute of the Su-shên bone armor in 262 A.D. is separated from the closing year of the Han period in 220 A.D. only by the brief span of forty-two years; that is, the average duration of a generation. If, accordingly, these two developments should have run parallel to each other in point of time in two widely different culture areas which otherwise had not a single point in common, the inference would have to be drawn that these two developments have taken place independently, and may have each been prompted by factors coming from a different quarter. In the present state of our knowledge it is safe to assume that bone armor in north-eastern Asia is as old as, or even older than, any iron plate armor in China or Korea.

If an outward impetus to the making of bone armor in that region must be assumed, I am disposed to believe that it came from the interior of Siberia.[1] In regard to ancient Siberian armor, our information is exceedingly scanty. Only traces of plates of armor have been discovered in graves on the Berel,[2] and a famous petroglyph on the Yenisei depicts to us a horseman armed with lance and mail-clad (Fig. 35). The long continuity of the iron age in Siberia renders it impossible at

---

[1] For evidence see below, p. 274.

[2] W. RADLOFF, Aus Sibirien, Vol. II, p. 130. Also in Siberia iron armor may have formed the exception, while hide, as the cheaper material, always maintained its place. MARCO POLO (ed. of YULE and CORDIER, Vol. I, p. 260) says concerning the Tartar (that is, Mongol) customs of war, "On their backs they wear armor of cuirbouly [boiled leather], prepared from buffalo and other hides, which is very strong."

the present time to fix a date for these antiquities with any degree of certainty; but a general deduction may be hazarded. There are good reasons for assuming that the Chinese derived their iron armor from Turkish and Iranian peoples,— first, because their knowledge of smelting and forging iron came from them; and, second, because their own inventiveness in defensive and offensive armor was rather poor, and because others of their weapons, like swords and daggers, were adopted from the same group (p. 215). The sudden appearance of iron armor in the Later Han dynasty speaks in favor of this view; and as only copper plate armor was known in the preceding period of the Former Han dynasty, it seems very likely that iron armor among the Turkish tribes was not much older than in China. As previously stated, the Su-shên sent iron armor along with skin and bone armor to China, but only the latter two types formed their national armor, according to the later report of the Annals of the Tsin Dynasty. The occasional introduction of iron armor, consequently, did not suppress among them the employment of skin and bone armor; and although iron armor was known to them at the end of the third century, they adhered, for several centuries downward, to bone and hide, that seem to have represented a more efficient means of defence at that time than iron armor, the making of which must still have been in a primitive and experimental stage. On the other hand, in opposition to this theory of a foreign influence, it must be emphasized that the culture types of north-eastern Asia, on the whole, have strong and pronounced characteristics which have hardly any parallels in the rest of the Asiatic world, and that owing to geographical conditions the entire area has remained purer and more intact from outside currents than any other culture group in Asia. The profound researches of Bogoras and Jochelson have shown us that in language, folk-lore, religion, and material culture, the affinities of the Chukchi, Koryak, Yukagir, and Kamchadal go with Americans, not with Asiatics. In fact, Turkish-Mongol influence on these tribes is exceedingly small; Chinese influence, if any, amounts to a minimum;[1] and the alleged Japa-

---

[1] While the Chinese, owing to political circumstances, were comparatively well acquainted with the tribes inhabiting Manchuria, Korea, and the Amur region, their knowledge of the tribes beyond has always been very limited. Their first acquaintance with the Ainu dates from the year 659 A.D., when some members of this tribe accompanying a Japanese embassy made their appearance at the Court of the Emperor Kao-tsung (650–683) of the T'ang dynasty; they are described on this occasion as "forming a small country on an island in the ocean, having beards four feet long, being clever archers, and sticking arrows through their hair; they have a man hold an arrow (according to another reading, a vessel) which they use as a target at a distance of ten paces, without missing their aim" (T'ang shu, Ch. 220, p. 11; and Yen kien lei han, Ch. 231, p. 47). They are called by their Japanese name Yemishi (Chinese, Hia-i). This embassy is mentioned under the same year also in the Japanese Nihongi (ASTON, Nihongi, Vol. II, p. 260), where it is said that the

nese influence is a chimera. Plate armor, if due in that region to a stimulus received from outside, would represent a somewhat isolated instance of historical contact in the line of warfare;[1] and whatever the psychology of this first stimulus may have been, — I venture to deny that it ever operated in the haphazard and purely external manner indicated by Ratzel and Hough, — a certain independent course of development in that area cannot be absolutely denied.

While I am very far from contesting that historical interrelations may have been at play in the dissemination of the plate idea in north-eastern Asia, I wish to maintain for the present an attitude of reserve toward this point. The downright failure of the Japanese hypothesis should put us on our guard; and, the imitation theory, I confess, be it formulated with reference to the Japanese, Chinese, or Siberians, does not strike me very favorably. Whatever we may now be inclined to assume in that direction, it will remain mere assumption in our present state of knowledge; and it must be upheld that no imitation theory, with whatever modifications, can be backed up by certain facts. In other words, the problem is not yet susceptible of a definite solution. There is, however, not only an historical, but also a technical side to this question, and we should not entirely lose sight of the technical point. We observe in various culture-groups that plate armor is never a primary type of armor, but occupies a secondary place in point of

---

Japanese took with them a Yemishi man and woman of Michinoku to show to the T'ang Emperor. In the Description of the Tributary Nations of the Ts'ing Dynasty (*Huang Ts'ing chi kung t'u*, Ch. 3), published under the patronage of the Emperor K'ien-lung, the Ainu are figured and briefly characterized under the name *K'u-ye*. This is the Gilyak designation *Kuhi* for the Ainu, identical with the *Huye* of Du HALDE (Description de l'empire de la Chine, Vol. IV, p. 15; compare also L. v. SCHRENCK, Reisen und Forschungen, Vol. III, p. 129). On some Chinese maps Saghalin is still designated as "Island of K'u-ye." The Gilyak came to the notice of the Chinese at a very late date; they do not seem to be mentioned earlier than in the *Se wên hien t'ung k'ao* (published in 1586) under the name *Ki* (or *K'i*)-*li-mi* (Gilāmi), the name given this people by its Tungusian neighbors (compare A. WYLIE, Chinese Researches, pt. 3, p. 249, who alludes to this passage without identifying the tribe). In the Chinese work previously quoted, the Gilyak are pictured and described under the term *Fei-ya-k'a* as inhabiting the country to the extreme east of the Sungari, the littoral of the ocean, and scattered over the islands (compare L. v. SCHRENCK, *l. c.*, pp. 100–103).

[1] A very interesting case was established by FRANZ BOAS in his study Property Marks of Alaskan Eskimo (*American Anthropologist*, 1899, pp. 601–613). Property marks are very frequently used by these tribes on weapons employed in hunting with the object of securing property-right in the animal in whose body the weapon bearing the mark is found. It is a remarkable fact that these marks occur only among the Eskimo tribes of Alaska, but are not known from any other Eskimo tribe. This fact, taken in connection with the form and occurrence of such marks among the north-eastern tribes of Asia, suggests to Boas that this custom, like so many other peculiarities of Alaskan Eskimo life, may be due to contact with Asiatic tribes. This case is very plausible, and would merit a more profound historical investigation in connection with the practice of *tamga* now disseminated throughout Siberia.

time; it is always preceded by plainer types, usually cuirasses of hide or
cotton, and scale armor. Cuirasses of rhinoceros-skin were utilized
in China for thousands of years, before any metal harness became
known. In China as well as in Egypt we clearly recognize the intermediary stages of hide and plate armor, the surface of the hide being
first reinforced by irregular, scale-like metal pieces (first of copper,
later of iron), which gradually assumed the standard rectangular plate
shape; and then, by removing the hide foundation, the pure metal
plate armor sprang up as a new and independent type. The history of
defensive and offensive weapons, moreover, is closely interrelated; the
eternal game of modern war industry — first inventing bullet-proof
naval armor-plates, and then the bullets to pierce them — was in full
swing even in the stages of primitive life. The growing perfection of
metal weapons constantly forced man to devise new means of increasing
the power of his defensive armor, and this accounts for the coming into
existence of ever-varying new types. I am certainly not competent
on any subject of American ethnology, and must leave it to our Americanists to reason out the case for themselves. But this much may be
said. Nearly everywhere in North America, even in the eastern area,
we generally find the type of hide armor, the indigenous development of
which is admitted by Dr. Hough and cannot seriously be challenged;
thus hide armor may have been the oldest form of body protection in
war also in this region.[1] We meet there also the intermediary stages,
as, for instance, the wooden cuirass of the Thompson River Indians,
covered with elk-hide, described by JAMES TEIT,[2] and the application of
wooden slats, of reeds, of bone plates to the exterior or interior of the
cuirass, to strengthen it more efficiently, — the secondary development.
Finally those materials were exclusively utilized in its construction,
leading up to pure plate armor as a tertiary and ultimate stage. No
fundamental difference can be found in the employment of wood and
bone, or ivory, which simply present purely technical changes of material; and American-Asiatic bone plate armor, after all, might be conceived as quite a natural development, which may have arisen independently, without the contact of an outside culture. Its coming into
existence could be explained by the trend of indigenous thought and the

---

[1] "The American savages were acquainted with body armor when they were
first encountered. Wherever the elk, the moose, the buffalo, and other great land
mammals abounded, there it was possible to cover the body with an impervious suit
of raw-hide" (O. T. MASON, The Origins of Invention, p. 390).

[2] The Thompson Indians of British Columbia (*Jesup North Pacific Expedition*,
Vol. II, p. 265). See also A. P. NIBLACK, The Coast Indians of Southern Alaska
(*Report U. S. National Museum*, 1888, pp. 268–270).

inventiveness of the aborigines, which may have resulted in a large variety of ingenious armor spread over an extensive area.[1]

There remain other considerations to be made which would seem to confirm this impression. The cut, the style, and the mode of wearing armor in the North-Pacific region are different from those in eastern Asia. The peculiar Chukchi fashion of having the left side covered up and the left arm and hand hidden in the armor, while only the right arm remains free for action,[2] is a striking feature, which is entirely lacking in any other part of Asia. At any rate, I am inclined toward the opinion that the type of bone plate armor under consideration is not exclusively due to an impact of foreign influence. In some form unknown to us it may have pre-existed, before any metal plate armor had reached the Far East; while I am quite willing to admit that at some later period the regular, rectangular shapes of the ivory plates, and the peculiar method of lashing them together, may be the outcome of an adaptation to some imported model.

The memorable passage in the Chinese Annals concerning the Su-shên may elucidate still another problem. Their gifts to China in 262 consisted not only of bone armor, but also of iron armor. BOGORAS [3] has shown that ancient iron armor, made of small pieces of iron with fastenings of narrow leather strips, was until recently very common among the Reindeer Chukchi; and he makes it probable that iron was known among them before the arrival of the Russians. And here the Su-shên come again to our assistance in dispelling the Japanese spectre; for the question of the origin and manufacture of Chukchi iron armor suggests to Mr. BOGORAS "a connection with the Japanese which does not exist at present,"— and which in all probability has never existed. Mr. BOGORAS is unable to furnish any evidence for such an alleged intercourse, which is certainly not proved by the occasional occurrence of a modern Japanese article of trade in that region.[4] The facts in the case

---

[1] I do not mean to say, of course, that the development has actually and objectively taken place that way, but only wish to point out that it may be thus construed in our minds.

[2] HOUGH, Plate V; BOGORAS, The Chukchee, p. 163 (shows also a suit of left-handed iron armor).

[3] The Chukchee (*Jesup North Pacific Expedition*, Vol. VII, No. 1, p. 54).

[4] The statement of Bogoras that the armor and helmet figured on p. 164 are Japanese seems to me to require further proof. It rather conveys the impression of being un-Japanese. Bogoras alludes to the advance of the Japanese to Kamchatka without citing sources in support of this opinion. I presume he must have had in mind the passages of G. W. STELLER (Beschreibung von dem Lande Kamtschatka, pp. 3, 249) saying that the Japanese were long known as traders to the inhabitants of the littoral of the Okhotsk Sea (on the Kamchadal name of the Japanese, see L. v. SCHRENCK, *l. c.*, p. 192). Kamchatka was vaguely known to the Japanese of the eighteenth century, as we see from KLAPROTH'S Aperçu général des trois royaumes

are that the Japanese never have penetrated much beyond Saghalin Island, where the southern portion inhabited by the Ainu was their main field of exploitation, while the northern part remained a *terra incognita* to them. The Japanese have exerted no influence on the culture of the Gilyak settled there,[1] nor is there any Japanese trace on the mainland in the region of the Amur. Even without such considerations, however, the point of view taken by Bogoras in this matter can no longer be upheld. The fact that the Su-shên possessed knowledge of iron armor in 262 goes to prove that iron armor around that time was within the boundaries of the North-Pacific culture-zone.[2] Again, it must be called to mind that the Su-shên iron armor cannot have been of Japanese origin, as iron armor was not then in existence in Japan; neither can it be set in relation with Chinese iron armor, as it would be absurd to suppose that the Su-shên should have sent Chinese iron armor as tribute to the Chinese Court. Their tribute certainly consisted of curious and valuable objects which were new and impressive to the Chinese. As the Su-shên were not able to make iron armor, not being acquainted with the technique of smelting and forging iron, they consequently must have received it in the channel of trade from an iron-producing region, such as we find in ancient times in the interior of Siberia, in Central Asia,[3] and in the beginning of our era also in

---

(p. 195, Paris, 1832). The Itālmen, the ancient Kamchadal, knew the Japanese chiefly as importers of iron needles, and styled these *šis* (plural *šisin:* I. RADLINSKI, Slownik narzecza Kamczadalów, p. 72, Cracow, 1892) after *Sisam*, the Ainu designation of the Japanese. But it is altogether the simple question of a superficial trading relation along the coast by way of the Kuriles; and there is no trace of Japanese influence whatever on the culture of the Kamchadal.

[1] Likewise L. v. SCHRENCK (Reisen und Forschungen im Amur-Lande, Vol. III, p. 570).

[2] This chapter, as it now stands, was in substance written in the autumn of 1912, an abstract of it having been read at the meeting of the American Anthropological Association held in Cleveland, December, 1912 (see *Science*, 1913, p. 342, or *Am. Anthr.*, 1913, p. 960). A confirmation of the above conclusion is now furnished by the highly interesting study of R. and K. TORII (*l. c.*, p. 72), who found in eastern Mongolia a metal (seemingly iron) plate of an armor (4 × 2.5 cm) with four apertures in the long sides. It is correctly diagnosed by the two Japanese authors, who remark that such plates are now dispersed among the ruins left by the Tung Hu ["Eastern Hu," a general Chinese designation for the populace of eastern Siberia], especially in the region of the Shira Muren. This archæological discovery bears out the fact that iron armor anciently did exist in eastern Siberia, and that it was of the type of plate armor. Thus the supposition is gaining ground that the iron harness in the possession of the Su-shên was iron plate armor, and existed in that region side by side with bone plate armor. Messrs. Torii, in this connection, remind us of the fact that the Wu-huan, according to the Annals of the Later Han Dynasty, are capable of making their bows and arrows, also saddlery, and turn out their own arms from forged iron.

[3] It is known that L. v. SCHRENCK (*l. c.*, Vol. III, p. 569) attributes to Japanese influence the knowledge of iron-forging among the Ainu and Gilyak. This being an affair of recent origin is certainly not a serious case; these tribes purchase Japanese pig-iron, and work it up into blades for knives. Schrenck's point of view that iron-

Korea.[1] These considerations are instructive also in that they reveal the baselessness of what might be styled "the Japanese mirage of American ethnology." Not only objects of material culture like plate armor, but also motives of myth and legend, have been traced from America directly to Japan, as, for instance, by the late PAUL EHRENREICH.[2] This method seems to me inadequate for historical reasons. The primeval culture type of Japan, as we know it, is a comparatively recent production, very recent when contrasted with the great centres of culture developed on the mainland of Asia, and recent even in comparison with all indigenous cultures found on the American Continent. I mean to say that most phenomena of culture, inclusive of myth and religion, are by far older on this continent, and still preserved in an older form, than any corresponding phenomena in Japanese culture, even if the latter are reduced to their oldest attainable condition. The Kōjiki and Nihongi, the main text-books of Japanese mythology, do not present a pure source of genuine Japanese thought, but are retrospective records largely written under Chinese and Korean influence, and echoing in a bewildering medley continental-Asiatic and Malayo-Polynesian traditions. But more than that, — it may be safely stated at the present time that the history of American cultures has never had, and never could have had, any relation with Japan, which always was beyond the pale of American-Asiatic relations, and that American ethnology offers no point of contact with Japan. The threads of historical connection running from America into Asia do not terminate in Japan, but first of all, as far as the times of antiquity are concerned, in a territory which may be defined as the northern parts of modern Manchuria and Korea. From ancient times the varied population of this region has shared to some extent in the cultural elements which go to make up the character-

---

forging among the Gold on the Amur is due to the adjoining Manchu-Chinese, however, is entirely erroneous, as this art doubtless is much older in that region than the rule and influence of the Manchu, and points decidedly in the direction of the Turkish Yakut. Many iron objects of an ornamental character in use among the Gold can be plainly recognized as Yakutan in origin, and Yakut are constantly living and trading in their midst. Neither the Japanese nor the Chinese need be invoked to explain iron-forging in eastern and north-eastern Siberia, as it is much older in the interior of Siberia, where there have been at all times better blacksmiths, forging better iron-work than was ever turned out in China.

[1] The Annals of the Later Han Dynasty (*Hou Han shu*, Ch. 115, p. 5 b) relate that the country Shen-han in Korea produced iron, that the Wei, Wo (Japanese) and Ma-han went there to purchase it on the market, and that iron was the means of barter in all business transactions. There was no iron in the country of the Shi-wei, and they received it from Korea in exchange for sable-skins (*Pei shi*, Ch. 94, p. 9 b). The considerable beds of iron ore in Kang-wun Province are still worked by the natives, who scrape it up from the surface of the ground, and smelt it in furnaces by means of charcoal (H. B. HULBERT, The Passing of Korea, p. 274).

[2] Die Mythen und Legenden der südamerikanischen Urvölker, pp. 77 *et seq.* (Berlin, 1905).

istics of the North-Pacific culture-province.   It does not suffice for the study of American-Asiatic relations to take into consideration only the present ethnological conditions, as has been done, but the ancient ethnology of that region must first be reconstructed.   From this point, the further contact, if any, may be given, and as our knowledge advances, may eventually be established at a future date (I speak only hypo-thetically) with ancient China on the one hand, and ancient Siberia on the other, — relations which would all refer to pre-Japanese times, and move outside of the current of Japan.   The early existence of bone armor is one of the examples proving that this view seems to be on the right track, and entitling us to speak of an historic antiquity in North-Pacific culture.

A pragmatic history of the development of plate armor cannot yet be written, as the subject has not been thoroughly investigated by specialists in the antiquity of western Asia, and as there are doubtless many missing links still unknown to us.   Meanwhile the following in-dications which I have been able to trace may be welcome.

In Assyria, plate armor is unmistakably represented on monuments of King Sargon (B.C. 722–705) in connection with foot-archers, whose coats consist of six or seven parallel rows of small rectangular plates.[1] It seems that in Assyria plate mail sprang up during that period, for in the reign of Salmanassar II (B.C. 860–825) the bowmen sculptured in stone are frequently clad with long coats reaching from the neck to the ankles and girdled below the chest, the coats being covered with an irregular checkered design, but not with rows of rectangles.[2]   Further, we find metal plate armor in ancient Egypt;[3] there a cuirass of thickly wadded material was covered with metal plates.   It is ascribed to the reign of Ramses II, who ruled in the thirteenth century B.C.

Also the Shardana armor described by OHNEFALSCH-RICHTER [4] — consisting of bronze plates, two of which are mutually joined by means

---

[1] P. S. P. HANDCOCK (Mesopotamian Archæology, pp. 350–2), who speaks only of coats of mail.

[2] Ibid., pp. 260, 350.

[3] An illustration of it may be seen in A. ERMAN's Life in Ancient Egypt (p. 545, London, 1894).   As a rule, the helmet and body armor did not consist there of metal, being more probably made, as many of the pictures seem to indicate, of thickly wad-ded material, such as is worn even now in the Sudan, and forms an excellent protec-tion.   In rare instances, however, defensive armor may have been covered with metal plates.   No special investigation of this subject has as yet been made in regard to the two culture zones of Assyria and Egypt; but these indications, however brief, will suffice to show that plate armor must have been widely distributed in ancient times, and that a mere consideration of present conditions alone, as attempted by Ratzel and Hough, cannot bring about the solution of the problem of its history.

[4] Zeitschrift für Ethnologie, Vol. XXXI, 1899 (Verhandlungen, p. 360).

of hinges, and sewed to a foundation of linen or leather — evidently belongs to this category.

The most valuable contribution to the question is presented by a number of single bone plates of rectangular shapes, found in barrows about Popovka on the Sula in southern Russia. Five of such plates are reproduced by E. H. MINNS.[1] As these have perforations (one, two, or three) only at the top and base, we must suppose that they were sewed on to a foundation of cloth or leather; they could not have been lashed together freely without such a background, as in the Chukchi and Eskimo plate armors discussed above.[2] Those with pointed top and a single perforation, having the one side curved and the other straight, formed the ends of a plate-row. This find attests the fact that bone plate armor anciently existed in the western part of the Old World among Scythian tribes; and this case shows that in regard to Northeast-Asiatic and American bone plate armor we need not resort to the theory of explaining it as an imitation of iron in bone. If imitation it is, it may have been Scythian (or Siberian) bone armor (a single piece or several), which by trade found its way to north-eastern Asia. In the territory of the Scythians we find plate armor not only of bone and horn, but also of bronze and iron; and it seems to me that the adoption, on the part of the Scythians, of the Iranian tactics of cataphracti (p. 220) gave the impetus to the introduction among them of this type of armor. The rock-carving of the mounted lancer on the Yenisei (Fig. 35) demonstrates that plate armor, presumably of iron, had penetrated into Siberia during the iron age. I suspect the institution of cataphracti of being largely responsible for the wide dissemination of this type of armor; it was peculiarly adapted to fighting on horseback, and the Iranian mode of tactics, as we saw in Chapter III, expanded into the Roman Empire, and was adopted by the Huns, to be continued by the Turks (T'u-küe) under the T'ang dynasty. When tactics and cavalry organization spread over the boundaries of Irān, the armature of the cavaliers was necessarily bound to migrate along the same path.

The fresco paintings discovered in Turkistan furnish many valuable contributions to the history of body armor, and particularly of plate armor. A. STEIN[3] was the first to correctly recognize this type of armor in a Buddhist statue excavated by him at Dandan-Uiliq. The figure, standing over the body of a prostrate foe, is clothed with a coat of mail reaching below the knees and elaborately decorated. "The gay colors

---

[1] Scythians and Greeks, p. 188 (Cambridge, 1913).

[2] In these, perforations likewise run along the long or vertical sides of the plates.

[3] Sand-buried Ruins of Khotan, p. 272 (London, 1904); and Ancient Khotan, Vol I, p. 252, Vol. II, Plate II (Oxford, 1907).

of the successive rows of small plates which form the mail, alternately red-blue and red-green, were remarkably well preserved, and not less so all the details of the ornaments which are shown along the front and lower edge of the coat and on the girdle around the waist. Even the arrangement of the rivets which join the plates of mail, and the folds of the garment protruding below the armor, are indicated with great accuracy. There can be no doubt that the artist has carefully re-produced here details of armor and dress, with which he was familiar from his own times." [1]

A rich material for the study of plate mail in the art of Turkistan is offered by the fascinating work of A. GRÜNWEDEL,[2] who himself has clearly recognized and pointed out this armor type.[3] The fact that the plates are painted blue clearly proves that they were wrought from iron. The coats are tight-fitting, and open in front; the sleeves are likewise bedecked with plates, and the shoulders with pauldrons. A further example will be found in the work of A. v. LE COQ.[4]

The T'ang period (618–906) is responsible in China for a far-reaching innovation in the line of armor, which has persisted at least down to the end of the eighteenth century,— the combination of armor with the military uniform, resulting in a complete armor-costume. Up to that time, armor and garment had been distinct and separate affairs. The ancient hide harnesses were worn over the ordinary clothing or uniform, and were naturally put on only when making ready for battle; while

---

[1] The comparison made by Stein (Ancient Khotan, p. 252) between this armor and that on a Gandhāra relief figured by GRÜNWEDEL (Buddhist Art of India, p. 96) is not to the point. The two suits of armor are of entirely different types, the former being plate armor; the latter, as correctly interpreted by Grünwedel, scale armor. Stein did not recognize this difference, nor did V. A. SMITH (History of Fine Art in India, p. 122), who copied him on this point. Among the finds made by A. Stein (Ancient Khotan, pp. 374, 411) at Niya, there is a single piece of hard, green leather, shaped and perforated very much like the metal plate of an armor. Stein suggests that "it probably belonged to a scale armor" (he means plate armor), and thinks that this supposition is confirmed by the metal plates of an armor coming from Tibet (p. XVI). This is possible; I do not believe, however, that an entire suit of armor was ever made in Turkistan in this manner, but that only certain parts of an armor suit were of this technique. There would be no sense in producing a complete suit by means of such separate leather laminæ, —a very toilsome and cumbrous process; any plain hide coat would probably present a more enduring protection than such an affair. Indeed, this technique is known to us from Japan: thus a shoulder-guard believed to date from prior to 1100 (BASHFORD DEAN, Catalogue of the Loan Collection of Japanese Armor, Fig. 12 B) is made from bands of laminæ of boiled leather interlaced with rawhide. Leather laminæ, of course, do not present any original state, but are a secondary development, being the outcome of an imita-tion of metal laminæ.

[2] Altbuddhistische Kultstätten in Chinesisch-Turkistan (Berlin, 1912).

[3] L. c., p. 201, and Figs. 451, 452, 456, 460, 512, 513, 628.

[4] Chotscho, Plate 48 (Berlin, 1913).

during the march they were rolled up and carried.[1] Scale, chain, ring, and plate armor were all a great burden on the body owing to their heavy weight, and a serious obstacle to the mobility of troops. The reform is attributed to Ma Sui, who was president of the Board of War under the Emperor T'ai-tsung of the T'ang dynasty, and who died in 796.[2] He conceived the idea of combining armor with the costume (styled k'ai i, "armor clothing") in three grades differentiated according to length; and the soldiers thus clad were enabled to run, and to advance comfortably. The helmets he made in the form of lions.[3] This innovation is illustrated by an interesting passage in the Ch'u hio ki,[4] where some new names for the parts of armor are given, derived from the names of clothing. "The skirt attached to the armor is called shang (No. 9734, "the clothes in the lower parts of the body"); the inner side of an armor is styled lei (No. 6843);[5] and the coat of the armor (kia i, No. 5385) is termed kao (No. 5949)."[6] The general expression for clothing, i-shang, finds here application to armor: the upper portion of the armor is directly styled i ("upper clothing"), and the term kao used with reference to it plainly indicates that a robe made of some textile material was worn over the mail to cover it all round.

This state of affairs is confirmed by the Wan hua ku,[7] where, besides cuirasses and six kinds of iron suits, are enumerated armor made from white cotton stuff (pai pu kia), that made of black silk taffeta (tsao chüan kia), and even wooden armor (mu kia).[8]

---

[1] As expressly stated by Sun-tse (see L. GILES, Sun Tzŭ on the Art of War, p. 58, London, 1910).

[2] GILES, Biographical Dictionary, p. 569.

[3] T'ang shu, Ch. 155, p. 1 b.

[4] Compiled by Sü Kien in the early part of the eighth century (BRETSCHNEIDER, Botanicon Sinicum, pt. 1, p. 143, No. 76).

[5] COUVREUR (p. 473 c) explains this word as mailles d'une cuirasse.

[6] Ordinarily "a quiver," but originally a case to place any arms in; hence COUVREUR (p. 304 a) enveloppe de cuirasse, de bouclier, de lance (see p. 176). In the above case, the costume worn over the armor is thus called, because, like a case, it envelops the armor.

[7] See above, p. 196.

[8] Wooden armor existed perhaps under the Later Han dynasty, though alluded to only in a metaphorical sense. In the Chapter Wu hing chi (Hou Han shu), ice-crusts covering trees (mu ping) are likened to wooden armor (mu kiai); and the commentary explains kiai as symbolizing military armor (P'ei wên yün fu, Ch. 69, p. 42); thus the existence of wooden armor at that time might be presupposed as being instrumental in this comparison. "Wooden armor" can be nothing but wooden slat armor, as described by W. HOUGH (Primitive American Armor, l. c., pp. 632, 636) among the North-American Indians. Another type is presented by the wooden armor of the Thompson Indians described by JAMES TEIT (The Thompson Indians of British Columbia, Jesup North Pacific Expedition, Vol. II, p. 265) as consisting of four boards an inch and a half thick, two for the front and two for the back, which reached from the collar-bone to the hip-bone; these boards were laced together with buckskin, and the whole covered with thick elk-hide; while the same tribe made also

We do not know from the literary records how the armor credited to Ma Sui was constructed in detail; but it was doubtless the forerunner of the armor-costumes, as we find them duly sanctioned by the emperors of the Sung, Ming, and Manchu dynasties; those, in my opinion, go back to types established in the T'ang period. Ma Sui's invention was a coat of cotton or silk, the exterior or interior of which was covered with rows of small iron or steel plates. Indeed, plate mail is well represented on Chinese clay statuettes of the T'ang period, in accordance with what we find in the art of Turkistan. The nearest approach to Ma Sui's contrivance may be recognized in the clay figure of a soldier (five of these are in our collection) on Plate XXX. These figures coming from graves of Shen-si Province are clad with an ordinary long-sleeved coat; in front and back, over the chest, and along the lower edge, we notice a row of plates emerging.[1] Plates, accordingly, strengthen the front and back of the coat, and are covered with the same material as the latter consists of. The whole affair is tightly held together by two bands adorned with bosses.

The two clay figures on Plate XXXI represent two identical specimens of the same type of warrior, coming from Shen-si Province. The left hand, which is raised as if brandishing a weapon (spear), is unfortunately broken off in both pieces. The expression of lively motion and the quality of modelling are remarkable. In the grim faces slightly bent and turned sideways, the demoniacal power of these armored knights watching over the grave is well represented. The helmet-mask is formed by a bird's head with a strong flavor of the Indian Garuḍa; a horn or crest in the centre of the head is broken off. The well-developed eyebrows of the bird's faces terminate in spirals arranged on the foreheads; the beak is strongly curved; the interval between the eyes is filled with a pigment of indigo. The helmet covers the back of the head, nape and chin. A shawl is elegantly draped around the shoulders, and tied in a knot over the chest, the two round iron breastplates being visible beneath it. An animal head is brought out in relief in the middle, apparently a metal clasp holding the two sheets of the armor together.[2] An apron, a sort of undivided braconnière, consisting of three horizontal rows[3] of long, rectangular iron plates is worn over

---

corselets from narrow strips of wood from half an inch to an inch in thickness or of rods, going entirely around the body; the strips of wood were placed vertically, and laced together with bark strings; such vests were generally covered with one or two thicknesses of elk-skin.

[1] Compare Plate XVIII.

[2] Sheet armor is discussed in Chapter VI.

[3] It is interesting to compare it with the clay statuette found by GRÜNWEDEL, *l. c.*, Fig. 460.

the coat (Plate XXXI, Fig. 1); the plates are distinctly represented by parallel rows of lines executed in black ink and continued on the back (Fig. 42); the lines are somewhat rounded at the top, and leave no doubt of the real shape of these armor-plates. In Fig. 2 of the same Plate these lines are omitted, or may have been worn out.

FIG. 42.
Back of Clay Statuette represented on Plate XXXI, Fig. 1.

As those two statuettes represent the typical armed warriors of Shen-si Province, so the pair on Plate XXXII illustrates the character-istic types current in Ho-nan, and is for this reason inserted here, though not vested with plate armor. Of powerful martial appearance, "armed at point exactly, cap-a-pie," these heroes valiantly lean on the hilts of their straight swords resting between their feet,— not dissimilar to a mediæval Roland.   They are protected by iron sheet armor,[1] over which a jerkin is thrown, two circular spaces being cut out on the thorax, and exposing the iron plastrons or breastplates.   The helmet envelops the occiput, nape, and cheeks, and is held by a broad leather mentonnière.   The baggy trousers are fastened with garters over the upper parts of the thighs.   Many of these figurines, as indicated by the remains of pigments, must originally have been well painted, the pigments being spread on a background formed by a thick coating of white pipe-clay.[2]   In the two figures in question, judging from the traces of pigments, the helmet was colored a crimson-red, the face pink, the eyeballs black, and likewise the big mustache with turned-up tips; the breastplates were vermilion, and the garment surrounding them light green.   The sleeves on the upper arms are still decorated with parallel black stripes; those on the lower arms are painted a crimson color, the hands pink.   Geometric ornaments that are but partially preserved were painted in red on the portion of the coat beneath the girdle.

Plate armor is met also on contemporaneous Chinese sculpture in stone.   There is in the Museum's collection a marble slab dug up in the environment of the city of Hien-yang, Shen-si Province (Plate XXXIII).   It represents a mock-gate which denoted the entrance to a tomb.   The two door-leaves countersunk in the slab are divided by a faint line in the centre, and kept closed by means of a bolt carved in relief.   On each leaf is delicately traced the figure of a guardian completely armored with plate mail, and holding a sword.   On the lintel two phenixes surrounded by rich foliage are chiselled out in flat relief.

Plate armor was officially adopted by the Sung dynasty.   In 1134, the Imperial Armory had four model pieces constructed, which were founded on the principle of the plate.   The first of these, an armor suit, consisted of 1825 plates (styled *ye*, "leaves," written without the classifier 'metal') polished and burnished on both sides; the épaulières (pauldrons) were protected on the inner side by 504 plates; each of these plates weighed one fifth of an ounce plus six *fên*.   The second, also a coat, was formed of 332 plates, each plate of the weight of two-fifths of an ounce plus seven *fên*.   The third piece, a lower garment, was composed of 679 plates of the shape of a tail-feather of a hawk, each plate weighing two-fifths of an ounce plus five *fên*.   The fourth piece was a helmet consisting of 310 plates, each weighing one-fifth of an ounce plus five *fên;* the total weight of the helmet, inclusive of its appurte-

[1] See Chapter VI.
[2] The same process is applied to T'ang pottery vessels, as will be seen in Part II.

nances sheltering the nape and the forehead, amounted to one catty and one ounce. The leather straps wound around the head weighed five catties, twelve ounces and a trifle more than a half. Each suit had a weight of forty-nine catties and twelve ounces. The weight of an armor naturally depends upon the weight of the individual wearer; in the army, however, concern about the individual would not be feasible, and would incur heavy expense as well as waste of material. It was therefore thought advisable to reach a compromise, and to standardize the weight of the armor at from forty-five to fifty catties, with the strict understanding that in no case should it exceed fifty catties.[1]

In regard to the Mongols, we mentioned the employment of hide and hide scale armor in their armies (pp. 190, 197). There are also accounts to the effect that plate mail was known to them. In the earliest European document regarding the Mongols, written by MATTHEW PARIS under date of 1240, giving the first description of this new people, they are described as "men dressed in ox-hides, armed with plates of iron, . . . their backs unprotected, their breasts covered with armor;" their backs remained unprotected so that they could not flee.[2] WILLIAM OF RUBRUCK, travelling from 1253 to 1255, makes us acquainted with sundry types of armor in use among the Mongols, — the haubergeon (chain mail), scale hide armor, and iron plate armor, the iron plates being introduced from Persia.[3] But the Franciscan Friar John of Pian de Carpine (or Latinized, Plano Carpini), who travelled to the Court of Kuyuk Khan (1245–47) as ambassador of Pope Innocent IV, is that mediæval writer who has left to us the clearest and most complete description of Mongol plate armor. At the same time he is the first European author to give any description of Eastern plate armor at all. In his "Libellus historicus" (Cap. XVII) [4] he describes the defensive armor of the Mongols, and states that the upper part of their helmet is of iron or steel, while the portion guarding the neck and throat is of leather. Whereas the majority wear leather armor, some have their harness completely wrought from iron, which is made in the following manner. They beat out in large numbers thin iron laminæ a finger broad and a full hand long. In each they bore eight small apertures, through which they pull three straight leather thongs. Thereupon they arrange these laminæ or plates one above another, as

---

[1] See *Sung shi*, Ch. 197, p. 6.

[2] W. W. ROCKHILL, The Journey of William of Rubruck, p. xv (London, 1900).

[3] *Ibid.*, p. 261. He mentions also iron caps from Persia.

[4] In the new edition of G. PULLÉ, pp. 86–88 (*Studi italiani di filologia indo-iranica*, Vol. IX, Firenze, 1913). C. R. BEAZLEY, The Texts and Versions of John de Plano Carpini, pp. 89, 124 (London, 1903, *Hakluyt Society*).

it were, ascending by degrees, and tie the plates to the thongs mentioned by means of other small and tender thongs drawn through the apertures. And in the upper part they fasten a single, small thong, doubled on each side, and sew it on to another, that the plates may be well and tightly connected. Thus a uniform protection is effected by these plates, and such-like armor is made for their horses as well as for their men. It is so highly polished that a man may mirror his face in it. In regard to shields, Carpini observes that they have them made of wickerware or small rods (*de viminibus vel de virgulis factum*), but that they carry them only in camp and when on guard over the emperor and the princes, and then only at night. The armament of the Mongols was not uniform; and this complex and expensive structure of plate armor was probably within the reach of but few. Their ordinary armor was a cuirass of boiled-leather scales. According to Carpini, the leather was that obtained from an ox or some other animal; and the scales were a hand broad.[1] Three or four of these were held together by means of pitch, and connected with one another by means of cords. In double or triple rows they were laid around the trunk. The complete set of armature consisted of four parts,—the front piece, reaching from the neck down to the lower part of the thighs, and well adapted to the form of the body; the back protector, and an apron encompassing the back and abdomen; and the brassards and cuishes. The back of the upper arm was guarded by two iron plates hinged together.

The plate idea has remained the basic principle of the officially recognized body armor down to the end of the eighteenth century. The changes were those of style and ornamentation only, while no fundamental innovations were added in the Ming and Manchu periods. The Statutes of the Ming Dynasty (*Ta Ming hui tien*) contain the following regulations relative to plate armor: "In 1374 it was ordered that instead of the threads, by means of which the armor-plates were held together, leather thongs should be used. In 1376 the General Staff was ordered to make war-suits of cotton (*mien hua chan i*), and to apply to them four colors, — red, purple, dark blue, and yellow; for Kiang-si and other places, to make war-coats with different colors on the exterior and interior, and to cause the officers and petty officers to change their uniforms accordingly. In 1383 orders were given for harness, each set to be made as follows: for the colletin (neck-guard) thirty plates, for the body armor two hundred and nine plates, for the plastron (breastplate) seventeen plates, for the pauldron (épaulière)[2] twenty plates.

---

[1] PULLÉ's complete text is followed here; this portion is lacking in the former editions of Carpini.

[2] In Chinese, "arm-pit plates" (*chi wo ye*).

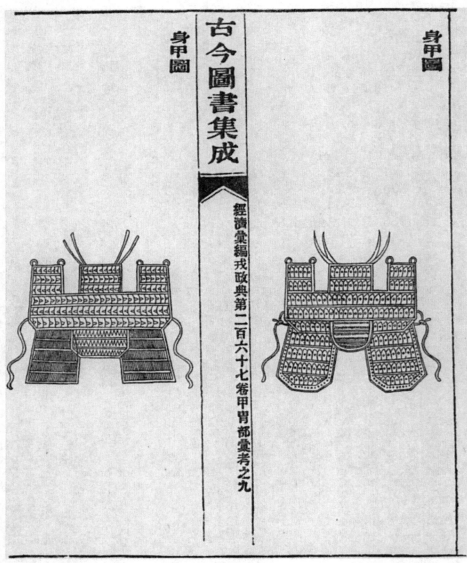

FIG. 43.
Illustrations showing the Conventional Chinese Style of Drawing Plate Mail (from *T'u shu tsi ch'êng*).

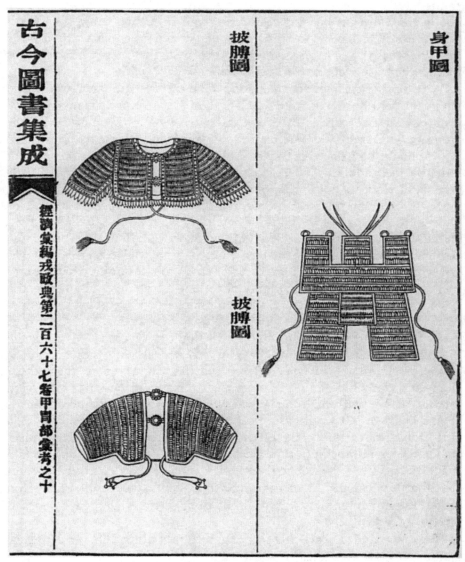

FIG. 44.
Illustrations showing the Conventional Chinese Style of Drawing Plate Mail (from *T'u shu tsi ch'êng*).

All these pieces are soaked with lime, and united by means of soft, tanned leather thongs passing through the perforations of the plates. Along the maritime coast of Chê-kiang and in Kuang-tung, the guards stationed there have to utilize black-lacquered iron plates perforated and connected by cotton strings; for the rest, however, their armor is made in the style of the 'brilliant armor' (*ming kia*).[1]

" In 1435 (tenth year of the period Süan-tê) the ordinance was issued that each coat had to be fixed at a length of four feet and six inches, with a supply of two catties of cotton and velvet; for the making of the trousers, half a catty of cotton and velvet should be used; the wadded boots should be from nine inches and a half up to one foot, or one foot and two inches long. Now, the regulation was provided to make wide coats and trousers, and to employ for these fine, closely woven, broad, and white cotton stuff dyed blue, red, or green; the sleeves should be wide and long; and the materials employed, like cotton and velvet, should be of solid quality. The wadded boots should be fine, thick, and strong. In the finished garment a written entry was to be made by the government officers who inspect the troops and examine their equipments; they shall enter the family name and surname of the tailor, the cost-price, the measurements in feet and inches, the weight, the number of strips of cloth used in the skirts, with seal attached. At fixed terms, every year before the seventh month, the uniforms were to be furnished.

"In the year 1496 (ninth year of the period Hung-chi under the Emperor Hiao-tsung) it was ordered that for the covers of the armor[2] thick and dark blue and white cotton stuff should be employed, that for the 'armor with nails' (*ting kia*) small studs with lacquered heads should be used. It was further settled that, for each set of a blue cotton stuff iron armor, iron to the quantity of forty catties and eight ounces should be required, and that each set of the finished armor should weigh twenty-four to twenty-five catties. In 1503 order was given that the guards stationed in southern China should exchange their iron armor for that made of water-buffalo skin sewed together by means of cotton ropes."

Figs. 43 and 44 are here inserted to illustrate the conventional Chinese style of representing plate mail.[3]

The Manchu dynasty adopted the military institutions of the Ming in their entire range, and in particular the defensive armor, without making any new additions in the line. Plate XXXIV illustrates a

---

[1] A technical term frequently employed in the Annals; it presumably refers to highly varnished and polished plates of iron or steel.

[2] In Chinese, "the face of the armor" (*kia mien*).

[3] Compare note 4 on p. 243.

horseman's suit of armor, as it was in vogue during the K'ien-lung period (1736–1795). It is complete with leggings and helmet. The lower garment is covered by four parallel rows of very thin, light and elastic steel laminæ of rectangular shape, 9 cm long and 1 cm wide, rounded at the upper end, perforated at both apex and base, and sewed on to a foundation of cloth, the lower ends being hidden in a fold, where they are riveted by means of studs with broad, gold-plated heads. They are not mutually joined, but one overlaps another to a slight degree. In the upper garment the steel plates are invisible, being inserted as an interlining (between the lining and the silk on the exterior), and fastened by means of rivets, so that their gilt heads appearing on the surface indicate the hiding-places of the plates.[1] Dragons, all together six, rising from the sea and standing erect, are embroidered with gold threads on the front and back of the coat, on the two separate shoulder-pieces, and on the two side-pieces underneath the arms. The casque, composed of two steel sheets and surmounted by a black velvet plume, has chased dragons in front, and is provided with silk protectors enveloping occiput, neck, ears, and chin.

The uniform of an artillery-man (Plate XXXV) consists of a coat, lower garment, and pair of leggings of wadded black satin lined with light-blue silk, and studded with gold-plated, riveted bosses. These bosses, of a merely decorative character, are the survivals of the iron or steel plates which, as in the preceding harness, are wrapped up in the interior of the garment or are fastened to the lining. The plates are retained in this specimen only for the protection of the shoulders, but have a decorative rather than a positive value. They are arranged in rows of three, two rows being in front and two at the back on each

---

[1] It is singular that the students of plate armor have never turned their attention to China, although it was very clearly described as early as by GERBILLON (in DU HALDE, Description of the Empire of China, Vol. II, p. 340, London, 1741): "All the soldiers who were in the camp, headed by their officers, repaired to the place appointed, armed with their casques and cuirasses. The Emperor put on likewise his cuirass and helmet, being accompanied with his eldest and third sons; but this latter was not armed, being too young to bear the weight of a Tartarian cuirass. This cuirass consists of two pieces; one is a sort of under petticoat which is girt about the body, and reaches below the knee when they are standing, but covers all their limbs when they are on horseback: the other piece is like the coats of armor of the ancients, but the sleeves are longer, reaching to the wrist. The outside of both these pieces is of satin, for the most part purple, embroidered with gold, silver, and silk of various colors. Next to this satin, lined with some pieces of taffety, are hammered plates of iron or steel, finely burnished, which are placed like scales on the body of a fish, whence they probably took the notion. Each plate, which is about an inch and half long, and a little more than an inch in breadth, is fastened to the satin by two small nails, the heads, being round and well polished, appearing without. Some few put another piece of taffety within-side, which covers the iron plates. These cuirasses have this conveniency that they do not deprive the body of the liberty of turning and moving easily; but then they are exceeding heavy."

shoulder, and connected by a broad, quadrangular plate resting on the shoulder.   Each lamina is of steel and gold-plated, and chased with a four-clawed dragon soaring in clouds.   From the lower ends of the plate rows project two gold-plated arms, likewise chased with figures of dragons and encircling a round metal plaque (of white copper or tootnague, with brass rim).   A plaque of the same material and size is fastened to the back.   Such circular plaques are known as *hu sin king* (No. 2170), literally "mirror guarding the heart;" that is, a protective amulet.   The helmet is identical with the one previously mentioned, and heavily lined with quilted material.

The archer's suit of armor (Plates XXXVI, XXXVII) is made of black silk, the interior being covered with broad steel plates, each secured by means of two rivets only, so that the plates are loose and movable. Their disposition on the shoulders is at variance with that in the preceding specimen.   There is but one row of three brass plates in front of each shoulder, extending in length as much as the two rows in the previous armor.   There are three narrow plates arranged side by side on the surface of the shoulders, and three on the back much shorter than those in front.   The three rows covering either shoulder are interlaced and riveted together.   Each of these shoulder-plates is decorated with two rampant dragons playing with a flamed ball.   The coat is embroidered with six dragons all together.

In 1901 I saw a very interesting and ancient suit of plate mail in the Mahākāla Temple, which is situated within the walls of the Imperial City of Peking.   The suit is of yellow silk, to which iron plates are attached both outside and inside, — those on the exterior being very narrow slips, those on the interior being four times broader and occupying the interval left by the outside plates; so that by this alternating process a complete plating is insured.

On Plates XXXVIII–XL is represented what may be styled a parade or ceremonial armor.   It is the uniform belonging to a guard-officer of the first rank, detailed on duty in the Imperial Palace.[1]   These military officers were divided into seven ranks, each distinguished by a special coat and helmet, and an equipment with appropriate insignia.   Their outfits are minutely described in the State Handbook of the Manchu Dynasty.   The cut, the style, and the main characteristics of body armor are well preserved in this costume, which is magnificently embroidered with heavy gold thread, and studded with gilt bosses.   Dragons', tigers', and lions' heads are the prevailing motives of ornamentation.   The disposition of the shoulder-plates is identical with that

---

[1] This is ascertained from the descriptions and illustrations of the official costumes given in *Huang ch'ao li k'i t'u shi* and *Ta Ts'ing hui tien t'u.*

in the suit of the artillery-man, except that the dragons are here embossed, and the clouds are treated in open-work, all metal pieces being heavily gilded. Five similar plates are suspended from the ends of the shoulder-pieces.

The steel helmet (Plate XXXIX) is a gaudy and elaborate affair of admirable workmanship. It is surmounted by a high crest terminating in a pair of eagle-feathers painted with dragons in gold, and is adorned with twelve black sable-tails,[1] seven of which are preserved. Dragons are lavished on it, being chased in the plated brass mountings, or cut out of the same material in full figure, or represented in inlaid feather-work.[2]

The bow-case and quiver figured on Plate XL belong to the accoutrements of the same official. They are of leather, dressed with red velvet; the upper corners and lower portion of the bow-case are finished with black leather. The metal fittings, of gilt bronze, fastened to the centre and corners of both objects, are of very elegant forms and delicate workmanship. The quiver, in addition to these ornaments, is decorated with three symbols meaning "longevity" (shou). The arrows are stuck into the folds in the interior formed by layers of brown felt.

Reference has been made above (p. 272) to the early mining of iron in Korea, and the barter carried on in this metal from there to the neighboring tribes. Metal armor (k'ai kia) seems to have prevailed in the kingdom of Kokurye (Kao-kü-li) at an early date.[3] The Annals of the Sui Dynasty[4] state in regard to the kingdom of Sinra in Korea that its defensive and offensive armor is identical with that of China, which would mean that Sinra had derived its armor from China. The Books of the T'ang Dynasty mention a kind of armor, seemingly peculiar to the state of Pek-tsi in Korea, under the name "armor of bright lustre" (kuang ming k'ai), which must have been iron armor. Such a suit was presented in 622 to the Emperor of China, and in 637 iron armor (t'ie kia), together with carved axes, was sent as tribute to the Emperor T'ai-tsung.[5] Metal armor is alluded to likewise in the Annals of Korea.[6] When the Japanese plundered the royal palace of Kokurye, in 562,

---

[1] This is the required number according to the official statement.

[2] From the blue plumes of the kingfisher, Halcyon smyrnensis (in Chinese, fei-ts'ui).

[3] Liang shu, Ch. 54, p. 9 b; Nan shi, Ch. 79, p. 1 b.

[4] Sui shu, Ch. 81, p. 4 (also Pei shi, Ch. 94, p. 7).

[5] T'ang shu, Ch. 220, pp. 4, 7.

[6] See, for instance, Ta tung ki nien (published at Shanghai, 1903), Ch. 1, p. 69 b. The Koreans possess a considerable literature on military art (M. COURANT, Bibliographie coréenne, Vol. III, pp. 63–89).

they obtained among other treasures two suits of armor.[1] We have no exact information as to what these ancient suits of armor were like, and can base our conclusions only on such specimens as we find in the country at present. Among these are some of considerable age; that is, ranging within the time of the last two centuries or so. We have two main types of harness from Korea, — padded armor[2] and plate mail.

A very interesting specimen of the latter type is in the Museum collection (Plates XLI, XLII). It is a rough-looking coat of strong twill, lined with blue cotton, and covered with hemp cloth of loose texture imprinted all over with charms by means of wooden blocks.[3] The designs are effaced to such a degree that the details can no longer be recognized: birds' heads, floral designs, trees, arabesques, are conspicuous; Sanskrit letters, which occur in other specimens, are absent. The buttons in front are of bone; the sides are open, and provided with rows of buttons. Both front and back are strengthened by seven parallel rows of rectangular steel plates (averaging 10.2×7.5 cm), very flexible, each coated on both faces with a black varnish. The plates are not mutually connected, but merely imbricated, — a feature not yet observed in Chinese plate mail. Each plate is clinched to the cloth foundation by means of two rivets with flat heads. They are driven through, and appear on the exterior as big iron nail-heads. A number of plates have additional perforations that are not utilized, but which show that the plates could have been tightly sewed on to the background had not the wearer of this armor preferred to have them loose and movable. The shoulders are covered on the interior by two rows of

---

[1] ASTON, Nihongi, Vol. II, p. 86.

[2] A Korean armor consisting of many thicknesses of coarse cotton cloth is figured by W. HOUGH (The Corean Collections in the U. S. National Museum, *Report U. S. Nat. Mus.*, 1891, Plate XXVIII, and Primitive American Armor, *l. c.*, p. 645); the surface of portions of the coat is printed with prayer formulas (*dhāraṇī*) in Sanskrit, and such are inscribed also on the helmet. This practice seems to be derived from China: the helmets used by the imperial house during the Manchu dynasty were chased with Sanskrit characters (see *Huang ch'ao li k'i t'u shi*, Ch. 13, or *Ta Ts'ing hui tien t'u*, Ch. 61). A modern Korean helmet is illustrated by E. ZIMMERMANN (Koreanische Kunst, Hamburg, no date, Plate VI). It is a leather helmet of conical shape, surmounted by a bunch of horse-hair and a metal ball in open-work, and adorned with dragons and a hydra about to attack, wrought in gilt metal; fur-lined ear-warmers covered with metal studs are attached to it, the whole style being that of the Manchu dynasty. The costume on Plate VII, explained as the official robe of a minister, is in fact a pseudo-armor, as shown by the rows of metal bosses and the two appliqué dragons playing with balls; it is similar to the one on our Plate XLIII. Generals' and soldiers' helmets are figured and briefly described by F. H. JENINGS (Korean Headdresses in the National Museum, *Smithsonian Miscellaneous Collections*, Vol. 45, 1904, pp. 161–163). Good specimens of these are also in the Field Museum.

[3] Much in the style of Tibetan cloth prints which are attached to flag-poles set up on the roofs of houses in order to bring luck to the inmates.

plates, and are reinforced on the outside by iron bars, one for either shoulder, each bar consisting of two parts hinged together, so that easy motion is secured. The casque (Plate XLII) consists of two sheets of iron riveted together, with a projecting visor and frontal covering the forehead. The couvre-nuque and the ear-protectors attached to the casque are made from the same hemp cloth as the harness; they are likewise printed with designs, and stuffed with iron plates, which are kept in place by means of the clinches appearing on the surface. The top of the helmet is surmounted by an iron trident and a tuft of red-dyed horse-hair. There is no doubt that this Korean armor represents a very primitive type of plate mail, and conveys to us an excellent idea of what the ancient Chinese plate mail may have been like.[1]

On Plate XLIII is shown the Korean court costume of a high official, which is a pseudo-armor in imitation of Chinese style. The cloak-like robe consists of red cloth trimmed with otter-fur, and lined with light-blue Chinese silk. It is strewn with regular rows of brass bosses representing purely decorative survivals or reminiscences of plate armor. Three globular buttons close the garment in front; the two lower ones are hidden under a broad sash of figured blue silk. Around the neck are laid twelve maple-leaves cut out of brass and riveted to the cloth (in the illustration hidden by the ear-protectors of the helmet). The epaulets are adorned with full figures of gilt, embossed dragons hunting for the flamed jewel; they are worked in sections, which are cleverly connected by hinges, so that the shoulders are not handicapped in any motion. The helmet is an elaborate affair, composed of strong, com-pressed, glazed leather, lined with soft leather. The surface is divided by means of four metal bars into four compartments, two of which are each adorned with a dragon, the two others each with a phenix on the wing, — all of gilt bronze. On the sides, silver phenixes filled with dark-blue enamel[2] are added. The most interesting point concerning our subject is the fact that the ear-muffs and nape-guard, likewise of red cloth trimmed with otter-fur, have thin copper plates concealed between the outside material and the lining. They are kept in place by copper nails with gilt heads. A quilted cap of blue silk is worn next to the skull,

---

[1] W. E. GRIFFIS (Corea, the Hermit Kingdom, p. 101) figures what he calls "a Korean knigl t of the sixteenth century." I have no judgment on the authenticity and alleged dating of this illustration, but in itself it is interesting in that the laminæ forming the plastron and reinforcing the sleeves and brassards are arranged in hori-zontal (not, as usual, vertical) position. "Many of their suits of armor," GRIFFIS says, "were handsomely inlaid, made of iron and leather, but less flexible and more vulnerable than those of the Japanese, which were of interlaced silk and steel on a background of tough buckskin, with sleeves of chain mail. The foot-soldiers on either side were incased in a combination of iron chain and plate armor."

[2] A process still extensively applied in China to silver jewelry.

under this helmet.   Below, on the same Plate, is illustrated the black-varnished wooden chest in which the suit is stored, with a special conical compartment for the casque.   This arrangement is also in imitation of a Chinese practice.   Japanese plate armor has so frequently been described[1] that it is not necessary to dwell on this subject.   What is important for the purpose of our investigation is the fact that it does not arise in Japan earlier than the first part of the ninth century; [2] that is, in the T'ang period, when it was perfectly known in China.   It is therefore certain that the idea has penetrated into Japan from China and Korea, whatever subsequent developments, changes, and improvements plate mail may have undergone in Japan.

Armor composed of horizontal rows of small iron plates, presumably of Chinese origin, seems to occur occasionally in Tibet.   A specimen recently presented by the Dalai Lama to the King of England is now preserved in the British Museum.[3]

Looking backward at the remarkably wide distribution of plate armor, we cannot fail to recognize in this fact a certain degree of historical coherence.   This coherence, without any doubt, exists in the T'ang period between Turkistan and China on the one hand, and between China, Korea, and Japan on the other hand.   But the T'ang epoch denotes only the culminating point in this development, — that period in which we observe plate mail wrought to its greatest perfection.   Metal plate mail is a complex affair of difficult and refined technique, a downright product of higher civilization, which is witnessed by the fact that it is conspicuously absent among all primitive cultures of Asia, Africa, and ancient Europe.   Certainly it did not come into existence all at once as a finished product of industry.   It ran through many experimental stages, and took time to develop and to mature.   The elegant specimens of the T'ang, granting the muscles free motion and aiming at æsthetic qualities, were preceded by those of coarser and cruder workmanship; as we see, for instance, in the Korean specimen on Plates XLI and XLII.   There is a great deal of probability in the supposition that such existed, both in China and among the Iranian and Turkish tribes of

---

[1] First by PH. H. V. SIEBOLD, Nippon, Vol. I, p. 333.

[2] J. CONDER, The History of Japanese Costume (*Transactions Asiatic Society of Japan*, Vol. IX, 1881, p. 256).   According to this author, the employment of plates and scales of iron in armor was finally established as late as the epoch Tenshō (1573–1592).   See chiefly BASHFORD DEAN, Catalogue of the Loan Collection of Japanese Armor.

[3] It is figured on Plate III of the Ethnographical Guide published by the British Museum.   See also A. STEIN, Ancient Khotan, Vol. I, p. XVI.   Armor of small steel plates riveted on red velvet appears also in Europe (see, for instance, BASHFORD DEAN, Catalogue of European Arms, p. 48), but this subject is not within the scope of the present investigation.

Central Asia, ages before the T'ang, presumably as early as the era of the Han (p. 214). Iranians surely were the mediators between the west and the east in this matter, in the same manner as they acted in the transmission of chain mail, caparisons for horses,[1] and the great principles of cavalry tactics. Up to this point the territory is fairly well reconnoitred. But thus far we are entirely ignorant of when and how plate mail may have arisen in Irān, nor do we positively know whether it existed there at all; if it did, the possible connection with the plate mail of ancient Egypt and Assyria remains a subject for investigation. Altogether the impression remains that plate armor, the last offshoots of which we encounter in the farthest north-east corner of Asia and the farthest north-west of America, took its origin from western Asia. This field is entirely beyond my competency; and it is the sole object of these notes to point out the existence of the problem, and to leave its final solution to the ambition of others.

---

[1] See Chapter VII.

VI. DEFENSIVE ARMOR OF THE T'ANG PERIOD

In the preceding notes we had occasion to refer repeatedly to defensive armor of the T'ang period (618–906). Mention has been made of the fact that cuirasses of rhinoceros-hide were then still in existence (p. 189), and also that those of buffalo-hide then came into vogue (p. 162). Plate mail reached its climax at that time (p. 277), and chain mail was introduced from Iranian regions (p. 246). The types of armor utilized under the T'ang must have been of a large variety. The Statutes of the T'ang Dynasty, drawn up by the Emperor Yüan-tsung (713–755),[1] enumerate thirteen classes of armor manufactured by the Imperial Armory (*wu k'u*): six of these were of iron, and of the types of plate, scale, and chain armor; others were of white stuffs, black silk, hide, and even of wood (p. 276). How the military uniform was then combined with armor has also been set forth (p. 275). Besides the means of protection officially recognized in the army, there were other plain and cheap contrivances for the use of the people, such as are still common in the country. Thus we hear in the Annals of the T'ang Dynasty in regard to a certain Ch'êng K'ien that he made defensive armor from layers of felt.[2] The most curious armor of which we read in that period was a kind made from sheets of paper laid in folds, which could not be pierced by the strongest arrows; this invention is credited to Shang Sui-ting.

Under the Sung dynasty, paper armor was officially recognized, for we hear that in the year 1040 the troops stationed in Kiang-nan and Huai-nan (in An-hui Province) were ordered to fabricate thirty thousand suits of paper armor, to be distributed among the garrisons of Shen-si Province. The localities mentioned are celebrated for their paper manufacture, and were accordingly obliged to contribute to a demand which could not be filled in Shen-si. The *Wu pei chi* (Ch. 105, p. 17) of 1621 has preserved for us an illustration of such paper armor (Fig. 45), arranged in triangular scales slightly rounded at the base. These suits were especially favored under the Ming in southern China by the soldiers fighting the Japanese, who then invaded the Chinese coasts.[3] The favorite brand of paper for this purpose in recent times

---

[1] See above, p. 189.

[2] *P'ei wên yün fu*, Ch. 40, p. 86. In 1286, according to *Yüan shi*, the country of Ma-fa sent a tribute of saddles, bridles, and felt armor.

[3] The same work illustrates also armor of plaited rattan; but it is not known at what time this type of armor sprang up in China.

292

was the famed Korean paper highly prized in China and Japan for its toughness and durability, and forming part of the annual tribute sent from Korea to Peking. In the treaty of 1637, concluded after the Manchu invasion, the figure was stipulated at five thousand rolls of large and small paper.[1] A good deal of Korean paper was utilized by

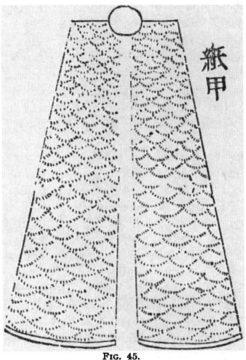

FIG. 45.
Paper Armor (from *Wu pei chi*).

the tailors of the Chinese metropolis as lining for the coats of officials and gentlemen. It served also for the covering of window-frames. A sewed wad of from ten to fifteen thicknesses of it made a protective armor for the troops. It is said to have resisted a musket-ball, but not a rifle-bullet.[2]

[1] W. W. ROCKHILL, China's Intercourse with Korea, p. 25 (London, 1905). A notice on Korean paper is contained in the *Wei lio* (Ch. 12, p. 1 b).

[2] W. E. GRIFFIS, Corea, the Hermit Nation, p. 153 (New York, 1904). Paper and cotton armor still exist in southern China. Consul BEDLOE (quoted above, p. 180) offers the following remarks on this subject: "Parallel to this alternating of leather and wool in the north was that of paper and cotton cloth in the south of China. It seems ridiculous to call such combinations armor, and yet they make an armor superior in many instances to steel. Thirty thicknesses of alternate calico and paper will resist a pistol bullet or one from a rifle at a distance of a hundred yards. A spearman who thrusts his weapon into a man clad in this kind of garment

The most interesting source for the study of T'ang defensive armor is naturally offered by the clay figures and figurines; and these reveal to us a new style of armor, that of sheet armor, which is thoroughly characteristic, not of the life, but of the art, of this period.

The type of clay image which comes here into question is of the greatest interest, as it originated in the Çivaitic worship of India, and became widely diffused over Tibet, Turkistan, China, and Japan. We may in general classify the manifold variations of this type among the so-called Dharmapāla ("Protectors of Religion"), guardian deities adopted by Buddhism, and more specifically designate it as Yama, the God of Death, who still plays such a prominent rôle in Tibetan Lamaism. J. EDKINS[1] holds that he may be pointed to as the most remarkable example of the influence of Hindu mythology on the popular mind of China.

Among the clay figures of the T'ang period we find two fundamental types of this Hindu god, — a zoömorphic and an anthropomorphic form. The zoömorphic form is doubtless the older one, and is closely associated with the Lamaist representation of Yama as Dharmarāja ("King of the Law"), figured with the head of a bull, and dancing on the back of this animal.[2] Old ZIEGENBALG, who wrote in 1713 at Tranquebar on the coast of Coromandel, gives the following description of his image as found in southern India: "Yama is represented as being quite black, with a horrible face, and a crown on his head, and altogether surrounded by fire. In his mouth he has a lion's teeth, and in his four hands he holds respectively a club, ropes, a trident, and a wine-jug, from which he gives wine to the dying to mitigate the bitterness of death. On the whole he is adorned like the king, and rides on a black buffalo. The poets have written many stories about him, which these heathens receive with undoubted credence."[3]

On Plate XLIV we see him modelled in clay, with most powerful

---

can neither wound his enemy nor extract his weapon, and if his enemy is an archer or is armed with a long sword or javelin, he is likely to lose his life for his mischance. The suit of a famous Yūn-nan bandit consisted of sixty thicknesses of cotton cloth and paper, and made him practically invulnerable. These suits are comparatively light, are very durable, and of course, extremely cheap." Heavy quilted cotton armors are still occasionally worn by Chinese in this country under their garments, when the members of secret societies are on the war-path. The writer was once shown a wonderful specimen in the Police Department of New York, which weighed so heavily upon the unfortunate Chinaman that he was unable to run, and was easily captured after a shooting-affair.

[1] Chinese Buddhism, p. 219 (London, 1893).

[2] PANDER and GRÜNWEDEL, Pantheon des Tschangtscha Hutuktu, p. 62; GRÜNWEDEL, Mythologie des Buddhismus, pp. 62, 168, 174.

[3] B. ZIEGENBALG, Genealogy of the South-Indian Gods (translated into English by G. J. METZGER), p. 192 (Madras, 1869).

expression and lively motion, standing on the body of a sow.[1] The animal is represented in the agony of death, with wide-open muzzle and with its facial muscles distorted, stretching forth its four feet. The terrific god has the head of a bull, exactly as in the corresponding Tibetan images, — with two curved horns, bushy eyebrows, and protruding eyeballs painted black; his mouth is wide agape, and shows the esophagus. Palate and face are coated with a red pigment. Hands and feet are provided with sharp eagle-claws. The head is surrounded by flames.[2] A projecting crest is attached to the spine, and there is a tail at the end of it.

Another representation (Plate XLV), likewise with horned bull-head, shows him in the same posture, standing over the back of a reclining bull, a snake winding around his left arm. In another clay figure (Plate XLVI) he is clad with a leopard-skin, and standing in the same attitude as the two preceding ones, but without a bull; the bearded face, though of human traits, bears a grim, demoniacal expression, and is painted red, beard and mustache being in black outlines. The erect ears are animal-like, as are the hands and feet; the head is surmounted by a long, slightly twisted horn, somewhat similar to that on the clay figures of sphinxes.

Between the animal and the human types, there is an intermediary form with some features borrowed from both. In Fig. 1 of Plate XLVII, his head is still modelled in the style of the bull-faced Yama, with horns and flames, but he is equipped with an armor in the same manner as the human forms; and the plume surmounting his head-dress is identical with the one in the figures of knights (Fig. 2 of the same Plate). The statuette on Plate XLVIII, belonging to the same intermediary type, displays all these features brought out still more clearly, — the two-horned bull-like head with a certain assimilation to human traits, the high plume and pommels of the elaborate head-dress, animal-heads protruding from the sleeves, breastplates, an apron, and a skirt consisting of two flaps; thus he is standing over the figure of a demon.[3] A demon of exactly the same type is modelled in the glazed statuette on Plate XLIX. The god, however, is here represented as a purely human form, a knight clothed with heavy armor, pressing his right hand on his hip, and raising his left. The figure, except the head, is coated with

---

[1] Why in this particular case a sow, and not as usual a cow, is represented, I do not know. The interpretation itself is indubitable, the animal being modelled in a most naturalistic style and thoroughly characterized by the anatomy of the head and the crest on the skull and spine.

[2] The tips of two of them are broken off.

[3] Compare in Indian art Kubera standing on a Yaksha (GRÜNWEDEL, Buddhist Art in India, p. 40; and Mythologie des Buddhismus, p. 15).

soft lead glazes in four colors,— green, blue, brown, and yellowish white; the demon is glazed yellowish white with brown hair. The plastron of the knight's armor is blue, the circular portions are white, the knobs in the centre are blue.[1]

Besides the god in the garb of a knight trampling down a demon, we meet again a similar type of knight standing on the back of a reclining bull (Plate L).[2] The positions of feet and hands are quite stereotyped. The right foot is set on the head of the bull, the left on its croup; the left arm is akimbo, and the right hand is raised as if throwing a weapon (Plates LI and LIII, Fig. 1). Or, the left foot rests on the bull's head, the right on its croup, while the left arm is akimbo, and the right hand raised for attack (Plates LIII, Fig. 2, and LIV). It will be noticed how the conventionalization of this type gradually advances. Somewhat more artistic features adhere to the statuette on Plate LII, which, with the exception of the head, is glazed in three colors,— green, brown, and yellowish white; the bull is lost, and may be supplemented from the preceding figure in Plate LI.[3] The bull, as previously pointed out, alternates with the demon (Plate LIII, Fig. 2). In Plate LIV, Fig. 2, a human body is plainly fashioned; so that in this case we have the same motive as in the Lamaist images, in which a human corpse serves as basis for certain Tantrik deities.

The flat miniature figure on Plate LV is very curious, in that it is cast from lead; it shows Yama in the same pose as the preceding ones, and standing on a bull. Finally we see the ultimate stage of develop-

---

[1] The method of glazing in the T'ang figures is very interesting: the idea underlying the application of glazes, if more than one glaze is enlisted, seems to centre upon the tendency of reproducing the colors of costume or armor. In the majority of cases, probably in all human figures, it is only the costume which receives the colored glaze, while head and hands remain uncoated. In the figurines of women it is sometimes merely the central portion which is glazed, the dresses usually being of green and brownish-yellow tinges, while the remaining portion is covered with a white plaster. In the case of monochromes, the glazing as a rule extends to the whole figure.

[2] A curious analogy to this type is offered in European mediæval art by the brasses of English lords in full armor standing on the back of a lion or another animal, and by the monument of Count Otto IV of Henneberg, and other German statues (for illustrations see, for example, BASHFORD DEAN, Catalogue of European Arms and Armor, Figs. 17–22; or Encyclopædia Britannica, Vol. I, p. 587).

[3] A type similar to this one is figured on Plate XIV of the Catalogue of Early Chinese Pottery, published by the Burlington Fine Arts Club (London, 1911), except that in this figure both feet are straight on the same plane. The modelling of the head, the position of the left arm, the armor, and the style and colors of the glazing, are identical in both figures. The pose of the right arm, however, must have been different in our figure, in accordance with the drawn-up right foot; it doubtless has to be supplemented correspondingly with the left arm in the figure on Plate XLIX; that is, the arm was raised, and the hand either formed into a clenched fist, or the palm stretched outward. Also in the specimen referred to, which is in the possession of Mr. G. Eumorfopoulos of London, the face and hands are unglazed, while the remainder is glazed in cream, orange-yellow, and green colors.

ment in clay figures without the mythological attributes of the bull or demon; these are purely armored knights or guardians. But the derivation of this type is unmistakable. The demoniacal expression in the face of the tall figure (Plate LVI)— the mouth is agape, as if he were represented shouting — reveals his affinity with the group of the God of Death. His style of hair-dressing is the same as that in the figure on Plate L, and he is armored in the same manner as the preceding images. Such a demon-like creature is disclosed also by the warrior on Plate LVII, with very elaborate body armor consisting of large plastron and dossière of metal, connected by leather straps running over the shoulders. It is plainly visible how the two breastplates join together in the middle. He wears a high collar and turned-up sleeves, animal-heads being brought out on the upper arms; the waist is narrow and tied by leather straps, and an apron of plate mail is hidden under the garment.

Finally we come to clay figures which are plainly knights or guardians armored *cap-à-pié*, without any mythological reminiscence (Plates LVIII–LX).

In Japan, types exist which are related to the Chinese clay figures already described. These are of highly artistic qualities, and show us that in the T'ang period a Buddhist school must have flourished, the tradition of which embraced the whole of eastern Asia. Two examples are here selected. The one is a clay figure, originally colored, in the Tōdai temple in Nara, founded in the middle of the eighth century (Fig. 46).[1] This remarkable statue is justly dated by the Japanese in the eighth century (T'ang period). Head-dress and armor, as well as pose of hands and feet, closely agree with those of the Chinese types; here we observe that the raised hand was indeed grasping a weapon. The Japanese name this figure Dhṛitarāshṭra, one of the four Mahārāja or Lokapāla of Hindu mythology guarding the world-mountain Sumeru. Another very similar statue (Fig. 47),[2] likewise and justly attributed to the eighth century, is named Virūpāksha, the third of the four guardians of the world. Both are posed on the bodies of demons.[3] The four Lokapāla are conceived as kings and heroes, and hence represented as

---

[1] The sketch is reproduced from the *Kokka*, No. 170, 1904.

[2] From the *Kokka*, No. 42. In the same manner Vajrapāṇi is represented (*Kokka*, No. 28, Plate V).

[3] The Japanese identifications are doubtless based on correct traditions, but I am not inclined to transfer these interpretations to the Chinese figures standing on demons as those mentioned before. We noticed that in some of these the bull-face of Yama is still preserved, and that consequently this figure is Yama: hence we may infer that also the anthropomorphic figures standing on demons are derived from the same type. Compare also the four wood-carved Lokapāla posed on crouching demons in *Kokka*, No. 165, 1904.

FIG. 46.
Japanese Colored Clay Statue of Dhrĭtarāshtra, Eighth Century (after *Kokka*).

FIG. 47.
Japanese Statue of Virūpāksha, Eighth Century (after *Kokka*).

armored; at the same time they are regarded as "protectors of religion" (Sanskrit *dharmapāla*), and for this reason are shown in so-called terrific forms.[1]

Analogous types of Lokapāla are met in the contemporaneous stone sculpture of China, for instance, in the caves of Lung-mên.[2] A marble relief (Plate LXI) in the Museum collection shows an armored Virūpāksha leaning on a two-edged sword, and holding a miniature Stūpa (tope) in his left hand.[3] The armor is very clearly represented: the breastplates tightly envelop the thorax, and are held in place by means of broad leather suspenders running over the shoulders and connecting with the dossière. The metal buckles fixed to the edge of the plastron are plainly visible, and tongues are passed through perforations of the straps. The ends of these straps reach the centre of either breastplate, and are strengthened at this spot by an additional piece of leather. The belt is a broad leather band starting in a rosette from the sternum, the end being turned upward from beneath the girdle.

It is of especial interest that similar clay figures representing Lokapāla (the term is perhaps too narrow, and should rather be Dharmapāla) have been discovered in Turkistan.[4] These are likewise enveloped by suits of armor much resembling those of the Chinese and Japanese clay statuettes. It is therefore obvious that in this case the question is not of any national type of armor which the Chinese applied to the clay figures, but that this armor was already peculiar to the latter when they were received in the channel of Buddhist art and reproduced by the potters of China. The art displayed in the caves of Tun-huang on the boundary of Turkistan and China may be made directly responsible for the transmission of this particular type from Turkistan to China; for there we find a statue of a Dharmapāla standing on a demon, and with exactly the same characteristics as our Chinese clay figures.[5] Was this armor ever a living reality in China, or did it merely remain an artistic motive? It is not very likely that it ever became of any practical use among the Chinese. It is not described in the official records of the T'ang dynasty; at least, in the records at our disposal no armor is

---

[1] Styled in Sanskrit *krodha*, in opposition to *çānta*, the mild forms. A mild form of Yama seated on the back of a bull was painted by the Buddhist monk Eri, who died in 935 (reproduction in *Kokka*, No. 133, 1902).

[2] CHAVANNES, Mission, No. 353. Besides the hero and warrior type of Lokapāla, we have in the same period a nude type clad only with an apron, and with fine modelling of strong, well-developed muscles (*ibid.*, Nos. 358, 359). An excellent marble of the latter type is in the collection of Field Museum.

[3] Styled in Chinese "King of Heaven lifting a Stūpa" (*T'o t'a t'ien wang*).

[4] A. GRÜNWEDEL, Altbuddhistische Kultstätten in Chinesisch-Turkistan, p. 205.

[5] A. MAYBON, L'art bouddhique du Turkestan oriental, p. 55 (*L'art décoratif*, 1910).

described that could freely be recognized in it.   Sheet armor, indeed, was never peculiar to China, but is plainly of western origin.   Above all, this type of armor, even if it should have sparsely existed here and there in China during the T'ang, has left no trace in any later period; it does not survive in any harness of the Ming and Manchu periods; and this is a signal fact, as otherwise the T'ang tradition in regard to armor was still alive in that recent age.[1]

Buddhism, however, may have influenced Chinese armature to a certain degree.   A peculiar kind of armor styled "lion-armor" (ni k'ai) is attributed to the T'ang period.[2]   The helmet and the coat are roughly figured in T'u shu tsi ch'êng (Fig. 48); but only the former is explained by a note to the effect that for each single piece five or six catties[3] of

---

[1] In Japan, however, specimens of such armor, though very rare, do occur. BASHFORD DEAN (Catalogue of Japanese Armor, p. 52) has figured one exactly corresponding to the sheet armor of our clay statuettes.   It is said to date about 1500, and "this form simulates the naked body and is known as the Hotoke-dō (saint's breastplate), an Indian saint being often represented with the body naked."   This term means "Buddha's breastplate (Hotoke=Chinese Fu, "Buddha"), and clearly indicates that this armor was made in imitation of that represented on Buddhist statues.   Among modern Indian armor, a very similar type is still found (W. EGERTON, Illustrated Handbook of Indian Arms, Plate XII, No. 587, and p. 124).   A somewhat different type of iron sheet armor is figured by W. GOWLAND (The Dolmens and Burial Mounds in Japan, p. 48, Westminster, 1897; the same also in YAGI SHŌZABURŌ, Nihon Kōkogaku, II, p. 153, Tōkyō, 1898; and N. G. MUNRO, Prehistoric Japan, pp. 396, 417, Yokohama, 1908).   It is likewise a harness composed of plastron and dossière which are formed of horizontal plates of iron skilfully forged and clinched together with iron rivets.   Gowland makes the interesting and correct observation that both body armor and helmet are entirely different in form and construction from those of historical times, but that they agree very closely with the armor represented on the terra-cotta figures called haniwa.   It is very interesting that the two TORII, in the publication previously mentioned (Etudes archéologiques, Journal College of Science, 1914, p. 73), figure such a haniwa with the description "cuirasse de style européen trouvée en Musashi, Japon."   The Japanese authors, accordingly, are struck by the "European" character of this armor.   It is now obvious that it has reached the East by way of Turkistan: consequently this haniwa adorned with this style of armor cannot be older than the age of the T'ang dynasty. Again we see in this example that the chronology of Japanese antiquities is in need of revision.

[2] AMIOT (Supplément à l'art militaire des Chinois, Mémoires concernant les Chinois, Vol. VIII, p. 373, Paris, 1782) was the first to describe this armor, but from a different source.   Amiot styles it "cuirass in imitation of the skin of the animal called ni (resembling, it is said, the lion)."

[3] The T'u shu tsi ch'êng, deviating from its ordinary practice, does not state the source of this passage, which is evidently not extracted from a contemporaneous record of the T'ang period, which, however, seems to go back to a tradition of that time.   The catty (kin) of the T'ang period is not identical with the present one. In the Museum collection there is a spherical bronze weight of the T'ang period (Cat. No. 116,892) inlaid with gold speckles and engraved with an inscription (the grooves of the characters being laid out with gold foil) yielding the date 672.   The weight is stated in this inscription as being 1 pound (catty) 8 ounces, while it is 2 pounds in our weight.   According to the present Chinese standard, it weighs 1 pound 11.32 ounces, or 27.32 ounces.   Consequently 1 ounce of the T'ang period is equal to 1.138 modern Chinese ounce, and 1 pound of the T'ang period is equal to 18.24 ounces modern.

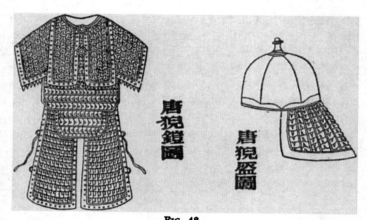

FIG. 48.
"Lion-Armor" and "Lion-Helmet" of the T'ang Period (from *T'u shu tsi ch'êng*).

FIG. 49.
"Lion-Helmet" of the T'ang Period (from *Wu pei chi*).

pure iron mixed with one catty of steel are required, and that a skin cut in five layers, to the weight of two catties, is laid around this foundation.[1] The term *ni k'uei* is not interpreted in this passage; but this word *ni* occurs only in the combination *suan-ni*, designating the lion. We noticed above (p. 276) that Ma Sui of the eighth century fashioned helmets in the shape of lions. A lion-helmet (*suan-ni mou*) is mentioned as having been in the possession of General Han Shi-chung, who died in 1151.[2]

A similar helmet with the same designation is illustrated also in the *Wu pei chi*[3] (Fig. 49); and the descriptive text there given is identical with that of *T'u shu tsi ch'êng*; nevertheless the illustration of the latter is not derived in this case from that book, as the knob of the helmet and the number of plate-rows in the attached coif of the helmet are different, being six in the *T'u shu*, and five in the *Wu pei chi*. It will be noticed that the triangles on the plates are alternately drawn point upward or downward, and that the *T'u shu* begins with points downward, the other book with points upward: the two sketches must therefore come from different sources.

Still more curious is the fact that the *Wu pei chi*[4] illustrates an armor of a different design under the same name, *T'ang ni k'ai* (Fig. 50). While the two drawings agree fairly well in the upper portions, the breast and sleeve coverings, they are considerably at variance in the middle and lower parts, though notwithstanding both evidently represent the same type of armor. The cut of the *Wu pei chi* is identical with the one figured by AMIOT;[5] and the quaint text supplied by him is found there also. It runs thus: "The lion-armor of the T'ang. First, five catties of the 'plant penetrating into the bones'[6] and three catties of radish-seeds are mixed into a pap which is placed in clear water to the quantity of a hundred catties, and boiled till it bubbles two hundred times. The residue is removed, and five scales of the pangolin[7] are added; further,

---

[1] Several designations for other kinds of helmets are added, and it is further said that in the south also old cotton is used in their making.

[2] GILES, Biographical Dictionary, p. 251. The passage alluded to above is contained in the biography of Han Shi-chung in the Annals of the Sung Dynasty (*Sung shi*, Ch. 364, p. 6 b). A "lion-armor" (*shi-tse kia*) is mentioned in the Annals of the Yüan Dynasty (*Yüan shi*, Ch. 79, p. 24 b, K'ien-lung edition).

[3] Ch. 105, p. 6.

[4] *L. c.*, p. 15.

[5] *L. c.*, Plate XXVIII.

[6] *T'ou ku ts'ao*, identified with *Mercurialis leiocarpa*, an euphorbiaceous plant (G. A. STUART, Chinese Materia Medica revised from F. Porter Smith's Work, p. 263, Shanghai, 1911).

[7] *Ch'uan shan kia*, the scaly ant-eater (*Manis tetra dactyla*). The word *ch'uan* is here written with the character 'river' (No. 2728) instead of No. 2739. This animal

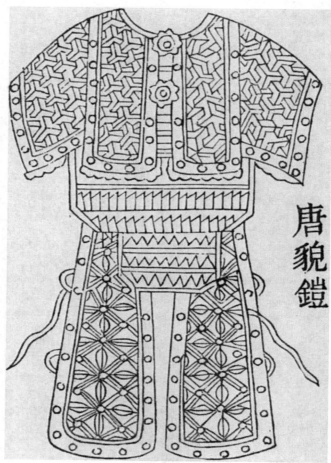

FIG. 50.
"Lion-Armor" of the T'ang Period (from *Wu pei chi*).

three catties of salt of Ta-t'ung, three catties of saltpetre, five ounces of stony nitre, and half a catty of sal-ammoniac. This mixture is tightly shut up in a kettle, and boiled for a day and night. Then the kettle is opened, and the mass is beaten with a leather ladle to secure various grades of thickness, and formed into the shapes of willow-leaves, fish-scales, square leaves, and rectangles. This armor has the advantage of being light in weight, and is much employed in the south."

This is apparently an alchemical recipe intended to produce a cut-proof body protection. The ingredients like the scales of the pangolin rest on sympathetic notions. Of course, it should not be understood with Amiot that the armor was manufactured from this substance; the illustrations show that the question is that of a substantial metal plate armor, although in the text it is a question of scales, and that the metal plates were covered with this essence. The idea of rendering the wearer invulnerable was perhaps responsible for the title of "lion-armor;" and this name, which conveys the impression of a rendering of Sanskrit *simhavarman*, savors of Indian-Buddhist influence. Indeed, on ex-amining closely the two designs of this armor, we cannot fail to notice that it is identical with the one represented in the late Buddhist art of China during the Ming period, especially in the statues of Wei-t'o (Veda) and the Four Heavenly Kings, the guardians of the world and armed defensors of the Buddhist religion. Numerous specimens of these in all dimensions, carved from wood or cast in bronze, are in the Museum's collection; whatever their artistic and scientific interest may be, they have no value for the study of body armor which is mechanically copied in various conventional and stereotyped designs not properly understood by the artists.

---

is an inhabitant of Fukien Province and Formosa, and has its trunk, limbs, and tail covered with large, horny, imbricated scales, which it elevates in rolling itself into a ball when defending itself against an enemy; the scales are medicinally employed (see J. H. EDWARDS, *China Review*, Vol. XXII, p. 714). Regarding the word "pan-golin" see YULE and BURNELL (Hobson-Jobson, p. 668), and A. MARRE (Petit Vocabulaire des mots malays que l' usage a introduits dans les langues d'Europe, p. 11, Rome, 1866).

VII. HORSE ARMOR AND CLAY FIGURES OF HORSES

Steeds shielded with armor are alluded to as early as the *Shi king*. It appears that horses harnessed to the war-chariots were sometimes covered at that period with a means of defence,[1] which, judging from the use of the word *kiai* (compare p. 195) in this connection, seems to have been of the type of scale armor, the scales being cut out of thin strips of hide or leather. During the Ch'un Ts'iu period, the horses of the war-chariots were likewise armored.[2] This horse armor of the archaic epoch was a plain caparison, and widely different from the complex and composite armor which, as we know with certainty, existed in the Mongol period.

As to metal armor for horses (*ma k'ai*), we hear it mentioned for the first time toward the end of or shortly after the Han, in two small compositions of the famed usurper Ts'ao Ts'ao, who died in 220 A.D., and of his son Ts'ao Chi (192–232). The latter says that the ancient emperors bestowed on their servants certain kinds of armor styled "shining like ink" (*mo kuang*) and "brilliant lustre" (*ming kuang*), an armor with double seat in the trousers, an armor with rings and chains, and a set of horse metal armor (*ma k'ai*). This passage is very suspicious because of its retrospective character: the metal armor (*k'ai*), while it existed at the author's time, had not yet appeared in the days of the early emperors; and the word is here used thrice consecutively with reference to them. The "ring and chain armor," as previously

---

[1] LEGGE, Chinese Classics, Vol. IV, pp. 131, 194. LEGGE translates in the one case "the chariot with its team in mail," and in the other case "his mail-covered team," explaining that the mail for the horses was made of thin plates of metal, scale-like. This interpretation is erroneous. The same misconception occurs in S. COUVREUR's translation of the *Shi king* (p. 136), "les quatre chevaux munis de minces cuirasses de métal," and is adopted by GILES (No. 1734); while in the other passage COUVREUR (p. 90) is correct in translating "les quatre chevaux munis de cuirasses," provided *cuirasses* is taken in its literal sense of "hide armor." It is impossible to assume that during a period when metal armor for the protection of the human body was entirely unknown, it should have been utilized in guarding a horse. Man of that age could conceive and employ no other armor for his horse than for himself; and since he was acquainted only with plain hide armor and hide scale armor, these two types must have served likewise for the horse, the term *kiai* being in favor of scale armor. The translations of the two passages of *Shi king* have to be corrected accordingly. The frontlets on the foreheads of the horses (*yang*, No. 12,882), once mentioned in *Shi king* (LEGGE, Chinese Classics, Vol. IV, p. 547) and once in *Tso chuan*, did not form part of an armor, but were metal ornaments which served for purely decorative purposes, and emitted pleasing sounds when the animal moved.

[2] LEGGE, *l. c.*, Vol. V, p. 345.

306

pointed out (p. 174), is an isolated instance in this period, and smacks of anachronism. For this reason also the metal horse mail must be looked upon with diffidence, and I am not inclined to attribute much importance to this text.

FIG. 51.
Armored Cavalier on Caparisoned Horse, Clay Figure in Collection of Mr. G. Eumorfopoulos, London (after Burlington Fine Arts Club, Exhibition of Early Chinese Pottery, Plate IV).

In 519 A.D., A-na-kuai, the King of the Juan-juan,[1] presented to the Emperor Su-tsung of the Wei dynasty one set of fine and brilliant[2] mail complete for man and horse (*jên ma k'ai*), and six sets of iron mail for man and horse.[3]

Caparisoned war-horses are repeatedly mentioned in the History of

---

[1] He committed suicide in 552, after having been vanquished by the Turks (HIRTH, Nachworte zur Inschrift des Tonjukuk, p. 110).

[2] This attribute is invariably used with reference to iron armor with varnished or polished plates.

[3] *Pei shi*, Ch. 98, p. 6.

the T'ang Dynasty. The rebel Kao K'ai-tao, who conquered Yü-yang in 618 and styled himself Prince of Yen, for example, was in possession of several thousand mail-clad horses and ten thousand men.[1] Among the types of armor officially established by the T'ang dynasty we find also "horse cuirasses" (*ma kia*); and a charger caparisoned in this manner appears in a contemporaneous clay figure (Fig. 51) coated with a yellow glaze. The armor covers the war-horse almost down to its knees; and as it appears as a solid mass without any divisions, it may be one of hide (also the rider apparently wears a hide armor); it is possible, however, that the hide is merely the exterior cover, and is placed over an armor of solid plate mail indicated by the row of laminæ along the lower edge.[2]

Under the Sung dynasty the horses received facial masks of copper.[3] According to *Ts'e fu yüan kuei*, Chang Yen-tsê, Governor of King-chou,[4] presented in 942, on his arrival at the capital, in order to show his gratitude for favors received, nine horses, and again fifty horses together with silver saddles and bridles, and iron armor for the protection of the faces of horses and men; at a later date he presented fifty horses with gold saddles and bridles, with complete armor for the horses and men.

The furniture of the horses of the Mongols is described by the Franciscan Plano Carpini in 1246.[5] It was of two kinds,— iron plate mail, as described in Chapter V, and leather scale armor. The latter consisted of five parts,— the body armor in two halves extending from the head to the tail, and fastened to the saddle, a protection for the croup, a neck-guard, a breastplate reaching down to the knees, and an iron lamina on the forehead (being the chanfrin).

In another passage the same writer says that many of the horses of Kuyuk had bits, breastplates, saddles, and cruppers, quite twenty marks' worth of gold.[6] The Armenian historian Haithon states that the horses of the Mongols, like their riders, were clothed with leather armor.[7]

Interesting illustrations depicting the single pieces making the complete furniture of the horse are preserved in the *Wu pei chi* (Figs. 52–54)

---

[1] *T'ang shu*, Ch. 86, p. 4 b.

[2] Also among the Moghuls the horses were first covered with mail, over which was put a decorated quilt (see H. BLOCHMANN, Ain I Akbari, Vol. I, Plate XIV, and the explanation on p. XI).

[3] *Sung shi*, Ch. 197, p. 2.

[4] In Kan-su Province (PLAYFAIR, Cities and Towns of China, 2d ed., No. 1112).

[5] Edition of G. PULLÉ, p. 87 (*Studi italiani di filologia indo-iranica*, Vol. IX, Firenze, 1913). This passage is lacking in the former editions of Carpini.

[6] W. W. ROCKHILL, The Journey of William of Rubruck, p. 20.

[7] G. ALTUNIAN, Die Mongolen und ihre Eroberungen, p. 81 (Berlin, 1911).

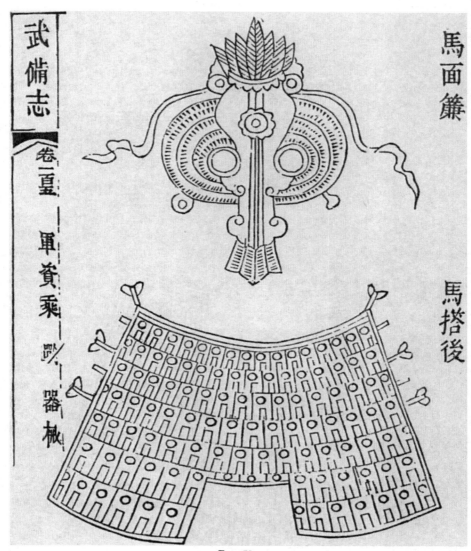

FIG. 52.
Chanfrin and Armor for the Croup of a Horse (from *Wu pei chi*).

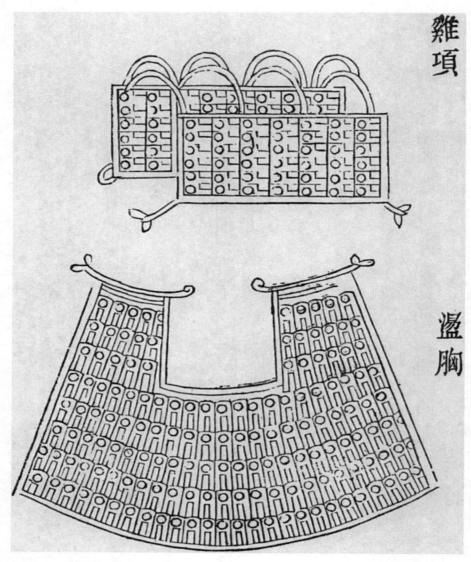

雞項

盪胸

FIG. 53.
Neck-Guard and Breastplate of Horse (from *Wu pei chi*).

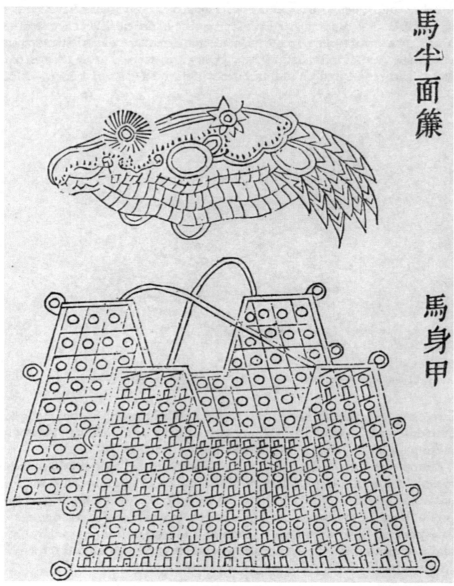

馬半面簾

馬身甲

Fig. 54.
Half-Chanfrin and Trunk Mail of Horse (from *Wu pei chi*).

of 1621, where no description of them, however, is given. The armor parts for the croup, neck, breast, and trunk, consist of plate mail; they represent the tradition of the Ming period, and may be identical with those of the Yüan.   It is not known to me whether horse armature was still employed under the Manchu dynasty.   Fig. 55 is here inserted after CIBOT; from what Chinese source this illustration is derived I do not know.   It is

FIG. 55.
Chinese Sketch of Caparisoned Horse (from L. P. Cibot, Lettre sur les caractères chinois, Brussels, 1773).

interesting as showing a horse with complete equipment, — a facial mask or frontal with chanfrin of scale armor, neck and shoulder guards of plate mail, and a chabraque enveloping the trunk.

From what has been set forth above in regard to the relations between Irān and China, it appears also that Chinese horse mail might have been influenced from the same direction. This influence is very probable; but the discussion of this matter may be left for the present, as it is preferable to wait until a thorough investigation of Iranian horse mail has been made by a competent specialist; ample material for such study is particularly furnished by the Persian miniatures.[1]

---

[1] In an illuminated manuscript of the Shāh-nāmeh preserved in the Royal Library of Munich, and representing the costume and arms of the Persians in the seventeenth century, according to Egerton, the combatants generally wear conical helmets with solid guards over the neck and ears. The horses as well as their riders have a complete covering of mail with alternate rows of gold and silver scales (W. EGERTON, Ill. Handbook of Indian Arms, p. 142). In ancient India, elephants and horses were protected by armor (G. OPPERT, On the Weapons, Army Organization, and Political Maxims of the Ancient Hindus, p. 8, Madras, 1880). The Chinese

Numerous clay figures of horses and cavaliers have been unearthed in recent years from the graves of Shen-si and Ho-nan, and a brief description of these may find a suitable place here. Particulars in regard to the history of the burial of such clay figures and their signifi-cance will be given in Part II. The observation of the local differentia-tions is an essential point of view to be pursued in the study of these clay figures.

The divergence between the grave-finds of Ho-nan and Shen-si is peculiarly manifest in the horses. Those of Shen-si usually represent the bare horse in a sober and mechanical conception;[1] those of Ho-nan illustrate more realistic types, always harnessed, in a variety of poses effected particularly by manifold turns of the neck. Most of the horses are posed on a flat rectangular clay base. Among seven clay horses of miniature size acquired by the writer at Si-ngan fu, six are almost identical, while the seventh is differentiated only in that the mane is coarsely fashioned. The horse on Plate LXII is an exception, being somewhat better shaped, and coated with soft lead glazes in three colors, — a deep brown, a light yellow, and a plant green; also saddle and sad-dle-cloth are represented (but not the stirrups); the saddle is padded with a textile material gracefully draped on both sides. The horse shown on Plate LXIII excels by its massive dimensions, but is other-wise the outcome of the routine work of an ordinary craftsman. The Ho-nan horses, on the other hand, appeal to us by the gracefulness of their motions, and the variety of actions in which they are represented (Plates LXIV, LXV); also the details of the harness are better and more efficiently worked out. In the horse on Plate LXVI, the trappings with their ornaments in metal, the tinkling bells on the breastband, as well as the lotus-flower designs on the crupper, are neatly moulded in relief.

The clay figure of the horse on Plate LXVII, found in fragmentary condition north of the city of Ho-nan fu in 1910, is notable for its un-usual dimensions and its perfect glazing.[2] The natural coloration of the animal is reproduced by a light-yellow soft lead glaze; the saddle, of the

pilgrim Hüan Tsang reports that the Indian war-elephants were covered with strong armature (S. BEAL, Buddhist Records of the Western World, Vol. I, p. 82). In Tibet the high officers sometimes clothe their horse with armor, and a set was cap-tured by the British expedition under Colonel Younghusband. A Tibetan cavalry-man whose horse is clad with chanfrin, neck and breast guard, is pictured in WAD-DELL'S Lhasa and its Mysteries (Plate opp. p. 168).

[1] Sometimes a mere saddle is represented without any other trappings; such a horse will be figured in Part II as forming part of a complete set of finds from the same grave.

[2] The technique and colors of these glazes are identical with those on the statue of the Arhat recently acquired by the British Museum, and ably described by R. L. HOBSON (Burlington Magazine, Vol. XXV, 1914, pp. 69–73). The excellent colored plate accompanying this article affords a good view of the T'ang potter's glazes.

same form as the one in use at present, is glazed a plant green; the double saddle-cloth underneath it, dark brown intermingled with green. The seat of the saddle is padded with a material arranged in graceful drapery. The mane is brown; the ornamental metal pieces attached to the head-stall, the breastband, and crupper are glazed green. The design which is brought out on these is characteristic of the T'ang period, and found also as relief decoration on coeval pottery vases.[1]

The horses on which human figures are mounted occupy a special place. Their significance in relation to the dead may be ascertained from their position in the grave: they were found either as preceding or as following the coffin. This seems to allude to the fact that they were regarded as the mounted escorts of the occupant of the grave, in the same manner as the living one, when on an official visit riding in a cart or in a sedan-chair, is accompanied by outriders in front and in the rear. As only persons of rank were granted this privilege, it seems certain that the same rule was observed in the grave, and that the clay statuettes of cavaliers appertain to dignitaries.

From Shen-si only figures of male riders are known to me (Plates LXVIII–LXX). The Shen-si horses are of somewhat stronger build, taller, and with more developed chests, than the Ho-nan breed. In the former, the curly hair on the forehead is parted and combed toward the sides, while in the latter it hangs straight downward. The men wear a pompon in the front of their round caps, and are strangely clad in long gowns. The cavalier on Plate LXVIII makes a poor figure as a horse-man, and shows that the Chinese of the T'ang period had as poor a knowledge of the art of riding as at present. The women of Ho-nan are better seated in the saddle than the men of Shen-si. The rider in question has his left foot pushed forward and his right foot backward; his hands come too near to the horse's neck, and seem to be in motion.

---

[1] An illustration of such a vase will be found in Part II. Chinese horse-trappings of the T'ang period may be viewed in Tōyei Shukō, Vol. III, Plates 196, 197. In none of the clay figures which have come to my notice is the saddle-girth represented. Judging from the clay figures, saddlery must have been almost the same in the T'ang period as at present. The frame of the modern saddle is carved from wood, frequently covered with shagreen and edged with metal-work, usually iron incrusted with silver wire forming geometric or floral designs. The seat is padded with a blue or red satin or velvet cover. There are, as a rule, two saddle-cloths, the lower one of wadded cotton cloth, the upper either of leather, ornamented with designs in color or appliqué patterns, or of wool or silk carpeting. A single bridle of cotton webbing is used. Headpiece, breastband, and crupper are usually decorated with brass work, or sometimes with silver gilt. A neckcollar fitted with small brass bells is occasionally added. Two tassels of red-dyed horse-hair are suspended, the one from the breastband, the other from the band under the chin. The stirrups are large and heavy with solid bases ellipsoid in shape, usually of iron damaskeened with silver, more rarely of brass. In Kan-su and north-eastern Tibet, wooden stirrups were also observed and collected by the writer; these are made as substitutes only when iron is lacking. Compare also Plate XXII.

Whoever has observed Chinese riding will have witnessed such perform-
ances; and in this case the potter must be granted all credit for his
power of observation. There is another type of mounted soldier from
Shen-si, whose left hand appears as if seizing the bridles, while he is
pressing his right hand against his chest (Plate LXIX, Fig. 2).

The figure on Plate LXX is curious in exhibiting a helmeted soldier
rising in the saddle in an upright position, in order to salute by lifting
his folded hands to the height of his face. The headstall of the horse is
decorated with floral ornaments, probably chased in metal.

In the Ho-nan types, the horses prick up their ears; their necks are
elegantly curved; the manes are either upright, or falling down to the
right side, and are carefully modelled. In all Ho-nan figures of riders
known to me, the stirrups are represented.[1] Fig. 1 of Plate LXIX
illustrates a female rider very well seated; the body of the clay is coated
with a yellowish-green glaze, and the mane of the animal is well treated;
but the form of the head is bad. In the figure on Plate LXXI the mane
of the steed is painted vermilion. The woman[2] wears male attire, a
girdled coat with triangular lapels (as in our man's clothing), trousers,
and boots; she is sitting straight and with arms crossed, the short sleeves
rendering the hands visible. The saddle-cloth is painted with small
circles in black ink, and thus is presumably intended for a panther's
skin. The reins and crupper likewise are so decorated, and there are
a few black circles on the neck of the animal. The stirrups are represented.

The horse illustrated on Plate LXXII is fairly well modelled. The
neck is painted red, and overstrewn with white spots. Headstall and
bridle are painted in black outlines, while the crupper is brought out in
relief. The muscles of the head, the nostrils, the jaws (agape), teeth,
and tongue are carefully modelled. The woman, almost Japanese in
expression, wears a flat cap, from which a long ribbon is floating down
her back. Her dress is painted a brown-red. Her right arm is hanging
down, her left hand is raised to seize the bridles. The saddle-cloth
seems to be a cotton quilt.

---

[1] As has already been shown by F. HIRTH (*Zeitschrift für Ethnologie*, 1890,
Verhandlungen, p. 209), stirrups were in vogue during the T'ang period; the people
availed themselves of iron stirrups, those of the dignitaries were made from the metal
alloy called *t'ou-shi*.

[2] Horseback-riding was a common exercise for women in the T'ang period.
Female equestrians were represented by pictorial art. Yang Kuei-fei was painted in
the act of mounting on horseback (GILES, Introduction to the History of Chinese
Pictorial Art, p. 50). In the Gallery of the Sung Emperors there was a picture by
Chang Süan, representing a Japanese woman on horseback (*Süan ho hua p'u*, Ch. 5,
p. 6).

PLATE IX.

DÜRER'S RHINOCEROS (see p. 83).

(After a Photogravure published by the British Museum, London.)

In 1515 Albrecht Dürer made the sketch of an Indian rhinoceros which had been shipped from India to King Emanuel of Portugal. The original is preserved in the British Museum, London, and bears from the hand of the artist the title "Rhinoceron 1515," and an explanation written in German along the lower edge of the picture.

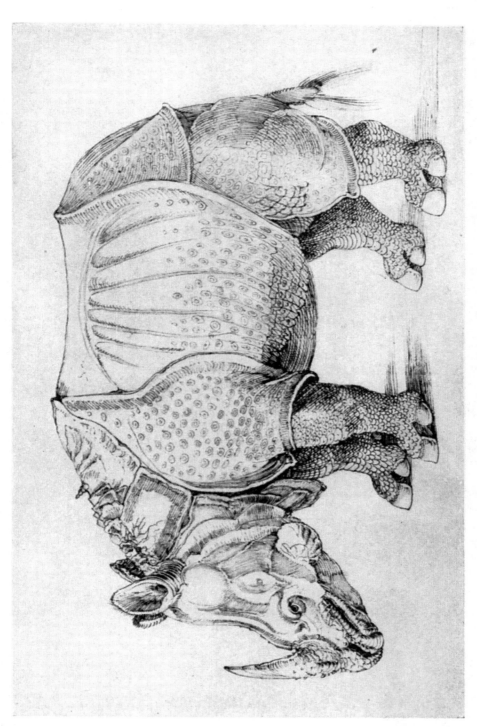

THE RHINOCEROS BY ALBRECHT DÜRER 1515.

PLATE X.

MASKS REPRESENTING INDIAN HERMIT (see p. 112).

Fig. 1. Japanese mask, representing the Indian hermit Single-Horn (in Japanese, Ikkaku-sennin). From *Nōgaku dai-jiten* ("Dictionary of Nō Plays"), by Masada Shōjirō and Amaya Kangichi (Tōkyō, 1908).

Fig. 2. Tibetan mask, representing the Indian hermit Single-Horn (in Sanskrit, Ekaçṛiṅga). Obtained in the Lama Temple of Bagme, western Sze-ch'uan. Cat. No. 120732.

JAPANESE AND TIBETAN MASKS OF THE HERMIT SINGLE-HORN.

PLATE XI.

AMERICAN HIDE ARMOR (see p. 183).

Made from hard, tanned moose-skin of two thicknesses, the two layers being tightly pressed together. From the Tlingit, Alaska. Presented by Mr. E. E. Ayer. Cat. No. 18165.

TLINGIT HIDE ARMOR.

PLATE XII.

ESKIMO HIDE ARMOR (see p. 183).

Two seal-skins are utilized in this hide armor, covering in width 155 cm. Wooden slats are encased between the two skins. From Asiatic Eskimo, East Cape, Chukotsk Peninsula. Obtained by Mr. A. M. Baber. Height, 57 cm. Cat. No. 34150.

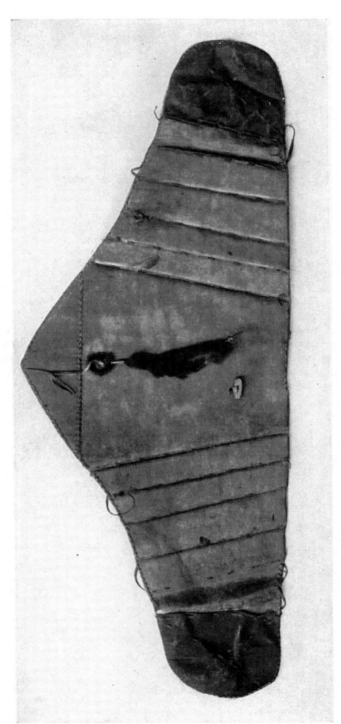

HIDE ARMOR OF ASIATIC ESKIMO.

PLATE XIII.

AMERICAN HIDE ARMOR (see p. 184).

Covered with about a thousand Chinese coins inscribed with the periods Shun-chi, K'ang-hi, Yung-chêng, and K'ien-lung, and procured in trade from the Russians, whose ships, exchanging the furs of the North Pacific with the Chinese for tea, plied constantly between the two countries, by which means many Chinese articles found their way to Alaska. Secured by Lieut. G. T. Emmons from the Tlingit, Tarku Tribe, on the Tarku River, Alaska. Cat. No. 78559.

TLINGIT CUIRASS COVERED WITH CHINESE COINS.

PLATE XIV.

COPPER SCALE ARMOR (see p. 196).

The copper scales are imbricated, and fastened by means of brass wire to a foundation of sackcloth. The collar consists of a single row of smaller scales. The coat folds over to the right side, and is fastened by three pairs of brass rings connected by cords. The epaulets are cut out of leather. Secured in Ch'eng-tu, Sze-ch'uan. Length, 80 cm; weight, 38¼ lbs. Cat. No. 118349.

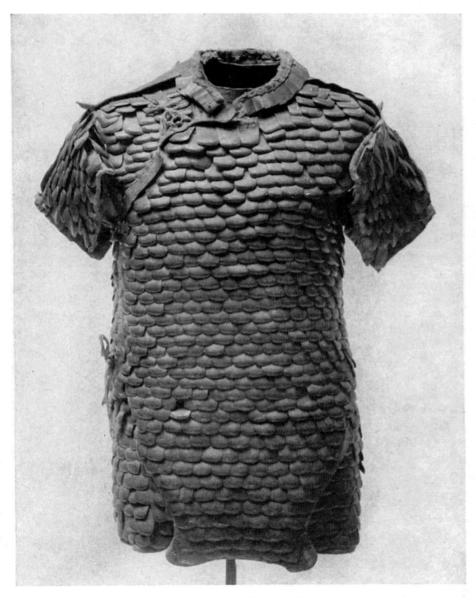

CHINESE ARMOR OF COPPER SCALES.

PLATE XV.

CLAY FIGURE REPRESENTING SHAMAN OF ARCHAIC PERIOD (see p. 199).

He is clad with sleeveless, tight-fitting scale armor, the scales being cut out of leather. They are outlined in black varnish over a coating of pipe-clay. The lines are so fine that they cannot be brought out. He wears a hide helmet surmounted by a high crest. Note the oblique and almond-shaped eyes. He is represented in the act of combating the demons and brandishing in his right hand a spear, which, being of wood, has rotted away under ground. The figure is hollow, and the clay walls are very thick and hard. Found in Ho-nan Province. Height, 51.2 cm. Cat. No. 117842.

CLAY FIGURE REPRESENTING SHAMAN.

PLATE XVI.

BACK OF CLAY FIGURE SHOWN ON PLATE XV.

The scales of the cuirass are modelled in the surface of the clay. A coif of hide scales is attached to the helmet, which envelops the head on all sides.

BACK OF CLAY FIGURE ON PRECEDING PLATE.

CLAY FIGURE REPRESENTING SHAMAN (see p. 199).

(see p. 199)

Archaic period. He is clad with a leather scale armor, the scales being painted in black outlines. He is in the act of throwing a spear during a struggle with demons. His hair is bound up in a snail-like chignon. His eyeballs protrude, and the cheek-bones are prominently accentuated. The tip of the nose is broken off. The figure is hollow, and the clay walls are very thick and hard. Found in Ho-nan Province. Height, 37.9 cm. Cat. No. 117841.

CLAY FIGURE REPRESENTING SHAMAN.

PLATE XVIII.

CLAY FIGURE OF A MAGICIAN (see p. 200).

Front view and profile. He wears a shirt of mail beneath his coat, a cape of tiger-skin around his shoulders, and a necklace. The hood-like helmet is worked into scales. T'ang period (618–906). Height, 36 cm. Cat. No. 118014.

CLAY FIGURE OF MAGICIAN.

PLATE XIX.

MILITARY WATCH-TOWER (see p. 208).

Model of green-glazed Han pottery, in the collection of Mr. Charles L. Freer of Detroit. It is here inserted to illustrate the military life of the Han period.

HAN POTTERY MODEL OF WATCH-TOWER.

PLATE XX.

TWO-EDGED BRONZE SWORDS OF THE HAN PERIOD (see p. 215).

Fig. 1. Much-worn blade, highly polished by means of an alloy of mercury and tin (such as is employed for metal mirrors), rhomboid guard, hollow handle. Length 45.6 cm. Cat. No. 116754.

Fig. 2. Unpolished blade, solid handle. Length, 45 cm. Cat. No. 116757.

Fig. 3. Blade, guard, and handle, made in one cast. Guard and knob of hilt show the same designs on the reverse side. Length, 71 cm. Cat. No. 116756.

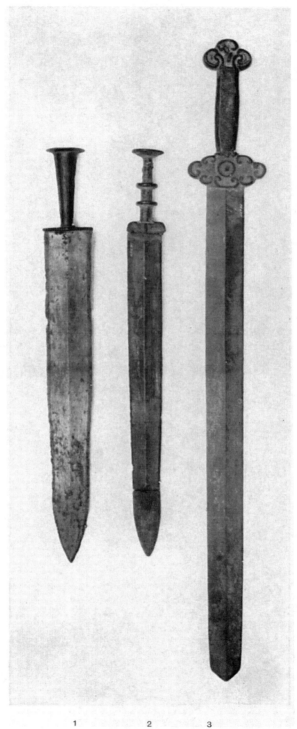

1   2   3

BRONZE SWORDS OF THE HAN PERIOD.

PLATE XXI.

CAST-IRON WEAPONS OF THE HAN PERIOD (see p. 216).

Figs. 1–2. Remnants of cast-iron spears. Length, 122.8 cm and 99 cm. Cat. Nos. 120995, 120996.

Figs. 3–4. Cast-iron swords with rhomboid bronze sword-guards. Length, 117.6 cm and 114.3 cm. Cat. Nos. 120993, 120994.

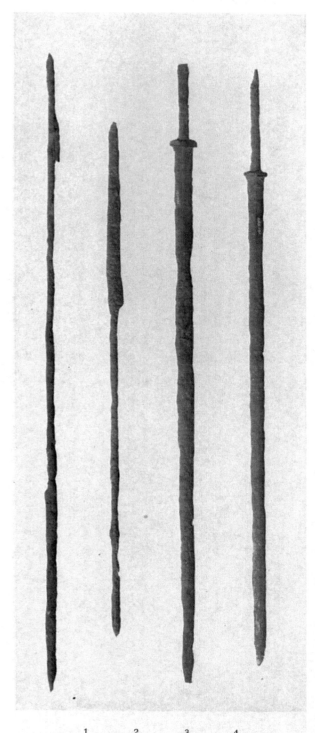

CAST-IRON WEAPONS OF THE HAN PERIOD.

1　　　　2　　　　3　　　　4

PLATE XXII.

IDEALIZED PORTRAIT OF A SOVEREIGN OF THE HUNS.

Painting, attributed to Han Kan, the famous horse-painter of the T'ang period. I have not seen the painting itself, and know it only from a photograph. I therefore have no opinion as to the period when it was executed. It is a good copy, presumably of the Yüan or Ming period, undoubtedly made after some T'ang production, which may have emanated from the school of Han Kan. It is an interesting piece of work, from a culture-historical point of view, and is here reproduced to give an idea of the Chinese conception of a sovereign of the Huns (see p. 224). He is represented on the hunt. His keen eyes have spied a bird in the branches of a tree, and he is going to fix the arrow to the bow-string. The string passes through the sleeve of his left arm, so that he does not have to hold the bow while trotting or galloping. He wears a turban, large ear-rings, and high boots. His long under-garment displays a checkered design, and may be composed of fur of alternately black and white squares, such as we still find among the tribes of eastern and central Siberia. His cloak is sleeveless, buttoned in front, and with girdle. The horse is furnished with a double saddle-cloth,— an ornamented rug, and a leather (or felt) cover. The saddle is mounted with shagreen.

The upper portion of the painting, taken up by scenery, is not here illustrated in order to insure a larger reproduction of the portrait. The entire composition may be viewed in L. Binyon, Painting in the Far East, 2d ed., Plate VIII (London, 1913).

In the collection of Sir William van Horne, Montreal, Canada. Secured through Mr. Stephan Bourgeois, to whom I am indebted for a photograph of the painting.

FIELD MUSEUM OF NATURAL HISTORY.

ANTHROPOLOGY, VOL. XIII, PL. XXII.

A SOVEREIGN OF THE HUNS.

Plate XXIII.

Persian Chain Mail (see p. 244).

Made of twisted iron wire, with helmet. Obtained at Tiflis by Mr. Charles R. Crane, Chicago, and now in the possession of Dr. Charles B. Cory, Chicago.

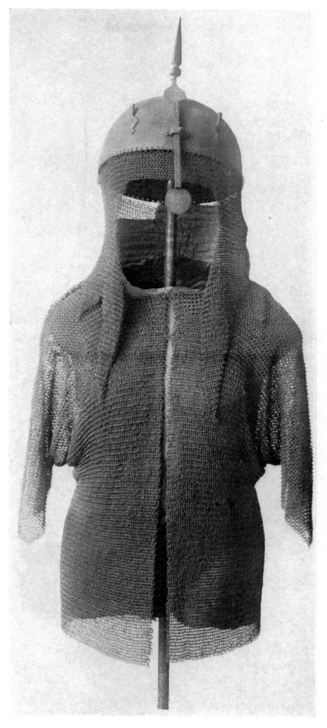

PERSIAN CHAIN MAIL, FRONT VIEW.

PERSIAN CHAIN MAIL, BACK VIEW.

Outfit belonging to Persian Chain Mail shown on
the two preceding plates (see p. 244).

(see p. 244)

Fig. 1.   Two-edged sword with steel blade and iron handle.   The ornaments on the blade are incrusted with gold; those on the handle, with silver.

Fig. 2.   Iron arm-guard, with representations of four scenes in Persian style.

Fig. 3.   Hauberk, consisting of a coif of mail, suspended from a wadded cotton quilt.   Width, 26 cm.

Fig. 4.   Gauntlet of mail.   The back is formed by red cotton stuff, lined with chamois leather.   The mail protecting the palm consists of a single layer of chain twisted from iron wire.   Length, 18 cm.

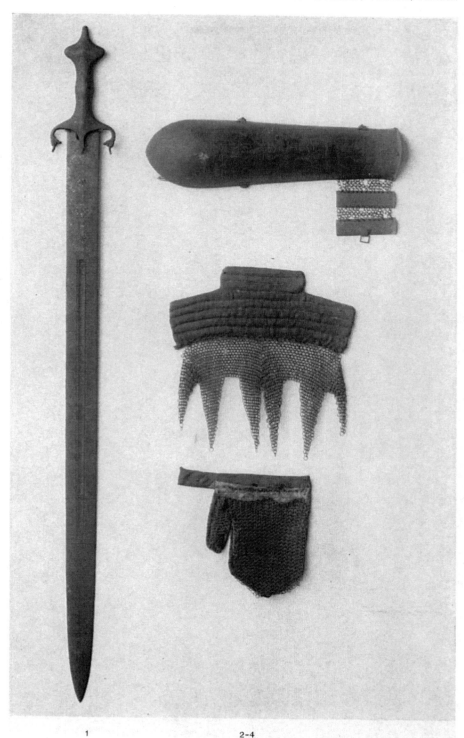

1   2-4

OUTFIT BELONGING TO PERSIAN CHAIN MAIL

Plate XXVI.

Chain Mail (see p. 249).

Fig. 1. Suit of chain mail consisting of riveted steel rings. Obtained at Si-ning, Kan-su Province, China, and said to have come from Tibet. Cat. No. 118348.

Fig. 2. Suit of chain mail composed of welded iron rings. Obtained at Si-ngan, Shen-si Province, China. Cat. No. 118347.

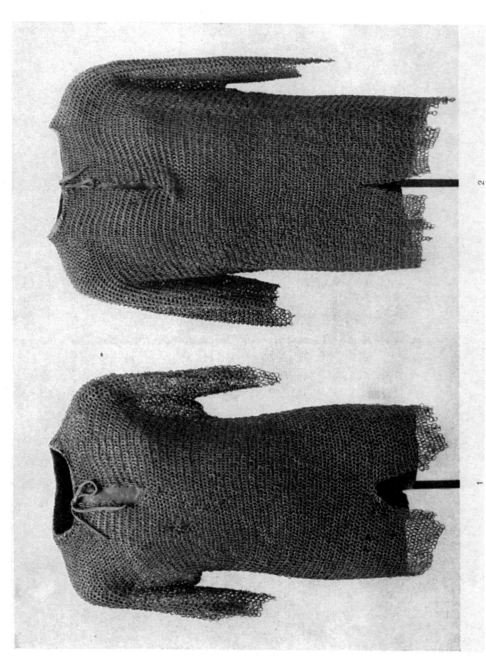

CHAIN MAIL FROM SI-NING AND SI-NGAN.

PLATE XXVII.

TIBETAN SHIELDS (see p. 257).

Fig. 1. Convex shield cut out of rhinoceros-hide, and ornamented with four brass bosses. Shields of this kind were manufactured in India and imported into Tibet. Cat. No. 122178.

Fig. 2. Shield of rattan, plaited in the basketry style of circular coils. This is the national shield of the Tibetans. Cat. No. 122179.

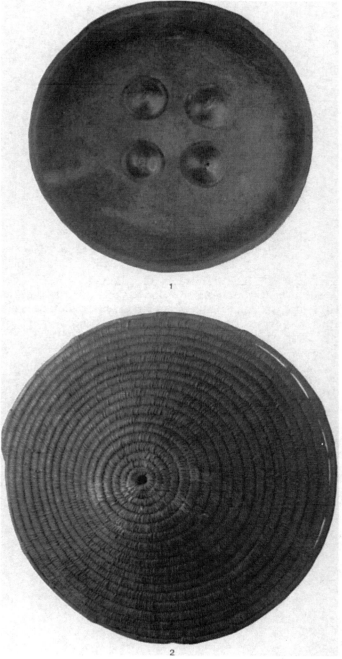

1

2

TIBETAN SHIELDS.

TIBETAN HELMET (see p. 257).

Composed of steel sheets, incrusted with gold and silver wire, forming floral designs. A coif of mail is attached to it for the protection of the nape. A nose-guard (nasal) in front, sliding up and down, serves for the protection of the nose; in the illustration it is down. Helmets of this type were manufactured in India and imported into Tibet. Cat. No. 122180.

TIBETAN HELMET OF INDO-PERSIAN STYLE.

PLATE XXIX.

AMERICAN PLATE ARMOR (see p. 263).

Composed of three rows of ivory plates, averaging 2.5 cm in width, and 15 cm in length. Each plate contains six holes, through which pass rawhide thongs, thus lashing the plates together. These plates are slightly imbricated, as are also the different rows, so as to ward off more effectually the weapons of the enemy. The lower contains forty-three plates; and the middle, thirty-eight. The upper row consists of two sections,— one containing ten plates, and protecting the breast; the other, eight, and protecting the upper part of the back. A rawhide strap passes over the shoulders and supports the armor.

Total width, 110 cm. From Eskimo of Cape Prince of Wales, Alaska. Obtained by H. R. Thornton, 1892. In the possession of the U. S. National Museum, Washington (Cat. No. 153491).

The smaller specimen in the upper right-hand corner represents the fragment of a plate armor consisting of nine iron plates bound together with three lashings of rawhide. This object was dug up in a bog at Cape Prince of Wales, near the locality where the ivory armor was obtained.

Length of each plate, 11.9 cm; width, 4.4 cm. Secured by H. R. Thornton, 1892. In the possession of the U. S. National Museum, Washington (Cat. No. 153492).

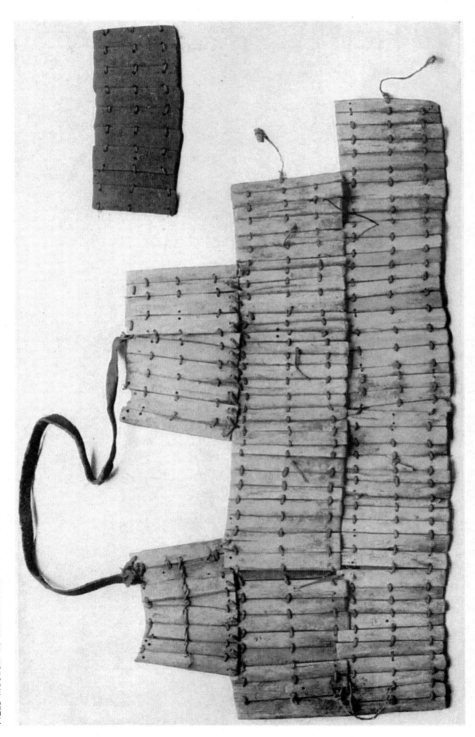

ESKIMO ARMOR OF IVORY PLATES AND FRAGMENT OF IRON PLATES.

CLAY FIGURE OF SOLDIER (see p. 277).

Both front and back views are shown. He is clad with armor in combination with costume. T'ang period (618–906).

Height, 20.5 cm. Cat. No. 117916.

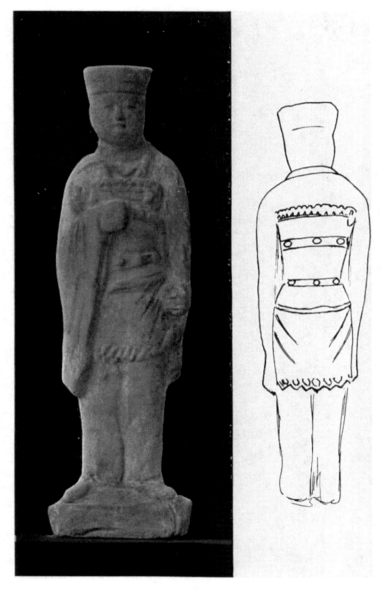

CLAY FIGURE OF SOLDIER, FRONT AND BACK VIEWS.

PLATE XXXI.

PAIR OF ARMORED KNIGHTS (see p. 277).

Clay figures of the T'ang period (618–906), from Shen-si Province. They are clad with sheet armor, — a clasp in the shape of an animal-head holding the plastron together, — and with apron consisting of metal plates (compare Fig. 42). The helmet-mask is formed by the head of an eagle. Compare the peculiar pose of these figures with that of the figures on Plates XLIX, LIII (Fig. 2), LIV, and LV.

Height, 34.3 cm. Cat. Nos. 11809, 11810.

CLAY FIGURES OF ARMORED GUARDIANS OF THE GRAVE.

CLAY FIGURES OF ARMORED GUARDIANS OF THE GRAVE.

PLATE XXXIII.

MARBLE MOCK-GATE (see p. 279).

    This formed the entrance to a tomb, and was dug up in the environment of the city of Hien-yang, Shen-si Province. Two soaring phenixes are carved in flat relief on the lintel. The gate is marked by lines and kept closed by means of a bolt, brought out in high relief. In each of the two wings is finely traced the figure of a guardian completely armored with plate mail, and handling a sword. Height, 52.5 cm; width, 34.5 cm; T'ang period (618–906). Thickness, 8.2 cm. Cat. No. 121623.

MARBLE MOCK-GATE OF A TOMB.

CHINESE PLATE ARMOR (see p. 284).

Horseman's uniform, of K'ien-lung period (1736–95). The skirt is covered with four parallel rows of light and elastic steel laminæ. In the coat, the steel plates are inserted as an interlining. Steel helmet, surmounted by velvet plume, dragons being engraved on the front, with silk covers for neck, ears, and occiput. The plume has not been represented on the Plate, in order that the suit might be reproduced on a larger scale. Obtained at Si-ngan. Cat. No. 118344.

HORSEMAN'S SUIT OF ARMOR.

CHINESE PLATE ARMOR (see p. 285).
(see p. 285)

Artillery-man's uniform, of K'ien-lung period (1736–95). The plates are retained only for the protection of the shoulders. Each lamina is of steel and gold-plated, and chased with a four-clawed dragon soaring in clouds. Steel helmet lined with quilt, and chased with gilt figures of dragons in pursuit of the flamed jewel. Obtained at Si-ngan. Cat. No. 118346.

ARTILLERY-MAN'S SUIT OF ARMOR.

ARCHER'S SUIT OF ARMOR, FRONT VIEW.

ARCHER'S SUIT OF ARMOR, BACK VIEW.

PLATE XXXVIII.

CEREMONIAL UNIFORM (see p. 286).

Belonging to guard-officer of the first rank detailed on duty in the Imperial Palace. The costume is magnificently embroidered with heavy gold thread, and studded with gilt bosses. The shoulder-plates are arranged in the same manner as in the suit of armor on Plate XXXV. Cat. No. 32853.

Helmet, bow-case, and quiver belonging to this uniform are represented on the following two Plates.

1　　　　　　　　　　　　　　2

UNIFORM OF PALACE GUARD-OFFICER, FRONT AND BACK VIEWS.

HELMET OF PALACE OFFICER.

BOW-CASE AND QUIVER OF PALACE OFFICER.

PLATE XLI.

KOREAN PLATE ARMOR (see p. 288).

Front and back of the coat are strengthened in the interior by seven parallel rows of rectangular steel plates, coated on both sides with a black varnish.

Length, 81 cm. Cat. No. 33281.

The following Plate illustrates the interior of this suit, with the iron casque.

KOREAN STEEL PLATE ARMOR, EXTERIOR.

SAME ARMOR, INTERIOR, WITH HELMET.

KOREAN PSEUDO-PLATE ARMOR (see p. 289).

It has no plates, but the rows of brass bosses on the surface of the coat are decorative survivals or reminiscences of plate armor. Thin copper plates are inserted as an interlining in the ear-muffs and nape-guard attached to the helmet.

Length, about 1 m.   Cat. No. 33263.

KOREAN COURT COSTUME OF HIGH OFFICIAL.

PLATE XLIV.

YAMA, THE GOD OF DEATH (see p. 294).

He stands on the body of a sow, and is represented with the head of a horned bull, and with eagle-claws on his hands and feet.   His head is surrounded by flames.

Clay figure from Shen-si, of mediæval times, probably T'ang period (618–906). Traces of red pigment; eyeballs painted black.

Height, 60 cm.   Cat. No. 117987.

ZOÖMORPHIC FORM OF THE GOD OF DEATH.

PLATE XLV.

YAMA, THE GOD OF DEATH (see p. 295).

He stands on the body of a bull, and is represented with the head of a horned bull, surrounded by flames.   A snake is winding around his left arm.

Clay figure from Shen-si, of mediæval times, probably T'ang period (618–906). Height, 34 cm.   Cat. No. 117985.

ZOÖMORPHIC FORM OF THE GOD OF DEATH.

ZOÖMORPHIC FORM OF THE GOD OF DEATH.

PLATE XLVII.

CLAY FIGURES FROM SHEN-SI AND HO-NAN (see p. 295).

Fig. 1.   Intermediary form of the God of Death.   His head is modelled in the style of the bull-faced Yama, as shown on Plates XLIV and XLV, but he is equipped with armor in the same manner as the human forms.   He stands over the figure of a demon, and seems to have grasped a weapon in his right hand, which is perforated.

Clay figure from Shen-si.   T'ang period (618–906).

Height, 45 cm.   Cat. No. 117998.

Fig. 2.   Fragmentary clay figure from Ho-nan, of armored knight with plumed head-dress,— a type evolved from Yama as triumphant warrior.   Here inserted for comparison of the head-dress with that in Fig. 1.

Height, 31.5 cm.   Cat. No. 117994.

INTERMEDIARY AND HUMAN FORMS OF THE GOD OF DEATH.

PLATE XLVIII.

FORM OF THE GOD OF DEATH (see p. 295).

Intermediary between the zoömorphic and anthropomorphic types. The attitude is that of a triumphant victor, standing over the figure of a crouching demon. Clad with armor and an elaborate head-dress, like the figures of knights shown on the following plates, he shares the two-horned bull-head with the purely animal forms of Yama, illustrated previously.

Clay figure from Shen-si. T'ang period (618–906).

Height, 68 cm. Cat. No. 117993.

THE GOD OF DEATH.

THE TRIUMPHANT GOD OF DEATH (see p. 295).

He is represented as a knight with complete armor, standing on the figure of a demon.   The figure is coated, except the head, with glaze in four colors,—green, blue, brown, and yellowish white.

Clay figure from Shen-si.   T'ang period (618–906).

Height, 52.6 cm.   Cat. No. 118000.

HUMAN FORM OF THE GOD OF DEATH.

HUMAN FORM OF THE GOD OF DEATH (see p. 296).

Posed on the back of a reclining bull, and clad with sheet armor. There are two identical specimens of this figure in the Museum collection, said to have been found in the same grave.

Clay figure from Shen-si. T'ang period (618–906).

Height, 67 cm. Cat. No. 118006.

THE GOD OF DEATH.

HUMAN FORM OF THE GOD OF DEATH.

HUMAN FORM OF THE GOD OF DEATH (see p. 296).

The figure of the bull is lost, but may be supplemented in accordance with the figure in the preceding Plate, with which it agrees in pose and general style. It is, however, much more artistic. The face is well modelled and very expressive. Note the mustache with turned-up tips. The clay piece, which appears dark on the Plate, is a recent supplement. The entire clay figure, with the exception of the head, is glazed in three colors,— green, brown, and yellowish-white.

From Ho-nan. T'ang period (618–906).

Height, 68.8 cm. Cat. No. 118069.

GLAZED FIGURE OF THE GOD OF DEATH.

PLATE LIII.

THE GOD OF DEATH (see p. 296).

Fig. 1.   Of the same type and style as the clay figure on Plate LI, only without helmet.   His hair is parted and bound up in a chignon.

Fig. 2.   In this figure, the pose of hands and feet is reversed, the right arm being akimbo, and the left one being raised.   He stands on the body of a demon.

Clay figures from Ho-nan.   T'ang period (618–906).

Height, 40 and 38 cm.   Cat. Nos. 117876, 117991.

1   2

CLAY FIGURES OF THE GOD OF DEATH.

PLATE LIV.

THE GOD OF DEATH (see p. 296).

Fig. 1. The God of Death trampling on the body of a demon, of the same style and pose as Fig. 2 on the preceding Plate.

From Ho-nan. T'ang period (618–906).

Height, 38.3 cm. Cat. No. 118065.

Fig. 2. The God of Death trampling on the figure of a human body (probably child), coated with a thick layer of white pipe clay,— eyes, brows, nose, and mouth being painted in black; so are also the boots of the God. Further, the outlines of his eyes are black (the eyeballs being red). The middle portion of the sleeve of his right arm is covered with a red pigment.

From Ho-nan. T'ang period (618–906).

Height, 29.3 cm. Cat. No. 117995.

1   2

CLAY FIGURES OF THE GOD OF DEATH.

PLATE LV.

THE GOD OF DEATH (see p. 296).

Represented as armored knight, standing on a bull or a demon (the figure is not sufficiently distinct to allow of positive identification).   Miniature figure, solid cast from lead, in high relief; the back is flat.

From Shen-si.   T'ang period (618–906).

Height, 11 cm; width, 4.3 cm; thickness, 2.2 cm.   Cat. No. 117091.

LEAD FIGURE OF THE GOD OF DEATH.

PLATE LVI.

GUARDIAN OF THE GRAVE (see p. 297).

Knight or warrior clad with sheet armor, animal-heads being brought out on the sleeves. In the point of armor, in the weird and demoniacal expression of his face (he is represented as shouting), and in the style of his chignon (compare Plates L and LIII, Fig. 1), he reveals his affinity with Yama, the God of Death.

Clay figure from Ho-nan, of unusual dimensions. T'ang period (618–906). Height, 79.7 cm. Cat. No. 118154.

CLAY FIGURE OF ARMORED KNIGHT.

GUARDIAN OF THE GRAVE (see p. 297).

This figure affords a good example for the study of sheet armor. Plastron and dossière are conspicuously represented, each consisting of two halves joined in the middle, and are connected by leather straps running over the shoulders. He holds a weapon in his right hand.

Well-modelled clay figure from Ho-nan, with traces of red pigment. T'ang period (618–906).

Height, 61.9 cm. Cat. No. 118008.

CLAY FIGURE OF ARMORED GUARDIAN.

CLAY FIGURE OF ARMORED GUARDIAN.

1   2

PAIR OF ARMORED GUARDIANS.

GUARDIAN OF THE GRAVE (see p. 297).

Upper portion of clay figure, representing shouting warrior clad with sheet armor, shoulder-guards, and hood-like helmet. His right fist has an aperture (made by means of a drill), in which a wooden spear seems to have been inserted.

From Ho-nan. T'ang period (618–906).

Height, 34 cm. Cat. No. 118011.

FRAGMENTARY CLAY FIGURE OF ARMORED GUARDIAN.

PLATE LXI.

GUARDIAN OF THE WORLD (see p. 300).

One of the four Lokapāla or Guardians of the World of Hindu mythology, who hold sway at the foot of the World-Mountain Sumeru. This is King Virūpāksha residing on the western side of the mountain, holding in his left hand a miniature pagoda, and seizing a sword with his right. Here inserted to illustrate the identity of sheet armor in Buddhist stone sculpture with that in the preceding clay figures of the same epoch.

Relief marble plaque, obtained from the temple King-ch'êng-se at Si-ngan, Shen-si. T'ang period (618–906).

Height, 38 cm; width, 21 cm. Cat. No. 121555.

MARBLE RELIEF OF GUARDIAN OF THE WORLD.

CLAY FIGURE OF SADDLED HORSE (see p. 313).

Coated with soft lead glazes in three colors,— a deep brown, a light yellow, and a plant green. Saddle-cloth and saddle are represented, the latter being padded with a gracefully draped textile material.

Excavated in Lung chou, prefecture of Fêng-siang, province of Shen-si. T'ang period (618–906).

Height, 27.5 cm. Cat. No. 118039.

GLAZED CLAY FIGURE OF HORSE.

CLAY FIGURE OF HORSE, FROM SHEN-SI.

PLATE LXIV.
CLAY FIGURE OF HORSE (see p. 313).

This horse turns its head sideways.   The muscles are brought out in its head. Headstall, saddle-cloth, and padded saddle are represented.

From Ho-nan.   T'ang period (618–906).

Height, 27.7 cm.   Cat. No. 118038.

CLAY FIGURE OF SADDLED HORSE, FROM HO-NAN.

PLATE LXV.

CLAY FIGURE OF HORSE (see p. 313).

Horse with complete harness and upright mane. The head is well modelled; and, though the pose is somewhat stiff, the potter seems to have attempted to represent the animal as though mourning for its deceased master.

From Ho-nan. T'ang period (618–906).

Height, 30 cm. Cat. No. 118060.

CLAY FIGURE OF SADDLED HORSE, FROM HO-NAN.

PLATE LXVI.

CLAY FIGURE OF HORSE (see p. 313).

Horse with complete harness, mourning for its dead master. The trappings with their metal ornaments, the tinkling bells on the breastband, as well as the designs of lotuses on the crupper, are neatly moulded in relief.

From Ho-nan. T'ang period (618–906).

Height, 32 cm. Cat. No. 118037.

CLAY FIGURE OF SADDLED HORSE, FROM HO-NAN.

PLATE LXVII.

CLAY FIGURE OF HORSE (see p. 313).

Fragmentary figure of horse, of unusual dimensions, and coated with lead glazes of light-yellow, plant-green, and brown tints.

From Ho-nan; found in the spring of 1910 during the cuttings for a railroad north of the city of Ho-nan fu. T'ang period (618–906).

Height; 80 cm. Cat. No. 118040.

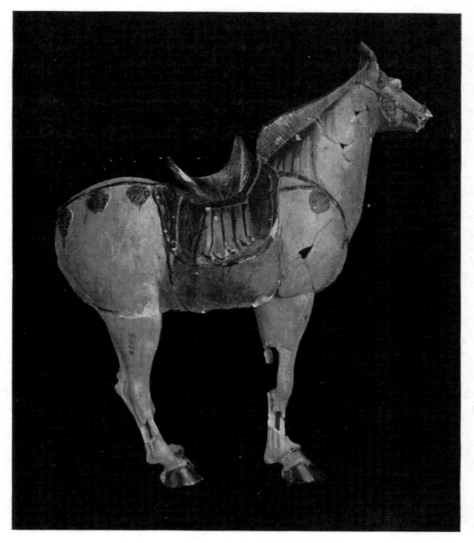

GLAZED CLAY FIGURE OF HORSE, FROM HO-NAN.

PLATE LXVIII.
CAVALIER (see p. 314).

Horseman, escort of the inmate of the grave. Such figures were placed in front of, or behind, the coffin. The hair on the forehead of the horse is parted and combed toward the sides.

Clay figure from Shen-si. T'ang period (618–906).

Height, 33 cm. Cat. No. 118049.

CLAY FIGURE OF CAVALIER, FROM SHEN-SI.

PLATE LXIX.

CAVALIERS (see p. 314).

Fig. 1. Horsewoman well seated in the saddle. The figure is uniformly coated with a lustrous, yellowish-green glaze.

Clay figure from Ho-nan. T'ang period (618–906).

Height, 28.7 cm. Cat. No. 118055.

Fig. 2. Clay figure of horseman, of same type as that in the preceding Plate.

From Shen-si. T'ang period (618–906).

Height, 33 cm. Cat. No. 118048.

CLAY FIGURES OF CAVALIERS.

PLATE LXX.

Cavalier (see p. 315).

Horseman, represented in the act of saluting by lifting his folded hands to the height of his face. The headstall of the horse is decorated with floral ornaments, probably chased in metal. When found, the feet of the horse were broken off. From Shen-si. T'ang period (618–906). Height, 30 cm. Cat. No. 118059.

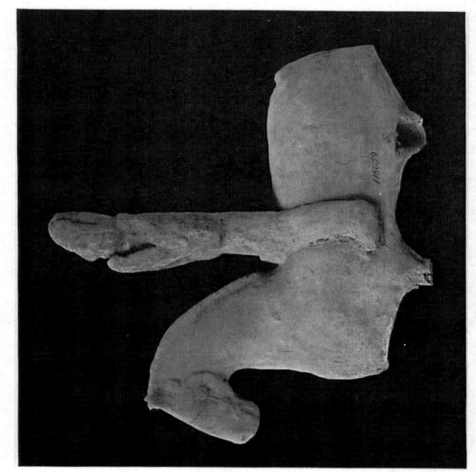

FRAGMENTARY CLAY FIGURE OF CAVALIER.

PLATE LXXI.

HORSEWOMAN (see p. 315).
Wearing male attire, a girdled coat with triangular lapels, trousers, and boots. The saddle-cloth is formed by a panther-skin.

Clay figure from Ho-nan. T'ang period (618–906).

Height, 30.2 cm. Cat. No. 118058.

CLAY FIGURE OF EQUESTRIAN WOMAN.

PLATE LXXII.

HORSEWOMAN (see p. 315).

In brownish-red dress, with flat cap from which a long ribbon is floating down her back. The neck of the horse is painted red and overstrewn with white spots. The muscles of the head, the nostrils, jaws, teeth, and tongue are carefully modelled.

Clay figure from Ho-nan. T'ang period (618–906).

Height, 36 cm. Cat. No. 118057.

CLAY FIGURE OF EQUESTRIAN WOMAN.

AMERICAN ANTHROPOLOGIST

NEW SERIES

ORGAN OF THE AMERICAN ANTHROPOLOGICAL ASSOCIATION,
THE ANTHROPOLOGICAL SOCIETY OF WASHINGTON,
AND THE AMERICAN ETHNOLOGICAL
SOCIETY OF NEW YORK

PUBLICATION COMMITTEE

VOLUME 19

LANCASTER, PA., U. S. A.
PUBLISHED FOR
THE AMERICAN ANTHROPOLOGICAL ASSOCIATION

1917

Japan, are represented by well selected material. A good sketch of the prominent figures of the Lamaic pantheon is inserted. The unexplained female statuette (plate XXIX and p. 91) appears to have formed part of a triad, and to represent one of the consorts of Padmasambhava, probably Mandārava; at any rate, it is the type of an Indian, not a Tibetan woman. A piece of evidence of Japanese relations with Indo-China is presented by a Japanese sword guard found at Angkor Vat. Some terra cotta statuettes discovered in the marshes or rice fields of villages in the proximity of Hanoi are said to be relatively modern and of Canton manufacture; but one representing a lion with human head is regarded as older than the others. This type is well known among the T'ang clay figurines from Shen-si Province, being doubtless modeled after an Iranian prototype.

From plate XIX we glean with some consternation that a fine set of sacrificial bronze vessels of the K'ien-lung period is not cased. Chinese bronze, it is true, in general possesses a much greater power of resistance to atmospherical influences than Egyptian, Greek, or Roman bronze; but even if sheltered in a fairly air-tight case, it demands the constant, watchful care of a museum curator as to possible formation of malignant patinas. In the humidity of the Hanoi climate any open exhibition of whatever character would seem unsafe. We entertain the best wishes for the future growth and prosperous development of the Hanoi Museum, to which this attractive guidebook will assuredly win many new friends.

B. LAUFER

Chinese Clay Figures. Part I. *Prolegomena on the History of Defensive Armor.* BERTHOLD LAUFER. (Field Museum of Natural History, Publication 177, Anthropological Series, vol. XIII, no. 2.) Chicago, 1914. Pp. 69–315, 64 pls., 55 figs.

The basis of this series of studies, which the author modestly terms prolegomena, is afforded by certain ancient clay figures from the provinces of Honan and Shen-si—the region, in other words, where the old Chinese culture first took on its historical form. As is the case with all of Dr. Laufer's writings, the subject is treated with the utmost thoroughness, and light is thrown on it from almost every conceivable angle. In doing this the author has availed himself of his very wide knowledge of Chinese records and customs, and has based his conclusions upon evidence which seems incontrovertible. It is fortunate that Dr. Laufer is an ethnologist of thorough training and wide experience, for his consistent adherence to the modern anthropological point of view gives his work a quality which is lacking in much that has been done in the same field.

The first chapter of the work under consideration, entitled a "History of the Rhinoceros," is an excellent example of the thoroughness with which this writer treats his subject. Because the earliest defensive armor possessed by the Chinese was, according to him, of rhinoceros hide, the whole question of the former existence of that animal in China is gone into most exhaustively. Evidence is drawn not only from the ancient Chinese writings themselves, but from the records of Greece and Rome and of the Mohammedan world and mediaeval Europe, while the latest conclusions of zoology and palaeontology are carefully sifted for any data which may bear upon the subject.

Dr. Laufer's argument, in brief, is this: that in the archaic period (*i. e.*, down to the third century B.C.) the Chinese used armor made from the hides of two animals designated by the words *se* and *si;* that further, these words referred not to some large bovine form, as had heretofore been taken for granted, but to the rhinoceros; and, finally, that the *si* is to be identified with the existing two-horned Asiatic rhinoceros (*R. sumatrensis*), and the *se* with the single-horned animal (whether *R. unicornis* or *R. sondaicus* is not clear).

Previous writers have apparently treated this question only incidentally, adopting without adequate examination the statements of relatively late Chinese historians and lexicographers. It was the most natural thing in the world for earlier writers to compare the rhinoceros of their day with some such large bovine as, let us say, the water-buffalo; anybody who has seen both creatures must have been struck by their likeness, particularly when viewed tail-on. It would be equally natural, after the rhinoceros had become extinct and when it was remembered only as an animal "resembling a wild ox," to transfer its now ownerless name to the still surviving bovine. This process, as we all know, has been going on the world over; an almost exact parallel to the fate of the words *se* and *si* is the transference of the name "auerochs" to the European bison after the extermination of *Bos primigenius*. Dr. Laufer's argument seems most cogent and conclusive. There is, however, one minor point which seems to demand revision. On page 158 the suggestion is thrown out that the Chinese *si* may have been a descendant of the well known woolly rhinoceros (*R. tichorhinus*) which inhabited Siberia in Pleistocene times, and which, Dr. Laufer suggests, may have retreated gradually southward. To accept this, however, would be to controvert the much more plausible previous identification (p. 93) of the *si* with the existing *R. sumatrensis*, inasmuch as the place of *R. tichorhinus* is with the modern African group (*Atelodus*), and not with the much more primitive Sumatran animal.

6

One gains from this chapter not only much that is new regarding the former distribution of the rhinoceros, but also a most attractive picture of the archaic Chinese world of the first millennium B.C. We see not the congested and intensively cultivated country of today, but a region still affording shelter in its forests and jungles to numerous species of big game animals, and occupied by a warlike and chivalrous race with a highly developed bronze culture in many ways strangely reminiscent of that of Europe.

In chapter II, entitled "Defensive Armor of the Archaic Period," Dr. Laufer points out that until the termination of the Chou dynasty, in the middle of the third century B.C., armor was made exclusively of hide, and that iron armor and helmets were unknown until later. Several pages are devoted to a study of the nature of this armor, and most interesting comparisons are made with hide armor occurring in other culture areas, and particularly North America. The writer also shows that rhinoceros hide by no means went out of use with the introduction of metal armor, but that on the contrary it persisted as late as the T'ang period (A.D. 618–907). He then describes the two types of hide armor, the one (*kia*) a cuirass made in the form of a coat, the other (*kiai*) consisting of leather scales arranged like those of a fish. This most interesting chapter concludes by pointing out the striking coincidences between the development of defensive armor in the archaic epoch of China and that found in other ancient culture groups of Asia as well as in some primitive societies of the present.

The third chapter, "Defensive Armor of the Han Period," deals with the introduction of metal armor in China and the striking development in military organization which accompanied it. The congeries of loosely federated states which formed the China of the archaic period had been consolidated by the short-lived but exceedingly energetic dynasty of the Ts'in into a true centralized empire. This passed, at the very close of the third century B.C., under the dominance of the great house of Han. Marked advances now took place along all lines of culture, and both the political boundaries and the intellectual horizons of the Chinese people underwent a very remarkable expansion. In the field of warfare, Dr. Laufer points out, the old bow was superseded by the much more powerful crossbow, possibly adopted from the non-Chinese aboriginal tribes in the regions south of the Yangtse. At the same time the old bronze sword gave way to one of cast iron traceable to Siberia, while the chariot force which had formed the most important part of the armies of former times was replaced by regular cavalry.

Unfortunately the sculptured reliefs of the latter Han period, which give us such extraordinarily vivid and varied representations of the life of the time, reveal little regarding the defensive armor.

After a discussion of the shields of the Han warriors, and a reference to the life of the frontier guards who protected the northwestern border of the empire against marauding nomads, the author goes on to speak of the way in which metal armor first came into use, and the causes for its adoption. The probability is shown that it was copper which was first utilized, iron replacing it only during the first two centuries of our era (the latter half of the Han period). The author then points out what so many have failed to grasp, or at least to emphasize adequately—the great historical connections linking all Asia in matters of military art. This point is so important, and exemplifies so well Dr. Laufer's point of view in all his work, that his exact words deserve quoting. He says,

No human invention or activity can be properly understood if viewed merely as an isolated phenomenon, with utter disregard of the causal factors to which it is inextricably chained. Every cultural idea bears its distinct relation to a series of others, and this reciprocity and interdependence of phenomena must be visualized in determining its historical position.

Dr. Laufer then goes on to show how the Persians, some time after their invasions of Greece, developed a true cavalry in the modern sense, utilizing shock tactics instead of distant skirmishing. The advantages and weaknesses of the new system are pointed out, as is the fact of its adoption by the Huns about the close of the third century B.C., just at the time, it will be recalled, when China passed under the sway of the powerful Han dynasty. Finally, the various stages are traced by which the Chinese themselves, out of the necessities of their secular strife with marauding neighbors, developed a body of true cavalry in emulation of their foes. The whole chapter affords a most interesting picture of certain aspects of the culture of the Chinese at a period when they had for the first time come in direct and conscious contact with the great civilized world of the west.

One hesitates to differ, even in minor points, with one who has so thoroughly mastered his subject as has Dr. Laufer. There is, however, a suggestion, early in this chapter (p. 201), which seems not quite convincing, although the point would not be worth dwelling upon did it not seem to disclose the existence of a certain convention among the artists of that day. Dr. Laufer remarks that many of the soldiers represented on the Han reliefs carry their shields in their right hands and their swords in their left, and suggests that this was done in order to rest themselves while fighting, by changing hands from time to time. Ambidexterity of this sort has certainly been known in the past; but its occurrence has invariably been so rare as to call for special mention. A careful study of the Han monuments suggests another explanation. In general, in these reliefs, movement seems to be from left to right; and when this direction is reversed, the attitudes of the personages seem also to be reversed. This is exactly what occurs, for example, in a relief (described by Dr. Laufer on p. 228) showing a battle with the Huns. Here in every case the archers advancing from the right hold the bow in the right hand and draw the arrow with the other, while their opponents, coming from the other direction, handle their bows in the normal manner. There can of course be even less question of changing hands with the bow than with the sword. This proposition appears to hold good for all the Han reliefs known to the present reviewer. Hence it would appear that this method of representation was due either to a feeling for symmetry, or else, perhaps more probably, to a desire to show the action of both hands; and the latter could not of course be accomplished if the arm held closest to the body were the one on the side away from the spectator.

The fourth chapter, devoted to a "History of Chain Mail and Ring Mail" begins with a definition of the two types; here, as always, Dr. Laufer's terminology is precise, logical, and consistent. The origin of chain mail is shown to have been in all probability in ancient Iran, appearing first during the Sassanian period. The evidence is then given for its introduction in China, where it seems first to have been made during the T'ang period (A.D. 618–907), from models derived from Samarcand. Ring mail, says Dr. Laufer, has never been much employed by the Chinese, but is ascribed by them to the Tibetans, who, as is known, had early relations with the Persians. The Tibetan sword and helmet, of Indo-Persian type, indicate the existence of this contact also.

Chapter v is a discussion of the "Problem of Plate Armor." This type is carefully differentiated from scale armor on the one hand and from sheet armor on the other, and is defined as consisting of flat rectangular laminae mutually lashed together, instead of being attached to a backing of leather or fabric, as is the case with scale armor. A generous tribute is paid to the work of Ratzel and Hough in investigating the distribution of plate armor in northeastern Asia and northwestern America, but it is shown that the conclusion arrived at by these investigators that Japan was the center of distribution of plate armor is untenable. The statement, however (p. 264), that there was no metal armor in

Japan prior to the end of the eighth century A.D., seems somewhat sweeping, although backed by the very great authority of Bashford Dean. At all events iron armor seems to have been associated in some instances with the dolmen type of burial, which tended to disappear after the introduction of Buddhism in the sixth century, and the discovery of blacksmiths' pincers of iron in deposits of that period would suggest that not all the metal objects found were importations from the continent. The point seems at least to call for further elucidation.

In the latter part of the chapter Dr. Laufer traces the history of plate armor during the T'ang period as illustrated by clay figures and carvings on stone. It is then followed through the Sung and Mongol periods, and the fact is pointed out that no fundamental change in plate armor has been made since the Mongol dynasty (A.D. 1280–1368), such alterations as have taken place pertaining only to style and ornamentation. A description follows of individual suits of recent date, and the chapter closes with a reference to the complexity of the technique of plate armor and the probability that a historical coherence exists between its various manifestations.

Chapter VI, upon "Defensive Armor of the T'ang Period," opens with a summary of the development of armor up to and including the T'ang epoch. Among the types mentioned is the curious paper armor which was then devised. The bulk of the chapter, a comparatively short one, is devoted to a discussion of the armor shown on certain clay figures of type originating in the Çivaitic worship of India and forming an unbroken series from early zoomorphic forms of Yama, God of Death, to figures of knights or champions of purely human character, without mythological connotations. It is pointed out that this type of image also occurs in Turkestan, and that therefore the armor which it wears, consisting essentially of metal plastron and dossière, is not Chinese in origin, but has some undefined relation to the sheet armor of the west. The chapter is brought to a close with a discussion of the so-called "lion armor" of the T'ang period, its probable nature, and the Indian-Buddhistic influence which it displays.

The closing chapter of the book, "Horse Armor and Clay Figures of Horses," gives us a historical résumé of horse armor, its Iranian provenance, and an account of the clay figures of horses and their riders from the provinces of Shen-si and Ho-nan, including a discussion of individual pieces and their significance.

Speaking of the work in its more mechanical aspects, there is very little fault to be found with it. Dr. Laufer's arrangement of his subject matter is logical and easy to follow, while his language is clear and makes pleasant reading. It seems, however, as though some of the matter dealt with in the footnotes might have been incorporated in the text with a distinct gain in the continuity of the thought. While the notes give an enormous amount of historical and other detail, all of great interest and having a direct bearing upon the question under discussion in the text, there is nevertheless a certain amount of distraction in referring to them so constantly. Also it would seem to have been better to have used the Chinese character along with the word transliterated, instead of merely referring to it by its number in Giles. The latter work is not always at hand, especially when one is traveling in the interior of China, and not the least of the many merits of Dr. Laufer's writings is their very great usefulness in field work.

The illustrations in the book are excellent, particularly the plates, which are carefully chosen and well executed and illustrate more adequately than is often the case the subject matter of the text. The proof reading has also been exceptionally well done, and only one or two small errors are to be noted. In the reference on p. 192, note 3, to Giles' *Biographical Dictionary*, p. 242 should be 212; while in that, p. 154, note, to Bushell, *Chinese Art*, vol. I, p. 119 should be 111.

On the opening page of this work Dr. Laufer says, "The second part of this publication will deal with the history of clay figures, the practice of interring them, the religious significance underlying the various types, and the culture phase of the nation from which they have emanated." It is to be hoped that this promised second part will not long be withheld.

C. W. BISHOP

OCEANIA.

The Mythology of All Races. Vol. IX: *Oceania.* ROLAND B. DIXON. Marshall Jones Co.: Boston, 1916. Pp. XV, 364, 24 pls., 3 figs., map.

Professor Dixon's volume is a scholarly contribution that must prove enlightening to the general reader interested in myth and will be of real service to the ethnologist who wishes to get his bearings amidst the chaos of Oceanian literature. Under each of the headings of Polynesia, Melanesia, Indonesia, Micronesia, and Australia, there is a chapter on Myths of Origins and the Deluge, and another devoted to Miscellaneous Tales; where the material gives warrant an additional chapter is added, and a Summary closes each section. Naturally it is only possible to refer to a few points of interest within the limits of a review.

THE

JOURNAL

OF THE

ROYAL ASIATIC SOCIETY

OF

GREAT BRITAIN AND IRELAND

FOR

1915

PUBLISHED BY THE SOCIETY

22 ALBEMARLE STREET, LONDON, W.

M DCCCC XV

instructive in more ways than one. We may look for good work from Dr. Nesbit.

T. G. PINCHES.

CHINESE CLAY FIGURES. Part I: PROLEGOMENA ON THE HISTORY OF DEFENSIVE ARMOR. By BERTHOLD LAUFER, Associate Curator of Asiatic Ethnology in the Field Museum of Natural History, Chicago. 1914.

Dr. Laufer has the pen of a ready writer, the equipment of a trained scholar, and the keenness of the scientific explorer. He has already made his mark in the field of Far Eastern history, art, and civilization, and seems destined to cut it deeper still there, though it should be borne in mind that sinologic topics form only a part of his professional studies, a fact which his growing competency in Chinese matters rather tends to obscure.

Before giving some account of this work there is one small bleat of discontent I am impelled to utter concerning the title. It is long, and so far as this part of the book goes it would more closely describe Dr. Laufer's treatment if it ran in some such terms as Early Armour, illustrated by Chinese clay figures and other plates. For other readers should be interested in these chapters besides those who devote themselves to the forbidding fruits of sinology.

The present part of the work consists of a volume of 315 pages, in seven chapters, followed by sixty-four plates, by no means all of which illustrate clay figures. There are also fifty-five text-figures. The plan of the whole cannot be better described than by its opening paragraph :—

" An extensive collection of ancient clay figures gathered in the provinces of Shen-si and Honan during the period from 1908 to 1910 is the basis of the present investigation. As the character of this material gives rise to research of manifold kinds, it has been thought advisable to publish it in two separate parts. Many of the clay statuettes which form the nucleus of our study are characterized by the wear of defensive armor, hence this first part is devoted to an inquiry into the history of defensive armor,—a task of great interest, and one which heretofore has not been attempted. It will be recognized that this subject sheds new light on the ancient culture of China and her relations to other culture zones of Asia. The second part of this publication will deal in detail with the history of clay figures, the practice of interring them, the religious significance underlying the various types, and the culture phase of the nation from which they have emanated."

This being the scheme, and the author being nothing if not thorough, we begin in chapter i, "History of the Rhinoceros," at the beginning, and at once find ourselves in an awkward place where no safety is, between Dr. Laufer on the one side, with a rhinoceros, or rather with two, *unicornis* and *sumatrensis*, and Professor Giles on the other, behind a vague but formidable "bovine animal". These three quadrupeds are claimants for the right to wear the Chinese names *ssŭ* and *hsi* (alias *se* and *si*), and therewith the honour of providing the ancient Chinese with the material of their first body armour, as described in the classical book, the *Chou Li*, or *Rites of Chou*. The chapter is very interesting and the longest of the seven, extending to no less than 173 pages. But just because it has raised a controversy it may unduly obscure the value of the remaining chapters, and I shall perhaps be of service to readers if I pass from it to an *aperçu* of the contents of those that follow.

Dr. Laufer's general view of Chinese civilization and its origins is summarized in chapter ii, "Defensive Armor of the Archaic Period," on p. 185. Speaking of the war-chariot, he says that like many other basic factors of ancient Chinese culture it is one of those acquisitions which ancient China has in common with Western Asia, and which go back to a remote prehistoric age. He proceeds in this chapter to consider what the most archaic armour of the Chinese was. Basing himself on the statements of the *Rites of Chou*, he concludes that contemporary armour was marked by the absence of any metal, and consisted only of a cuirass and a helmet, both of rhinoceros hide. (He uses "cuirass", however, to include a corselet and a short skirt, as I understand him.) He argues that the crucial passage in the work just named has been misunderstood by all the Chinese commentators, and in their wake by the French scholar Biot in his translation, and that the text does not mean, as they supposed, that a suit of armour consisted of seven, or six, or five pieces sewed together by the edges, but of that number of superposed layers of rhinoceros hide, cut up into large and thin sheets, first cured, and afterwards tightly pressed and sewn together.

The archaic helmets (*chou*) were, he thinks, only round caps of the same hide, corresponding to the Roman *galea*. I may add, however, that if we may judge by one ancient example of the character for *chou* which has survived, the latter would seem to have resembled the German *pickelhaube*, for a pointed spike appears, springing from a spherical or thickened base. Dr. Laufer maintains that the use of rhinoceros hide persisted in Chinese armies down to the T'ang period (A.D. 618–906), but not to the exclusion of metal, as the centuries passed. Naturally, no specimens of such hide armour have survived.

But the author thinks the archaic period knew also a scale-armour of hide, a type in which horizontal rows of scale-shaped leather pieces were fastened on a backing or foundation, also of hide. This type was known as *kiai* (*chieh* in Pekinese), a scale. Examples have been found in Japan, though not in China, but some of the curious clay figures illustrated in the plates, and representing Shamans, may be, as the author suggests, wearing such scale jackets, while brandishing spear and shield in their

exorcising dances.   Among the Khalkha Mongols the late
Captain Binsteed, of whose recent death at the Front we
have heard with deep regret, witnessed and described
in the last October Number of this Journal the modern
counterpart.   In general, the chapter concludes, this
archaic armour agrees closely with that of other primitive
populations in Asia, as, for instance, the Scythians described
by Strabo.

With chapter iii, "Defensive Armor of the Han Period,"
we pass to the introduction of metal armour into China,
and the fact occasions an interesting and suggestive
discussion of the reasons for the change.   Metal suits,
helmets, brassards, and neck-guards now appear, but
Laufer points out that the documents discovered by Stein
show that both the old type and the new hide reinforced
with metal were in use in this period among the
Turkestan garrisons.   It is here argued that the first
metal thus applied was copper, replacing the earlier
leather scales, and gradually developing a type of uniform,
oblong, rectangular "plate".   Ultimately iron ousted
copper armour, and was usual in the time of the T'ang
dynasty.   A corresponding change took place under the
Han, from copper to cast iron, for *offensive* weapons also.

In a most interesting passage on p. 217 Dr. Laufer
broaches a theory to account for these and other military
developments in China.   Briefly, it propounds the view
that ancient Irān evolved far-reaching military reforms
deeply affecting the entire ancient world, and, among
others, the Turkish peoples of Central Asia and of
Siberia.   One of these reforms was the institution of
a regular cavalry armed with metal-plated armour and
with sword and shield—the *cataphracti* of Xenophon's
day.   This mode of fighting and these weapons were
adopted, Laufer argues, by the Huns, the perpetual
enemies and scourge of the Chinese, but by the time
of the Han dynasty the latter had been wise enough, in

their turn, to imitate both the tactics and the equipment of their predatory neighbours.

Chapter iv, "History of Chain-mail and Ring-mail," illustrates the previous general thesis by an examination of the appearance of a special type (in the two varieties just mentioned) in China and other Eastern countries. The type was widely prevalent on the Volga and in Siberia, but observers agree that it was of foreign origin there, Persian in our author's view. He points out that the monuments show that both scale-armour and chain-mail were in use in the time of the Arsacides and their successors, the Sassanides. It was from Persia the author believes that both the Moghuls and the Arabs derived chain-mail.

Dr. Laufer obtained and illustrates two suits of such armour in China, one from Kansu and one from Shensi province, but cannot find that this type was ever in use by Chinese. But both chain-mail and the simpler scale-armour were worn by Tibetan soldiers, and as it is difficult to believe the latter country could have had the skill to produce chain suits, these must have been imported from the West, leaving the scale coat as an indigenous manufacture.

Chapter v is devoted to "The Problem of Plate Armor". By plate-armour, Dr. Laufer is careful to point out, is meant not that which consists of large surfaces of metal enveloping the front and back of the wearer (such suits he terms "sheet-armor"), but a defensive dress of horizontal rows of narrow, rectangular laminæ mutually lashed together, and each row similarly secured to the one above and below. This type, he insists, must be distinguished from scale-armour, for which a backing is indispensable, for in his opinion the two types are of independent origin.

Such plate-armour was worn in Japan, in North-West America, among the Eskimo, and by the curious tribe known to the Chinese historians as the Su-shen in North-East Asia. In this region the material was of bone, and it was in use in the third century, apparently long before the Japanese made any armour at all, even of plain leather. The author thinks from the available evidence that bone plate-armour in North-East Asia was as old as, perhaps older than, any of iron in China or Korea. The Scythians used bone armour of this kind, some plates of which have been found in South Russia. Lastly it existed in Assyria, and in the Egypt of Rameses II. Dr. Laufer suspects that such a wide dissemination is due to fitness for use by a cavalry of *cataphracti*. But when, where, and how the type first arose, and in what manner it spread to the widely separated regions in which it has been found, these are the unknown points which remain for the ambition of others to solve.

"Defensive Armor of the T'ang Period" is the heading of chapter vi. In this period, besides the armours previously described, we find figures of guardian deities clad in sheet-armour. The type, Laufer says, originated in the Sivaitic worship of India, and became widely diffused over Tibet, Turkistan, China, and Japan. The figures given in plates 46–61, excavated in Honan and Shensi, are remarkable. Especially perhaps plate 49, which is called "The Triumphant God of Death", who is represented as a knight with complete armour and has a strangely Western and mediaeval appearance.

Chapter vii, on "Horse Armor and Clay Figures of Horses", concludes this part of the work. It is mainly a description of the clay figures of horses, with and without riders, recently dug up from graves in Shensi and Honan. The figures are on the whole only indifferent, but plates 64 and 67 show really well-modelled forms, especially the latter. Both are from Honan.

I suppose I ought to take a few exceptions on points

of detail in taking leave of this solid contribution to knowledge until part ii is published. The alligator is not extinct in the Yangtze River, as the author supposes, p. 156. I have seen a living specimen myself, and the species was fully described by the late M. A. Fauvel. On p. 187 the expression *kan ko* is translated "shield and spear", but the *ko*, as numerous examples prove, was a kind of halberd. On p. 208 the author refers to pl. xix as illustrating "a three-storied watch-tower rising from the bottom of a round bowl; on the two parapets and roofs the sentinels are engaged in showering from their cross-bows a volley of darts at an advancing column of scouts". This same model is illustrated by R. L. Hobson in his recently published *Chinese Pottery and Porcelain*, where, on p. 13, it is described as a "fowling-tower". On p. 209 the left-hand entry from the Shuo Wên dictionary is not that intended by the author, who meant to cite the word *yeh*, but has inadvertently inserted the Shuo Wên's previous entry *tsi*.

L. C. HOPKINS.

I. DAVIDSON. SAADIA'S POLEMIC AGAINST HIWI AL-BALKHI. A fragment edited from a Genizah MS. 8vo; 104 pp. with a facsimile. New York: The Jewish Theological Seminary of America, 1915.

Among the fragments found in the Genizah in Cairo Dr. Davidson was fortunate enough to discover one more of the lost works of Seadyah, the great scholar, philosopher, and polemical writer of the tenth century. The object of this newly discovered treatise was the refutation of the anti-Biblical theses of a certain Hiwi of Balkh. Up to now only scattered allusions to this writer had been found in various books. The nature of his objections had practically remained obscure. It was more a guess than real knowledge which led the Jewish scholars

The
Burlington Magazine

for Connoisseurs

Illustrated & Published Monthly

Volume XXVII—No. CXLV—CL
April to September 1915

LONDON
THE BURLINGTON MAGAZINE, LIMITED
17 OLD BURLINGTON STREET, W.

NEW YORK: JAMES B. TOWNSEND, 15-17 EAST FORTIETH STREET
PARIS: BURLINGTON MAGAZINE, LTD., 10 RUE DE FLORENCE, VIIIe
BRUSSELS: LEBÈGUE & CIE., 36 RUE NEUVE
AMSTERDAM: J. G. ROBBERS, SINGEL 151-153
FLORENCE: B. SEEBER, 20 VIA TORNABUONI
BASLE: B. WEPF & CO,

Reviews

CHINESE CLAY FIGURES, Part I ; Prolegomena on the History of Defensive Armour ; BERTHOLD LAUFER ; 64 pl., 55 fig. (Field Museum of Natural History, Publication 177), Chicago, N.P.

As an anthropologist nothing comes amiss to Dr. Laufer, and it is not so much the diversity of his interests which amazes us as the profound erudition which he displays on every branch of his extensive work. Those who eagerly devoured his learned books on Han pottery and jade will welcome the first part of his study of Chinese clay figures ; and many others besides these, for in point of fact only a small proportion of the volume is actually concerned with the subject of the general title. The clay figures are here considered from the point of view of the armour which is indicated on them either by pigment or in the modelling, and the chief interest in Part I lies in the prolegomena on the history of defensive armour. That the history of the rhinoceros should occupy a hundred pages is at first sight disconcerting, but Dr. Laufer is concerned to prove that the ancient body armour of the Chinese was largely composed of the hide of this pachyderm. Hence the lengthy disquisition on the words *ssŭ* and *hsi*, which undoubtedly mean rhinoceros in the later Chinese writings, though sinologues have thought that in the archaic periods—the Chou dynasty, for instance—they referred to some animal of the genus buffalo. Existing Chinese illustrations of the *ssŭ* and *hsi* only serve to complicate the question, for they are no more like any living animal than the Chinese porcelain lion is like the king of beasts. However, Dr. Laufer is satisfied that the words in question mean rhinoceros in the Chou writings, and that rhinocerous hide was the armour of that period. Whether he has proved his point or not can only be determined by those who are able to follow him through the original Chinese texts. Other critics must accept his conclusions or suspend judgment. Incidentally we learn many interesting things about rhinoceros horn, how it was regarded as a protection against poison and numerous other evils, ranging from disease and dust to bad temper. Apart from this question the reader will follow with unflagging interest Dr. Laufer's history of armour in the archaic period, in the Han dynasty, and in the T'ang dynasty, as well as his researches on chain and ring mail, and on plate and sheet armour ; and in our present frame of mind we can especially enjoy his story of the evolution of military tactics and the never-ending struggle between attack and defence in which Persians, Scythians, Parthians, Romans, Huns Mongols, and Crusaders pass in review. Collectors of Chinese pottery and those who are interested in the clay figures of the Han and T'ang and intermediate periods will await with impatience the completion of Dr. Laufer's studies of these remarkable objects. In the first volume he illustrates a few types only. These are armed figures of the Chou period which he describes as *shaman* or exorcists in the act of attacking demons, besides similar figures of the well-known T'ang type and a few armed grave guards. The casual observer will group all these as warriors, and will not readily distinguish the *shaman* from the guard. Next to these is a very interesting series which presents similar difficulties. At one end is the zoomorphic form of Yama, god of death, with ferocious bull head with horns and bristling flames, a bear-like body and eagle's claws, the fore paws raised in a threatening attitude, and the hind paws resting on a prostrate figure of a sow, bull or demon. At the other end is the familiar figure of a fierce warrior with crested helmet, armed *cap à pie* and standing in triumphant attitude on a demon or a bull. This latter figure is generally regarded as a Lokopala or one of the four Buddhistic guardians of the world, and Dr. Laufer admits this attribution in the case of two Japanese examples in wood and a Chinese stone carving. But the series is carefully graduated, and it must be confessed that the one type passes into the other in a way that suggests that the individuality of these two divinities may have been merged by the Chinese when they had adopted them. On the other hand, there seems to be a connexion between the animal-headed Yama and the strange sphinx-like figures which are found in T'ang burials. But no doubt Dr. Laufer will deal with the sphinxes in his second volume. The remaining illustrations represent equine and equestrian figures, interesting for harness and caparison. The T'ang horses are well known and much admired, and Dr. Laufer makes some interesting generalisations from those in his collection. The horses found in Honan, he notices, are more spirited in modelling than those from Shensi, and the riders found in the latter province are distinguished by a pompom head-dress. All the figures except the Chou *Shaman* are assigned to the T'ang period. Dr. Laufer's conclusions throughout are backed by an overwhelming array of references and quotations, and his narrative, while teeming with erudition, is relieved by a vein of quiet humour. There are seventy-two excellent half-tone plates, besides numerous line blocks in the text, and the book is as worthy of Dr. Laufer's reputation as it is characteristic of his thoroughness and superhuman industry.   R. L. H.

(1) DECORATION IN ENGLAND FROM 1660 TO 1770 ; (2) FURNITURE IN ENGLAND FROM 1660 TO 1770 ; FRANCIS LENYGON. (" The Library of Decorative Art ", Batsford.) £2 each.

198

108

贝克《柏林皇家图书馆的藏文手稿索引》书评

THE

JOURNAL

OF THE

ROYAL ASIATIC SOCIETY

OF

GREAT BRITAIN AND IRELAND

October — — — — — — **1914**

OCTOBER 15

PUBLISHED BY THE SOCIETY

22 ALBEMARLE STREET, LONDON, W.

Price Twelve Shillings

wood which is stained black and which the author
mentions (p. 553) is Chinese ebony. The two Manchu
characters on the reverse of the Chinese coins give the
mint at which they were cast and the word "currency"
(p. 306). The Mandarin speech of Nanking is Southern
Mandarin, that of Peking is the northern variety,
and the latter city being the capital, the Northern or
Pekingese Mandarin has been the Court language
(p. 392). The Execution Ground in Canton is about
a couple of miles further down the river than the old
"Thirteen Factory Section". The Chinese language
possesses pronouns (p. 359) and they are in constant
use. If "London Mission" (p. 387) is meant for the
London Missionary Society, it should be struck out, as
that society is not represented in Fuchau. Sir Matthew
Nathan is still alive (p. 269). The author must have
confused him with someone else.

J. DYER BALL.

DIE HANDSCHRIFTEN-VERZEICHNISSE DER KÖNIGLICHEN
BIBLIOTHEK ZU BERLIN. Band XXIV: Verzeichnis
der Tibetischen Handschriften, von Dr. HERMANN
BECKH. 1. Abteilung: Kanjur. pp. 192 (22 × 30 cm.).
Berlin, 1914.

The Royal Library of Berlin has certainly deserved well
of Orientalists through a continuous accumulation of
manuscripts and books in almost all Oriental languages
and through the careful preparation of catalogues which
render these treasures accessible to the scientific world.
The works of Weber, Steinschneider, Dillmann, Pertsch,
Ahlwardt, and Sachau have achieved great reputation,
and become household books in the hands of specialists.
The cause of the literatures of Central and Eastern Asia
has heretofore been somewhat neglected, though the older
stock of Chinese and Manchu books was registered by

J. Klaproth as early as 1822, and published again in a more thorough manner by W. Schott in 1840. These lists, excellent as they were for their time, no longer satisfy the demands of modern sinology; and, with the great increase of books, a new catalogue of East Asiatic literature has become imperative. Professor H. Hülle, on the staff of the Library, who has had the advantage of several years' study in Peking, is now actively engaged in drawing up the catalogue of Chinese and Manchu books, the appearance of which is anticipated with great interest; while Dr. H. Beckh has devoted his energy during the last years to a survey of the Tibetan manuscripts. Dr. Beckh is no novice among students of Tibetan literature; in 1907 he brought out his painstaking study of the Tibetan version of Kālidāsa's Meghadūta, and in 1911 he gave us a critical edition of the Tibetan text of the Udānavarga. The Royal Library did well in choosing such a circumspect and persevering worker for the arduous task of examining the 108 volumes in its possession, constituting a magnificently written copy of the Kanjur; this was secured in 1889 from the Lama temple Yung-ho-kung at Peking,[1] through the good offices of Mr. v. Brandt, then German Minister in China. The inventory of this Kanjur edition forms the contents of the present volume, being the first of the entire series that is to deal with the Tibetan manuscripts and books of the Royal Library.

Our previous knowledge of the Kanjur was based exclusively on the analysis made by Alexander Csoma (*Asiatic Researches*, vol. xx), a very meritorious work, considering the trying circumstances under which the author laboured. The summaries of contents which he

---

[1] Dr. Beckh might have added that the merit of having obtained the copy is due to E. Pander, who at that time was Professor at the Peking University and on very friendly terms with the Lamas (see *Zeitschrift für Ethnologie*, vol. xxi, p. (203), 1887).

added to every work are still our main guidance in the labyrinth of this vast collection of religious literature, being based on faithful translations from Sanskrit. Sanskrit, however, was one of Csoma's weak points; and the Sanskrit titles, as recorded by him, suffer from numerous defects. Beckh is the first to inaugurate in this respect a sound reform, and to reconstruct by means of efficient critical methods both the Indian and Tibetan titles in their original and correct readings. In view of the host of copyists' errors by which the Berlin text is marred, this enterprise was by no means easy; and the author's patience and care merit the highest praise. He has indeed presented us with the first critical and accurate catalogue of Tibetan literature, which is extremely useful, not only to the librarian, but to the student of Buddhism as well.

Under each volume the miniatures adorning the covers are listed with their designations, and this has special importance for the iconography of Lamaism. The titles are given in Sanskrit and Tibetan, the latter in Tibetan types followed by a romanization, and, what is very gratifying, are accompanied by a translation. Then, the introductory salutation formula is given; and the locality of India where the plot of the story is laid has been added. The names of authors and translators, as far as they are on record, are pointed out; and colophons, with the exception of a few lengthy ones, have usually been translated in full. As the number of the folio and even the number of the line where each treatise begins is recorded, it is possible for the reader in the Royal Library to lay his hands on any desired work at a moment's notice. The two excellent indexes of the Sanskrit and Tibetan titles [1] are a most valuable addition, and

---

[1] Dr. Beckh (p. viii) asserts that an alphabetic list of the Sanskrit titles of the Kanjur had not previously existed. But there is one by L. Feer appended to his "Analyse du Kandjour" (*Annales du Musée*

simultaneously afford a concordance with the Index published by the Imperial Academy of St. Petersburg and prefaced by I. J. Schmidt, and with the analysis of Csoma. Beckh's index shows at a glance that many works of the Berlin edition (I noted fifteen of these) are not contained in the Kanjur utilized by Csoma and in the Index of St. Petersburg; and it is this peculiar feature that lends the Berlin version its special scientific importance, which will be discussed presently. Beckh has numbered the single treatises under each volume; in the opinion of the reviewer it would have been preferable to number the treatises right through from the beginning to the end of the work, irrespective of the volumes, in accordance with the Index of St. Petersburg and Bunyiu Nanjio's *Catalogue of the Chinese Tripiṭaka*. This procedure would have considerably simplified the work of indexing; it would enable one to recognize at once the total number of treatises embodied in the Berlin copy, and would facilitate the quotation of a treatise by referring to the individual number. This, however, is a purely technical point of minor importance.

As there are many editions and recensions of the Kanjur at variance with one another in the arrangement of the subject-matter and in contents, the main question to be raised is, What edition is represented by the Berlin copy, and what is its specific importance? Dr. Beckh has but partially responded to this query. The problem as to the date of the Berlin Kanjur is naturally one of consequence. In his brief discussion of this matter in the prefatory notice, Dr. Beckh has correctly seen two points, first that a recension of the Kanjur in 108 volumes, like the Berlin edition, differs from one in 100 volumes, as for instance represented by the edition of Csoma and the one underlying the *Index Schmidt*, and second that the

Guimet, vol. ii, pp. 499–553) ; Beckh's list, however, is far superior and represents the first complete and accurate inventory of the Sanskrit titles.

manuscript copies now current in China, Tibet, and Mongolia have no independent value, but are traceable to printed editions from which they have been copied. Misled by an unfounded statement of Colonel Waddell, Dr. Beckh is inclined to trace his copy to an alleged Derge print of the Kanjur in 108 volumes, and tentatively conjectures the date of the Berlin copy as being some time after the year 1731, which is an alleged date for the printing of the Kanjur at Narthang.[1] This argumentation is inadmissible, and the result conflicts with the facts as disclosed by the Berlin Kanjur itself. The edition of Derge published in 1733 cannot come into question as its prototype, for the technical reason that it consists in fact of only 100 volumes (this statement refers to a copy of this edition in the Library of Congress, Washington, examined by the writer), and for the inward reason that the Berlin version contains a number of works which are lacking in the Derge edition. But more than that—the Berlin copy, as revealed by inward evidence, cannot have been made from any edition of the Kanjur issued in the age of the Manchu dynasty (1644–1911); it can only have been copied from a print published under the preceding Ming dynasty (1368–1643), and its scientific importance rests on the fact that it reflects the tradition of the Kanjur, as it was established in the Ming period.

[1] Dr. Beckh credits this date to Waddell (*Buddhism of Tibet*, p. 159); but Waddell in this passage exactly copies a statement of A. Csoma (*Asiatic Researches*, vol. xx, p. 42), and even repeats the misnomer " wooden types " (instead of " wooden blocks "). As has been demonstrated by M. Pelliot, Csoma's dates are all unreliable. In fact, the preface of the index volume of the Narthang edition is dated 1742 (*c'u p'o k'yii lo*, " water male dog year "). If Beckh had consulted Csoma's work in its original issue, he would have been saved from the incorrect statement (p. vi) that Csoma gives no information as to the Kanjur edition utilized by him. Csoma, indeed, states that his study is based on the Narthang edition in the possession of the Asiatic Society of Calcutta and procured by Hodgson. Moreover, his analysis is in harmony with the index of Narthang. Koeppen also (*Lamaische Hierarchie*, p. 280) had already pointed out this fact.

Statements in regard to the date of the Berlin Kanjur have previously been made, and these should have been mentioned by Dr. Beckh. E. Pander, when he reported on the acquisition of the copy in Peking, left the chronological point undecided.[1] Grünwedel[2] first made the general statement that this copy is said to come down from the end of the sixteenth century; and on another occasion referred to the Wan-li period (1573–1620) of the Ming dynasty as the time when the work was copied, presumably from the Yung-lo edition of the Kanjur printed in 1410.[3] Of the latter, the Royal Library possesses thirty - seven volumes procured in Peking, likewise by E. Pander,[4] among which the twenty-four volumes of the Tantra are said to be complete.

From these statements I suspected that the manuscript Kanjur might contain a colophon making a reference to the Wan-li period, or at least imparting some information as to its date, and therefore appealed to my friends in Berlin. Professor Grünwedel, in a letter of April 22, was good enough to write me that his notice previously quoted was based on an entry in the Accession Documents of the Pander Collection: "In a letter of Minister H. von Brandt, dated Peking, March 18th, 1888, the Berlin Kanjur is styled 'a unique copy made at the time of the Ming Emperor Wan-li (1573–1620) after a printed edition from the time of Yung-lo, accordingly, prior to the reforms of bTson-k'a-pa'." Professor H. Hülle of the Royal Library, in a letter of May 15, very courteously replied in detail to all my queries regarding the Berlin Kanjur editions, and stated that the colophon suspected by me in the beginning or end of the work is not to be found, nor

[1] *Zeitschrift für Ethnologie*, vol. xxi, p. (203), 1889.
[2] *Mythologie des Buddhismus*, p. 178.
[3] "Die orientalischen Religionen" (in *Kultur der Gegenwart*), p. 161.
[4] Loc. cit., p. (201). Pander gives the date as 8th year of Yung-lo, but wrongly identifies it with the year 1411 instead of 1410.

does it contain any Chinese prefaces or postscripts; but Mr. Hülle found in a catalogue of the Library the same observation as made by Professor Grünwedel, entered by the hand of L. Stern, the late Director of the Department of Manuscripts. This statement, ultimately, must have emanated from E. Pander, who acquired the copy at Peking, and seems to have received this communication from his Lama friends. The question, then, pivots around the point whether Pander's information is trustworthy. Pander was exceedingly well posted on Lamaist affairs, and at the outset I can see no reason why his information should be discredited. There are, in fact, several circumstances which conspire to prove that his opinion is well founded. Professor Hülle has been so kind as to place at my disposal photographs of the imperial preface and postscript in Chinese and Tibetan accompanying the printed Ming Kanjur, and these are indeed written by the Emperor Yung-lo and dated in the manner that Pander had indicated.[1] Pander, accordingly, was quite right in this assertion; and the Yung-lo Kanjur of Berlin is a fact. Further, Dr. Beckh, to whom I submitted my opinion on the date of the manuscript Kanjur, advised me that the latter closely agrees with the Yung-lo edition, as far as he had collated the two, and that for this reason he too had formed the opinion that the manuscript edition is copied from the Yung-lo print. This, then, forms the second point in which Pander is correct. Of course, the assumption that on account of this agreement the copy was actually made in the Ming period is not cogent; in theory it may have been executed at any posterior time. But such a theory is not very probable. Each historical period has had its standard

---

[1] The exact date is 永 樂 八 年 三 月 初 九 日 ; in Tibetan ཡུང་ལོ་བརྒྱད་པའི་ལོ་ཟླ་བ་གསུམ་པའི་ཚེས་དགུའི་ཉིན ॥ "On the ninth day of the third month of the eighth year of the period Yung-lo (1410)."

printed edition of the Kanjur, from which the manuscript copies were written out; and the issue of a reprint is always the symptom of a previous edition being extinct. When the K'ang-hi Kanjur was out in 1700, when in the reign of K'ien-lung several editions were published in Peking as well as in Tibet, it is not very likely that at that time a copyist should have resorted to the Yung-lo edition as his model. The greater probability, at any rate, is that a Kanjur copy coinciding with the latter is also the work of the Ming period. It is hoped that the further researches of Dr. Beckh will settle this point positively. Meanwhile I wish to call attention to the fact that the Kanjur analysed by Dr. Beckh does not contain any traditions relating to the Manchu dynasty, and does contain works which are not extant in the editions coming down from the age of the Manchu.

In like manner, as works were added to the Chinese Tripiṭaka under the Manchu dynasty, so also books translated from Chinese into Tibetan were joined to the Kanjur in the same epoch. The titles of such books are to be found in the *Index Schmidt*, for instance No. 199 (p. 33): *byaṅ=c'ub sems-dpa dga-ldan gnam-du skye-ba blaṅs-pai mdo*, "Sūtra as to how to be reborn as a Bodhisatva in the Tushita Heaven," translated from Chinese by Bab-toṅ (Tibetan transcription of a Chinese name) and Šes-rab seṅ-ge (Skt. Prajñāsiṁha). This work, however, is not contained in the Berlin edition.

In the *Index Schmidt*, No. 446 (p. 67), we find a work translated by Bu-ston, with a colophon saying that a new translation of it was made after the Chinese version by mGon-po skyabs, professor at the Tibetan school of Peking.[1] This personage is well-known, and lived during the K'ang-hi period (1662–1722). Turning to Beckh's catalogue (p. 86 *b*), we notice that the fact of this revision is lacking in the colophon of the Berlin edition, which

[1] The colophon is translated in the writer's *Dokumente*, i, p. 52.

likewise attributes the translation to Bu-ston, but refers
to a revision through Rin-c'en rgyal=mts'an. The tradition
of the age of the Manchu is therefore unknown to the
Berlin Kanjur. A similar case occurs in No. 502 (p. 76)
of the *Index Schmidt*, where a brief *dhāraṇī* is cited as
having been translated afresh by the same mGon-po
skyabs, who on this occasion is characterized as "the
great translator of the present great Ts'ing dynasty".[1]
Hence it follows that the *Index Schmidt* represents the
tradition of the time of the Manchu dynasty. Again,
this colophon is absent in the Berlin recension (Beckh,
p. 97, xii, 1). The last eminent scholar who laid his
hands on the Kanjur was Tāranātha, born in 1575; he is
expressly named as such in the index volume of the
Kanjur printed at Derge in 1733 (fol. 97*b*). Several
separate issues of Kanjur treatises are known to me
which, according to the colophons, were revised by
Tāranātha.[2] His name, however, is not mentioned in any
colophon of the Berlin copy.

As already stated, at least fifteen treatises of the latter
are wanting in the subsequent editions. It would be
very interesting to study these, and to ascertain by what
reasons their elimination under the Manchu may have
been prompted. Another task that remains to be done is
to draw up a list of those treatises found in the Manchu
prints of the Kanjur, and which are lacking in the Ming
edition.

The greatest surprise offered by the Berlin Kanjur is
the fact that the uncanonical "Sūtra of the Dipper",
which has never before been pointed out in any other
edition, is embodied in that collection (Beckh, p. 70). This
work has been discussed by the writer in *T'oung Pao*,
1907 (pp. 391–409, with an additional note of S. Lévi,
ibid., p. 453), together with the texts and translations of

---

[1] Tib. *da-lta c'en-po C'iñ gur-gyi lo-tsts'a-ba c'en-po.*
[2] In particular an edition of the *dhāraṇī* styled *Vajravidāraṇā.*

the two interesting colophons, which yield the date 1337 for the Tibetan translation. Dr. Beckh has overlooked this contribution; he leaves the first colophon untouched, and gives but a brief abstract of the second,[1] omitting all difficult points and any dates, and without discussing the important fact that this Sūtra is lacking in other editions, and is for the first time here revealed in the Kanjur. Bu-ston, the editor and first publisher of the Kanjur, according to the chronological table *Reu mig*, lived from 1290 to 1364; and his edition of the Kanjur was printed at Narthang, at the time of the Emperor Jên-tsung (1312–1320).[2] The translation of the Sūtra of the Dipper made at Peking in 1337, therefore, cannot have been embodied in his collection. Indeed, it has never been included in any of the subsequent re-editions of Narthang as shown by the index volume of the last edition of 1742 and the analysis of Csoma based thereon, where this work is omitted. It is likewise absent from the *Index Schmidt*. We are thus bound to conclude that the Sūtra of the

---

[1] This has been done by him in several cases; and he observes that the context, as in many colophons, remains obscure (p. 68), or that the entire colophon is very difficult and obscure (p. 136). In such cases it would have been advisable to publish the texts of the colophons *in extenso* in order to enable future students to make the best use of them.

[2] Huth, *Geschichte des Buddhismus*, vol. ii, p. 165, and Laufer, *Dokumente*, i, p. 53. From the text of a Jigs-med nam-mk'a it follows that the first Narthang edition was printed in black by means of Chinese ink. Also the later edition of 1742 was printed in black. This point is mentioned here because Schmidt and Boehtlingk (*Verzeichnis der tibetischen Handschriften*, p. 4) speak of a Narthang edition of the Kanjur in St. Petersburg as printed in red (that is, vermilion); but no such vermilion print has ever been issued from the press of Narthang. Technically it is impossible to print from the same blocks a copy in black and another one in vermilion; the same blocks can be utilized for impressions either in black only, or in vermilion only. From the summary of contents given by Schmidt and Boehtlingk it follows that the Kanjur in question cannot be an edition of Narthang, for in the latter the section Nirvāṇa (Tib. *myaṅ-adas*) occupies a separate department (No. vi); while in their edition this section is joined to the Sūtra class. It is therefore probable that this edition is the one printed at Derge, which is, indeed, in vermilion.

Dipper was inserted during the Ming period in an edition of the Kanjur somewhat deviating from the orthodox line; certainly, this cannot have been any Ming Narthang edition, nor in all likelihood any edition issued in Tibet. The Sūtra of the Dipper having first been published in Peking, the greater probability is that also the first Kanjur print containing it may have been one executed in China. The Yung-lo edition to which the Berlin copy is supposed to go back is therefore the one suspected, and it would be interesting if the Sūtra in question could be traced in one of the thirty-seven volumes of that edition extant in Berlin. On the other hand, there is also a reason militating against this supposition; and this is that, as formerly remarked by me, the Sūtra is not contained in the Ming edition of the Chinese Tripiṭaka. It has been adopted, as observed by M. Lévi (loc. cit.), in the new Japanese edition of the Tripiṭaka published in Tōkyō. I have been searching unsuccessfully through the index of the K'ien-lung Tripiṭaka of 1738, but I am not positive in asserting that the work should not be contained therein. Thus the Berlin Kanjur raises a problem which remains to be investigated.

A catalogue of the Kanjur and Tanjur should furnish us with the material with which to build the most important chapter of the literary history of Tibet. This, in fact, remains a task to be desired. The chief sources from which we have to draw for this purpose are the colophons appended to the individual works, and giving names of authors, translators, patrons, localities, etc., with greater or less fullness. There are very simple and brief colophons, easy to grasp at first sight; there are complex and lengthy ones of problematic nature, and exacting hard study. Tibetan is not an easy language, and we are all liable to err in translating from it: those who fancy themselves to be infallible are usually those with the largest quota of mistakes.

The correct understanding of proper names and separation of personal from clan names and titles, especially, is one of the difficult points in Tibetan, and there is no reference book to assist the student. Frequent misconceptions occur in linking two names into one, or in taking as the name of one man a compound in which two names are abridged. On p. 33 Dr. Beckh imparts a name in the form " der an der Grundlage des Dharma festhaltende mit ausgezeichnetem Verständnis begabte Lotsava rGya-mts'oi-sde ". In fact, two names are here intended, namely, dGe-bai Blo-gros with the title *Dharmai - gži adsin* (that is, " comprehending the foundation of the Dharma ") and rGya-mts'oi-sde with the title *Locāva* (" translator "). On p. 75 (and similarly on p. 90) we read " C'os-kyi dbaṅ-p'yug, der Übersetzer aus Mar ", instead of Mar-pa, the translator, with the title *c'os-kyi dbaṅ-p'yug* (" the lord of the doctrine," Skt. *dharmeçvara*). In lieu of " Pad-ma-ka-ri ", which occurs in the same colophon, read " Padmākara " (see *Index Schmidt*, No. 366). Instead of " the Tibetan translator Bande Zla-bai od-zer from Gyi-jo " (p. 73) read " the Tibetan translator Bande Zla-bai od-zer, the lord of Gyi ". The phrase *žal sña-ṅas* in the same colophon does not mean " in the presence of ", but is an honorary appendix frequently attached to the names of clericals in the Kanjur and Tanjur, as well as in historical records, and may approximately correspond to our Rev. So-and-so. On p. 75 (ii, 1) the same phrase has been rendered " with the assistance of " ; for " Dpal-šes ", etc., read " dPal Ye-šes ", etc. " C'os-kyi blo-gros from Mar-pa-lho-brag " (p. 87), in my opinion, must be altered into " C'os-kyi blo-gros and Mar-pa from Lho-brag ", the latter being a province of Southern Tibet bordering on Bhūtan, from which Mar-pa, the teacher of the famous mystic and poet Mi-la ras-pa, hailed. On p. 63 Dr. Beckh correctly annotates that Žaṅ Ye-šes sde (that is, Ye-šes sde from Žaṅ, the latter being a locality and at the same time his clan name) is identical with the personage, otherwise styled plainly Ye-šes sde; but then how can Dr. Beckh (p. 31) derive the same man from Samarkand by believing that sNa-nam Ye-šes sde means " Ye-šes sde von Samarkand " ? How should a Tibetan translator who worked in the first part of the ninth century at the time of King K'ri-lde sron̄-btsan have originated from Samarkand ? His full name is Žan̄ sNa-nam Ye-šes sde; that is, Ye-šes sde from sNa-nam in Žan̄, Žan̄ being a district in the province of gTsan̄ in Central Tibet. Consequently, sNa-nam in this case is a locality in Tibet.[1] Me-ñag (p. 95) is not the name of a king, but of a country, usually styled Mi-ñag. As regards the Bhikshu Çiladharma from Li (p. 53), Dr. Beckh is inclined to identify the name Li with

---

[1] Cf. *T'oung Pao*, 1914, p. 106. The same error of taking sNa-nam for Samarkand in connexion with a purely Tibetan name is committed by P. Cordier (*Cat. du fonds tibétain*, ii, p. 84, No. 46). These translations are based on the fact that Jäschke in his Dictionary, with reference to *rGyal rabs*, assigns to sNa-nam the meaning of "Samarkand"; Chandra Das, without adducing any proof, has merely copied Jäschke. The question is whether Jäschke is correct, and on what evidence his opinion is founded. In *rGyal rabs* we find mention of a queen from sNa-nam, married to King Mes 'Ag-ts'om. I. J. Schmidt (*Geschichte der Ost-Mongolen*, p. 349), translating from the *Bodhi-mör*, the Kalmuk version of the Tibetan work, styles her the chief consort from a clan of Samarkand, without advancing evidence for this theory. In his *Forschungen* (p. 231, St. Petersburg, 1824), however, the same author states that the Kalmuk original has in this place *Samardshen*. The date of the Kalmuk work is not known; since Kalmuk writing was framed as late as 1648, the Kalmuk translation, as a matter of principle, cannot be earlier than the latter part of the seventeenth century. The case therefore hinges on the point whether the Kalmuk rendering of recent date is correct in its understanding of the Tibetan word. Neither Kovalevski nor Golstunski (in their Mongol Dictionaries) has recorded the word *Samardshen*. Whether sNa-nam ever had such a meaning remains to be proved, if indeed it can be proved. For the time being the matter is open to doubt, and it seems more than doubtful that the Tibetans ever had relations with Samarkand. But the supposition that Tibetan authors living and working on Tibetan soil were born in Samarkand, which would presuppose the existence there of a Tibetan colony in the T'ang period, is somewhat adventurous.

Li-t'ang. This is very improbable, as the monastery of Li-t'ang was founded as late as 1580,[1] and owing to its location in the western part of Sze-ch'uan, is far remote from Central Tibet, where most of the translations took place; no translators from this monastery are otherwise known in the Kanjur, and it is always styled Li-t'ang, and never Li. *Li* is very familiar to us in the Kanjur and Tanjur as a designation of Khotan, as has convincingly been proved by Mr. Rockhill.[2]

The colophon on p. 136 (No. 5) is difficult, but the text is evidently corrupt, and must be collated with another edition. It seems hardly possible that a work could have been translated " auf der Spitze des Turmes des Klosters Byams-sprin in Maṅ-yul in der Verborgenheit ". The word *dbu-rtse* does not mean " Spitze ", but designates the chief temple-building or hall in a lamasery ;[3] *ya-t'og* is not a tower, but the upper story of a building. The sense of the passage therefore seems to be that the translators withdrew into and kept themselves in retirement in the main hall on the upper floor in the monastery Byams-sprin (Skt. *Maitrīmegha*).

A very attractive task it is to pursue the gradual growth of the Kanjur and Tanjur through the course of many centuries, and to establish the chronology of the translations. This task is not entirely insuperable now, especially when we avail ourselves of historical literature, like the *dPag bsam ljon bzaṅ* and other works. Dr. Beckh should not have wholly neglected the historical question of the translations ; only through an attempt at

---

[1] By the third Dalai Lama bSod-nams rgya-mts'o (1543–88) ; see Huth, *Geschichte des Buddhismus*, vol. ii, p. 224. The foundation of the monastery, accordingly, falls within the Wan-li period (1573–1620), during which the Berlin Kanjur was presumably copied. It is therefore impossible to assume that a translator named in this edition could have come from Li-t'ang.

[2] *The Life of the Buddha*, p. 230.

[3] See *T'oung Pao*, 1908, pp. 20, 22.

determining the time of the translators and translations may we hope to correctly understand the colophons and the proper names.   Kun-dga rgyal-mts‘an, for instance, mentioned as translator on pp. 93–4 and 128, is nobody but the celebrated aJam-mgon Sa-skya Paṇ-c‘en (1182–1251).[1]   His collaborator bDe-bar gśegs-pai dpal (Skt. Sugataçrī) is apparently identical with the Paṇḍita Saṁghaçrī, who instructed him in logic, *pāramitā*, grammar, poetry (*kāvya*), metrics, lexicography, and dramatic art.[2]   Hence he receives in the Kanjur the title " the great grammarian ".[3]

Whatever these matters bearing on details may be, they do not detract from the great value of Dr. Beckh's thorough work, for which we have every reason to be grateful to him.   In the most disinterested manner he has presented us with a handbook of practical and

[1] Huth, loc. cit., p. 118.   In the index volume of the Kanjur of Derge (fol. 97 *b*), where he has the attribute *sNar-t‘an-pa*, " the man from Narthang," he is expressly listed among the collaborators of the Kanjur.

[2] Ibid., p. 122.   Tib. *zlos-gar* is not " art of dancing ", as translated by Huth, but " dramatic art " (*nāṭaka*).

[3] Tib. *brda sprod-pa c‘en-po*, which does not mean " der grosse Erklärer von Symbolen ", as Dr. Beckh (p. 128) translates.   A title which has greatly embarrassed the author occurs in the same colophon, in the form *sgrai gtsug lag lam rmoṅs-pa*, tentatively translated by him " one who obscures the road of linguistic science ", and accompanied by a note to the effect that this might possibly be a proper name, though somewhat strange.   It is not, however, a proper name but rather a title.   The word *rmoṅs-pa* was indicated as an epithet of Tāranātha by A. Schiefner (*Tāranātha's Geschichte des Buddhismus in Indien*, p. vii, n. 2), and *rmoṅs-pai gñen-po*, " the adviser of the ignorant," is a title bestowed on members of the clergy (*Dokumente*, i, p. 61).   Thus the above title apparently means " one who is a guide along the dark points in the science of language ".   For the rest, *rmoṅ-ba, rmoṅs-pa* is not a transitive verb, and never means " to obscure " (which is *rmoṅs-par byed-pa*), but " to be obscured, obscurity ", etc. The monastery Sar-sgreṅ on p. 9 is to be corrected into Ra (or Rva)-sgreṅ, as shown by a Peking print of the work in question containing the same colophon.   This follows also from the historical context of the passage, owing to the mention of aBrom-ston, who was the founder of the monastery Ra-sgreṅ.

permanent utility which will be warmly appreciated by all present and future students of Tibetan literature. The next volumes which are promised us are eagerly anticipated. The clear printing of this volume in two columns reflects much credit on the printing-house of Unger Bros., as well as on the munificence of the Royal Library, which deserves sincere congratulations on this enrichment of its catalogues.[1]

B. LAUFER.

[1] After the above was written, I had meanwhile an opportunity of exactly collating the Index of the Kanjur of Derge with that of Berlin, and may now positively state that the two editions are independent, and that the Berlin version cannot be traced to that of Derge. There are treatises in the latter wanting in the former and vice versa; above all, the arrangement of the works in the section Tantra is widely different in Derge from the Berlin copy and other editions of the Kanjur. I hope to come back to these questions in detail in a future bibliographical study of the Kanjur. The collation with other editions bears out the fact that many colophons of Berlin are sadly deficient, and especially that numerous proper names are disfigured. A few examples may suffice. On p. 76a (below) we read of a monastery Yu-tuṅ-lhan in Nepal; the real name is Yu-ruṅ, while *lhan* is an error for *lhun*, which does not belong to the name, but to the following *-gyis grub-pai gtsug lag k'aṅ* ("miraculous monastery"). On the same page, and again on p. 77, we meet the wrong name of a translator in the form La-bciṅs-yon-tan-ạbar; it should read Yon-tan-ạbar from C'iṅs (written also aC'iṅs). Byai gdod-pa-can (on p. 95b) should be *gdoṅ* ("the Bird-faced one"). The name of the translator K'u-ba-lha btsas (p. 126a) is correctly K'ug-pa lhas-btsas; instead of Klogs-skya (ibid.) read Glog-skya; instead of Ḍo-ma-bi (p. 87a) read Ḍombi. In many cases the Berlin colophons are incomplete, or there are none at all where they can be supplied from other editions. It is therefore unsafe to found a study of the translators on the work of Beckh. The colophon on p. 106 (No. 29) has been entirely misunderstood by the author: he distils from his corrupted text a monastery Dbe-rñid in Kashmir, and makes it the place where the translation took place. Neither, however, is the case; *dbe rñid* is an error for *dpe rñiṅ* ("old book"), and the passage means, "The Paṇḍita Parahitaprabha [thus written in the Index of Derge] and the Locāva gZu-dGa-rdor have translated the work, and edited it on the basis of an ancient book hailing from the monastery Amṛitasambhava (Tib. *bdud-rtsi ạbyuṅ-gnas*) in the country of Kaçmīra." A wrong translation occurs on p. 67 in the colophon of *mDsaṅs blun*, which does not mean "seems to be a translation from Chinese", but "it has been translated from Chinese". The verb *snaṅ-ba* never assumes the significance "to seem".

109

一些关于中国文化的基本理念

附：书评一则

THE JOURNAL

OF

RACE DEVELOPMENT

EDITED BY

GEORGE H. BLAKESLEE AND G. STANLEY HALL

CONTRIBUTING EDITORS

PROFESSOR DAVID P. BARROWS,
University of California
PROFESSOR FRANZ BOAS,
Columbia University
PROFESSOR W. I. CHAMBERLAIN,
Rutgers College
PROFESSOR W. E. B. DuBois,
New York
GEORGE W. ELLIS,
Chicago
WM. CURTIS FARABEE,
University of Pennsylvania
PRESIDENT A. F. GRIFFITHS,
Oahu College, Honolulu
ASS'T-PROFESSOR FRANK H. HANKINS,
Clark University
ASS'T-PROFESSOR ELLSWORTH HUNTINGTON,
Yale University
PROFESSOR J. W. JENKS,
New York University

GEORGE HEBER JONES,
Seoul, Korea
JOHN P. JONES,
Madura, India
ASS'T-PROFESSOR A. L. KROEBER,
University of California
PROFESSOR GEORGE TRUMBULL LADD,
Yale University
PROFESSOR EDWARD C. MOORE,
Harvard University
K. NATERAJAN,
Bombay, India
PROFESSOR HOWARD W. ODUM,
University of Georgia
JAMES A. ROBERTSON,
Manila
PROFESSOR WM. R. SHEPHERD,
Columbia University
ASSOC. PROFESSOR PAYSON J. TREAT,
Stanford University
ASS'T-PROFESSOR FREDERICK W. WILLIAMS,
Yale University

VOLUME 5
1914-1915

CLARK UNIVERSITY
WORCESTER, MASS.
LOUIS N. WILSON, *Publisher*

SOME FUNDAMENTAL IDEAS OF CHINESE CULTURE

By Berthold Laufer, Ph.D., Associate Curator of Asiatic Ethnology, Field Museum, Chicago

Of all the numerous problems with which the scientific research of China is concerned, the problem of the early origin and development of Chinese civilization is the most important, and at the same time the most fascinating. In former times, when the exploration of China was still in its infancy, two main theories, in strong contrast with each other, were advanced in regard to the origin of the Chinese. In the eighteenth century, when both China and Egypt were imperfectly known, it was almost inevitable that the two should be linked together by a common source of origin; and in more recent times the romantic school of sinologues, headed by T. de Lacouperie, stamped the Chinese as emigrants from Babylonia, bringing from there all the essential elements of West-Asiatic civilization. The French Count Gobineau is responsible for the not very serious hypothesis that the culture of China in its total range may have been derived from India. Other scholars endowed with a lesser degree of imaginative power insisted on the independence and originality of Chinese culture, and vigorously stood on the platform of a Monroe doctrine, "China for the Chinese." But this theory of perfect seclusion and isolation of ancient Chinese culture can no longer be upheld; for we begin to recognize more and more its historic and prehistoric connection with other culture-groups of Asia, and to understand that also the Chinese were a people among peoples.

Indeed, no culture on this globe was ever exclusive or singled out, or had a purely internal development prompted by factors wholly within itself. The growth and diffusion of culture are due to historical agencies, and must be

160

comprehended in connection with the universal history of mankind. No historical problem can be understood and solved with any hope of success by limiting the attention to one particular culture-sphere to the exclusion of all others, and even in the minutest specialization of our work we must never be forgetful of the universalistic standpoint. Aside from the lack of critical methods, the principal error of those who simply reduced Chinese culture to a loan received from the west, was that the antiquity of the fundamental elements of civilization was far undervalued, and that a purely imaginary drama of migration of tribes was staged which has no basis in fact. Beyond any doubt, the foundations of civilization are far older than the period to which the oldest extant documents of the Egyptians, Sumerians, and Chinese, carry us back; and the impression even prevails at present that they are still older than we are now inclined to assume on the ground of archæological facts and internal evidence. The acquisition of cultivated plants, their wide distribution over immense geographical areas in Asia and Europe, the introduction of agriculture, the domestication of animals, the mining and working of metals, the conception of the important technical inventions, in order to come into being, must have taken, even within the boundary of reasonable calculation, not centuries but millenniums of human labor and exertion, and are removed far beyond the bounds of all historical remembrance. As to the question of migrations, it is not tribes but the very ideas of culture which have constantly been on the path of migration, which were transmitted from people to people and fertilized and advanced the life of nations. In the earliest records of the Chinese we meet no tradition pointing to an immigration from abroad. All that the conservative historian may safely assert is, that they inaugurated their career in the fertile valley of the middle and lower course of the Yellow River and its affluents, and gradually expanded from this centre of their early habitat eastward toward Chili and Shantung, and in a southerly direction toward the Yangtse. In their onward march they encountered a large stock of an aboriginal population of most varied tribes, partly related to them in language, with whom they struggled many centuries for the supremacy in China. The comparative study of Indo-Chinese languages has brought out the fact that the Chinese are a member of an extensive family of peoples, the best-known representatives of which are the Siamese, the Burmese, and the Tibetans. In early historical times all these peoples lived in close proximity to and relationship with the Chinese, in the western and southwestern part of China; and we are able to trace from their records and tradition the history of their migrations into the countries which they now occupy. The Tibetans designate themselves Bod (Sanskrit Bhota), and Ptolemy knows them by the name Bautai inhabiting the river Bautisos, identified with the Upper Yellow River. The present territory of Western Kansu and Szechuan was the cradle of the Tibetan branch which moved from there westward into the present territory of Tibet, probably during the first centuries of our era. The province of Yünnan is the home of the forefathers of the modern Siamese formerly known as Shan or Ai-lao (the modern Laos), who formed the highly organized kingdom of Nan-chao. Their state was destroyed by the Mongols in 1252, and the Mongol invasion gave the incentive to an emigration of the Shan from Yünnan down into the peninsula, where they founded the Kingdom of Siam in about 1350.

In the extreme southeast of Asia, scattered over the mountains and littorals of Indo-China, we meet another large group of peoples whose languages show no affinities with Chinese, and who form a distinct family. The most prominent members of this stock are the Annamese, the Khmer of Cambodja, the Mon of Pegu in the delta of the Irawaddy, the Khasi, and the Colarians, whose remnants are dispersed over the hill tracts of Central India. In prehistoric times this group extended also into southern China, and it is due to the expansion of the Chinese that they were subsequently driven back farther toward the south. These ethnical movements render it clear that the present Chinese territory is in the main composed of two distinct culture-areas,—a northern one, decidedly Chinese;

and a southern one, originally non-Chinese, but later colonized, absorbed by and assimilated to Chinese rule. Present-day China is a political, not a national or ethnical unit. The antagonism that still prevails between the people of northern and southern China, and which nearly resulted in a partition of the country during the recent revolution, has come to the notice of everybody. It amounts not only to a question of racial differences, but to a far-reaching divergence of culture and economy as well. The farmer of the north grows wheat, barley, and various species of millet, and tills the ground with the ox as the draught-animal of his plough. The south is engaged in the cultivation of rice, and the peasant avails himself of the water-buffalo, an animal domesticated in south-eastern Asia. His method of farming, corresponding to the subtropical flora characterized by palms, evergreen shrubs, fragrant woods, and tropical fruits, consists essentially in gardening, where that primitive system of hoe-culture still partially survives in which not the plough, but only the hoe, is employed. The north is traversed by highways, and the two-wheeled cart drawn by mules is the usual means of conveyance; besides, the horse, the donkey, the camel, are in evidence as pack-animals and for riding. The south is densely intersected by rivers and a net of skilfully laid out canals connecting rivers and lakes, so that boats are the favorite method of journeying and transporting goods; on land, the sedan-chair carried on the shoulders of bearers is the means of transportation, whereas horses and mules are almost absent or scarce. The northerners are typical children of the soil, conservative, and somewhat heavy; the southerners, more alert and quick tempered, are sons of the watery element, river boatmen, bold seafarers, enterprising merchants, emigrants and colonists. The Chinese, of course, are by origin a purely continental race; and one of the most attractive chapters of their history is the one telling how they gradually extended from their inland seats toward the sea-coast, how their naïve astonishment at the grandeur of the ocean produced a marine mythology and legends of distant lands and blessed isles, and how they learned and acquired the art of navigating from the seafaring nations along the shores of Indo-China. The north, in close contact with central and northern Asia, was constantly engaged in perpetual defensive wars against the restless hordes of Turkish and Tungusian nomads, and subject to influences coming from that direction; the horse, the donkey, the camel, the tactics of mounted archers and cavalry, felt and rug weaving, are due to this contact. The south was always deeply influenced by currents of thought pouring in from Malayan and Indian regions, and still visible in the laying-out of settlements, in domestic architecture, in every-day implements, and in certain industries and products.

The knowledge of the geographical distribution of such culture-elements as are here pointed out is naturally the basis for the understanding of their origin and historical development. For this reason let us now turn our attention to that northern culture-province which represents the original culture of the Chinese, and which was subsequently welded with the south into that unit which is now included under the name "China." The main question to be raised is, What relation did that culture hold to the other cultures of Asia? When we attempt to reconstruct by comparative and intense methods the oldest accessible primeval forms of the ancient civilizations of Asia, we are ultimately led to the result that in an undefinable pre-historic age a great universal and uniform culture-type must have existed in the northern or central hemisphere of the Old World, in strong contrast with the cultures of all primitive tribes which we encounter in the rest of Asia, in Africa, and in America. In the earliest stages of Sumero-Babylonian, Indo-Iranian, and Chinese cultures leaving aside the manifold subsequent differentiations due to indigenous development, we are confronted with a number of traits which most strikingly coincide, and which cannot be attributed to a chance accident. Conspicuous among these is the economic system of the three peoples, which was founded in like manner on agriculture and cattle-breeding; that is to say, it was then already on the same basis as our modern system of economy. Their agricultural implements were highly developed, they tilled their fields by means of the plough drawn by an ox that was regarded as a sacred animal, and cultivated several cereals, chiefly wheat and barley. Methods of artificial irrigation were perfectly known, and elaborate agrarian laws were in force. Cattle were exclusively employed as draught-animals, particularly in connection with the plough, and were originally raised, not for their milk, but for their meat only. Carts and chariots built on the principle of the wheel are found alike in the three groups; while it is a notable fact that transportation by means of wheels is obviously absent with all primitive tribes of Asia, Africa (except ancient Egypt), in the South Sea, Australia, and America. It is remarkable also that the cart appears everywhere hand in hand with the plough, and consequently must be an invention made in the agricultural stage of civilization. The peoples of Babylonia, India, and China, in the same manner, employed chariots for making war, to which horses were harnessed. As the ox was domesticated in the interest of agriculture, the camel in the interest of commerce, so the horse was essentially an animal of war, and in its military capacity was employed as the draught-animal of war-chariots. Horseback-riding is a much later art conceived by the nomadic tribes of Scythia and inner Asia. Of other domestic animals, dog and swine were reared. As to the importance of swine, ancient China offers a striking analogy to the prehistoric cultures of central Europe, with which it has also millet and the water-chestnut in common. The offensive arm of that period was the composite bow,—a very complex affair, consisting principally of flexible wood combined with horn, to which are bound layer upon layer of pliable sinew,—while most of the primitive tribes of Asia know only the simple wooden bow. Of metals, only copper and bronze were employed. In the manufacture of ceramics, the potter's wheel has been utilized since early times in the East as well as in the West. Homer, in a verse of the Iliad (XVIII, 600), compares the movements in the round of a dance to the whirling

motion of the disk turned by the potter's hand. The
ancient Chinese philosophers likened the action of Heaven
in evolving the universe and its beings to a potter fashioning
the objects of clay by the revolution of his wheel; and
Heaven, for his creative power, is directly styled a moulder
or a potter's wheel. Again, this contrivance is unknown
wherever pottery is worked in the primitive stages of culture. Furthermore, we find in ancient Babylonia and China
a highly-developed stage of knowledge of astronomy combined with an intelligent chronological sense, time-reckoning, and a calendar system.

The time and locality in which this reconstructed primeval culture common to Western Asia, India, and China,
was developed, certainly escape our knowledge, and I must
forbear on this occasion discussing this side of the problem.
The point which should be emphasized is, that the characteristic features, particularly the system of economy, are
fundamental principles and factors of civilization, and
exactly those which still form the fundament of our own
modern life. We still depend for our subsistence, in the
same manner as the prehistoric culture-groups of Asia
did ages ago, upon the products of the soil, the cultivation
of cereals, the breeding of cattle; and in principle, our
methods of farming, despite all technical improvements,
are still the same. We may reform almost everything
in our life; we may change our language, our manners and
customs, our political institutions, or our philosophy;
but we cannot change that one most stable and persistent
factor of our culture which we may briefly sum up in
the words "cereals," "cattle," "plough," and "wheel."
Whatever our modern progress in the perfection of land
transportation may be, whether we consider our steam-
engines or motor-cars, they all depend upon the basic
principle of the wheel,—that wonderful invention of prehistoric days, of the time, place, and author of which we are
ignorant.

The main contribution, however, to the problem under
consideration, is that ancient Chinese culture in its earliest
stage cannot be the product of an isolated seclusion, but

has its due share and its root in the same fundamental ideas as go to build up the general type of Asiatic-European civilization. This opinion certainly does not imply that the basis of primeval Chinese culture is merely derived from the West, but only that it has a substratum of ideas to be met alike in the other great culture-groups of Asia from which we may reconstruct the common ancestral form of culture that must have once prevailed in most ancient times.

While, thus, the place of China is determined in the general history of civilization, there are, on the other hand, visible symptoms in existence which warrant the belief that as early as prehistoric times the Chinese must have undergone a development during several thousands of years entirely independent of any Western influence. And here we touch on one of the most interesting problems in the oldest history of Asia. Ancient Asia with its European annex is split into two large, sharply-defined economic camps, as regards the production and consumption of milk and other dairy products. The entire East-Asiatic world, inclusive of China, Korea, Japan, Indo-China, and all Malayans, does not take animal milk for food, and evinces a deep-rooted aversion toward it; and this was the state of affairs even in remotest times. On the other hand, all Indo-European peoples, the Semites, the ancient Scythians, and all nomadic tribes of northern and central Asia, as Turks, Mongols, and Tibetans, are all milk-drinkers, and were so in early historical times. The remarkable feature about this case certainly is not the bare fact that the East-Asiatics abstain from milk,—for the aboriginal tribes of America and Australia and others, simply for the lack of milk-producing animals, do exactly the same,—but the essential point is that the Chinese and their followers adhere to this practice, despite an abundance of milk-furnishing domestic animals in their possession, and despite long-enduring intercourse with neighboring milk-consuming peoples, whose habits and mode of life were very familiar to them. They rear cows, buffalo, mares, camels, sheep, goats, all animals from which milk could be derived, but they do not even understand how to milk them. They were at all times surrounded by Turkish and Mongol peoples, whose daily sustenance depends upon milk and kumiss, butter and cheese. This fact has been perfectly known to the Chinese, but, notwithstanding, they never acquired the habit. In India and Indo-China we face the same striking fact, in that the aboriginal inhabitants, though willing to submit to the higher civilization of the Aryan Hindu, never adopted from them the custom of milk-drinking. It follows, therefore, that our consumption of animal milk cannot be looked upon as a self-evident and spontaneous phenomenon, for which it has long been taken, but that it is a mere matter of educated force of habit As natural as it appears to us, owing to time-honored practice and tradition, so just as unnatural, tedious, and barbarous does it strike the Chinese and other peoples of eastern Asia, who uphold that it is cruel to deprive the calf of its mother's milk. This ethical opinion, surely, does not give the true reason for their abstinence from milk, but is no more than a speculative after-thought. No less remarkable is it that no religious taboo is placed on milk in any of the Eastern religions, and that the aversion is not prompted by motives of any religious character; it is purely a matter of social and economic life. Thus we are led to distinguish in the history of the domestication of cattle two main and fundamental stages. In the primary stage, the milking faculty of the cow was unknown to man, and the ox was exclusively the sacred animal of agriculture, drawing the plough; and the invention of the plough and the cultivation of cereals are events closely affiliated with the taming of the ox. This is the very point which Chinese society has in common with the rest of Asia and Europe. At the close of this initial period, the western portion of the Old World subsequently advanced to a further stage of development, from which the Chinese were debarred,— the acquisition of dairy economy. This was an exceedingly complex, slow, and long process, moving along two lines,— one in the producer, the animal, in which the productive power was gradually trained; the other in the consumer, man, who just as slowly acquired the habit of taking to milk.   It should be understood that the obvious advantages which we derive from domesticated animals are not the reasons which prompted their domestication.   These material advantages are but the effect and result of prolonged activity in matters of domestication, and could not have been anticipated by primitive man when he first conceived the idea of rearing and training animals.   Wild fowl, e.g., when they are first being taken care of by man, do not propagate to a large extent, nor do they lay eggs in great numbers.   The egg-laying habit of our chickens, to such an extent that it was of some advantage to man, was only attained in the course of the gradual process of domestication.   Consequently other reasons than material considerations must have led to the first step in this direction.   Likewise the productive power of our milk-animals is only the consequence and the ultimate result of long-continued domestication.   This development, which must have been in operation for millenniums, ages before any recorded history, remained confined solely to the western part of Asia, while the East was never affected by this movement.   This effect is still obvious in the division of labor brought about between cow and ox in the West, the ox performing duty as working-beast, the cow being the milk-animal.   The same is reflected in language: all Semitic, Indo-European, and Ural-Altaic languages have separate words for "bull" and "cow," while the Chinese express the same notions with one word only.

There are other such negative criteria of peculiar character which conspire to prove a lengthy prehistoric period of Chinese independence.   The ancient Chinese raised sheep and goats, but never utilized the wool of these animals for the making of material for clothing, as was done in the West.   The employment of wool for felt and rugs is an idea of nomadic peoples of inner Asia, and was taught by them to the Chinese in historical times.   The latter always used for their garments vegetable fibres obtained from various kinds of hemp, and silk, which most clearly stands out as a pre-eminently brilliant example of their power of nature-observation and of their technical genius. The art of baking eavened bread, which was first applied by the Egyptians and adopted by the Greeks and Romans, has always remained unknown in China, where no leavening or fermenting agent is employed for bread-making.

Speaking of mental achievements, we observe that the Chinese, like all other peoples of eastern As·a, have never produced any epic poetry. Epic poems are met with among all Indo-European nations, among the Finno-Ugrians, the Turkish, Tibetan, Mongol, and Tungusian tribes; and it is a peculiar coincidence that all these peoples of epic songs are also milk-consumers, while those abstaining from milk are deficient in epic poems. I do not mean to say that there is an interrelation between milk and epics, but merely wish to point out the fact of this curious coincidence.

Thus the conviction is gaining ground, on the one hand, that Chinese culture, in its material and economic foundation, has a common root with our own; and, on the other hand, that it independently marched along its own way and evolved its own ideas for numberless ages, until at the time when the nation emerged from prehistoric life it had grown to full maturity. The keynote of its rapid progress in historical times is chiefly signalled by the sound development of all social and civic virtues, finally culminating in the political and ethical system expounded by the sage Confucius. The sane family organization based on the religious institution of ancestral worship, the high conception of the sacredness and purity of family life, filial devotion, and the subjection of the individual to the ideal of the family and the state, must be regarded as the principal manifestations accounting for the racial and national continuity of the Chinese, that indestructible vital power and tenacity of their culture and institutions. No nation has ever presented a more sensible and effectual solution of the problem of the sexes than China by her common-sense marriage-laws, which enjoin marriage on every one as a moral obligation due to the ancestors. Despite its religious function, it has always been strictly a matter of civil law, and was never usurped by a Church or bound to ecclesiastic sanctions, as has so long been the case among ourselves. Early marriages were always made possible in consequence of a just economic system, with an almost equal distribution of landed property among single owners, in which a large, probably the largest, portion of the population, enjoys a share; while great real-estate owners are few, resulting in a levelling process of economic and social equality. Husbandry was at all times upheld as the bone and sinew of society, and encouraged and promoted by Government. In the social division the farmer ranks next to the scholar or official, and precedes the laborer and merchant. Wholesome principles in matters of nutrition, frugality, and temperance—in general, a good share in the knowledge of that greatest of all arts, the art of living—have contributed to the stability and persistency of Chinese society. An extraordinary capability for passive resistance, and an unlimited power of absorption, are prominent characteristics of this civilization, whose vitality has been tested many times. Military defeats and even widespread conquests have never been able to make a deep impression on this people. By dint of intellectual force and superior diplomatic tactics they have usually overcome the most serious conditions. The Huns could overrun Europe; but the very same Huns, knocking at the gates of China for many centuries, were unable to bring about her downfall. The Mongols trampled the Occident under the feet of their horses; but under their sway in China they became converted into Chinese, and made art, literature, and commerce flourish. In the same manner they absorbed the Khitan, the Manchu, and others of their foreign rulers.

We are all familiar with the fact that we are indebted to the inventive genius of the Chinese for the mariner's compass, gunpowder, fire-works, rag-paper, wall-paper, paper money, silk, porcelain, the goldfish, tea and many other valuable cultivated plants. All these things, being exceedingly useful and practical, come within the daily reach and experience of every one. They have also enriched the field of our popular games and entertainments by the addition of kite-flying, shuttlecock, playing-cards, dominoes, checkers, and that jolly theatrical performance the shadow-play. But all these advantages sink into insignificance when we come to consider the intellectual gain which has accrued to us from the wonderful development of the mental and moral forces alive in the Chinese nation. They follow suit with us in their eminently chronological and historical sense, one of their most striking and admirable intellectual traits. Next to the Greeks, they are that nation which has furnished the most solid and extensive contributions to our scientific knowledge. While ancient India revels in mythology but gives us no clew to her history, the Chinese have recorded for us with minute accuracy and painstaking conscientiousness, and above all with objective impartiality, every event in their internal history and in their relations with foreign peoples. The continuity of their traditions laid down in their twenty-four national Annals may be styled, more justly than many other things, one of the great wonders of the world, and this stupendous work is the most permanent monument that they have built to themselves. They are born philologists and students, and there is no domain of human thought which their fertile literature has not efficiently cultivated.

While the knowledge of China which the Greeks and Romans possessed was vague and hazy, the Chinese have bequeathed to us invaluable notes on the conditions and commerce of the Roman Orient. Their generals, diplomats, and Buddhist pilgrims, who journeyed across Central Asia to India, or visited Korea, Japan, and the ports of the Indian Ocean, have proved as keen and trustworthy observers in the memoirs which they have jotted down on the geography, manners, and customs of foreign countries. But for the writings of many heroic Chinese explorers, we should still be groping in the dark as to the ancient history, topography, and archæology of India and Turkistan; and a serious study of India and Buddhism, as well as of Tibet and Mongolia, is no longer possible without the staff of sinology. The first Chinese traveller was the famous

General Chang K'ien, who in 138 B.C. started on a diplomatic mission to the west; he was held in captivity for ten years by the Hiung-nu, the ancestors of the Huns of Attila, and finally reached Ferghana and Bactria, advancing as far as the Oxus. In 126 B.C. he returned to his fatherland with the seeds of a number of new cultural plants, and submitted to his astounded countrymen a glowing account of the new world which he had discovered, and which was nothing less than the Hellenic-Iranian civilization inaugurated in those regions by the successors of Alexander the Great. He was the first Chinese who had learned, and brought the idea home, that behind the thick veil of roving nomads threatening the empire in the north and west, there were sedentary nations in the farthest west, of enormous wealth and civilized forms of government, with whom a profitable trade might be opened. The report of this undaunted pioneer left an indelible impression upon the minds of his countrymen. In 115 B.C. he set out again for his former goal, but died the next year from the fatigue of his journey. About two centuries afterwards, in A.D. 73, General Pan Ch'ao, after a struggle of sixteen years centring around Kashgar, established a protectorate over Central Asia, and heard people speak with much praise about a great empire in the west, Ta Ts'in; that is, the Roman possessions in anterior Asia. In 97 he commanded Kan Ying, an officer of his military staff, to proceed to Syria in search of this mysterious land. Kan Ying reached the shores of the Persian Gulf, where he was to embark on one of the boats circumnavigating Arabia, and sailing up into the Red Sea to the point from which the caravans started for Syria. There he was deterred from his plan by natives, who warned him of the perils of navigation—presumably on account of the jealousy of the Parthians, who were the middlemen in the silk trade between Serika, as the ancients called China, and the Roman Orient, and who may have feared lest the Chinese might open direct trade with Rome, if Kan Ying should have succeeded. There is ample food for reflection in speculating as to how far Chinese culture would have

been affected or modified by an immediate intercourse
with the antique world.   We may regret that it was deferred
and that the seed first sown by Chang K'ien has reached
fruition only in our day. A common tie links Chang
K'ien, the first modern Chinese, with Kang You-wei,
the reformer, though the two men are separated by a space
of two thousand years.   Chang K'ien was the first to experience, that, besides his own country, there was another
great sunny world.   The ancient Chinese doctrine of the
State came to its climax in the axiom that China and her
civilization were identical with the world, and represented
the universal unlimited empire surrounded by barbarians.
This political philosophy was maintained by the Government until recent times, when K'ang You-wei was the first
to strike the death-blow to this theory, and to prove to his
countrymen that it is the individual national State only
which has a right of existence and a guaranty of enduring
under modern conditions. Under the guidance of this
fertile idea, the renaissance of the nation has taken place,
and its national rejuvenation will doubtless result also in
the reshaping of a new national culture.

AMERICAN
ANTHROPOLOGIST

NEW SERIES

ORGAN OF THE AMERICAN ANTHROPOLOGICAL ASSOCIATION, THE ANTHROPOLOGICAL SOCIETY OF WASHINGTON, AND THE AMERICAN ETHNOLOGICAL SOCIETY OF NEW YORK

PUBLICATION COMMITTEE

F. W. HODGE, Chairman ex-officio; PLINY E. GODDARD, Secretary ex-officio;
HIRAM BINGHAM, STEWART CULIN, A. A. GOLDENWEISER,
GEORGE BYRON GORDON, WALTER HOUGH, A. L. KROEBER,
BERTHOLD LAUFER, EDWARD SAPIR, MARSHALL H. SAVILLE,
JOHN R. SWANTON, ALFRED M. TOZZER.

PLINY E. GODDARD, *Editor*, New York City
JOHN R. SWANTON and ROBERT H. LOWIE, *Associate Editors*

(F. W. HODGE, Editor for pp. 1–221)

VOLUME 17

LANCASTER, PA., U. S. A.
PUBLISHED FOR
THE AMERICAN ANTHROPOLOGICAL ASSOCIATION
1915

generally synonymous of our word 'half' (*i. e.*, half of two = one). According to this rule Dr Sapir's *hwä* must not mean foot, but feet, in the same way as his *la* should be translated hands, not hand.

Dr Sapir's analysis of the short text which closes his essay is simply admirable, and betrays an insight into the morphology of his material which one is at a loss to know where, or how, it was acquired. Scarcely more than one criticism have I to offer on this part of his paper. The last syllable of the compound *dō-at-ṭi* is not a "noun suffix," as he believes (30). It is a regular verb, or rather a verbal stem, since the pronominal element of the same has disappeared through the process of word formation. *Dō* is the Carrier negation *au*, the Babine *so˙*, Chilcotin *tla*, Sekanais *ussé*, Nahanais *ætû*. These particles or words can, in the North, conveniently be omitted in many cases. They are the equivalents of the French *ne . . . pas*. *'At*, as we have seen, means "wife"; *-ṭi* is the root of the verb *æṭi*, "he has."

This is about the sum total of the criticisms I have to make on that author's rendering and interpreting of the southern dialect he introduces to the philological world.

ADRIAN G. MORICE

ASIA

Some Fundamental Ideas of Chinese Culture. By BERTHOLD LAUFER, PH.D., Associate Curator of Asiatic Ethnology, Field Museum, Chicago. (Reprinted from *The Journal of Race Development*, Vol. 5, No. 2, October, 1914, pp. 160–174.)

It is a matter of great regret that Dr Laufer's writings are so little known among Americanists and ethnologists generally. However specialized researches into the civilization of nations possessing a written literature may have become, there is surely no difference in principle between studying the culture of primitive and of civilized populations, and the ethnologist might reasonably expect valuable suggestions from the historian, the Egyptologist, the sinologue, *et al.* The sources of information of these scholars appear of higher authenticity than the oral traditions recorded in the course of ethnological field-work, and at all events promise a better chronological insight into the actual growth of cultures. Unfortunately we are often disappointed in these hopes through the fact that the student of higher cultures is not imbued with the anthropological point of view: too often he naïvely assumes "theories that are now gracing only the refuse heaps of the modern anthropologist's laboratory"; too often he remains ignorant of avenues of approach successfully trodden by modern ethnologists. As a striking illustration we may cite the lack of criticism with which speculative afterthoughts are frequently taken at their face value,—as trustworthy accounts of the actual origin of institutions. The distinctive value of Dr Laufer's work lies in the fact that he combines the erudition of the Orientalist with the spirit of latter-day ethnology, which he applies with originality, rare judgment, and unusual psychological insight to the problems of his chosen field. Readers of the *American Anthropologist* need only be reminded here of Dr Laufer's independent "Theory of the Origin of Chinese Writing" that appeared in this Journal in 1907 (pp. 487–492); of the pregnant general remarks scattered through the monograph on *Jade* (Chicago, 1912); and the interesting discussion of convergence in his *Dokumente der indischen Kunst*, 1 (Leipzig, 1913).

In the paper now under consideration Dr Laufer briefly outlines the general characteristics of Chinese culture. It would be difficult to compress more solid and suggestive information within the narrow compass of fifteen pages. Dr Laufer, dismissing the "Monroe doctrine" of sinologues, that Chinese culture is a purely indigenous product, establishes its relations with other cultures. He clearly distinguishes a northern and a southern culture area. The former is characterized by the use of the plough and ox in the cultivation of wheat, barley, and millet, while the southern farmer plants rice with the aid of the hoe and water-buffalo characteristic of southeastern Asia. In the north there are highways, and travel is by two-wheeled carts drawn by mules, while the horse, donkey, and camel are found as pack-animals and for transportation; in the south there are rivers and canals, and travel is by boat and sedan-chairs. The northern populations show cultural contact with Tungus and Turkish tribes, while the south was influenced by the Indian and Malayan cultures. It is the northern Chinese culture that shares traits of the greatest significance with the early Sumero-Babylonian and Indo-Iranian cultures. The common possession of certain cereals, of the plough and the ox, of wheeled vehicles, of the composite bow, and of the potter's wheel, indicates "that in an undefinable pre-historic age a great universal and uniform culture-type must have existed in the northern or central hemisphere of the Old World, in strong contrast with the cultures of all primitive tribes which we encounter in the rest of Asia, in Africa, and in America" (p. 164). This, the author hastens to assure us, does not imply that primeval Chinese culture was simply a wholesale importation from the West, but does prove that it cannot be considered "the product of an isolated seclusion." On the other hand, there is strong evidence that "as early as prehistoric times the Chinese must have undergone a development during several thousands of years entirely independent of any Western influence" (p. 167). In this connection Dr Laufer emphasizes a negative trait,—the fact that the Chinese, like the Koreans, Japanese, Indo-Chinese, and Malayans, do not use animal milk for food, which constitutes a significant difference from the Semites, the ancient Scythians, the Indo-Europeans, Turks, Mongols, and Tibetans (pp. 167–169). Another significant negative feature consists in the non-utilization of sheep and goat wool for clothing, this being an art taught to the Chinese in historical times by the nomadic peoples of inner Asia. A curious adhesion, in Tylor's sense of the term, is that of epic poetry and milk-consumption, those tribes abstaining from milk being likewise deficient in epic literature (p. 170).

However, it is impossible to cite all the interesting data in Dr Laufer's paper without quoting or paraphrasing it in its entirety.

ROBERT H. LOWIE

Aberglaube und Volksmedizin im Lande der Bibel. By T. CANAAN. Hamburg: L. Friederichsen & Co., 1914, pp. xii, 153, large 8°. With 6 plates and 50 text figures. (Abhandlungen des Hamburgischen Kolonialinstitutes, vol. xx.) Price 6 m.

The author of this work is a physician of Jerusalem who was born and lived among the people whose beliefs, modes of thought and practices he depicts. At the same time he is fully abreast of Western science and familiar with modern methods of observation and analysis, having received his education in Europe. He had therefore the equipment and opportunities of studying the habits of the natives and penetrating the innermost recesses of their motives and thoughts rarely, if ever, granted to an outsider.

After an introductory chapter which briefly describes the domestic conditions of the peasants (fellahins) and their utter disregard of hygienic rules, the views of the natives on the causes of sickness and its handling are treated in eight chapters under the heads of etiology; diagnosis; prognosis; the healers; prophylaxis; vows, and the treatment of disease. The three principal causes of sickness are: (1) Spirits or demons who are everywhere, in fact they fill out the space between heaven and earth and are organized in several hierarchies with princes at their head; (2) the evil eye, which again lurks everywhere, as its baneful potency is due to a poisonous substance inherent in all men which emanates through the eye, working its mischief unwittingly and unconsciously, even animals not being